Hydrogen Diplomacy

SEYED EHSAN HOSSEINI
Department of Mechanical Engineering, Arkansas Tech University,
Russellville, AR, United States

Future Publishing LLC

Future Publishing LLC

Notices
The field's understanding and best practices continually evolve. As our knowledge deepens through new research and experiences, adjustments in research methodologies, professional protocols, or medical interventions may become necessary.
Practitioners and researchers should always rely on their own expertise and knowledge when assessing and utilizing any information, techniques, substances, or experiments mentioned here. When doing so, they should prioritize their safety and that of others, especially those under their professional care.
To the maximum extent permitted by law, neither the Publisher nor the authors, contributors, or editors assume responsibility for any harm or damage to individuals or property arising from product liability, negligence, or any other cause, or from the use or implementation of any methods, products, guidelines, or concepts discussed in this material.

How to cite this document?
Hosseini, Seyed Ehsan, Hydrogen Diplomacy. Future Publishing LLC, 2024. DOI: https://doi.org/10.55670/fpll.book/1

For Information on all Future Publishing LLC publications, visit our website at: https://fupubco.com

Future Publishing LLC

Contents

Preface

The book "Hydrogen Diplomacy" delves into the intricate connections between scientific advancements, technological growth, strategic policy formulation, and international collaboration, all aimed at creating a sustainable energy landscape. It sheds light on the multifaceted role of hydrogen as a clean, renewable energy source with immense transformative potential. Through the lens of diplomatic efforts, it explores how these elements converge to pave the way for a more environmentally responsible and economically prosperous future. The urgent necessity to tackle environmental issues stemming from fossil fuel usage has become increasingly evident. The adverse impacts of greenhouse gas emissions on our planet have spurred global demand for change. This sense of urgency is propelling the search for alternative energy sources, with hydrogen emerging as a leading contender due to its environmentally friendly and sustainable characteristics. Hydrogen presents the opportunity to decrease carbon emissions, tackle climate change, stimulate technological advancement, and promote economic expansion-key elements for a sustainable energy future. Nevertheless, the widespread adoption of hydrogen fuel encounters substantial obstacles that demand attention. These challenges encompass technological barriers, infrastructure expansion, economic feasibility, and policy frameworks. Overcoming these hurdles necessitates a unified, global endeavor characterized by strategic cooperation and alignment.

As we transition to hydrogen production methods, a spectrum from fossil fuel-based processes to renewable energy-driven technologies is encompassed. The advantages and disadvantages of each method are scrutinized, with an emphasis on the imperative of transitioning to sustainable production approaches for hydrogen. In this journey of exploration, the domain of electrolyzers is examined, highlighting them as the often-overlooked champions propelling the hydrogen revolution forward. The evolution of hydrogen storage and delivery on a global scale is emphasized as playing a pivotal role in facilitating a smooth shift towards a hydrogen-based economy. The revolution in hydrogen transport is noted as another focal point of discourse. Advancements in fuel cell technology, the proliferation of hydrogen-powered vehicles, and the transformative potential of hydrogen in revolutionizing transportation across diverse sectors are discussed. Specific regions are then analyzed, with a closer look at the opportunities and challenges in the United States, the European Union, Asia-Pacific, the Middle East, and other key players in hydrogen technology. Each region's unique perspectives, strategies, and initiatives are considered in harnessing hydrogen's potential as a catalyst for sustainable development. Central to this discourse is the notion of hydrogen diplomacy. This intricate and multifaceted strategy involves navigating geopolitical complexities, fostering international collaboration, and harmonizing policies to establish a conducive environment for hydrogen-related endeavors and investments. Successful hydrogen diplomacy is imperative for unlocking hydrogen's full potential as a clean energy solution.

Seyed Ehsan Hosseini

This page intentionally left blank

CHAPTER 1

The urgency of hydrogen: environmental issues and the need for change

1.1 Environmental issues caused by fossil fuels

The world's economy heavily relies on energy from fossil fuels, constituting approximately 80% of the overall energy requirements. This carbon-centered economic model has given rise to multiple issues of concern due to the release of diverse pollutants through the combustion of these fuels. However, the most critical challenge tied to fossil fuel usage pertains to the global climate change crisis, which arises from the emission of carbon dioxide (CO_2). Measuring the rate of economic advancement and industrial development often relies on energy consumption as a benchmark. Energy usage is pivotal in driving rapid economic growth in both developed and developing nations. However, it is essential to note that the consumption of non-renewable energy sources in developing economies not only facilitates economic progress but also leads to significant emissions of harmful gases and heightened environmental concerns. Carbon emissions and environmental degradation are significant challenges faced globally in the 21st century, impacting both developed and developing nations. Developing countries, particularly, are more vulnerable to these issues due to their increased reliance on fossil fuel mixes to drive industrial growth and value-added manufacturing. Using fossil fuels contributes to environmental issues like global warming and air pollution, resulting in adverse health impacts and diminishing the quality of life for communities. In addition to the environmental and health challenges they pose, the uneven distribution of fossil fuels raises concerns about energy security, given their crucial role in present-day energy production systems [1]. Furthermore, as non-renewable resources, their limited availability becomes a concern for current and future generations. The instability of markets and prices associated with fossil fuel usage also carries negative economic repercussions. The issues mentioned above contribute to the unsustainable nature of current energy systems and drive the search for new

DOI: https://doi.org/10.55670/fpll.book/1

solutions to mitigate these systems' economic, social, and environmental drawbacks, thereby promoting sustainability. Despite the efforts made by governments, international agencies, and other stakeholders, achieving sustainable energy systems remains a significant challenge for most countries. Over the past few decades, various policies and incentives, such as feed-in tariffs, have been implemented to support renewable energy. Additionally, strategies like demand-side measures and smart grids have been employed to address the challenges of transitioning to low-carbon energy systems [2].

Environmental pollution is a major global issue characterized by its adverse effects on human well-being, ecological balance, wildlife, air quality, natural resources, and the overall ecosystem. The emission of CO_2, a significant contributor to pollution, has witnessed a substantial increase over the years. In 1990, CO_2 emissions amounted to 22,670,893 kilotons, which escalated to 36,240,721 kilotons in 2015, reflecting a significant 59.86% surge in three decades. This upward trajectory in global CO_2 emissions stems from the growing demand for energy consumption.

In 2017, global energy demand experienced a 2.1% growth rate, surpassing the 0.9% increase observed in 2016, while the average growth rate over the previous five years stood at 0.9%. Notably, global electricity demand rose by 3.1%, exceeding the overall production growth. In 2022, global energy-related CO_2 emissions experienced a growth of 0.9% or 321 million tons, resulting in a new record high of over 36.8 billion tons. While the protective function of CO_2 against freezing in space is unavoidable, the escalating frequency of droughts, storms, floods, and the overall increase in global temperatures are significant outcomes stemming from the elevated levels of CO_2 emissions. Greenhouse gases (GHGs) possess the capability to ensnare heat radiation emanating from the Earth's surface within the atmosphere, directly leading to the phenomenon of global warming. Nevertheless, it is believed that the energy crisis did not lead to the significant rise in global emissions that was originally anticipated. This outcome can be attributed to the remarkable expansion of renewable energy sources, electric vehicles (EVs), heat pumps, and energy-efficient technologies. Without the presence of clean energy, the growth in CO_2 emissions would have been almost three times greater [3]. Undoubtedly, the susceptibility of human civilization to global warming is a highly intricate environmental concern in today's world (Figure 1.1). There is solid evidence that increasing the concentration of CO_2 in the atmosphere, primarily caused by the combustion of carbon-based fuels like coal, gasoline, and natural gas, has a significant impact on the Earth's environment [4].

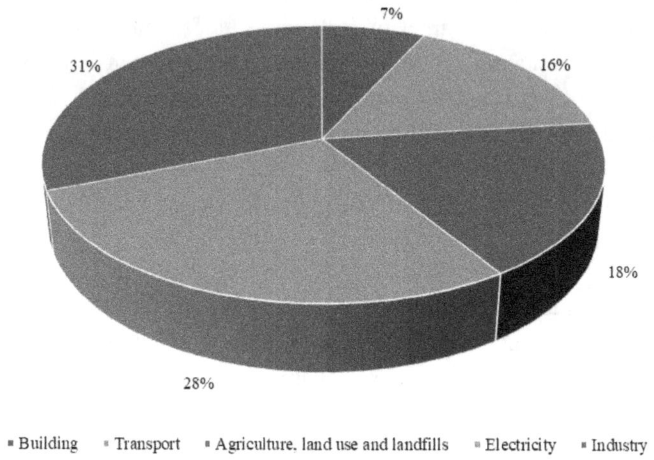

Figure 1.1 Global greenhouse gas emissions by sector

The majority of worldwide air pollution is attributed to energy-related fossil-fuel combustion in higher and middle-income nations, as well as biomass burning in lower-income countries. This activity contributes to approximately 85% of airborne particulate pollution that can be inhaled and nearly all emissions of sulfur dioxide (SO_x) and nitrogen oxide (NO_x) into the atmosphere. Additionally, emissions encompass substances like black carbon, polycyclic aromatic hydrocarbons (PAHs), nitrogen and sulfur dioxides, mercury, and volatile compounds that lead to the formation of ground-level ozone (O_3). The majority of electricity production in the United States is sourced from non-renewable energy. In 2022, the country generated approximately 4.24 trillion kilowatt-hours (kWh) of electricity by utility-scale electricity generation facilities. Of this total, around 60% of the electricity generation came from fossil fuels such as coal, natural gas, petroleum, and other gases. Nuclear energy accounted for approximately 18%, while renewable energy sources contributed to about 22% of the electricity generation [5].

The proportion of renewable energy in U.S. power generation is projected to rise to 44% by 2050. This growth in renewable energy primarily stems from adopting new wind and solar power technologies. The role of hydropower is expected to remain relatively steady through 2050, while other renewable sources of energy generation, like geothermal and biomass, are anticipated to collectively account for less than 3% of the total power generation. The Paris Agreement represents a critical international environmental accord focused on climate mitigation. Its core objective involves substantial reductions in human-caused carbon emissions. This entails a significant overhaul of the current system, incorporating widespread adoption of emissions-free energy generation facilities and alternatives to traditional fossil fuels. Meeting the Paris Climate target of restraining global temperature rise to either 1.5°C or 2°C necessitates swift and extensive decarbonization across all sectors of our economies [6]. Given the limited availability of fossil fuel reserves and the environmental issues linked with burning them, there is a pressing need to explore fresh renewable energy sources. These alternatives would serve as cleaner options in comparison to fossil fuels and contribute to the attainment of sustainability objectives.

1.2 The importance of reducing global greenhouse gas emissions

The issue of GHG emissions and their impact on the environment has become a pressing concern on a global scale. GHGs are critical in regulating the Earth's temperature by trapping atmospheric heat. However, the excessive accumulation of these gases due to human activities, primarily the burning of fossil fuels, deforestation, and industrial processes, has led to a range of detrimental effects on our planet. At its core, the greenhouse effect is a natural process that maintains the Earth's temperature at levels suitable for life. Sunlight passes through the Earth's atmosphere and reaches the surface, absorbed and re-radiated as heat. GHGs, in turn, absorb and re-radiate some of this heat back toward the surface, creating a natural balance. However, the problem arises when human activities release unprecedented quantities of GHGs into the atmosphere. These gases accumulate and amplify the greenhouse effect, leading to an increase in global temperatures- a phenomenon commonly known as global warming. The consequences of global warming are far-reaching and multifaceted [7]. One of the most immediate and visible effects of rising GHG emissions is the disruption of climate patterns. This disruption manifests in more frequent and intense extreme weather events, such as hurricanes, droughts, heat waves, and heavy rainfall. These events can devastate ecosystems, agriculture, infrastructure, and human lives. The warming of the planet also contributes to the melting of glaciers and polar ice caps, resulting in the rise of sea levels. As glaciers and ice sheets melt, they contribute to the increased volume of water in oceans, leading to coastal erosion, submergence of low-lying areas, and the displacement of coastal communities. These changes not only impact human settlements but also disrupt marine ecosystems and biodiversity.

Rising temperatures and changing climate patterns disrupt ecosystems, affecting the distribution and behavior of species. Many plants and animals are struggling to adapt, leading to shifts in habitats and potential extinctions. Coral reefs, for example, are highly sensitive to temperature changes, resulting in coral bleaching and a decline in marine biodiversity. Many species rely on specific temperature and climate conditions for migration, breeding, and feeding. As these conditions change, animals may find it challenging to navigate and locate suitable habitats, potentially disrupting ecosystems' delicate balance. Rising temperatures can force ecosystems to migrate poleward or to higher altitudes, impacting local flora and fauna. Forests may move upslope, endangering species at higher elevations, while coastal habitats can be threatened due to rising sea levels. Ecosystems provide valuable services like clean water, pollination, and natural resources. Global warming can disrupt these services, leading to water scarcity, crop failures, and reduced availability of vital resources [8]. The repercussions of unchecked greenhouse gas emissions extend beyond environmental changes. Economic sectors dependent on stable climate conditions, such as agriculture and tourism, face significant challenges. The costs associated with adapting to these changes and mitigating their effects can strain economies and societies worldwide. Governments need to address the changing availability of resources like water and arable land due to climate change. Droughts, heatwaves, and changing rainfall patterns can challenge water management systems and agricultural practices. Increasingly severe weather events, such as hurricanes and flooding, can damage critical infrastructure, disrupt transportation, and threaten public safety. Governments must invest in resilient infrastructure and disaster response mechanisms. Global warming contributes to the spread of diseases and heat-related illnesses. Governments must adapt healthcare systems to handle increased health risks and protect vulnerable populations. Climate change can exacerbate existing tensions and conflicts, particularly in regions with resource scarcity. Governments may need to collaborate on climate initiatives and adopt policies to address the resulting geopolitical challenges.

On the other hand, climate change affects crop yields and food production, leading to supply chain disruptions and food price volatility. Changing climate conditions can threaten the livelihoods of small-scale farmers and impact global food security [9]. Increased demand for cooling during heatwaves strains electricity grids, while changes in precipitation patterns can impact hydropower generation. Transitioning to clean energy sources becomes crucial to mitigate economic losses. Climate-related events, such as hurricanes and wildfires, can disrupt supply chains, impacting manufacturing and distribution. This can lead to increased costs and reduced efficiency. Recognizing the severity of the situation, governments, organizations, and individuals across the world are actively working toward reducing GHG emissions. International agreements like the Paris Agreement aim to unite nations in their efforts to limit global warming and prevent the most catastrophic consequences of climate change. This involves transitioning to renewable energy sources, improving energy efficiency, reforestation, and adopting sustainable practices across industries.

1.3 The current state of the energy industry

Currently, the industrial segment accounts for over 33% of all energy usage in the United States. This proportion is anticipated to grow further unless the ongoing challenges within industrial operations are promptly tackled. The total energy consumption by the industrial sector in the U.S. is projected to experience a 36% increase between 2020 and 2050 if the existing trends persist [10]. Similar trends can be observed globally, with developing countries experiencing rapid industrialization and increased energy needs (Figure 1.2). This growth in energy consumption presents both challenges and opportunities for sustainable development. Energy costs have been increasing since 2021 due to a combination of factors, including the swift rebound of the global economy, weather-related events affecting different regions, postponed maintenance activities resulting from the pandemic, and prior choices made by oil and gas firms and exporting nations to curtail investment [11].

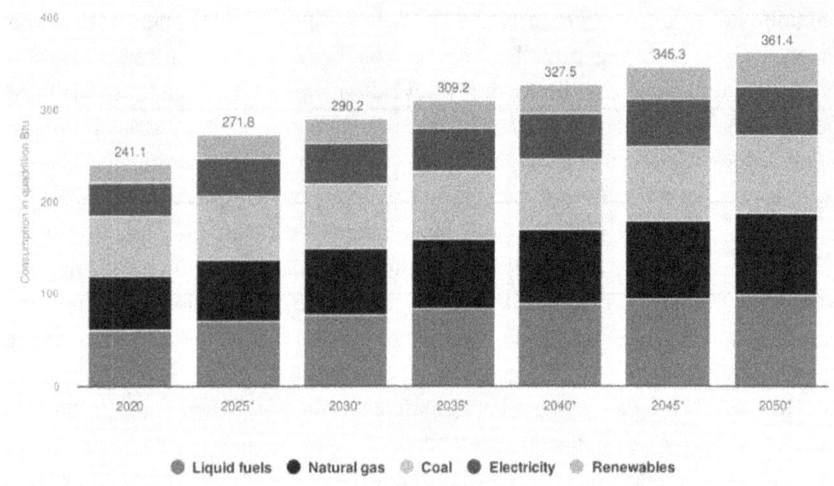

Figure 1.2 Global industrial energy consumption from 2020 to 2050 by energy source (In quadrillion British thermal units) [12]

In 2021, Russia initiated a halt in gas deliveries to Europe due to its invasion of Ukraine, resulting in further strain on already constrained supplies. The global energy crisis triggered by Russia's invasion of Ukraine is causing profound and long-lasting changes that potentially hasten the transition to a more sustainable and secure energy system [13]. During times of crisis, governments' actions come under intense scrutiny. In addition to immediate, short-term responses, many governments are embarking on more enduring strategies. Some explore options to enhance or broaden their oil and gas sources, while others focus on expediting structural transformations. Given the persistently vulnerable state of energy markets, the current energy shock is a stark reminder of the fragility and lack of sustainability inherent in our existing energy system. Policymakers face a critical dilemma: will this crisis impede the progress of clean energy transitions or accelerate the urgency for action? While some quarters attribute the surge in energy prices to climate policies and net zero commitments, there is limited evidence to support this claim. Notably, in the regions most heavily impacted, a correlation exists between higher proportions of renewables and lower electricity costs.

The realm of clean energy presents a substantial opportunity for growth and employment and emerges as a significant arena for global economic competition. By 2030, propelled notably by the US Inflation Reduction Act, the yearly additions to solar and wind capacity in the United States experience a two-and-a-half-fold increase compared to current levels, while electric vehicle sales surge sevenfold. Ambitious new targets continue to drive an extensive expansion of clean energy initiatives in China, leading to a peak in both coal and oil consumption before the close of this decade. The accelerated deployment of renewables and enhancements in efficiency within the European Union brought about a 20% reduction in natural gas and oil demand during this decade, accompanied by a 50% decline in coal demand. This drive is spurred by an added sense of urgency to identify fresh sources of economic and industrial advantage beyond dependence on Russian gas. Japan's Green Transformation initiative contributes a significant financial boost to various technologies, including low-emissions hydrogen, ammonia, and nuclear power. Similarly, South Korea is in the process of elevating the role of nuclear power and renewables in its energy mix. India, meanwhile, continues its stride towards achieving its domestic renewable capacity goal of 500 gigawatts (GW) by 2030, with renewables fulfilling nearly two-thirds of the country's rapidly escalating electricity demand [14]. As we look toward the future, the trajectory of global energy consumption is a critical concern. Projections indicate that energy consumption worldwide is expected to continue growing, primarily driven by factors such as population growth, economic development, urbanization, and technological advancements. This growth in energy demand is anticipated to persist across various sectors, including residential, commercial, transportation, and especially industrial sectors. Governments, industries, and policymakers around the world recognize the need to address these energy consumption trends in a sustainable manner. Efforts are being made to improve energy efficiency, develop cleaner technologies, and transition towards renewable energy sources to mitigate the impact of increasing energy demand on the environment and overall energy security. By taking proactive measures to address energy consumption, it is possible to manage the trajectory of global energy use and work towards a more sustainable future.

1.4 The potential of hydrogen as a clean and sustainable energy source

In contrast to petroleum fuels, hydrogen stands as a clean, renewable, and environmentally friendly energy source. Hydrogen energy uniquely offers both clean commercial electricity and fuel for transportation (Figure 1.3). The utilization of hydrogen in fuel cells through cold combustion generates electrical power, while in internal combustion engines (ICEs) of vehicles, hot combustion functions similarly to conventional fossil fuels. Both processes yield water as a byproduct. Hydrogen can be generated from various methods involving water dissociation, effectively returning to water upon combustion. This native production capability makes hydrogen less reliant on imports compared to petroleum. The primary advantage of hydrogen as a fuel is its absence of CO_2 emissions. Recognized as a renewable and sustainable option, hydrogen holds the potential to diminish global fossil fuel consumption and address climate change. Hydrogen has the potential to be derived from a variety of fossil-based sources, including coal, natural gas, and hydrocarbons (Figure 1. 4). This is achieved through various production methods like gasification, reforming, and pyrolysis. However, it's important to note that, in the current century, the energy needed for processes like hydrogen production, storage, distribution, and transportation primarily originates from fossil fuel resources.

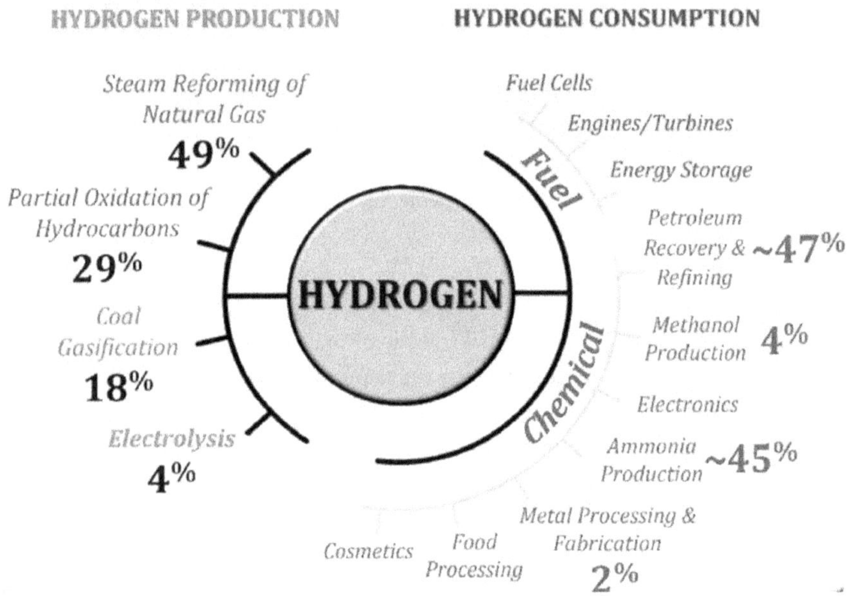

Figure 1.3 Clean hydrogen policy priorities [15]

It's noteworthy that fossil fuels hold certain advantages, such as widespread availability, comparatively lower costs, and existing infrastructure for delivery and distribution. As a result, these factors are likely to significantly influence hydrogen production in the immediate to foreseeable future. In this scenario, the extensive consumption of fossil fuels will lead to substantial environmental harm by emitting CO_2, NO_x, SO_x, and other harmful pollutants. As a solution to curbing the reliance on fossil fuels for hydrogen-related processes such as production and storage, efforts should be directed toward advocating sustainable energy-driven methods. This entails encouraging sustainable energy-based hydrogen production, storage, distribution, and transportation. Simultaneously, there should be a concerted effort to enhance and diversify the utilization of sustainable energy resources essential for various hydrogen-related activities.

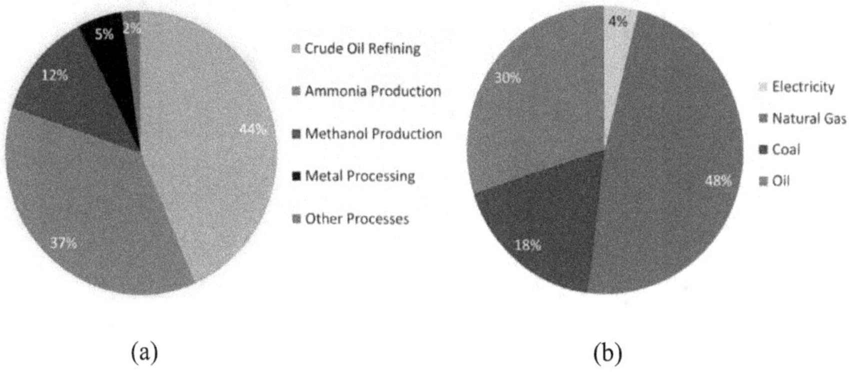

Figure 1.4 (a) Global hydrogen consumption (b) Global hydrogen production based on the source [16]

If implemented, hydrogen production from non-fossil fuel origins, such as water dissociation using a variety of sustainable energy sources and process heating methods, including those based on nuclear processes, could emerge as a primary fuel solution. This has the potential to address a multitude of objectives, including energy supply and security, facilitating the transition to a hydrogen-based economy, environmental improvement, as well as sustaining social, sectoral, technological, industrial, economic, and governmental aspects [17]. The prospects and obstacles associated with hydrogen arise from its energy characteristics. Hydrogen boasts a higher specific energy, indicating greater energy content per unit of mass compared to most hydrocarbon fuels. However, its volumetric energy density is the lowest, necessitating pressurization or liquefaction to render hydrogen suitable as a fuel (Figure 1.5). These fundamental properties not only determine the intrinsic value of hydrogen but also influence its suitability for diverse potential end-use scenarios.

The potential of hydrogen as a clean and sustainable energy source is garnering increasing attention in the global pursuit of mitigating climate change and transitioning to a more environmentally friendly energy landscape. Hydrogen, an element abundant in nature, holds remarkable promise due to its ability to produce energy without emitting greenhouse gases, making it a potential game-changer in combating the detrimental impacts of fossil fuel consumption. This potential lies not only in its capacity to produce electricity but also in its role as a versatile energy carrier with applications across various sectors [18].

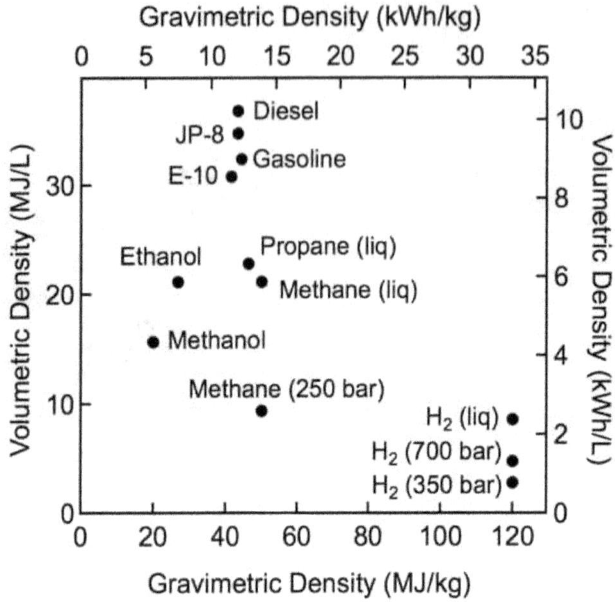

Figure 1.5 Volumetric and gravimetric density of various fuels compared to hydrogen [19]

As per the International Energy Agency (IEA), the attainment of Net-Zero Emissions (NZE) by 2050 will heavily rely on hydrogen as a clean energy carrier and as a precursor to hydrogen-based fuels, also referred to as electrofuels. Envisioning this scenario, the worldwide utilization of hydrogen would grow from under 90 million metric tons in 2020 to over 200 million metric tons in 2030, while the portion of low-carbon hydrogen would escalate from 10% in 2020 to 70% in 2030. Projected for 2050, it is expected that approximately 30% of the electricity consumption will be allocated towards the production of green hydrogen and its related products, including electrofuels like e-ammonia and e-methanol [20]. As indicated by the International Renewable Energy Agency (IRENA), hydrogen and its derivatives are projected to constitute approximately 12% of the overall global final energy consumption by 2050. This objective entails the need for nearly 5000 GW of hydrogen electrolysis capacity, a substantial increase from the current capacity of 0.3 GW [21]. Hydrogen has the potential to contribute to environmental sustainability due to its attributes as a high-quality and environmentally friendly energy source, serving as a non-toxic and clean carrier of energy that yields non-toxic exhaust emissions. Moreover, hydrogen holds the capacity to enhance environmental stability through its secure transport capabilities via pipelines, minimizing risks associated with its movement. Furthermore, Hydrogen is well-fitted for tapping into renewable energy sources and maintaining energy reserves for extended periods. This is achieved by utilizing diverse production techniques to derive hydrogen from non-fossil fuel sources. Additionally, hydrogen's ability to be stored for relatively prolonged periods, compared to electricity, furthers its role in ensuring long-term energy resource sustainability. Hydrogen contributes to establishing various new industrial sectors, thereby promoting industrial sustainability. This is attributed to the versatility of hydrogen, which finds advantageous application as a chemical feedstock across industries such as petrochemicals, food, microelectronics, metallurgy, and polymer synthesis, while also serving as a crucial energy carrier within clean and sustainable energy systems. Additionally, hydrogen facilitates the transition towards a hydrogen-based economy, effectively supporting both economic and social sustainability. Its adaptability allows it to be integrated across all facets of the economy and daily life, reinforcing its role as a driving force in ensuring holistic economic and societal stability.

Hydrogen's journey as an energy source begins with its production, which can be achieved through different methods, such as the electrolysis of water using renewable sources like solar or wind energy, and the use of sustainable bioenergy feedstocks. These methods hold the key to unlocking hydrogen's true potential as they provide a pathway to decarbonize its production process, rendering it truly clean from end to end. This stands in stark contrast to the conventional method of deriving hydrogen from fossil fuels, which produces carbon emissions, defeating the purpose of transitioning to cleaner energy sources. The advantages of hydrogen go beyond its environmentally friendly characteristics. Its versatility enables it to be utilized in a variety of applications, from powering vehicles to providing electricity and heat for industrial processes and homes. Hydrogen fuel cells have the potential to revolutionize the transportation sector by offering zero-emission alternatives to internal combustion engines, thereby reducing air pollution and dependence on fossil fuels. Furthermore, the potential of hydrogen is not confined to its direct applications; it can play a pivotal role in energy storage, enhancing the reliability and stability of renewable energy sources like solar and wind by storing excess energy and releasing it when needed. The size of the worldwide hydrogen generation market was assessed at USD 155.35 billion in the year 2022. This valuation reflects the total value of activities within the hydrogen production sector. Looking ahead, it is projected that the market will undergo significant growth, with a compound annual growth rate (CAGR) of 9.3% from 2023 to 2030. This indicates a consistent and considerable annual expansion of the market size over the specified period. The growth is expected to be driven by a range of factors, including increasing demand for clean and sustainable energy sources, advancements in hydrogen production technologies, and the adoption of hydrogen in various industries, including power generation, transportation, and industrial processes. This expansion signifies a shifting global focus toward hydrogen as a vital component of the future energy landscape [22].

While the potential of hydrogen is promising, the establishment of a hydrogen infrastructure, including storage, transportation, and distribution, poses a significant hurdle, and global collaboration and investment are crucial to overcome these challenges and maximize hydrogen's potential. Governments, industries, and research institutions around the world are increasingly focusing on research and development efforts to enhance hydrogen production, storage, and utilization technologies. Policies and incentives that encourage the adoption of hydrogen technologies, along with international cooperation in standardization and regulatory frameworks, are instrumental in fostering the growth of the hydrogen economy. As technological advancements continue to unfold and collaborative efforts gain momentum, the realization of hydrogen's potential is not only plausible but imperative for a greener and more sustainable future.

1.5 The potential benefits of hydrogen

In an era of increasing environmental consciousness, shifting geopolitical landscapes, and growing concerns over energy security, hydrogen emerges as a dynamic force that has the potential to reshape our world (Figure 1.6). With its unique properties and versatile applications, hydrogen offers a myriad of benefits that extend far beyond its immediate environmental advantages. For decades, nations around the globe have been at the mercy of volatile oil markets, often leading to economic instability and geopolitical tensions. One of the core advantages of hydrogen lies in its diverse range of production methods and feedstocks. While fossil fuels like oil are typically sourced from specific regions, hydrogen can be produced using domestic resources such as natural gas, biomass, and renewable energy sources like solar and wind.

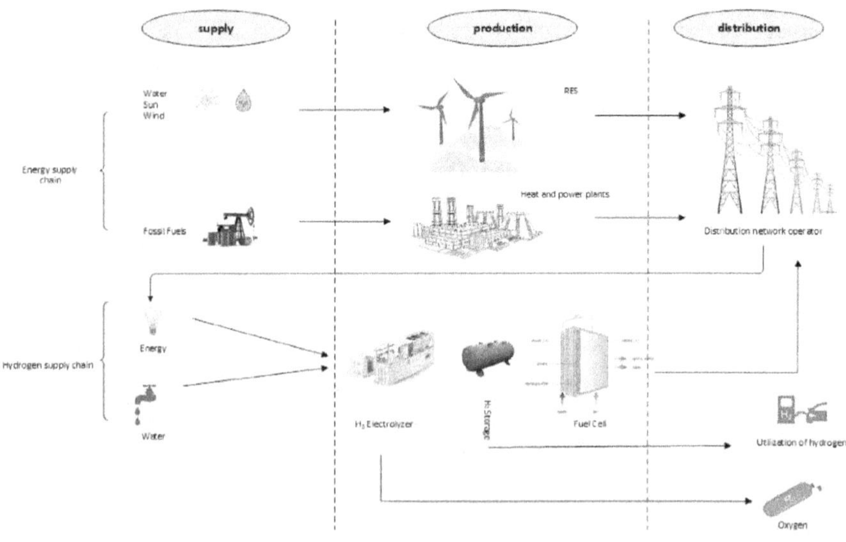

Figure 1.6 The hydrogen supply chain vs energy supply chain [19]

Hydrogen, with its capacity to be derived from various domestic sources, presents a promising solution to reduce the reliance on imported fossil fuels. The geopolitical dynamics associated with oil supply and demand can create vulnerabilities for nations heavily reliant on oil imports. Shifts in political alliances, conflicts, and trade tensions can disrupt oil supplies and lead to economic instability. By embracing hydrogen technologies, countries can decouple their energy needs from geopolitical considerations, reducing exposure to these risks and creating a more stable energy environment. The volatility of oil prices has a significant impact on national economies and industries. Fluctuations in oil prices can lead to inflation, trade imbalances, and reduced economic growth. Hydrogen offers an avenue to stabilize energy costs, especially when produced from locally available and diverse feedstocks. This stability in energy prices can provide economic predictability and resilience, shielding economies from the roller-coaster of oil price fluctuations. On the other hand, the global oil market is susceptible to sudden disruptions caused by conflicts, natural disasters, or technical failures. These disruptions can have cascading effects on energy supplies, transportation, and industries. Hydrogen, with its capacity for decentralized production and distribution, offers a buffer against such disruptions. Local hydrogen production facilities can continue functioning even when global oil supply chains are disrupted, ensuring a more secure energy supply during times of crisis. The transportation sector is a major consumer of oil, and by transitioning to hydrogen-powered vehicles and technologies, countries can alleviate the pressure of oil dependency, thus mitigating the adverse effects of market fluctuations and political uncertainties. Hydrogen fuel cell vehicles offer zero-emission mobility while utilizing domestically produced hydrogen. This transition can have a compounding effect on reducing oil dependence by curbing the demand for gasoline and diesel.

In an interconnected world where disruptions in one region can reverberate globally, energy security becomes a paramount concern. Energy security is a critical aspect of a nation's stability and development, as it ensures the availability, affordability, and reliability of energy sources. Historically, the global energy landscape has been marked by vulnerabilities stemming from reliance on limited and geopolitically sensitive fossil fuel resources. Traditional energy systems often rely on centralized production facilities and lengthy supply chains, making them vulnerable to disruptions caused by geopolitical conflicts, natural disasters, or technical failures. Hydrogen, with its potential for distributed production and local generation facilities, provides an avenue to diversify energy sources and enhance energy security. Unlike conventional fossil fuels, hydrogen can be produced using various methods and feedstocks, ranging from natural gas to water electrolysis powered by renewables or nuclear energy. This diverse production capability not only mitigates supply chain vulnerabilities but also reduces the potential impact of supply disruptions caused by geopolitical conflicts or natural disasters. In other words, the diversification provided by hydrogen enables countries to create a more resilient energy mix that reduces the risk associated with relying on a single source (Figure 1.7).

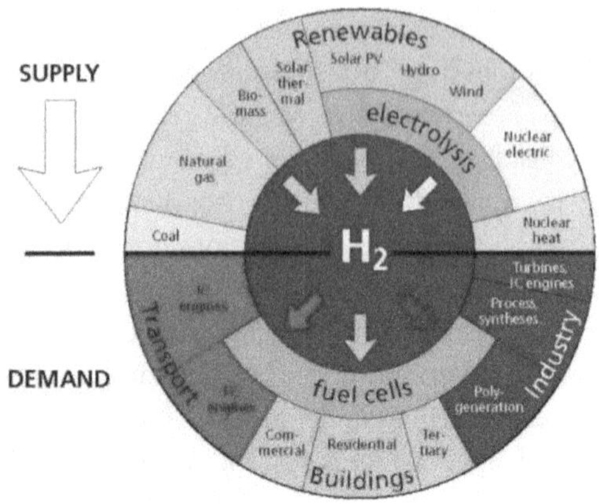

Figure 1.7 Hydrogen supply and demand [23]

The decentralized nature of hydrogen production and distribution lends itself to the creation of resilient and robust energy systems. Communities, industries, and even nations can establish local hydrogen production facilities, reducing their susceptibility to external shocks and fostering a more resilient energy infrastructure [24]. Hydrogen technologies align with climate goals, making them valuable assets in a world facing the impacts of climate change. By transitioning to cleaner energy sources like hydrogen, nations can enhance their resilience to climate-related challenges (Figure 1.8) [25]. Hydrogen can be utilized for power generation, heating, and industrial processes, reducing greenhouse gas emissions and contributing to a more climate-resilient energy system (Figure 1.9).

Embracing hydrogen technologies offers countries the opportunity to become self-sufficient in energy production and catalyze economic growth on multiple fronts. Hydrogen technology development opens up new markets and industries, leading to job creation across the value chain. From research and development to manufacturing, installation, and maintenance of hydrogen infrastructure, a robust hydrogen sector generates employment opportunities. Moreover, the growth of hydrogen-related industries can contribute to upskilling the workforce and providing specialized technical jobs, supporting a skilled labor pool, and driving economic growth. The development of hydrogen technologies requires innovation and collaboration across various sectors, including energy, transportation, manufacturing, and materials science.

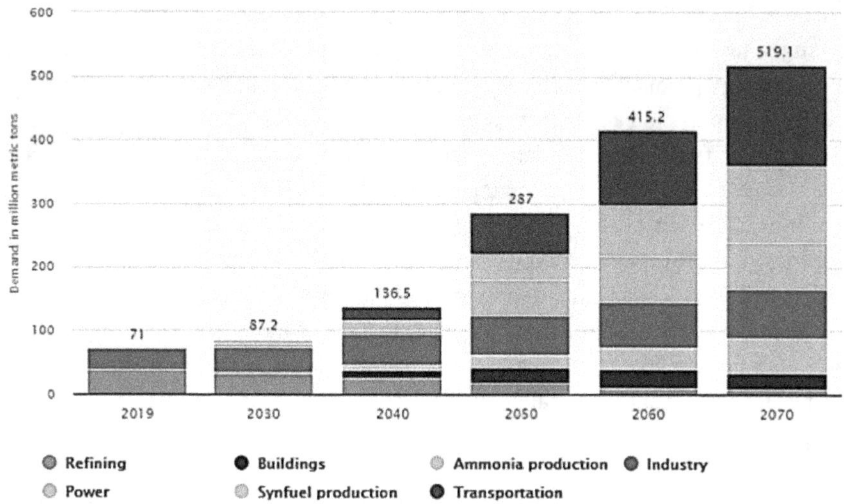

Figure 1.8 The anticipated global hydrogen demand by sector by 2070 [19]

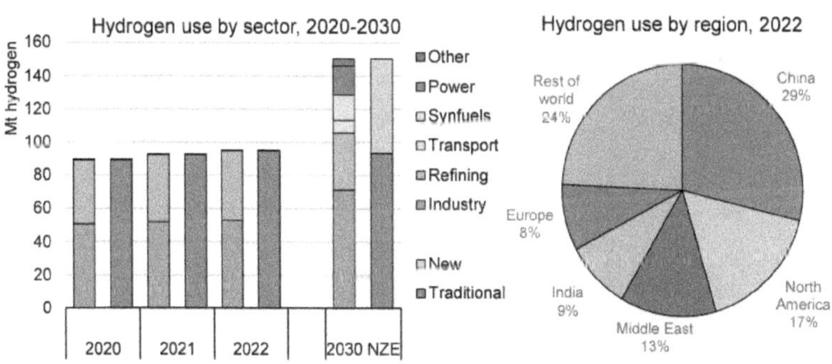

Figure 1.9 Hydrogen use by sector and by region [26]

As nations invest in research and development of hydrogen solutions, technological breakthroughs are likely to emerge. These advancements have spillover effects, potentially leading to the creation of new technologies and intellectual property, driving economic growth, and enhancing competitiveness on the global stage. Countries with abundant renewable resources, such as solar and wind energy, can leverage hydrogen production for export. Green hydrogen produced from renewable sources can be transported and traded globally, providing a new revenue stream for hydrogen-exporting nations. This can improve trade balances, create export-oriented industries, and enhance economic resilience. As hydrogen technologies mature and scale, they can contribute to reducing the overall cost of energy production and consumption. The use of hydrogen in various sectors, such as transportation and industrial processes (Figure 1.10), can lead to energy cost savings over time. This affordability can stimulate economic growth by reducing businesses' energy expenses, allowing them to allocate resources to other productive activities. The establishment of hydrogen infrastructure, including production facilities, transportation networks, and distribution systems, requires substantial investment. This investment not only creates jobs during the construction phase but also stimulates economic activity in the long term. Moreover, the deployment of hydrogen infrastructure can catalyze complementary infrastructure developments, such as renewable energy installations and smart grids.

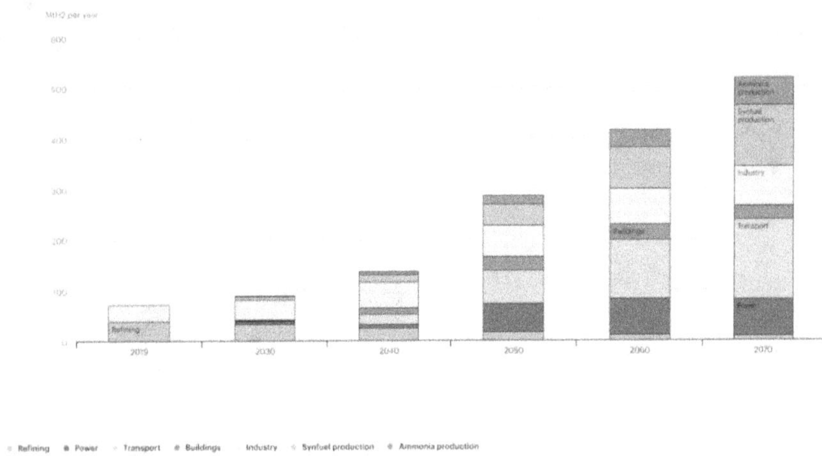

Figure 1.10 Global hydrogen demand by sector [25]

In terms of regional development, hydrogen projects can be strategically located in regions that have faced economic challenges or experienced a decline in traditional industries. These projects can act as anchors for economic revitalization, attracting investments and driving economic activity in areas that require rejuvenation. As hydrogen gains prominence as a clean and sustainable energy solution, countries that position themselves as leaders in the hydrogen economy can attract foreign direct investment. Investors seeking opportunities in renewable energy, technology, and infrastructure can be drawn to countries with favorable hydrogen policies and conducive business environments. Hydrogen technologies can support the transition to a circular economy by enabling the recycling of waste materials and the conversion of waste streams into valuable resources. This circular approach not only contributes to environmental sustainability but also presents economic opportunities by generating value from previously discarded materials. As nations embark on the journey towards a sustainable energy future, hydrogen stands as a driving force that not only addresses environmental challenges but also acts as an engine for economic prosperity and improved living standards [27].

1.6 Challenges facing the widespread adoption of hydrogen fuel

While the world looks to hydrogen fuel as a transformative solution for sustainable energy, the journey toward widespread adoption is not without its challenges. From technical intricacies to economic complexities and political considerations, the exploration of hydrogen technology is accompanied by a landscape of obstacles that require careful navigation. Efficient hydrogen production (Figure 1.11) is a critical puzzle piece in the broader picture of sustainable energy transition. The cost-effective production of hydrogen, particularly green hydrogen from renewable sources, remains a challenge. Electrolysis, a vital method for producing green hydrogen, requires advancements in electrolyzer technology to enhance efficiency and reduce costs. Safe storage and transportation are pivotal components of a successful hydrogen economy. Hydrogen, while a clean and efficient energy carrier, possesses unique properties that necessitate careful handling to mitigate potential risks. Given its low density, hydrogen requires specialized storage solutions to achieve adequate energy density for practical use. Challenges arise in selecting materials that can safely store hydrogen at high pressures or cryogenic conditions. Additionally, transporting hydrogen over long distances demands robust infrastructure to prevent leaks and ensure public safety. Addressing these challenges involves advancing materials science, engineering secure storage vessels, and developing leak-detection technologies.

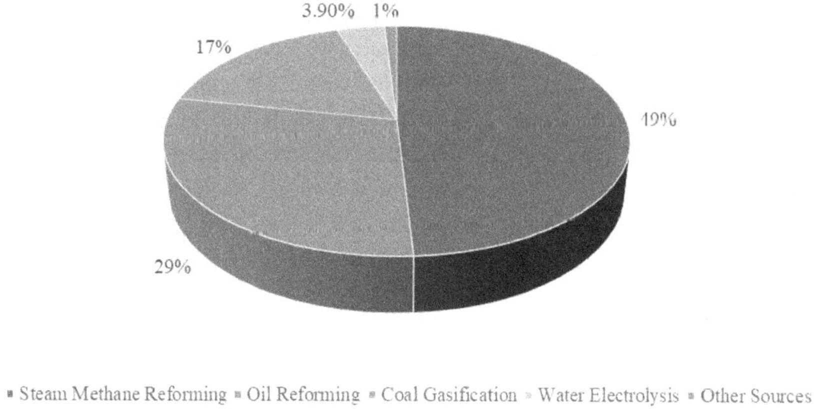

Figure 1.11 Global hydrogen production by method

A comprehensive approach encompassing rigorous safety standards, regulations, and public awareness campaigns is essential to building public confidence and realizing the full potential of hydrogen as a safe and sustainable energy source. Infrastructure development is a critical factor in the successful adoption of hydrogen technology. The transition to a hydrogen-based economy requires the establishment of a comprehensive network encompassing production, distribution, and refueling or recharging facilities. This entails upgrading or building new hydrogen production plants, pipelines, storage tanks, and refueling stations to ensure a seamless supply chain [28]. Moreover, integrating hydrogen infrastructure with existing energy systems and grids is essential for maximizing its benefits across sectors such as transportation, industry, and power generation. The complexity of this task is underscored by the need to adhere to safety standards, regulatory frameworks, and urban planning considerations. A coordinated effort involving governments, industry stakeholders, and research institutions is imperative to create a robust and interconnected hydrogen infrastructure capable of supporting widespread adoption while fostering economic growth and environmental sustainability. Besides these technical issues, the challenge of high initial costs presents a significant economic complexity in hydrogen adoption. While hydrogen holds immense potential as a clean and versatile energy carrier, the establishment of production, storage, and distribution infrastructure demands substantial upfront investments. The development of advanced technologies for efficient hydrogen production and storage, along with the establishment of refueling stations, can lead to elevated costs compared to traditional energy sources. This financial hurdle could deter private sector investments and hinder rapid deployment. Moreover, the economics of hydrogen systems are intricately linked to scale, as achieving cost competitiveness often relies on achieving economies of scale through large-scale production and utilization. In addition, the establishment of a hydrogen-based economy has encountered obstacles due to elevated expenses, intricate supply chain dynamics, regulatory frameworks, and policies. The production cost of green hydrogen exceeds that derived from fossil fuels, although the decreasing costs of renewables and the realization of economies of scale offer the potential for future economic viability. The complexity of hydrogen production processes, its applications across various sectors, and the intricacies of its sourcing and supply chain present challenges when compared to the more streamlined oil and gas industry. Additionally, the transportation and storage of hydrogen pose significant challenges, demanding substantial investments in infrastructure upgrades. Traditionally, governmental policy support for hydrogen has been minimal on a global scale, with preferences leaning towards alternative technologies and applications despite hydrogen's potential for substantial impact. Finally, existing standards for hydrogen use either lack comprehensive frameworks or have not been updated to meet evolving needs.

Addressing these challenges requires innovative financial mechanisms, government incentives, and collaborative efforts among industry players to drive down initial costs and accelerate the transition toward a hydrogen-based energy landscape. Cost competitiveness is a central economic complexity that shapes the widespread adoption of hydrogen technology. While hydrogen holds promise as a clean and versatile energy solution, its current production, storage, and distribution costs can often be higher than conventional energy sources. Achieving cost competitiveness is essential for hydrogen to become a viable alternative on a global scale. High initial investments, technology development, infrastructure setup, and operational expenses contribute to these higher costs. However, advancements in research, development, and commercialization can lead to innovations that lower the overall cost of hydrogen production and utilization. Moreover, increasing demand and scaling up hydrogen-related industries can create a positive feedback loop by driving cost reduction. Governments, industries, and research institutions must collaboratively address these economic complexities through targeted policies, incentives, and technological breakthroughs to accelerate the cost-competitive integration of hydrogen into our energy landscape. Investment uncertainty represents a significant economic complexity that hampers the broad acceptance of hydrogen technology. The development and implementation of hydrogen-related projects require substantial upfront investments, often involving unproven technologies and infrastructural changes. This uncertainty can deter private and public sector investors as they grapple with the potential risks and returns associated with these endeavors. Fluctuating energy prices, policy shifts, technological advancements, and market dynamics further contribute to this uncertainty. To address this challenge, it is crucial for governments, financial institutions, and industries to collaborate and establish clear and consistent regulatory frameworks, incentives, and funding mechanisms that mitigate investment risks. Creating a stable and favorable atmosphere for investors is essential to encourage the required financial dedication that can propel the advancement of hydrogen technology and its seamless integration into the worldwide energy sector [29]. Beyond the technical, economic, and political dimensions, public perception and societal acceptance of hydrogen technology play a pivotal role in its adoption. Awareness and education play a pivotal role in the domain of public perception and acceptance of hydrogen technology. Lack of adequate understanding and familiarity with hydrogen as an energy source can contribute to skepticism or misconceptions among the general population. Effective awareness campaigns, educational initiatives, and transparent communication about the benefits and safety aspects of hydrogen technology are essential to build public trust and facilitate its broader acceptance [30].

Convincing consumers to embrace hydrogen-powered products, such as fuel cell vehicles, necessitates addressing concerns about availability, affordability, and refueling infrastructure. Passenger fuel cell vehicle (FCV) deployment is primarily concentrated in OECD nations (The Organization for Economic Co-operation and

Development), which tend to be wealthier and more developed compared to their non-OECD counterparts. As illustrated in Figure 1.12, approximately 74% of the projected deployment by 2050 is attributed to OECD nations, with notable contributions from strong hydrogen-use countries like Japan, North America, and Western European nations. China and Southeast Asia are the predominant contributors from non-OECD nations. Expanding on the hydrogen penetration findings, it becomes feasible to estimate regional vehicle deployment figures for 2050, factoring in average vehicle occupancy and annual travel distances for each respective region [31].

The successful adoption of hydrogen technology by consumers requires not only technological advancements but also effective marketing strategies, clear communication of benefits, and addressing any perceived barriers. Moreover, the establishment of a supportive infrastructure, such as refueling stations for hydrogen vehicles, is crucial to encourage consumer confidence and facilitate the seamless incorporation of hydrogen-based solutions into their daily routines [32]. Environmental concerns encompass the apprehensions and considerations related to the ecological impact of hydrogen technology adoption. While hydrogen is touted as a clean and low-emission energy source, its production, storage, and transportation can still generate environmental challenges. These concerns range from potential greenhouse gas emissions during hydrogen production to the energy and resource intensity of its manufacturing processes. Additionally, questions arise about the sustainability of hydrogen production methods and the potential impacts on local ecosystems and water resources. Addressing these environmental concerns requires thorough lifecycle assessments, the implementation of sustainable practices, and the integration of renewable energy sources in hydrogen production. Achieving a balance between the promised benefits of hydrogen technology and its potential environmental consequences is a pivotal aspect in fostering public acceptance and advancing the widespread adoption of this technology [33]. The production of hydrogen through the electrolysis of water serves as a means to support the widespread generation of renewable electricity and functions as an energy storage solution. As the costs of renewables continue to decrease, there is a prospect for a corresponding reduction in hydrogen production costs, enhancing its competitiveness. Additionally, hydrogen plays a role in diminishing a nation's dependence on oil imports and contributes to the growth of the domestic job market. Furthermore, it positions countries to actively participate in the ongoing global energy transition, unlocking economic opportunities associated with this transition.

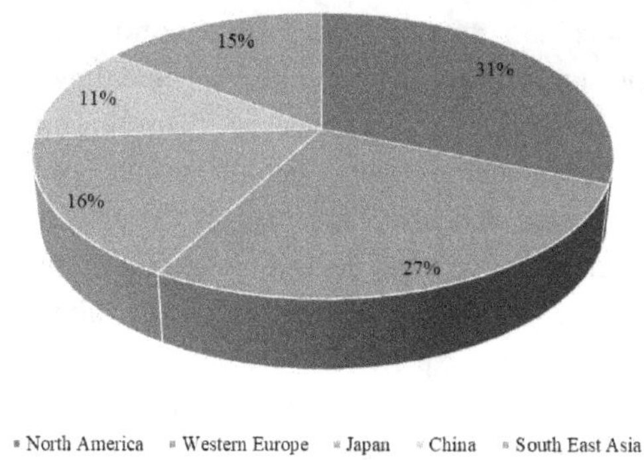

(a)

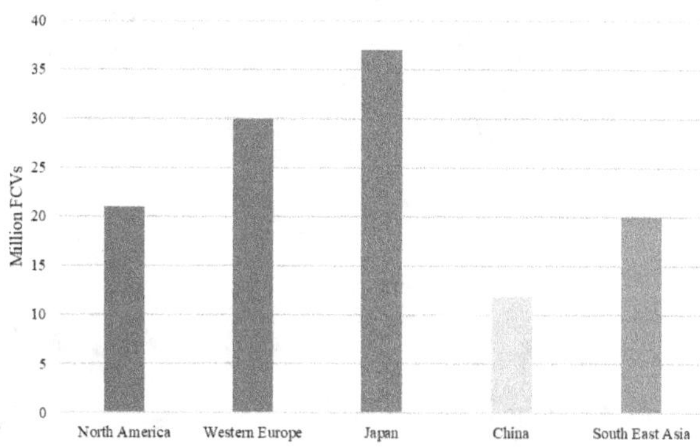

(b)

Figure 1.12 (a) Distribution of hydrogen regionally (% of the total global MTOE) for transportation, (b) Estimated regional deployment figures for FCVs in the year 2050

There is a renewed momentum observed as countries and companies increasingly commit to net-zero emission targets, recognizing hydrogen's potential to decarbonize challenging sectors. Currently, at least 43 countries have either established or are in the process of formulating strategies or roadmaps for a hydrogen-centric economy. The majority of government-backed research and development (R&D) funding for hydrogen is concentrated in Europe, the United States, Japan, and China. Table 1.1 summarizes hydrogen roadmaps and strategies of leading countries.

To date, more than 30 nations have unveiled their hydrogen strategies, and many others are in the process of developing their own (Figure 1.13). The hydrogen industry, once on the periphery, has now risen to prominence, sparking global enthusiasm and interest in the potential opportunities offered by hydrogen. Global hydrogen markets have undergone significant transformations, leading to a better understanding of the applications, both existing and emerging, where hydrogen can make the most substantial contributions to decarbonization and economic advancements. Numerous regions with established hydrogen strategies are now putting in place financial incentive mechanisms with the aim of expediting investments and fostering the development of the hydrogen industry. In light of the increasing support, the global supply and demand for hydrogen, particularly green hydrogen, are anticipated to experience substantial growth (Figure 1.14). The evolving economics of green hydrogen production contribute to sustaining this policy momentum. While the current hydrogen production predominantly relies on fossil fuels, there has been a notable surge in the number of electrolyzer projects.

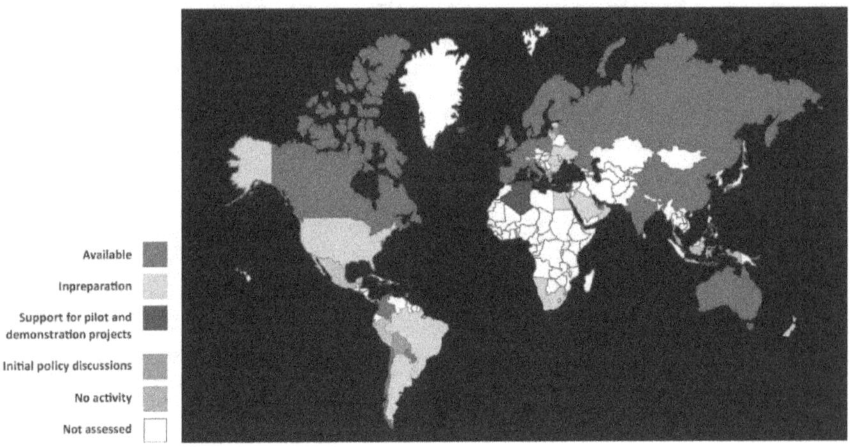

Figure 1.13 National hydrogen strategies as of August 16, 2022 [34]

The remarkable decline in renewable prices, coupled with the anticipated reduction in electrolyzer costs as technology advances and deployment scales up, leads to a consensus that green hydrogen production will become economically viable. Figure 1.15 shows low-emission hydrogen production projects throughout the world. The green hydrogen market could reach an annual value of US$120–US$175 billion by 2050 based on a range of projected prices. With the ongoing global push for decarbonization, current policy momentum, and improvements in the economics and durability of end-use technologies like fuel cells, hydrogen could potentially meet 7%–18% of global final energy demand by 2050.

Table 1.1 Hydrogen roadmaps and strategies of leading countries

	Current Hydrogen Demand	Policy Target Demand	Capital Allocated (US$)	Focused Hydrogen Source	Demand Focus			Export/Import Focus
					Industry	Transport	Others	
European Union		6 GW capacity by 2024; 40 GW by 2030; 10 MMTPA green H2 by 2030	609 billion	Low Carbon-Blue / Green	1.Chemical feedstock 2. Refining	1.Medium and Heavy Duty 2. Buses 3. Rail		
Germany	1.65MMTPA	2.7-3.3 MMTPA by 2030	15-20 billion	Carbon free - Blue / Green	1. Iron and Steel 2. Chemical feedstock 3. Refining	1.Medium and Heavy Duty 2. Buses 3. Rail		Import
France	0.9MMTPA	6.5 GW via electrolysis by 2030	> 7 billion	Low Carbon - Blue	1. Iron and Steel 2. Chemical feedstock 3. Refining 4. Others	1.Medium and Heavy Duty 2. Buses 3. Rail 4. Aviation		Export
Netherlands	1.5MMTPA	Not Available	40-55 million/yr	Blue / Green	1. Passenger Vehicle 2. Medium and Heavy Duty 3. Buses 4. Rail	1. Passenger Vehicle 2.Medium and Heavy Duty 3. Buses 4. Rail	Heating	EU Export/ Import Hub
Hungry	160ktpa	36 ktpa (low carbon) + 138 ktpa (grey) by 2030	450 million	Low Carbon - Grey / Blue	1. Chemical feedstock 2. Refining	1.Medium and Heavy Duty 2. Buses	Heating	
Portugal	150 ktpa	2-2.5 GW via electrolysis by 2030 400 ktpa overall by 2030	No dedicated capital	Green	1. Iron and Steel 2. Chemical feedstock 3. Refining 4. Others	1. Passenger Vehicle 2. Medium and Heavy Duty 3. Buses	Heating	Export
Spain	0.5MMTPA	4 GW via electrolysis by 2030	No details	Green	1. Chemical feedstock 2. Refining			Export
United Kingdom	0.7MMTPA	5 GW electrolysis capacity by 2030	2 billion	Blue/Green	1. Chemical feedstock 2. Iron and Steel	1. Medium and Heavy Duty 2. Buses 3. Rail 4. Aviation 5. Shipping		Export
Norway			23 million	Clean	1. Chemical feedstock	1. Maritime		
Japan	2MMTPA	3 MMTPA by 2030 and 20 MMTPA by 2050 (5-30 by 2050)	935 million / yr	Blue		1. Passenger Vehicle	1. Heating 2. Power	Import
South Korea	220 ktpa	3.9 MMTPA by 2030 and 27 MMTPA by 2050	653 million/ yr	Grey / Blue / Green		1. Passenger Vehicle 2. Medium and Heavy Duty 3. Buses	Power	Import
United States	10MMTPA		> 15 billion	Low Carbon-Blue/ Green / Others	1. Refining 2. Others	1. Passenger Vehicle 2. Medium and Heavy Duty 3. Buses 4. Aviation	1. Heating 2. Power 3. Energy storage	
Canda	3MMTPA	20 MMTPA	1.2 billion	Low Carbon Intensity-Grey / Blue	1. Iron and Steel 2. Chemical feedstock 3. Refining 4. Others	1. Passenger Vehicle 2.Medium and Heavy Duty 3. Buses 4. Rail	Heating	Export
Australia	650 ktpa		278 million (annual support)/ yr	Clean-Blue / Green	1. Chemical feedstock	1.Medium and Heavy Duty 2. Buses	Heating	Export
Chile	58.5 ktpa	5 GW/a (2025) 25 GW/a (2030)	50 million	Green	1. Chemical feedstock 2. Refining	1.Medium and Heavy Duty 2. Buses	Heating	Export
China	22MMTPA	35 MMTPA (by 2030); 160 MMTPA (by 2050)	13 million	Green (long-term)		1. Passenger Vehicle 2.Medium and Heavy Duty 3. Buses	Power	
Russia	2-3.5MMTPA	7 MMTPA by 2035 and 33 MMTPA by 2050 (export only)	1.2 billion	Low Carbon-Blue/Nuclear	Refining	Rail		Export

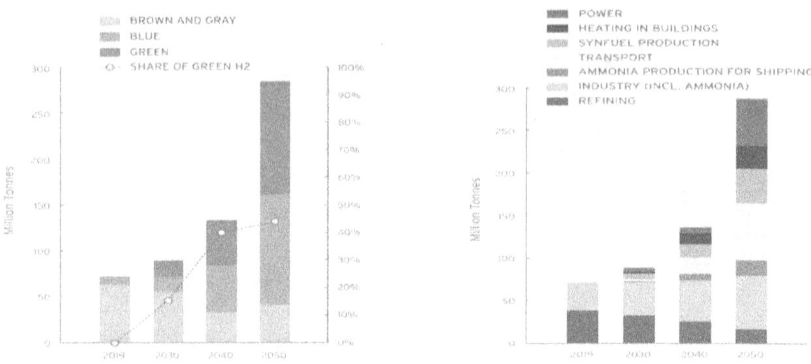

(a) (b)

Figure 1.14 (a) Global hydrogen production outlook (b) Global hydrogen demand outlook

If pursued diligently, net-zero targets could lead to a significant upside, with the IEA projecting a potential hydrogen demand of 528 million tons under their net-zero scenario by 2050, resulting in the mitigation of 1.6-3.5 gigatons of greenhouse gas emissions annually. The near-term demand for hydrogen is being driven by industrial decarbonization in both energy and feedstock, while longer-term opportunities are expected in transport, power, and even in decarbonizing the shipping and airline industry. Anticipating the projected growth in green hydrogen, there is an expectation of exponential growth in electrolyzer capacity. The electrolyzer market reached gigawatt-scale in 2022, driven by increased installations in China. Approximately 40 GW of electrolyzers are already proposed by 2030, with the potential for significant growth if a substantial decline in green hydrogen prices enables the replacement of blue hydrogen with the green counterpart. Extrapolating the upper global demand for hydrogen projected in national strategies for the year 2050 indicates a potential demand of up to 9,000 terawatt-hours (TWh) or approximately 270 million tons of hydrogen per year. This aligns with the current global supply of primary energy from renewable sources. Nevertheless, numerous countries may be unable to fulfill their hydrogen needs with their domestic hydrogen capacities, often due to restricted renewable energy potential. Countries like Germany, Japan, and South Korea anticipate a significant need for importing hydrogen in the future. This necessitates the development of suitable production capacities and the corresponding transport infrastructure. Typically, the substantial investments required for this development will not be solely undertaken by the exporting countries but will be realized through international collaborations. The establishment of international hydrogen relationships is presently mainly driven by bilateral and trilateral agreements between potential exporting and importing nations. It is anticipated that the number of hydrogen partnerships will escalate rapidly in the future.

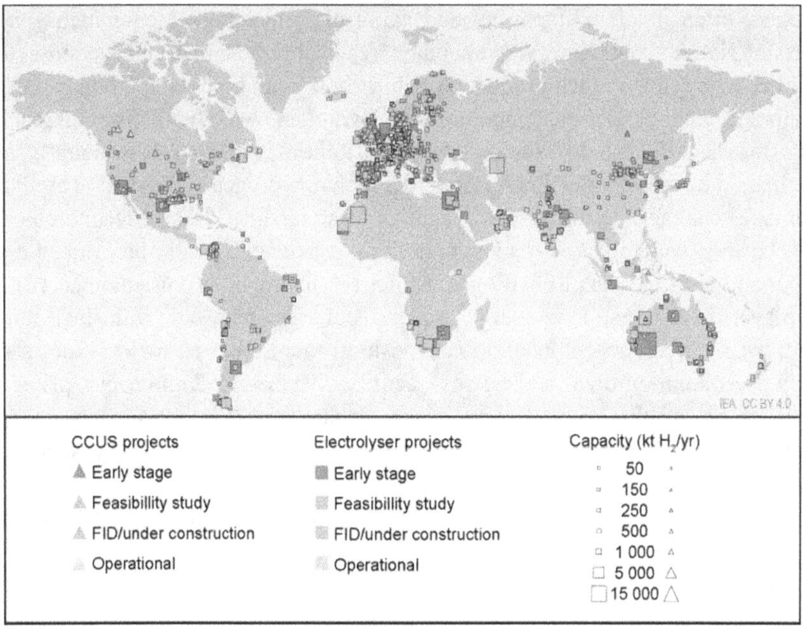

Figure 1.15 Low-emission hydrogen production projects throughout the world [26]

1.7 An overview of the hydrogen diplomacy concept

In the wake of pressing environmental challenges and the imperative to transition towards cleaner and sustainable energy sources, the concept of "Hydrogen Diplomacy" emerges as a crucial paradigm in the international arena. As nations grapple with the complexities of addressing climate change, reducing carbon emissions, and securing energy independence, the need for collaborative efforts on a global scale becomes increasingly evident. "Hydrogen Diplomacy", as explored in this book, transcends national borders, emphasizing the necessity for countries to engage in strategic alliances, negotiations, and cooperative ventures to harness the full potential of hydrogen as a clean energy carrier.

Countries worldwide are crafting unique hydrogen policies, outlining visions and strategies for integrating hydrogen into their energy landscapes. These policies serve as blueprints for domestic actions, encompassing investments in research and development, infrastructure development, regulatory frameworks, and incentives to stimulate the growth of the hydrogen sector. However, the global nature of the hydrogen economy necessitates a shift from unilateral endeavors to collaborative diplomacy. "Hydrogen Diplomacy" becomes the linchpin for fostering international cooperation, creating synergies, and addressing challenges that extend beyond national borders. One compelling reason for countries to engage in "Hydrogen Diplomacy" lies in the diverse and interconnected nature of the hydrogen value chain. Hydrogen, produced through various methods such as green, blue, or grey hydrogen, often involves importing and exporting technologies, equipment, and expertise. Nations endowed with abundant renewable resources may choose to specialize in green hydrogen production, while those with access to carbon capture and storage technologies may opt for blue hydrogen. A collaborative approach allows for the efficient distribution of responsibilities, enabling each country to capitalize on its strengths and contribute to a global hydrogen ecosystem. Through diplomatic channels, nations can negotiate partnerships, share technological advancements, and collectively overcome hurdles associated with the production, storage, and transportation of hydrogen. Moreover, hydrogen is poised to become a key player in international energy trade. Countries endowed with favorable conditions for hydrogen production can position themselves as major exporters, fostering economic growth and energy security. "Hydrogen Diplomacy" plays a pivotal role in establishing trade agreements, standardizing protocols for hydrogen transportation and storage, and addressing geopolitical considerations associated with the emergence of hydrogen as a global commodity. Collaborative frameworks not only enhance economic interdependence but also contribute to global energy security by diversifying energy sources and reducing dependence on fossil fuels.

The geopolitics of hydrogen resources further underscores the importance of diplomatic initiatives. Countries possessing significant renewable energy potential, such as solar or wind resources, may emerge as major players in green hydrogen production. However, the distribution of these resources is often uneven across the globe. "Hydrogen Diplomacy" becomes instrumental in bridging these resource gaps through international partnerships and collaborations. By fostering inclusivity and equitable access to hydrogen resources, nations can collectively work towards a more balanced and sustainable energy landscape. As the global community grapples with the urgency of mitigating climate change, achieving carbon neutrality has become a shared goal. Hydrogen, particularly green hydrogen produced using renewable energy, stands out as a viable solution for decarbonizing various sectors such as industry, transportation, and power generation. "Hydrogen Diplomacy" assumes a critical role in aligning national efforts with international climate goals, facilitating the exchange of best practices, and harmonizing standards for measuring the environmental impact of hydrogen production. Through diplomatic channels, nations can negotiate emission reduction targets, share insights on successful strategies, and collectively contribute to the global fight against climate change.

"Hydrogen Diplomacy" plays a pivotal role in shaping international collaborations focused on key aspects such as electrolysis, steam methane reforming, and the utilization of hydrogen in fuel cells or internal combustion engines. Electrolysis, as a clean method for hydrogen production using renewable energy, necessitates diplomatic efforts to facilitate knowledge exchange, joint research initiatives, and the establishment of international standards. Diplomacy becomes crucial in negotiating agreements that encourage the deployment of electrolysis technology on a global scale, fostering a harmonized and efficient transition to green hydrogen. Similarly, with steam methane reforming being a prevalent method for hydrogen production, diplomatic initiatives can focus on encouraging the implementation of carbon capture and storage technologies to mitigate emissions. Furthermore, the utilization of hydrogen in fuel cells for various applications, from transportation to stationary power, requires international cooperation to standardize regulations, share advancements, and ensure interoperability. By fostering diplomatic ties in these technological domains, nations can collectively address challenges, drive innovation, and accelerate the global adoption of hydrogen technologies, contributing to a sustainable and interconnected energy future.

"Hydrogen Diplomacy" is also essential in navigating the intricate web of regulatory frameworks and standards governing the hydrogen sector. Divergent national regulations can create barriers to the development of a seamless and integrated global hydrogen market. Through diplomatic negotiations, countries can work towards harmonizing regulatory approaches, developing international standards, and creating a conducive environment for cross-border investments. By fostering a regulatory landscape that encourages innovation and investment, "Hydrogen Diplomacy" facilitates the emergence of a robust and globally interconnected hydrogen market. Furthermore, the energy transition towards hydrogen offers an opportunity for countries to enhance their soft power and global influence. Those at the forefront of hydrogen innovation can leverage their expertise to shape international discourse, influence global agendas, and foster diplomatic relationships based on shared energy goals. By positioning themselves as leaders in "Hydrogen Diplomacy", countries can strengthen their diplomatic ties, attract foreign investments, and play a central role in shaping the future of the hydrogen economy.

Abbreviations

CAGR: Compound annual growth rate
EV: electric vehicles
FCV: Fuel cell vehicle
GHG: Greenhouse gas
ICE: Internal combustion engine
IEA: International Energy Agency
IRENA: International Renewable Energy Agency
NZE: Net-Zero Emissions
OECD: Organization for Economic Co-operation and Development
R&D: Research and development

References

[1] Narula K. Global energy system and sustainable energy Security. Lecture Notes in Energy 2019;68:23–49. https://doi.org/10.1007/978-981-13-1589-3_2/COVER.

[2] Hossain MS, Madlool NA, Rahim NA, Selvaraj J, Pandey AK, Khan AF. Role of smart grid in renewable energy: An overview. Renewable and Sustainable Energy Reviews 2016;60:1168–84. https://doi.org/10.1016/J.RSER.2015.09.098.

[3] Global CO2 emissions rose less than initially feared in 2022 as clean energy growth offset much of the impact of greater coal and oil use - News - IEA n.d. https://www.iea.org/news/global-co2-emissions-rose-less-than-initially-feared-in-2022-as-clean-energy-growth-offset-much-of-the-impact-of-greater-coal-and-oil-use (accessed June 28, 2023).

[4] Global Greenhouse Gas Emissions: 1990-2020 and Preliminary 2021 Estimates | Rhodium Group n.d. https://rhg.com/research/global-greenhouse-gas-emissions-2021/ (accessed August 17, 2023).

[5] Frequently Asked Questions (FAQs) - U.S. Energy Information Administration (EIA) n.d. https://www.eia.gov/tools/faqs/faq.php?id=427&t=3 (accessed June 28, 2023).

[6] United Nations Treaty Collection n.d. https://treaties.un.org/ (accessed August 14, 2023).

[7] Jin Y, Lenzen M, Montoya A, Laycock B, Yuan Z, Lant P, et al. Greenhouse gas emissions, land use and employment in a future global bioplastics economy. Resour Conserv Recycl 2023;193:106950. https://doi.org/10.1016/J.RESCONREC.2023.106950.

[8] Darzi Ramandi M, Khakzar Bafruei M. Effects of government's policy on supply chain coordination with a periodic review inventory system to reduce greenhouse gas emissions. Comput Ind Eng 2020;148:106756. https://doi.org/10.1016/J.CIE.2020.106756.

[9] Kamran M, Yan Z, Ahmad I, Jia Q, Ghani MU, Chen X, et al. Assessment of greenhouse gases emissions, global warming potential and net ecosystem economic benefits from wheat field with reduced irrigation and nitrogen management in an arid region of China. Agric Ecosyst Environ 2023;341:108197. https://doi.org/10.1016/J.AGEE.2022.108197.

[10] Patterson M, Singh P, Cho H. The current state of the industrial energy assessment and its impacts on the manufacturing industry. Energy Reports 2022;8:7297–311. https://doi.org/10.1016/J.EGYR.2022.05.242.

[11] Hosseini SE. An outlook on the global development of renewable and sustainable energy at the time of COVID-19. Energy Res Soc Sci 2020;68:101633. https://doi.org/10.1016/j.erss.2020.101633.

[12] Homepage - U.S. Energy Information Administration (EIA) n.d. https://www.eia.gov/ (accessed August 17, 2023).

[13] Hosseini SE. Transition away from fossil fuels toward renewables: lessons from Russia-Ukraine crisis. Future Energy 2022;1:2–5. https://doi.org/10.55670/fpll.fuen.1.1.8.

[14] World Energy Outlook 2022 – Analysis - IEA n.d. https://www.iea.org/reports/world-energy-outlook-2022 (accessed August 17, 2023).

[15] Hassan Q, Sameen AZ, Salman HM, Jaszczur M. A Roadmap with Strategic Policy toward Green Hydrogen Production: The Case of Iraq. Sustainability 2023, Vol 15, Page 5258 2023;15:5258. https://doi.org/10.3390/SU15065258.

[16] Jaradat M, Alsotary O, Juaidi A, Albatayneh A, Alzoubi A, Gorjian S. Potential of Producing Green Hydrogen in Jordan. Energies 2022, Vol 15, Page 9039 2022;15:9039. https://doi.org/10.3390/EN15239039.

[17] Midilli A, Dincer I. Hydrogen as a renewable and sustainable solution in reducing global fossil fuel consumption. Int J Hydrogen Energy 2008;33:4209–22. https://doi.org/10.1016/J.IJHYDENE.2008.05.024.

[18] Hosseini SE. Hydrogen fuel, a game changer for the world's energy scenario 2023. https://doi.org/10.1080/15435075.2023.2244050.

[19] Noyan OF, Hasan MM, Pala N. A Global Review of the Hydrogen Energy Eco-System. Energies 2023, Vol 16, Page 1484 2023;16:1484. https://doi.org/10.3390/EN16031484.

[20] Ibrahim OS, Singlitico A, Proskovics R, McDonagh S, Desmond C, Murphy JD. Dedicated large-scale floating offshore wind to hydrogen: Assessing design variables in proposed

typologies. Renewable and Sustainable Energy Reviews 2022;160:112310. https://doi.org/10.1016/J.RSER.2022.112310.

[21] Renewable Energy Statistics 2021 n.d. https://www.irena.org/publications/2021/Aug/Renewable-energy-statistics-2021 (accessed August 22, 2023).

[22] Global Hydrogen Generation Market Size & Share Report 2030 n.d. https://www.grandviewresearch.com/industry-analysis/hydrogen-generation-market (accessed August 18, 2023).

[23] Mohanty US, Ali M, Azhar MR, Al-Yaseri A, Keshavarz A, Iglauer S. Current advances in syngas (CO + H2) production through bi-reforming of methane using various catalysts: A review. Int J Hydrogen Energy 2021;46:32809–45. https://doi.org/10.1016/J.IJHYDENE.2021.07.097.

[24] Aguado R, Baccioli A, Liponi A, Vera D. Continuous decentralized hydrogen production through alkaline water electrolysis powered by an oxygen-enriched air integrated biomass gasification combined cycle. Energy Convers Manag 2023;289:117149. https://doi.org/10.1016/J.ENCONMAN.2023.117149.

[25] IEA – International Energy Agency n.d. https://www.iea.org/ (accessed March 11, 2022).

[26] Global Hydrogen Review 2023 – Analysis - IEA n.d. https://www.iea.org/reports/global-hydrogen-review-2023 (accessed November 27, 2023).

[27] Hosseini SE. Hydrogen has found its way to become the fuel of the future. Future Energy 2022;01:11–2. https://fupubco.com/fuen/article/view/28/19 (accessed October 3, 2022).

[28] Yazdi M, Zarei E, Pirbalouti RG, Li H. A comprehensive resilience assessment framework for hydrogen energy infrastructure development. Int J Hydrogen Energy 2023. https://doi.org/10.1016/J.IJHYDENE.2023.06.271.

[29] Abadie LM, Chamorro JM. Investment in wind-based hydrogen production under economic and physical uncertainties. Appl Energy 2023;337:120881. https://doi.org/10.1016/J.APENERGY.2023.120881.

[30] Emodi NV, Lovell H, Levitt C, Franklin E. A systematic literature review of societal acceptance and stakeholders' perception of hydrogen technologies. Int J Hydrogen Energy 2021;46:30669–97. https://doi.org/10.1016/J.IJHYDENE.2021.06.212.

[31] Chapman A, Nguyen DH, Farabi-Asl H, Itaoka K, Hirose K, Fujii Y. Hydrogen penetration and fuel cell vehicle deployment in the carbon constrained future energy system. IET Electrical Systems in Transportation 2020;10:409–16. https://doi.org/10.1049/IET-EST.2020.0014.

[32] Wang W, Li J, Li Y. Consumer willingness to purchase hydrogen fuel cell vehicles : A meta-analysis of the literature. Int J Hydrogen Energy 2023. https://doi.org/10.1016/J.IJHYDENE.2023.07.256.

[33] Harichandan S, Kar SK. An empirical study on consumer attitude and perception towards adoption of hydrogen fuel cell vehicles in India: Policy implications for stakeholders. Energy Policy 2023;178:113587. https://doi.org/10.1016/J.ENPOL.2023.113587.

[34] National Hydrogen Strategy Review. Australia Government | Department of Climate Change, Energy, the Environment and Wat 2023. https://storage.googleapis.com/files-au-climate/climate-au/p/prj277759bb48cda6df2c230/public_assets/National%20Hydrogen%20Strategy%20Review%20-%20Consultation%20Paper%20-%20July%202023.pdf (accessed February 1, 2024).

This page intentionally left blank

CHAPTER 2

Electrolyzers: enabling the hydrogen revolution

2.1 Introduction

An electrolyzer is an apparatus utilized to separate water into its constituent elements, hydrogen, and oxygen, by applying electrical energy. To comprehensively analyze the functioning of an electrochemical process, various principles, including mass and energy conservation, thermodynamic equilibrium fundamentals, transport phenomena within electrolyzers, and electric charge conservation, must be considered. Additionally, a thorough understanding of the electrochemical process necessitates an analysis of its economic and environmental impacts, as well as a materials assessment.

The chemical reactions involved in an electrochemical process are facilitated by an electric field generated by electrodes, typically composed of semiconductors or metals, which are connected to an external circuit supplied with electrical power. These electrodes transfer electrons to or from the atoms and molecules participating in the reaction. The electrolyte acts as a conductive medium for ionic species. A complete electrochemical process comprises two half-reactions, namely the oxidizing reaction (redox) and the reduction reaction, which occur at distinct locations. Figure 2.1 illustrates the fundamental mechanism of an electrolysis cell, where anode and cathode electrodes are immersed in the electrolyte and connected to a direct current (DC) power supply. The external circuit facilitates the flow of electrons to the negative pole (cathode), balancing the positive electric charge of ionic species passing through the electrolyte. As the required electrical power for water splitting is supplied to the electrodes, hydrogen and oxygen ions are generated at the cathode-electrolyte interface through the dissociation of water molecules. The produced hydrogen is collected from the cathode flow channel, while the oxygen ions (referred to as anions or negatively charged ions) migrate from the cathode to the anode through the electrolyte to form oxygen molecules. The electrolyte conducts the ions but does not conduct electrons. Under the influence of the electric field, cations (or positive ions) migrate from the anode to the cathode.

DOI: https://doi.org/10.55670/fpll.book/1

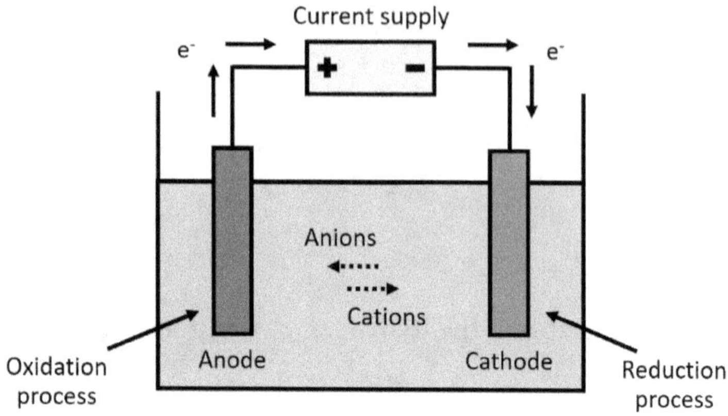

Figure 2.1 Fundamental mechanism of an electrolysis cell

Presently, the most promising electrolyzer technologies garnering significant attention can be broadly categorized into six main groups: alkaline, acidic, acidic/alkaline amphoteric, solid oxide, microbial, and photo-electrochemical [1]. While each of these categories has its own set of advantages and disadvantages (Table 2.1), not all are currently viable for industrial application (Table 2.2) [2]. Although the academic literature extensively covers alkali (alkaline) and PEM (acidic) systems that have already been commercialized, the current surge in interest is directed towards solid oxide and AEM (alkaline) solutions, which are reported to be closer to commercialization compared to other electrolyzer technologies [3]. Moreover, not all these technologies are well-suited for integration with intermittent renewable energy sources to establish a power-hydrogen nexus. Indeed, the performance characteristics of electrolyzers involve trade-offs among various factors such as cost, utilization of noble metals, technology readiness, and flexibility in operation.

Renewable hydrogen production through electrolysis relies on water as its primary feedstock, with approximately 10 liters of purified water typically necessary to generate 1 kg of hydrogen. Compared to alternative hydrogen production methods, electrolysis boasts the smallest water footprint. The production of hydrogen from natural gas with carbon capture, utilization, and storage (CCUS) consumes between 13 kg and 18 kg of water per kilogram of H_2. In contrast, coal gasification utilizes between 40 kg and 86 kg of water per kilogram of H_2, contingent on water consumption associated with coal mining [4]. Electrolyzers demand high water purity levels, posing potential challenges in water-stressed regions. However, in coastal areas, seawater desalination could serve as an alternative.

Table 2.1 Advantages and disadvantages of various electrolyzers

AWE	PEM	AEM	SOSE
Advantages			
• Well-established technology with gigawatts of historical deployment • Multi-megawatt stacks facilitate systems with substantial capacities already at present • Ability to utilize readily available and cost-effective materials	• High-power densities of over $2A/cm^2$, allowing for compact designs and a small physical footprint • Quick cold start-up time and rapid load change capabilities • Considered relatively mature, with capabilities reaching megawatts • Offers high intrinsic purity of product gas • Able to surpass its rated power for short durations, providing a wider range of flexibility	• Does not require costly or crucial minerals • Features compact designs and a small footprint similar to PEM systems • Capable of operating at high pressures (>30 bar) • Can utilize readily available and affordable materials	• Minimal standby power consumption • Inherently efficient thanks to rapid reactions and excellent conductivity at elevated temperatures • Ability to utilize external heat sources for stack electrical efficiencies exceeding 100% • Certain configurations can function reversibly as both fuel cells and electrolyzers • Capable of electrolyzing carbon-based compounds and co-electrolyzing them with water • Possibility to employ readily available and affordable materials
Disadvantages			
• Substantial material investment at the system level using strongly alkaline liquids as electrolytes • Lower power densities leading to a larger physical footprint • Additional measures needed to ensure gas purity • Lengthy cold start-up period • Elevated minimum load due to membrane gas permeability	• Application of costly materials like titanium and essential platinum group metals (PGM) at the cellular level • Requirement to validate long-term durability at the megawatt scale • Dependence on "forever chemicals" like PFAS that are not naturally found and exhibit minimal degradation in natural settings	• Stack sizes currently range in the low kilowatts (kWs), resulting in increased balance of plant expenses • Membrane degradation leads to shorter stack lifetimes compared to alkaline and PEM systems.	• Occupies the largest physical space, 1/3 larger than AWE and AEM systems and 3.5 times larger than PEM systems. • Elevated operating temperatures accelerate aging. • Hydrogen produced requires separation from steam and purification. • Low output pressure necessitates external compression. • Cold starts require several hours. • Historically known for its poor durability.

Table 2.2 Operational characteristics of commercially available electrolyzers

Technology	AWE	PEM	AEM	SOSE
Electrolyte	Liquid: 25 – 40% KOH	Solid: proton exchange membrane (Nafion)	Liquid-solid hybrid: 1% KOH/anion exchange membrane	Solid: ceramic-zirconia or ceria based
Operating temperature (°C)	70-90	50-80	40-80	500-900
Operating pressure (bar)	Conventional: atmospheric Modern: up to 30 bar (50 among startups)	Up to 80 (350 among startups)	Up to 35 with potential for much higher in the future	0-2 bar
Typical current densities today (A/cm^2)	0.4-1.0	0.2-4.0	0.2- 2.0	0.5- 1.5
System energy consumption ($kWh/kg\ H_2$)	50-78	50-83	57-69	38 (with steam import) 48(without steam import)
Stack cost (2020 $/kW)	70- 450	400-870	200	250- 2,000
Electrolyzer system cost with BoP (2020 $/kW)	800- 1,500	1,400- 2,100	3,333	917- 4,000
Stack lifetime (full load hours)	60,000- 100,000	50,000- 90,000	5,000- 40,000	20,000- 50,000
Degradation (% / 1000 hours)	0.13	0.25	0.4	0.55-1%
Ramp up time hot idle to nominal power	60s	10s	30 minutes	10 minutes
Cold ramp-up time	30 – 60 minutes	5 minutes	20 minutes	>600 minutes
Minimum load	10- 40%	5- 10%	10-20%	>3%

The use of reverse osmosis for desalination requires 3 kWh to 4 kWh of electricity per cubic meter of water, with a negligible impact on the overall costs of water electrolysis, resulting in an increase of less than USD 0.1/kg H_2 in total hydrogen production costs [5]. The production cost of electrolytic hydrogen is mainly determined by the expenses associated with electrolyzers and electricity. Presently, the cost of an electrolyzer, encompassing equipment, gas treatment, plant balancing, engineering, procurement, and construction, ranges from USD 1,400 per kilowatt (kW) to USD 1,770/kW [6]. Anticipated reductions in costs by 2030 stem from the widespread deployment of electrolyzers, with factors including economies of scale in manufacturing, technological innovation, increased proficiency, and more competitive markets playing significant roles. This relationship between deployment and decreasing equipment costs mirrors the trend observed in the past decade for other modular energy technologies like solar photovoltaic (PV). From 2010 to 2021,

PV module costs plummeted nearly 90%, dropping from USD 2.5/W to under USD 0.5/W, driven by a roughly 60-fold surge in cumulative installed utility-scale capacity, from less than 8 GW to 500 GW. Future estimates for electrolyzer costs can be made using a learning curve approach, where costs are modeled as a function of cumulative capacity deployment. Based on the current global pipeline, electrolyzer installed capacity is projected to rise from 0.5 GW in 2021 to 35 GW in 2025 and 134 GW in 2030. Consequently, capital costs for electrolyzers are expected to decrease by 64-67% by 2027 and 68-72% by 2030, ranging from USD 440/kW to USD 500/kW. Accessing an affordable, low-emission electricity source is crucial for achieving economically viable hydrogen production. This is because electricity constitutes a substantial portion of the overall costs, as a decrease in electricity costs from USD 35/MWh to USD 25/MWh results in a 30% reduction in the levelized cost of hydrogen, decreasing from an estimated USD 2.4/kg to USD 1.9/kg by 2030. According to data from S&P Global, in 2023, approximately 240 MW of electrolysis capacity are either under construction or permitted to be operational in the United States, with an additional 136 megawatts of capacity announced or in early planning stages. Collectively, these figures represent roughly 7% of the anticipated global electrolyzer production capacity added in 2023, compared to Europe's 16%. Nevertheless, projections suggest that the United States will make significant strides in the following years. By 2024, the US is expected to see 3,169 megawatts of electrolyzer capacity come online, constituting about 17% of the global added capacity, compared to Europe's 35% [5]. The preparations are underway, and it is expected to start manifesting in the overall capacity figures for the United States beginning in 2025 and throughout the remainder of the decade (Figure 2.2).

2.2 Alkaline water electrolyzer (AWE)

Alkaline water electrolysis stands out as the most mature technology among those mentioned, often regarded as the standard for large-scale industrial hydrogen production. Its long history dates back over a century, with the launch of the first significant water electrolysis plant in 1939, boasting a capacity of 10,000 normal cubic meters per hour (Nm^3/h) [6]. This technology's popularity stems from several key advantages. Firstly, AWEs predominantly utilize cost-effective components for electrodes and porous transport layer materials, typically nickel (Ni). Moreover, while platinum (Pt) and ruthenium (Ru) serve as catalysts, these electrolyzers can also employ non-noble and thus more economical metals like manganese (Mn), contributing to cost reduction [7]. Secondly, equipped with robust cell separators, alkaline electrolyzers demonstrate high durability and tolerance to impurities, leading to their longest estimated lifetime compared to other electrolyzer technologies currently available. Consequently, maintenance costs are significantly minimized [8]. Lastly, alkaline electrolyzers enable the largest scale of green hydrogen production at the lowest investment costs. With AWE sizes ranging from 1.8 to 5,300 kW, investment costs fluctuate between 800 and 1,500 USD/kW, approximately 2-2.5 times lower than the typical investment costs of the subsequent commercially available hydrogen production technology-proton exchange membrane (PEM) electrolyzers. The total installed costs for a 1-gigawatt (GW) green-hydrogen facility utilizing AWE technology are estimated at 730 €/kW, whereas for PEM technology, the estimate stands at 830 €/kW. Consequently, this equates to approximately 1580 €/(kg/day) for AWE and 1770 €/(kg/day) for PEM in terms of hydrogen production costs. The anticipated decrease in hydrogen production costs from 7.26 \$/kg to 2.60 \$/kg primarily stems from reductions in capital costs. Furthermore, advancements in AWE technology and its scaling up contribute significantly to reducing total installed costs compared to previous

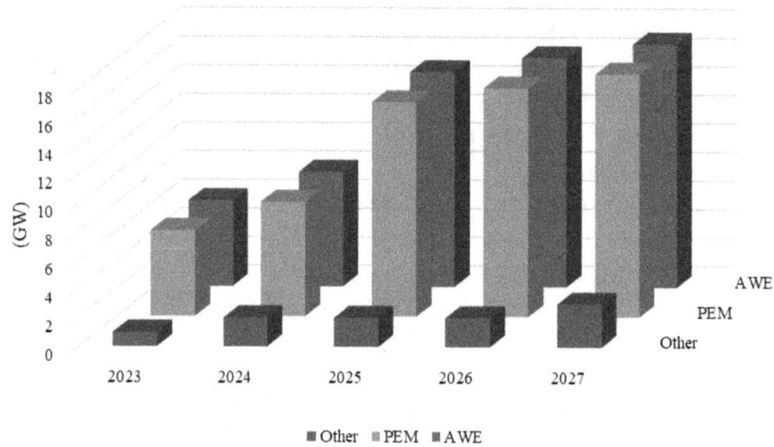

designs [9].

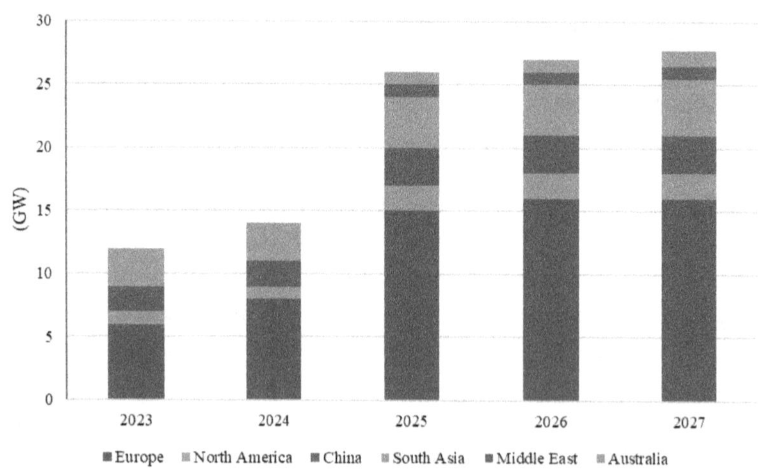

Figure 2.2 Global electrolyzer manufacturing capacity 2023-2027

While AWEs remain popular and widespread, they have drawbacks that may limit their applicability for hydrogen production purposes. One notable disadvantage is their high lower limit of minimum load, which restricts their compatibility with certain types of renewable energy sources. Moreover, their current density adjustments must be made gradually, as they are engineered for operation under fixed process conditions. Consequently, despite attempts to address this limitation through integrating batteries and control systems [10], alkaline electrolyzers may not be the optimal choice for effectively balancing variable renewable energy sources like wind and solar PV [7]. Additionally, AWEs operate at low current density and efficiency due to the significant internal resistance incurred across their thick porous diaphragm and electrolyte [11]. Consequently, these systems typically necessitate a bulky stack design at larger scales, with stack costs being major contributors to overall system capital costs, substantially increasing them. Furthermore, most conventional AWEs available on the market tend to produce hydrogen of lower purity compared to other power-to-hydrogen options, such as PEM electrolyzers, while consuming more energy [12]. Moreover, due to the corrosive nature of AWEs and the consequent risk of leakage, careful design and maintenance of the entire AWE system are imperative to prevent the mixing of the produced gases (hydrogen and oxygen), as their mixtures have high flammability. To address some of the challenges encountered in traditional AWEs, emerging polymer technologies like anion exchange membrane (AEM) and molten carbonate electrolysis cell (MCEC) are currently undergoing development. While they are still in the relatively early stages of research and development, they hold promise as a more effective means of integrating green hydrogen production with variable renewable energy sources.

2.2.1 Anion exchange membrane

AEM electrolyzers, unlike conventional AWEs, which utilize concentrated potassium hydroxide (KOH), employ a less concentrated alkaline solution, reducing susceptibility to corrosion [13]. Additionally, AEM setups eliminate the risk of electrolyte and gas leaks by replacing diaphragms with anion exchange membranes. Furthermore, besides lowering energy consumption and improving efficiency, AEM technology enhances the purity of generated hydrogen [14]. Moreover, similar to other alkaline and alkaline polymer electrolyzers, AEM systems can utilize inexpensive catalysts and core materials for electrodes, thereby reducing investment costs significantly. However, despite representing progress over conventional AWEs, AEM electrolyzers have notable drawbacks. Chief among them is their relatively high minimum load requirement, similar to their alkaline counterparts, making them less compatible with variable renewable generation, typically necessitating around 10-20 percent of design capacity to operate [15]. Moreover, while designed to be more efficient than AWEs, AEM systems still exhibit significantly lower water-splitting performance than PEM facilities [16]. Additionally, the scalability of available AEM electrolyzers is currently limited, and their performance deteriorates over time due to decreasing membrane ion conductivity and sensitivity to CO_2 intrusion, resulting in a relatively short lifetime of around 30,000 hours [17].

2.2.2 Molten carbonate electrolysis cell

MCEC technology is under development to address certain challenges encountered by AEM electrolyzers while enhancing the benefits associated with alkaline electrolysis. Similar to AEM electrolyzers in terms of producing high-purity hydrogen gas and not requiring expensive catalysts, MCECs offer distinct advantages over AEMs. Operating at significantly higher temperatures (600–700°C compared to 50–80°C for AEMs), MCECs are resistant to carbon monoxide (CO) poisoning and CO_2 intrusion, resulting in longer lifespans [18]. Moreover, MCECs can be larger in size, enabling them to produce more hydrogen per stack (up to 1,400 Nm^3/h). Consequently, once matured, this technology may be deemed superior to AEMs in terms of performance. However, similar to AEMs, MCECs are still in the early stages of development. Additionally, the estimated investment per stack (approximately $1,400) is already comparable to that of PEM electrolyzers and the most advanced conventional AWEs, which may have even greater capacities. Consequently, with a similarly high minimum load requirement (10–20% of design capacity), MCECs do not currently appear to be an ideal choice for green hydrogen production using intermittent renewable energy sources unless further technological advancements are made.

2.3 Acidic electrolyzer

Electrolyzer technologies functioning within an acidic setting can generally be categorized into acid electrolyzers and acid polymer electrolyzers. While both mentioned systems have been commercially implemented, acid polymer systems, exemplified by electrolyzers employing PEM technology, are typically regarded as significantly more advanced and better suited for integration with variable renewable energy sources. Acid electrolyzers typically utilize lithium-ion membranes and sulfuric acid (H_2SO_4) as the electrolyte, offering various advantages alongside drawbacks. With a 5-10% design capacity, acid electrolyzers generally surpass all categories of alkaline electrolysis facilities, making them more suitable for intermittent energy sources. Furthermore, they produce high-purity hydrogen and boast a relatively wide system size range, reaching up to 1,500 kW, comparable to PEM and surpassing AEM equivalents. However, the use of expensive noble metals such as platinum (Pt) and iridium (Ir) for electrodes and catalysts in acid electrolyzers may not always justify the investment, given their relatively low efficiency [19]. Additionally, presently available acid electrolyzers seem to have a shorter estimated lifespan compared to most other electrolysis alternatives.

2.3.1 Acid polymer electrolyzers

During the 1960s, efforts to address the major limitations of alkaline electrolyzers culminated in the development of initial systems based on a solid polymer electrolyte concept. Evolving over subsequent decades, this concept has evolved into PEM facilities, which are currently recognized as the most promising option for integrating hydrogen production with solar and wind power sources [20]. Despite the significant advantages of PEM technology over alkaline alternatives, these electrolyzers have certain shortcomings hindering their widespread industrial adoption. A primary advantage of PEM electrolyzers is their low level of partial load (0–10% of design capacity), a distinguishing feature from other available alternatives [7]. Additionally, the non-porous nature of the PEM separator enables rapid cycling in flexible operational conditions and offers a response speed suitable for electricity grid-balancing services. Consequently, PEM systems are better suited for operation with intermittent renewables, making them among the most appealing technological options for the transition to green hydrogen. Moreover, PEM electrolyzers exhibit higher efficiency compared to alkaline ones, typically offering better hydrogen purity and comparable stack and system lifetimes. Operating at higher temperatures and current densities than their alkaline counterparts, they feature a compact design and can be containerized even at larger scales, providing significant advantages for manufacturing, on-site development, and reducing transportation and maintenance costs. Despite these advantages that position PEM electrolyzers as potential key drivers of industrial growth for green hydrogen production, this technology still faces several challenges. One major barrier is the high cost of catalysts and electrodes, primarily due to the expense associated with noble metals, notably platinum. Additionally, the perfluorinated membranes and titanium current collectors used are relatively complex to produce and are costly [21]. Furthermore, in acidic environments, most non-noble metal oxides are prone to corrosion and are thus unsuitable as catalysts for PEM electrolysis. Without platinum alloy catalysts, the anode reaction of PEM electrolyzers may be susceptible to CO poisoning, resulting in a gradual decline in membrane performance [22]. Consequently, these systems face challenges in finding substantial cost-reduction options.

2.4 Acidic/alkaline amphoteric

As existing electrolysis technologies consume significant amounts of energy, there has been recent interest in developing acidic/alkaline amphoteric electrolyzers that could potentially decrease power usage. Although this technology is still in its early stages, it offers several potential advantages over current electrolysis systems, particularly alkaline ones. Acidic/alkaline amphoteric electrolyzers integrate both alkaline and acidic environments, each dedicated to producing different gases from water splitting. Specifically, these electrolyzers are divided by a membrane that centralizes neutralization, allowing hydrogen to be generated in an acidic solution chamber while oxygen evolves in an alkaline chamber. Pilot tests have shown a significant reduction in energy consumption (approximately 30%) and increased hydrogen production rates compared to AWE and AEM facilities of similar size and design capacity. Moreover, by avoiding Pt-group catalysts, which are costly in acidic electrolysis, acidic/alkaline amphoteric setups could potentially be more cost-effective while demonstrating higher efficiency than MCEC solutions [23]. However, despite the promising advantages of this technology, further development is required to address its limitations comprehensively. Improving the purity of produced H_2, increasing system size, and enhancing overall system lifetime is crucial for developing acidic/alkaline amphoteric electrolyzers. Additionally, substituting expensive Pt-based electrode coatings with more affordable alternatives could reduce the overall system cost.

2.5 Solid oxide steam electrolyzer

Besides AWE and PEM hydrogen production methods, solid oxide steam electrolyzers (SOSE) are anticipated to emerge as the third primary water electrolysis technology in the near future (Figure 2.3).

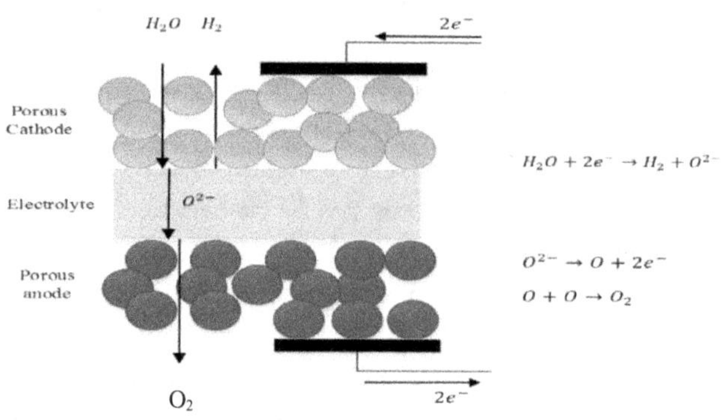

(a)

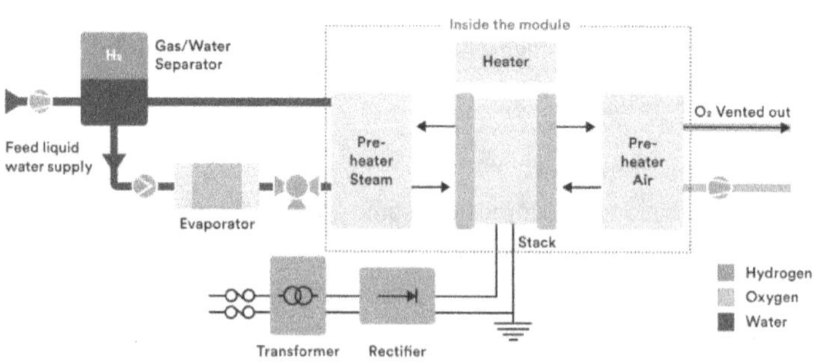

(b)

Figure 2.3 (a) A schematic of a SOSE function (b) A SOSE balance-of-plant components [24]

Although still in the research and development phase and far from maturity, SOSE technology is not new. Since its initial development in the late 1960s, it has garnered increasing interest due to its potential to significantly enhance water electrolysis efficiency by utilizing high operating temperatures ranging from 800 to 1000°C [11]. Consequently, SOSEs are essentially steam electrolyzers, leveraging high temperatures to split water into hydrogen and oxygen. One notable advantage of SOSEs is their exceptionally high efficiency, estimated at around 80–90.8%, potentially surpassing all existing alternative electrolysis methods. This elevated efficiency is primarily attributed to the fact that solid oxide-based cells are not constrained by Carnot efficiency [25]. With an anticipated minimum load level of around 3%, SOSEs may eventually compete with PEM systems regarding variable generation operations [7]. Additionally, they exhibit long-term stability and operate at low costs using inexpensive catalysts [26]. Despite the impressive benefits of solid oxide technology for hydrogen production, some drawbacks hinder its commercialization and wider adoption. The high operating temperature, while advantageous, also poses a significant challenge by increasing start-up and break-in times. This could potentially limit the system's compatibility with intermittent renewables despite the relatively low minimum load. High temperatures also introduce mechanical compatibility issues like thermal expansion mismatch and material diffusion within the cell layers, leading to rapid degradation of components, especially brittle ceramics. Consequently, SOSEs currently exhibit relatively low durability and overall lifetime. Coupled with high investment costs, bulky system design, limited production capacity, and a small production scale, these factors highlight the need for improvement before SOSEs can effectively compete with alkaline and PEM facilities [27].

2.6 Photoelectrochemical (PEC) water splitting

PEC involves using semiconductor materials to convert solar energy into hydrogen directly. These semiconductors are placed in a water-based electrolyte, where sunlight initiates the process of splitting water molecules into H_2 and O_2. Since this method relies on sunlight, its compatibility with other renewable energy sources is limited. By integrating a photovoltaic system and PECs, the size and capital costs are reduced. This integration also has the potential to decrease energy consumption due to the direct conversion of photons into chemical energy, consequently boosting the overall system efficiency by around 30%. Moreover, the absence of Pt-group catalysts in PECs makes them more cost-effective. However, for PEC technology to be commercially viable, enhancements in efficiency, durability, and cost are deemed necessary. While improvements have been made in efficiency through better sunlight absorption and surface catalysis, pilot models have shown a limit of around 40% efficiency [28]. To compete with other hydrogen production methods, the longevity of PEC systems must be significantly extended, which could be achieved by using more robust materials and protective coatings. Lastly, reducing material and processing costs could further lower the overall hydrogen production costs using PECs.

2.7 Microbial electrolysis cells (MECs)

MECs have the ability to not only produce hydrogen but also various organic chemicals by harnessing the chemical energy of biomass to support electrical energy [29]. This allows water splitting at relatively low temperatures, up to 30°C. While this technology shows promise for hydrogen generation, it is still in the early stages of development and requires further refinement. MECs can theoretically produce hydrogen from a low energy input ranging from 0.6 to 1 kWh/m^{3}H$_2$, which is significantly lower compared to the 4.5–50.6 kWh/m^{3}H$_2$ required by other systems. Therefore, operating costs for such facilities could be greatly reduced. With advancements in design, operation, and electrochemical characterization, pilot MEC projects have managed to reduce overpotentials and enhance overall efficiency [30]. Moreover, microbial electrolysis technologies also show promise for wastewater treatment [31]. However, MECs typically use precious metal catalysts, electrodes, and other materials, which hinders their commercialization, similar to the limitations faced by PEM electrolyzers. Additionally, current MECs have limited production capacity and potentially lower hydrogen purity compared to alternative technologies, necessitating significant technological advancements to compete commercially or complement existing alternatives.

2.8 Evaluation of electrolyzers for green hydrogen production

Electrolysers span a technological spectrum from well-established to emerging. Among the existing types, only AWEs and PEMs have reached commercialization (Figure 2.4). While AWEs have a century-long history, PEM ones have gained popularity since the latter half of the twentieth century. Currently, PEM electrolyzers stand out as the most suitable option for integrating intermittent renewable energy sources into the hydrogen production process. However, their high cost, attributed to platinum group catalysts and membrane production expenses, remains a significant challenge for electrolyzer manufacturers in the coming years. On the other hand, AWEs, despite benefiting from low-cost catalysts, have a relatively high minimum load and prefer stable operating conditions, making them less ideal for variable renewable energy use. Efforts are underway with SOSE and AEM technologies to bridge this capacity gap and leverage the advantages of both AWE and PEM systems without major obstacles. However, these technologies require further research and development before achieving widespread availability. AEM installations, in particular, face challenges with scalability and remain primarily low-capacity systems, while SOSEs are closer to commercialization but still have limited production capacity and durability issues due to brittle ceramics. As a result, no single electrolyzer technology currently excels in all essential features for efficient green hydrogen production. Looking ahead, alkaline, PEM, solid oxide, and AEM electrolyzers are expected to garner significant attention from researchers and industries in the global energy transition. Therefore, evaluating the costs associated with each technology and comparing them could provide insights into opportunities for cost reduction and help map future business development scenarios.

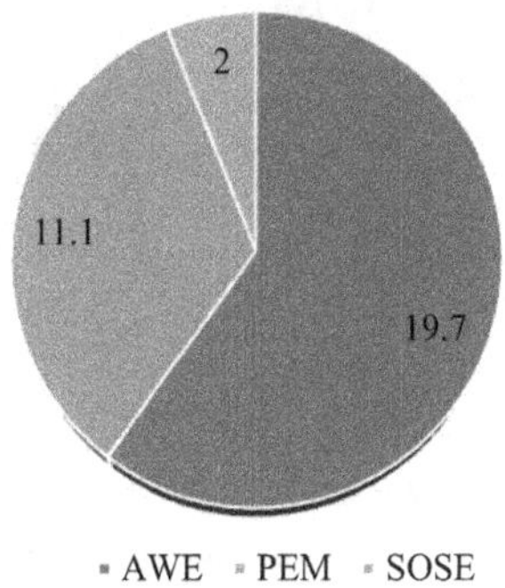

Figure 2.4 Global electrolyzer manufacturing capacity in 2023 (GW)

2.9 Cost assessment of electrolyzers

Assessing the precise costs linked with electrolyzers is challenging due to several factors. Firstly, some technologies are relatively new, leading to limited disclosure of cost-related data by companies aiming to maintain a competitive edge. Secondly, there's often ambiguity or inconsistency in defining cost boundaries, making cross-technology comparisons less accurate. However, a comparison can still be made between the major technologies: AWE and PEM systems. Currently, the commercially available PEM systems are approximately 50–60% more expensive than well-established AWEs. Both technologies have the potential for cost reduction through factors like automation, economies of scale, and component availability. This cost-reduction journey could align with commercializing SOSE and AEM alternatives, offering comparable cost benefits. While some companies have initiated the commercialization of AEM and SOSEs, many components remain in the laboratory phase, leading to small-scale pilot projects. Therefore, while these technologies promise cost-effective green hydrogen production, their widespread market presence is still evolving. Nonetheless, conventional alkaline, AEM, PEM, and solid oxide electrolyzers are seen as the most promising options by manufacturers for green hydrogen production. Figure 2.5 illustrates the detailed comparison of key cost drivers like a stack, power electronics, gas conditioning, and balance of plant across these four types of electrolyzers. The main cost component of electrolyzer technologies, which include cell units, porous transport layers, bipolar plates, end plates, and various smaller components like spacers, seals, frames, and bolts, are known as stacks.

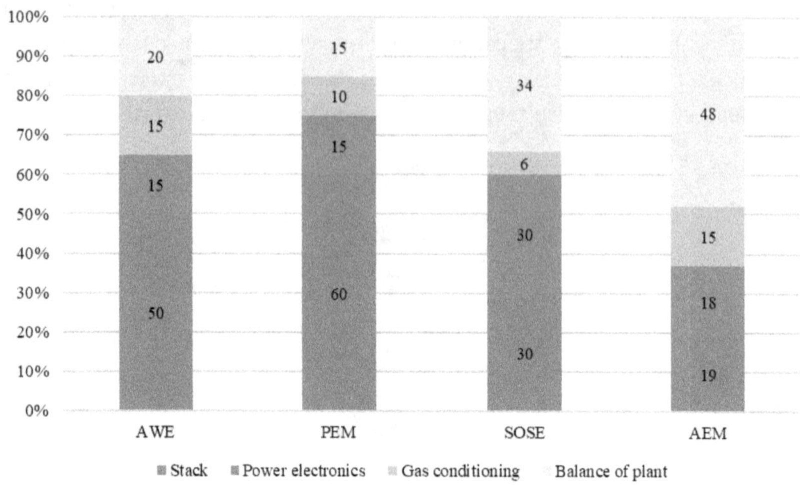

Figure 2.5 Cost breakdown for 1 MW electrolyzers

These stacks are crucial in both alkaline and PEM systems and account for a significant portion of the total expenses associated with these technologies. In PEM systems, the cell unit comprises the membrane with catalyst layers coated as electrodes, while in alkaline systems, it includes electrodes and diaphragms. Stacks typically comprise 40 to 60% of the total electrolyzer costs, with around half the cost for alkaline systems and about 60% for PEM systems [32]. The higher cost of PEM systems is primarily due to the use of precious platinum-group metals, which hinders their widespread commercial adoption. To address this issue, solid oxide and AEM electrolyzer technologies are exploring cheaper catalysts and materials, leading to a reduced share of stack costs of approximately 30% for SOSEs and 19% for AEM installations. While this reduction appears significant in percentage terms, the limited scalability of these newer systems, coupled with their current low maturity, still makes the stack component relatively expensive, especially compared to more established alkaline and PEM systems.

Power electronics refers to using solid-state electronics to manage and convert electric power. Its application is crucial for integrating renewable energy systems with the grid, as it converts the DC from the power source into the alternating current (AC) needed by the grid. In electrolyzer systems, the primary role of power electronics is to convert AC from the grid into the DC required by the electrolysis cell stack. This conversion is typically achieved using rectifiers, devices that convert AC, which reverses direction periodically, into DC that flows in a single direction. Additionally, most electrolyzers incorporate choppers, static power electronic devices that modify fixed DC voltage or power into variable DC voltage or power, essentially acting as switches controlled by another signal. Transformers are also commonly used to regulate voltage levels within a power grid. While the cost of power electronics constitutes around 15% in both alkaline and PEM systems, it is expected to be about 18% higher for AEM. For SOSEs, the cost of power electronics is even higher, at 30%, comparable to the stack cost and several times higher than in AWEs and PEMs. This difference in cost is primarily due to the complexity of hydrogen production in these technologies, including factors like high operating temperatures, mechanically unstable electrodes, brittle ceramics, and sealing challenges. Consequently, power electronics and the balance of plant components play a critical role in such circumstances.

Gas conditioning in electrolysis involves mechanically compressing gas to a specified storage pressure and drying it to enhance the purity of raw H_2. Depending on the level of technological development, hydrogen is purified to levels ranging from 99.5 to 99.9999% and compressed to over 50 bar for efficient storage. Despite its critical role in the overall hydrogen production process, gas conditioning is typically the least costly component among the four main ones discussed for each type of electrolyzer. Specifically, gas conditioning accounts for approximately 15% of costs in AWEs and AEM systems, drops to about 10% in PEM installations, and reduces further to only around 6% in SOSEs due to their different technological foundations. However, despite this lower percentage, the absolute cost of gas conditioning in SOSE and AEM systems is generally comparable to or higher than in commercialized alkaline and PEM electrolyzers, reflecting the significant cost gap between these technologies.

The remaining portion of essential electrolyzer costs is attributed to the balance of the plant system. This system typically encompasses components like control and safety systems, heat recovery and rejection equipment, piping, valves, pumps, and other material transport systems, waste management systems, maintenance and repair equipment, as well as facilities for ventilation, air conditioning, heating, cooling, power supply, and distribution, among others [33]. While the percentage of this cost component varies from 15 to 20% in traditional alkaline and PEM electrolyzers, it constitutes a larger share of expenses for more advanced systems currently in development and early commercialization stages. The balance of plant costs for SOSEs is estimated to be more than double that of PEM installations, accounting for 34% compared to 15%. In absolute figures, the cost difference is even more significant, with the balance of plant components in SOSEs being at least eight times more expensive. Additionally, for AEM electrolyzers, the balance of the plant is expected to represent nearly half of the total costs at 48%, which is two to four times higher than in AWEs and PEM systems in actual expenses. Much like power electronics, the balance of the plant is crucial in enhancing the longevity of SOSE or AEM membranes, which is currently constrained by the relative novelty of these technologies [34].

2.10 Electrolyzer's startup time

Alkaline electrolyzers, being the oldest among the analyzed water electrolysis systems with a history spanning about a century, were initially designed to align with the predominant power supply of their era. Consequently, they were primarily intended for stationary applications with grid connections rather than adapting to intermittent power sources from variable renewables that were not integrated into the grid. As renewable energy became more prevalent, alkaline electrolyzers underwent further enhancements, particularly with the introduction of pressurized versions, which are noted as the most efficient and cost-effective options for water electrolysis within the alkaline category. Despite technological progress and increased flexibility in handling variable power inputs, AWEs still lag behind PEM electrolyzers in terms of suitability for low-load operations. When these systems experience insufficient power and need to be stopped and flushed, they typically require a considerable amount of time, ranging from half an hour to an hour, for a cold start-up. AEM electrolyzers exhibit a similar cold start-up time as well. On the contrary, SOSEs have the potential for rapid electric load adjustments due to their wide load range. However, their start-up and shut-down procedures must be carried out gradually, and they need to maintain a temperature above 600 °C. If these electrolyzers are shut down, their cold start-up usually takes more than an hour. In contrast, PEM electrolyzers stand out for their ability to reach optimal operating temperature in just several minutes to half an hour, making them particularly noteworthy in terms of operational efficiency [35].

2.11 Electrolyzers lifetime

Improving the longevity of electrolyzer systems can effectively distribute the facility's cost across a greater volume of hydrogen production, thereby lowering the expenses per unit of produced hydrogen. This underscores the importance of enhancing the durability of power-to-hydrogen production systems to achieve overall cost reduction for their environmentally friendly output. Presently, AWEs, as the most mature systems, boast the longest operational lifespans. They are recognized for their robustness, with stack lifetimes ranging from 60,000 to 100,000 hours, and demonstrated system durability exceeding 30 years. While other electrolyzer technologies strive to match the durability of AWEs, they have yet to reach that level. The lifetime of electrolyzers is influenced by different factors that have various impacts. For AWEs, factors such as nickel alloys, gas permeation, electrodes, and water purity or impurity play crucial roles. Specifically, continuous gas permeation exposes the diaphragm in AWEs to increasing pin-hole failures, leading to gas contamination and ultimately diminishing the overall stack and system lifetime. Additionally, nickel alloys employed to handle highly caustic KOH at elevated concentrations must be free from chromium and iron to prevent electrode contamination and leaching, which would also significantly reduce their lifespan [36]. Moreover, water impurities adversely affect key components of AWEs, such as the diaphragm and catalysts. Addressing these factors would notably enhance the durability of such systems, subsequently benefiting the marginal cost of producing green hydrogen.

PEM electrolyzers currently have a lifespan ranging from 50,000 to 90,000 hours, which still falls short of the longevity seen in AWEs. This discrepancy is partly due to the relative newness of PEM technology and is influenced by several key factors affecting their durability. These factors typically encompass water quality, operating conditions, anode corrosion, fluctuating loads, and gas permeation. Since PEM systems generally operate at higher current densities, temperatures, and pressures compared to AWEs, these conditions can negatively impact their lifespan. Consequently, newer generations of PEMs often address these issues by incorporating thicker membranes, increased catalyst loadings, protective coatings on porous transport layers, and bipolar plates. However, these measures also lead to higher overall costs for PEMs. Variable load is another factor that adversely affects the lifespan of PEM electrolyzers. Despite being more suitable for intermittent power supplies than other systems, voltage fluctuations can accelerate the corrosion of stack components, reducing durability and increasing costs. Additionally, under certain temperature and voltage conditions, the iridium oxide commonly found on the anode can dissolve, prompting the need for higher amounts of catalysts and precious metals in protective layers, further adding to the cost. The same as AWEs, PEM electrolyzers are susceptible to gas permeation and water impurities. High differential pressure can compromise the mechanical stability of PEM membranes, leading to hydrogen permeation and degradation issues. Water impurities can also cause stack failures by affecting membranes, porous transport layers, ionomers in catalyst layers, and the catalysts themselves. These challenges are typically addressed through additional catalyst use for hydrogen conversion and water purifiers, but these solutions contribute to increased marginal costs. Improving the lifespan of PEM electrolyzers in a cost-effective manner is crucial for the wider adoption and advancement of PEM technology [37].

SOSEs have a variable estimated lifespan, which depends on the information source and the stage of commercialization of this technology. Current data suggest that their stack can last around 20,000 hours, with an experimental maximum of about 90,000 hours. However, these figures are often based on solid oxide systems powered by stable energy sources under specific operating conditions. It is anticipated that connecting SOSEs to intermittent renewables could accelerate their degradation due to thermochemical cycling during shutdown or ramping periods. Therefore, besides enhancing their lifespan, there's a need to improve their compatibility with intermittent renewables [2]. On the other hand, AEMs exhibit the shortest estimated lifespan, reaching only about 30,000 hours. Limited commercialization of AEM systems has resulted in scant information on key aspects. One major challenge lies in the stability of the AEM polymer used for membrane and catalyst layers. The polymer is prone to degradation from hydroxide attack, causing the membrane to collapse and the catalyst layer to dissolve within days. Efforts to address these chemical and mechanical stability issues have reduced efficiency and increased costs for AEM systems [38].

2.12 Policies for electrolyzers development

Despite the widespread interest among major global economies in advancing power-to-hydrogen technologies, there remains a significant level of uncertainty regarding the future demand for green hydrogen. Many countries lack defined hydrogen strategies, and among those that do, blue hydrogen (produced from fossil fuels with carbon capture and storage) is often considered a viable alternative for decarbonization efforts. Moreover, the current demand for hydrogen is largely agnostic about its source, resulting in limited recognition and valuation of green hydrogen's lower CO_2 emissions. This lack of market and valuation hampers the development of downstream uses and constrains demand for green hydrogen. Given the nascent stage of green hydrogen production, government policies are essential to foster a growing supply chain, mitigate investor risks, promote innovation, and deploy necessary infrastructure. These policies are crucial for addressing key cost categories such as electrolyzers and decarbonized electricity inputs. Many experts emphasize the importance of developing a robust hydrogen infrastructure to enable the industrial-scale delivery of both blue and green hydrogen. Without stimulating demand from industries and sectors, power-to-hydrogen technologies risk remaining expensive novelties rather than practical components of energy transition strategies. To comprehensively develop green hydrogen, supportive policies should not only promote renewable energy and electrolyzers but also focus on building hydrogen infrastructure and fostering demand among end-users. The primary strategies for supporting electrolyzer development include [39]:

- Establishing clear long-term goals for electrolyzer capacity to align with renewable energy growth, often outlined in national hydrogen strategies targeting 2030.
- Providing financial incentives and support to encourage investment in electrolyzer manufacturing, logistics, and deployment. This may involve subsidies, tax incentives, loan guarantees, and other measures to address the higher costs of green hydrogen compared to fossil fuel alternatives.
- Introducing policies that stimulate demand for green hydrogen, such as green procurement mandates, carbon pricing mechanisms, and emissions standards. These measures can drive investments in electrolyzer capacity.
- Developing essential infrastructure like hydrogen storage, transportation systems, and refueling stations, as the lack of infrastructure is a significant hurdle for green hydrogen adoption. Policies should focus on bridging this infrastructure gap.
- Establishing standards, certification programs, and knowledge infrastructure to ensure the environmental sustainability of green hydrogen production and utilization. This helps build trust and confidence in the market.

The primary obstacle hindering the widespread adoption of green hydrogen is its higher cost compared to many competing alternatives. While the expenses related to renewable power generation are recognized as a significant barrier, the excessive costs associated with electrolyzers represent the second most crucial factor contributing to the limited market appeal of green hydrogen (Figure 2.6). Therefore, while advancements in technology are expected to reduce these costs over time naturally, policy support is essential to expedite this process. Additionally, given the varying maturity levels among available electrolyzer technologies, tailored assistance is needed to propel their progression toward becoming competitive energy solutions.

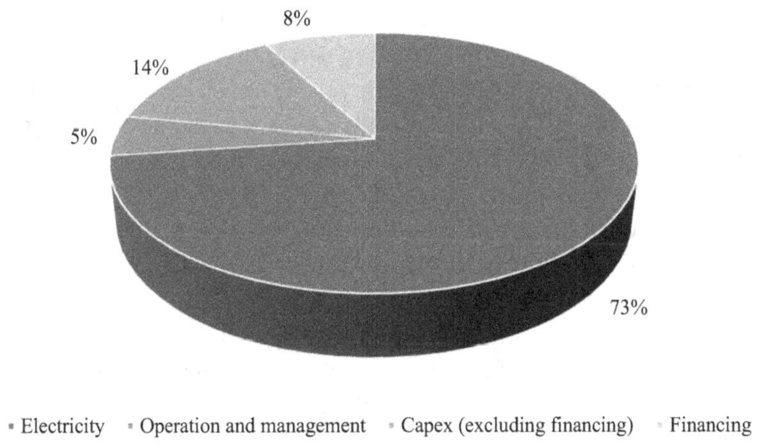

Figure 2.6 Main cost drivers for green hydrogen production

Countries and regions that are actively promoting the development of green hydrogen are offering fiscal incentives to electrolyzer manufacturers. An example of this is California, where projects combining solar PV with power-to-hydrogen can benefit from a 3.9 percent tax exemption through programs like the Sales and Use Tax Exclusion Program (with up to USD 20 million per project per calendar year), the California Research Credit, and the 'California Competes' Tax Credit (with a minimum of USD 20,000) [40]. While these incentives are unique due to the novelty of the issue, it is widely anticipated that more governments will recognize the crucial importance of such supportive fiscal policies for the success of 'green' hydrogen development. Additionally, direct support policies in the form of grants and loans have been implemented by leading economies. For example, Germany allocated €7 billion ($8.1 billion) to accelerate the implementation of hydrogen technology within the country as a component of its "Package for the Future" pandemic recovery strategy. This financial support aims to establish a production capacity of 5 GW for green hydrogen by 2030, with a further 5 GW targeted for 2035-2040 [41]. Australia granted USD 52 million for green hydrogen development targeting electrolyzers with capacities of 5 MW and above in 2020 and France allocated USD 2.4 billion for these purposes from 2020 to 2022.

The policy can also influence manufacturing and production capacity. The EU plans to install 40 GW of renewable energy hydrogen electrolyzers by 2030 to make green hydrogen a major fuel in the energy transition by mid-century [42]. Such an increase in capacity may require specific stimulating measures even within the world's largest economy. Global electrolyzer manufacturing capacity is expected to rise from around 135 MW/year in 2018 to about 54 GW/year by the end of 2027 [43]. However, to achieve the carbon neutrality target by 2050, annual manufacturing rates need to be up to 50 times the 2021 capacity, highlighting the necessity for dedicated supportive policies to meet capacity and energy transition goals. Currently, industrial policies aimed at scaling up and enhancing the efficiency of electrolyzer manufacturing primarily involve establishing long-term targets and allocating dedicated funds to improve production processes and technological advancements. For example, in March 2024, the US Department of Energy allocated $1 billion for research, development, demonstration, and deployment (RDD&D) activities to reduce the cost of clean hydrogen produced via electrolysis and $500 million for development and demonstration of improved processes and technologies for manufacturing and recycling clean hydrogen systems [43]. Similarly, Germany allocated EUR 700 million from the country's research ministry to three projects focused on developing gigawatt-scale serial production of electrolyzers for potential export [44]. This illustrates that major economies recognize the importance of expanding electrolyzer manufacturing and are dedicating specific funds for this purpose.

Besides supporting electrolyzer manufacturing, direct financial assistance should target projects in their early stages of development to stimulate research and development (R&D) activities. Currently, the emphasis is on improving PEM technology as it is considered the most suitable for integrating intermittent renewables. Although PEM systems share or exceed key characteristics of AWEs, their reliance on platinum-group materials remains a significant hurdle to cost reduction. Therefore, reducing the usage of scarce materials in PEM electrolyzer production is a crucial focus for further R&D. Similarly, SOSEs, while using scarce materials that elevate their costs, face challenges beyond material dependence. Efforts in R&D should also target increasing stack size and reducing cold start-up time to enhance their commercial viability [45]. Achieving these goals could lead to significantly lower electricity consumption and cost reduction for green hydrogen production due to their higher efficiency (around 90%, as reported). Regarding AEM electrolysis, R&D policies should concentrate on enhancing stack size, lifetime, and overall production scale. Although AEM systems currently seem furthest from commercial deployment among the mentioned technologies, their market entry is anticipated to lower the overall cost of green hydrogen [46]. This is attributed to the relative affordability of their components, expected high energy efficiency, and low operating costs.

Abbreviations

AC: Alternating current
AEM: Anion exchange membrane
AWE: Alkaline water electrolyzer
CCUS: Carbon capture, utilization, and storage
DC: Direct current
MEC: Microbial electrolysis cell
PEM: Proton exchange membrane
PV: Photovoltaic
Redox: Oxidizing reaction
R&D: Research and development
SOSE: Solid oxide steam electrolyzers

Reference

[1] Patonia A, Poudineh R. Cost-competitive green hydrogen: how to lower the cost of electrolysers? Oxford Institute for Energy Studies 2022. https://lutpub.lut.fi/bitstream/handle/10024/104326/MScThesis_JKK.pdf?sequence=2 (accessed March 10, 2024).

[2] Gniewomir Flis GW. Solid Oxide Electrolysis: A Technology Status Assessment. Clean Air Task Force 2023. https://cdn.catf.us/wp-content/uploads/2023/11/15092028/solid-oxide-electrolysis-report.pdf (accessed March 15, 2024).

[3] Hosseini SE. Integrating a gas turbine system and a flameless boiler to make steam for hydrogen production in a solid oxide steam electrolyzer. Appl Therm Eng 2020;180:115890. https://doi.org/10.1016/J.APPLTHERMALENG.2020.115890.

[4] Fairley P. Sizing up hydrogen's hydrological footprint. Nature 2023. https://doi.org/10.1038/D41586-023-03884-9.

[5] Commodities 2023: US clean hydrogen sector may see "start of exponential growth" | S&P Global Commodity Insights n.d. https://www.spglobal.com/commodityinsights/en/market-insights/latest-news/energy-transition/123022-us-clean-hydrogen-sector-may-see-start-of-exponential-growth-in-2023 (accessed March 11, 2024).

[6] Santos DMF, Sequeira CAC, Figueiredo JL. Hydrogen production by alkaline water electrolysis. Quim Nova 2013;36:1176–93. https://doi.org/10.1590/S0100-40422013000800017.

[7] Mittelsteadt C, Norman T, Rich M, Willey J. PEM Electrolyzers and PEM Regenerative Fuel Cells Industrial View. Electrochemical Energy Storage for Renewable Sources and Grid Balancing 2015:159–81. https://doi.org/10.1016/B978-0-444-62616-5.00011-5.

[8] Joonas Koponen. Review of water electrolysis technologies and design of renewable hydrogen production systems . Lappeenranta University of Technology 2015. https://ispt.eu/media/Public-report-gigawatt-advanced-green-electrolyser-design.pdf (accessed March 10, 2024).

[9] Hans van 't Noordende PR. A One-GigaWatt Green-Hydrogen Plant. Hydrohub Innovation Program 2022.

[10] Becker M, Brauns J, Turek T. Battery-Buffered Alkaline Water Electrolysis Powered by Photovoltaics. Chemie Ingenieur Technik 2021;93:655–63. https://doi.org/10.1002/CITE.202000151.

[11] Ursúa A, Gandía LM, Sanchis P. Hydrogen Production from Water Electrolysis: Current Status and Future Trends. Proceedings of the IEEE 2012;100:410–26.

[12] Sheffield JW, Martin KB, Folkson R. Electricity and hydrogen as energy vectors for transportation vehicles. Alternative Fuels and Advanced Vehicle Technologies for Improved Environmental Performance: Towards Zero Carbon Transportation 2014:117–37. https://doi.org/10.1533/9780857097422.1.117.

[13] Vincent I, Bessarabov D. Low cost hydrogen production by anion exchange membrane electrolysis: A review. Renewable and Sustainable Energy Reviews 2018;81:1690–704. https://doi.org/10.1016/J.RSER.2017.05.258.

[14] Mandal M. Recent Advancement on Anion Exchange Membranes for Fuel Cell and Water Electrolysis. ChemElectroChem 2021;8:36–45. https://doi.org/10.1002/CELC.202001329.

[15] Stanislaw LN, Gerhardt MR, Weber AZ. Modeling Electrolyte Composition Effects on Anion-Exchange-Membrane Water Electrolyzer Performance. ECS Trans 2019;92:767–79. https://doi.org/10.1149/09208.0767ECST/XML.

[16] Faid AY, Barnett AO, Seland F, Sunde S. Highly Active Nickel-Based Catalyst for Hydrogen Evolution in Anion Exchange Membrane Electrolysis. Catalysts 2018, Vol 8, Page 614 2018;8:614. https://doi.org/10.3390/CATAL8120614.

[17] Vega JA, Chartier C, Mustain WE. Effect of hydroxide and carbonate alkaline media on anion exchange membranes. J Power Sources 2010;195:7176–80. https://doi.org/10.1016/J.JPOWSOUR.2010.05.030.

[18] Lan Hu. Molten carbonate fuel cells for electrolysis. KTH Royal Institute of Technology 2016. https://kth.diva-portal.org/smash/get/diva2:920605/FULLTEXT01.pdf (accessed March 11, 2024).

[19] Khan MA, Zhao H, Zou W, Chen Z, Cao W, Fang J, et al. Recent Progresses in Electrocatalysts for Water Electrolysis. Electrochemical Energy Reviews 2018;1:483–530. https://doi.org/10.1007/S41918-018-0014-Z/FIGURES/23.

[20] Carmo M, Fritz DL, Mergel J, Stolten D. A comprehensive review on PEM water electrolysis. Int J Hydrogen Energy 2013;38:4901–34. https://doi.org/10.1016/J.IJHYDENE.2013.01.151.

[21] Aricò AS, Siracusano S, Briguglio N, Baglio V, Di Blasi A, Antonucci V. Polymer electrolyte membrane water electrolysis: Status of technologies and potential applications in combination with renewable power sources. J Appl Electrochem 2013;43:107–18. https://doi.org/10.1007/S10800-012-0490-5/METRICS.

[22] Baschuk JJ, Li X. Carbon monoxide poisoning of proton exchange membrane fuel cells. Int J Energy Res 2001;25:695–713. https://doi.org/10.1002/ER.713.

[23] Lei Q, Wang B, Wang P, Liu S. Hydrogen generation with acid/alkaline amphoteric water electrolysis. Journal of Energy Chemistry 2019;38:162–9. https://doi.org/10.1016/J.JECHEM.2018.12.022.

[24] Green Hydrogen Cost Reduction: Scaling up Electrolysers to Meet the 1.5^0C Climate Goal. IRENA 2020 2020. www.irena.org/-/media/Files/IRENA/Agency/Publication/2020/Dec/IRENA_Green_hydrogen_cost_2020.pdf (accessed March 15, 2024).

[25] Ishihara T, Jirathiwathanakul N, Zhong H. Intermediate temperature solid oxide electrolysis cell using LaGaO3 based perovskite electrolyte. Energy Environ Sci 2010;3:665–72. https://doi.org/10.1039/B915927D.

[26] Patrick T. Moseley, Jürgen Garche. Electrochemical Energy Storage for Renewable Sources and Grid Balancing. Elsevier, DOI: https://doi.org/10.1016/C2012-0-01253-7; 2015.

[27] Hosseini SE. Design and analysis of renewable hydrogen production from biogas by integrating a gas turbine system and a solid oxide steam electrolyzer. Energy Convers Manag 2020;211:112760. https://doi.org/10.1016/j.enconman.2020.112760.

[28] Fountaine KT, Lewerenz HJ, Atwater HA. Efficiency limits for photoelectrochemical water-splitting. Nature Communications 2016 7:1 2016;7:1–9. https://doi.org/10.1038/ncomms13706.

[29] Ogawa T, Takeuchi M, Kajikawa Y. Analysis of Trends and Emerging Technologies in Water Electrolysis Research Based on a Computational Method: A Comparison with Fuel Cell Research. Sustainability 2018, Vol 10, Page 478 2018;10:478. https://doi.org/10.3390/SU10020478.

[30] Badwal SPS, Giddey SS, Munnings C, Bhatt AI, Hollenkamp AF. Emerging electrochemical energy conversion and storage technologies. Front Chem 2014;2. https://doi.org/10.3389/FCHEM.2014.00079.

[31] Arun J, SundarRajan PS, Grace Pavithra K, Priyadharsini P, Shyam S, Goutham R, et al. New insights into microbial electrolysis cells (MEC) and microbial fuel cells (MFC) for simultaneous wastewater treatment and green fuel (hydrogen) generation. Fuel 2024;355:129530. https://doi.org/10.1016/J.FUEL.2023.129530.

[32] Böhm H, Goers S, Zauner A. Estimating future costs of power-to-gas – a component-based approach for technological learning. Int J Hydrogen Energy 2019;44:30789–805. https://doi.org/10.1016/J.IJHYDENE.2019.09.230.

[33] Bradley Ward. Balance of Plant Requirements for a Nuclear Hydrogen Plant . Idaho National Laboratory 2006. https://core.ac.uk/download/pdf/71313328.pdf (accessed March 14, 2024).

[34] Li D, Motz AR, Bae C, Fujimoto C, Yang G, Zhang FY, et al. Durability of anion exchange membrane water electrolyzers. Energy Environ Sci 2021;14:3393–419. https://doi.org/10.1039/D0EE04086J.

[35] Scheepers F, Stähler M, Stähler A, Rauls E, Müller M, Carmo M, et al. Improving the Efficiency of PEM Electrolyzers through Membrane-Specific Pressure Optimization. Energies 2020, Vol 13, Page 612 2020;13:612. https://doi.org/10.3390/EN13030612.

[36] Guandalini G, Campanari S, Valenti G. Comparative assessment and safety issues in state-of-the-art hydrogen production technologies. Int J Hydrogen Energy 2016;41:18901–20. https://doi.org/10.1016/J.IJHYDENE.2016.08.015.

[37] Krishnan S, Koning V, Theodorus de Groot M, de Groot A, Mendoza PG, Junginger M, et al. Present and future cost of alkaline and PEM electrolyser stacks. Int J Hydrogen Energy 2023;48:32313–30. https://doi.org/10.1016/J.IJHYDENE.2023.05.031.

[38] Vega JA, Chartier C, Mustain WE. Effect of hydroxide and carbonate alkaline media on anion exchange membranes. J Power Sources 2010;195:7176–80. https://doi.org/10.1016/J.JPOWSOUR.2010.05.030.

[39] Policy Priorities to Spur the Green Hydrogen Economy - RMI n.d. https://rmi.org/policy-priorities-to-spur-the-green-hydrogen-economy/ (accessed March 28, 2024).

[40] Josh Eichman MKOJG and BM. Optimizing an Integrated Renewable Electrolysis System. National Renewable Energy Laboratory 2020. www.nrel.gov/docs/fy20osti/75635.pdf (accessed March 28, 2024).

[41] Germany's Hydrogen Industrial Strategy n.d. https://www.csis.org/analysis/germanys-hydrogen-industrial-strategy (accessed January 17, 2024).

[42] European Commission, official website n.d. https://commission.europa.eu/index_en (accessed February 18, 2023).

[43] Global electrolyzer manufacturing capacity could outpace demand by 2030 – pv magazine India n.d. https://www.pv-magazine-india.com/2024/02/14/electrolyzer-manufacturing-capacity-could-outpace-demand-by-2030/ (accessed March 28, 2024).

[44] Germany to spend €700M on 3 hydrogen research projects | S&P Global Market Intelligence n.d. https://www.spglobal.com/marketintelligence/en/news-insights/latest-news-headlines/germanyto-spend-8364-700m-on-3-hydrogen-research-projects-62113936 (accessed March 28, 2024).

[45] Wang L, Chen M, Küngas R, Lin TE, Diethelm S, Maréchal F, et al. Power-to-fuels via solid-oxide electrolyzer: Operating window and techno-economics. Renewable and Sustainable Energy Reviews 2019;110:174–87. https://doi.org/10.1016/J.RSER.2019.04.071.

[46] 'Our unique AEM electrolysers will produce cheaper green hydrogen than any rival tech' | Recharge n.d. https://www.rechargenews.com/energy-transition/-our-unique-aem-electrolysers-will-produce-cheaper-green-hydrogen-than-any-rival-tech-/2-1-1042573 (accessed March 28, 2024).

CHAPTER 3

From fossil fuels to renewables: methods of hydrogen production

3.1 Introduction

The transition from fossil fuels to renewable energy sources has become a paramount pursuit in the quest for a sustainable and cleaner energy future. As the world grapples with the environmental, economic, and societal challenges posed by traditional energy systems, the imperative to adopt innovative methods of hydrogen production has gained unprecedented significance. Exploring a diverse array of sources for hydrogen production (hydrogen rainbow), ranging from conventional fuels like natural gas, propane, biomass, and coal to unconventional sources like methane from renewable processes such as wind, solar, hydropower, and even uranium, this chapter delves into the multifaceted landscape of methods that underpin the evolution from fossil fuels to renewables (Figure 3.1) [1]. Each source carries its distinct attributes and challenges, contributing to the intricate tapestry of hydrogen production methods that shape our journey towards a sustainable energy future (Figure 3.2).

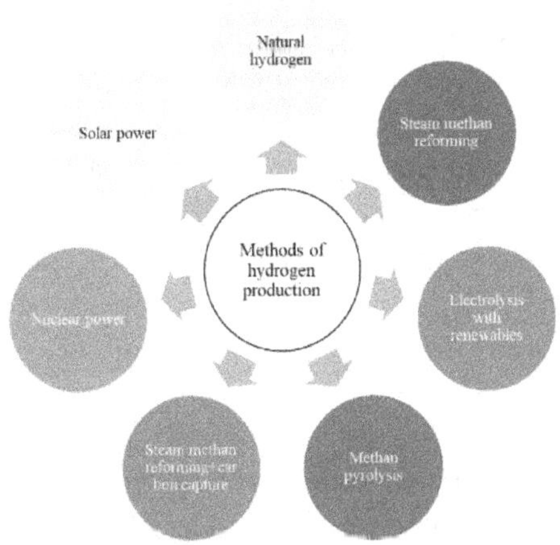

Figure 3.1 Hydrogen rainbow

DOI: https://doi.org/10.55670/fpll.book/1

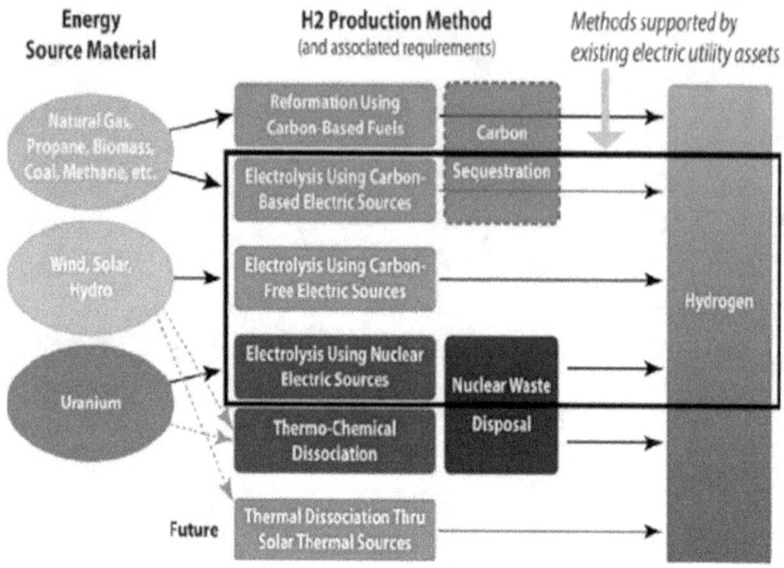

Figure 3.2 Diverse array of sources for hydrogen production [2]

As the call for reduced greenhouse gas emissions grows more urgent, the spotlight turns to hydrogen, an element poised to transform our energy landscape. A range of hydrogen production methods, from fossil fuel-based techniques to cutting-edge renewable approaches, are scrutinized, unraveling the intricate fabric of an evolving energy paradigm. The nuances, advantages, and challenges of these methods are aimed to be unearthed through meticulous examination, thereby illuminating the path to a hydrogen-powered future that not only promises enhanced sustainability but also resilience in the face of pressing global energy demands.

In the midst of this exploration of hydrogen production methods, it becomes clear that the journey toward a hydrogen-powered future is not just a technological pursuit but a realm interwoven with diplomatic intricacies. Within this section, the foundation is laid for the broader theme by illuminating the pivotal contribution of global collaboration, strategic policy structures, and cooperative ventures in influencing the trajectory of hydrogen technology. As the various sources of hydrogen production are investigated, a journey is embarked upon that transcends geographical boundaries, where the implications of each method ripple across nations and global energy landscapes. From North America to Europe, Asia Pacific to the Middle East, each region's distinct characteristics and contributions shape the broader context of the hydrogen economy. Illustrated in Figure 3.3, the primary sources that underpin the landscape of commercial hydrogen production exhibit a diversified composition. A substantial 48% emanates from the abundant reservoirs of natural gas, while another noteworthy 30% stems from the intricate processes of petroleum reforming. A further 18% emerges from coal gasification, marked by its distinct characteristics and implications. The remaining 4% originates from the innovative domains of electrolysis and renewable energy sources, representing the promising potential of sustainable pathways [2]. Beyond these prominent avenues, there exists a realm of supplementary possibilities. Biomass gasification and solar thermochemical hydrogen production, for instance, present themselves as emerging contenders, expanding the repertoire of hydrogen origins. Additionally, certain hydrocarbon processes, exemplified by the steam cracking of natural gas liquids to yield light olefins and associated co-products, contribute an auxiliary stream, accounting for a substantial 3.5 million metric tons annually. However, it's important to note that a considerable proportion of these auxiliary pathways exhibit lower energy efficiencies, coupled with elevated costs and the imperative for carbon capture technologies, as underscored in prior research [3,4].

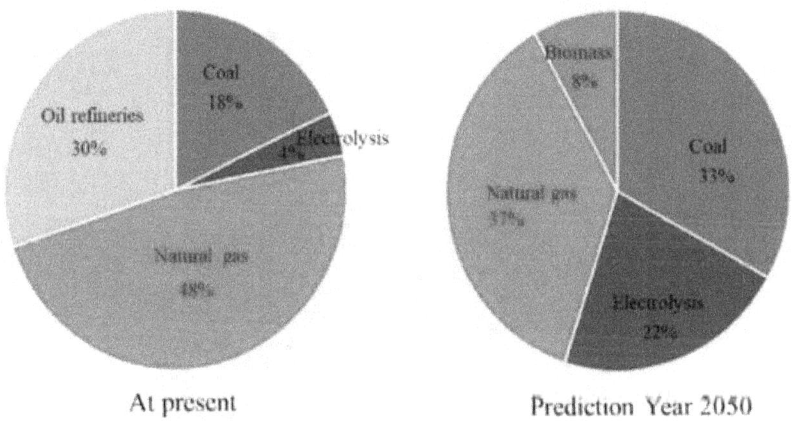

Figure 3.3 Global hydrogen production from different resources [2]

Global hydrogen production currently amounts to approximately 8 exajoules (EJ) annually, with the potential to reach 10 EJ per year by 2050. The hydrogen market is anticipated to experience substantial growth in the near future, with an annual increase of 5-10%, primarily driven by its use in refineries to process heavy oil fractions and the projected demand in the transportation sector, as well as its role as an energy carrier [5]. The vast majority of hydrogen production methods, accounting for 96%, rely on non-renewable sources. Among these methods, the most prominent are natural gas reforming (comprising 48% of production) and oil reforming (making up 30%), with coal gasification contributing to 18% of the total production. In contrast, a mere 4% of hydrogen production is achieved through water electrolysis. Nevertheless, it is anticipated that approximately 48% of hydrogen demand will be produced by renewables (green hydrogen) by 2050 (Figure 3.4) [6].

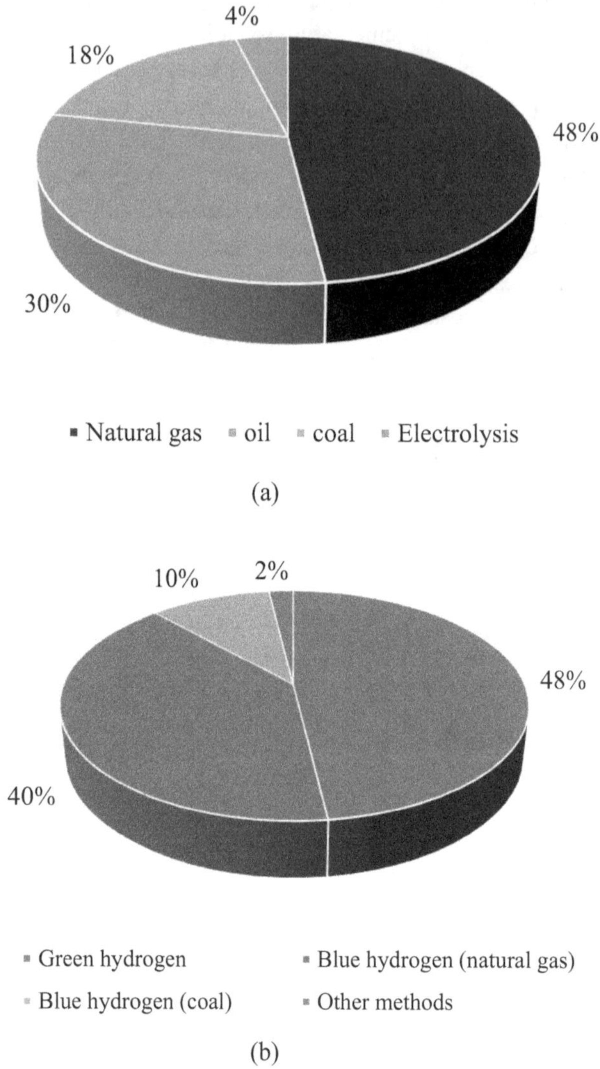

Figure 3.4 (a) Current hydrogen sources (b) anticipated hydrogen production methods by 2050

Table 3.1 outlines the various hydrogen color codes, detailing their sources, energy origins, production techniques, and associated carbon footprints. Moreover, Figure 3.5 visually represents the carbon footprint attributed to each hydrogen color, enabling a comparative analysis of their environmental impacts. Notably, black (or brown) hydrogen exhibits the highest carbon footprint, and it's essential to recognize that electrolysis, despite being a common method, may not always result in the cleanest outcome. For example, yellow hydrogen, which relies on grid electricity to power electrolyzers, can generate emissions surpassing those of gray hydrogen.

3.2 Fossil fuel-based hydrogen production

The journey from fossil fuels to renewables in hydrogen production marks a significant transition in the energy landscape. Historically, hydrogen was derived from fossil fuels, primarily natural gas, through processes like steam methane reforming (SMR) and partial oxidation. These methods involve subjecting natural gas to high temperatures and steam, resulting in the release of hydrogen molecules. While this approach has been widely used due to its efficiency and established infrastructure, it raises concerns about carbon emissions.

Table 3.1 Hydrogen sources and their carbon footprints (ATR: autothermal reforming, CCS: carbon capture and storage, NGD: natural gas decomposition, SMR: steam methane reforming)

Hydrogen color	Hydrogen source	Energy source	Technique	CCS	Carbon footprint (kg CO_2eq/kgH_2)	Reference
Black or brown	Coal	Thermal	Gasification	No	21.8-51.9	[7,8]
Gray	Natural gas	Thermal	SMR	No	10.9-18.4	[9,10]
Blue	Natural gas	Thermal	SMR, ATR, NGD	Yes	2.6-16.2	[9,11]
Turquoise	Methane	Thermal	Pyrolysis	No	4.4	[12]
Green	Water	Electricity (Renewables)	Electrolysis	No	0.63-6.6	[13,14]
Yellow	Water	Electricity (grid mix)	Electrolysis	No	28.6-32	[15,16]
Purple	Water	Electricity (nuclear)	Electrolysis	No	2	[17]
Pink	Water	Thermal+ Electricity (nuclear)	High-temperature electrolysis	No	0.4-2	[17]
Red	Water	Thermal (nuclear)	Thermochemical cycle	No	0.3-1.8	[17]
Aqua	Oil reservoirs	Thermal	Water-gas shift reaction	No	Not available	[18]
White	Water	Thermal (solar)	Thermochemical water splitting	No	Not available	[19]

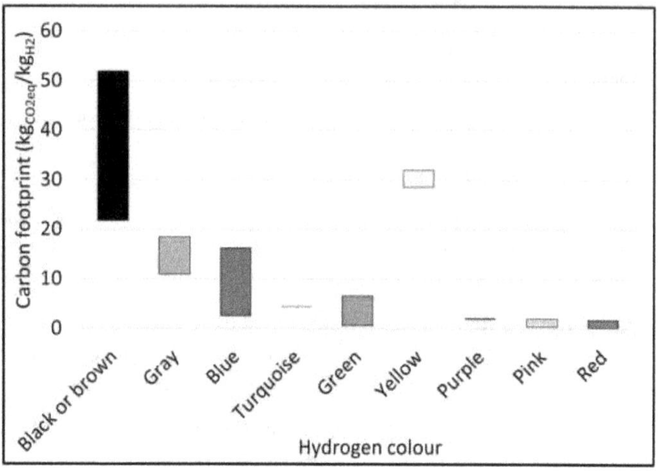

Figure 3.5 Carbon footprints of different hydrogen colors [20]

The carbon content in natural gas leads to the release of CO_2 during the production process, contributing to greenhouse gas emissions. This approach also perpetuates the reliance on fossil fuels and limits the progress toward a sustainable energy future. On the other hand, coal gasification technology has firmly established itself as a pivotal player in the hydrogen generation market, primarily due to its widespread adoption and alignment with strategic energy policies in several nations. The method involves the conversion of coal into syngas ($CO+H_2$) through a series of chemical reactions, offering a versatile platform for producing hydrogen and other valuable byproducts. This technology has particularly gained traction in countries like India, where the abundance of coal resources has driven its utilization as part of broader energy strategies. The adoption of coal gasification aligns with energy security and diversification objectives in these regions, as it allows for domestic hydrogen production and reduces dependence on fossil fuel imports. However, while coal gasification offers economic advantages and supports local industries, it also presents challenges related to carbon emissions and environmental impact. As governments aim to balance economic growth with environmental sustainability, policies are being crafted to incentivize cleaner approaches to coal gasification, including carbon capture and storage (CCS) initiatives. This proactive approach ensures that coal-derived hydrogen contributes to hydrogen economies while adhering to carbon reduction goals. This dual focus on economic growth and environmental responsibility underscores the role of coal gasification as a dynamic component of hydrogen strategies, harmonizing policy goals with market demands and sustainability imperatives. Fossil fuel-based hydrogen production is not only associated with environmental concerns but also poses challenges related to energy security and geopolitical stability. The dependency on fossil fuel resources, which are subject to price fluctuations and supply disruptions, can compromise the stability of hydrogen supply chains. Furthermore, the geopolitical complexities involved in fossil fuel trade can introduce uncertainties that impact energy security. Despite these challenges, fossil fuel-based hydrogen production has played a crucial role in advancing hydrogen technology and creating a foundation for the broader hydrogen economy. However, the need for cleaner and more sustainable alternatives has spurred research and development efforts to transition toward renewable sources for hydrogen production.

3.2.1 Steam methane reforming

The SMR process, a widely used method for hydrogen production, involves the conversion of natural gas into hydrogen and CO_2 through a series of chemical reactions. The process begins with the introduction of natural gas, typically methane (CH_4), into a reformer chamber, where it undergoes a high-temperature reaction with steam (H_2O). The endothermic reaction is catalyzed by a nickel-based catalyst and results in the production of hydrogen gas and carbon monoxide (CO). This mixture of hydrogen and carbon monoxide, known as synthesis gas or syngas, is the primary intermediate product in the SMR process. In the subsequent step, a shift reaction, specifically the water gas shift (WGS) reaction, is employed to increase the hydrogen yield and reduce the carbon monoxide content. In this reaction, the carbon monoxide reacts with water vapor to form additional hydrogen gas and carbon dioxide. The resulting gas mixture consists of hydrogen, carbon dioxide, and a reduced amount of carbon monoxide. The shift reaction is typically performed in two stages: the high-temperature shift (HTS) and the low-temperature shift (LTS). Catalysts are utilized to optimize the reaction conditions in each stage.

$$\text{Reformer:} \quad CH_4 + H_2O \rightarrow CO + 3H_2 \tag{3.1}$$

$$\text{WGS reaction:} \quad CO + H_2O \rightarrow CO_2 + H_2 \tag{3.2}$$

The final step involves the separation of hydrogen gas from the gas mixture, often using a pressure swing adsorption (PSA) process or a membrane separation technique. This separation is necessary to obtain high-purity hydrogen suitable for various applications, including fuel cells, industrial processes, and transportation. A widely recognized and economically viable approach for generating hydrogen fuel is SMR, exhibiting an efficiency range of 74% to 85%. The SMR technique accounts for over 80% of global hydrogen production, surpassing the output from the electrolysis process by a significant margin [21]. SMR represents the most cost-effective means of obtaining hydrogen, with costs potentially falling below the US$1 per kilogram mark. However, the so-called “grey hydrogen” produced through this method presents considerable environmental concerns, generating more than 10 kilograms of carbon dioxide per kilogram of hydrogen gas, making it the least environmentally friendly variant of hydrogen production (Figure 3.6).

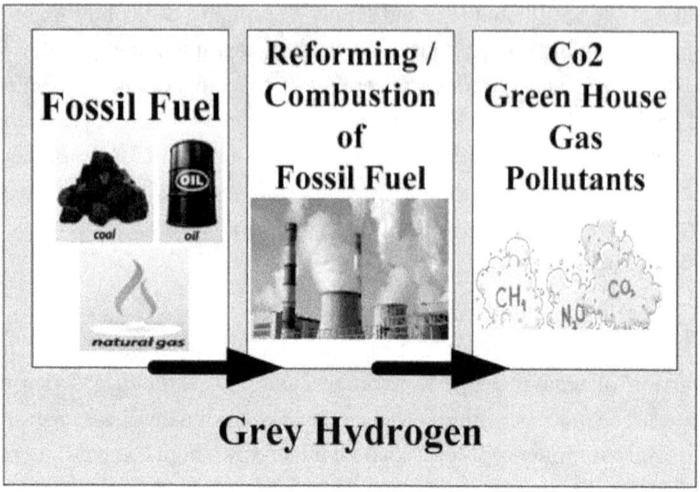

Figure 3.6 Schematic of grey hydrogen production [22]

Blue hydrogen (Figure 3.7) employs identical chemical procedures to gray hydrogen. Nevertheless, with the incorporation of carbon capture, the CO_2 generated during the procedure is prevented from escaping and is confined within dedicated facilities. Advocates assert that this approach is financially efficient and environmentally beneficial. According to calculations by the International Energy Agency (IEA), the cost of blue hydrogen in 2019 was limited to \$2.6 per kilogram, three times less than the highest costs associated with hydrogen derived from renewable electricity. Although this disparity is projected to diminish by 2050, blue hydrogen is still anticipated to remain more economical. Notable blue hydrogen initiatives have been unveiled in the past few years. A blue hydrogen plant valued at \$4.5 billion in the United States is being hailed as the world's largest carbon dioxide sequestration facility. Air Products, a gas company, has indicated that the facility will be operational by 2026 and capture 95 percent of carbon emissions. By 2030, blue hydrogen is projected to encompass 85 percent of North America's entire low-carbon hydrogen capacity. The SMR process has emerged as a pivotal cornerstone in the global landscape of hydrogen production, finding adoption and relevance in various countries due to its cost-effectiveness and efficiency. This process has been particularly attractive to countries with abundant natural gas reserves, such as the United States, Russia, and Middle Eastern nations. These countries have capitalized on their natural gas wealth to establish themselves as major players in the global hydrogen market. For instance, the United States, endowed with substantial natural gas resources, has embraced SMR as a means to not only produce hydrogen for its domestic industries (contributing to a substantial 95% of the country's total hydrogen output) but also to tap into the growing international hydrogen trade.

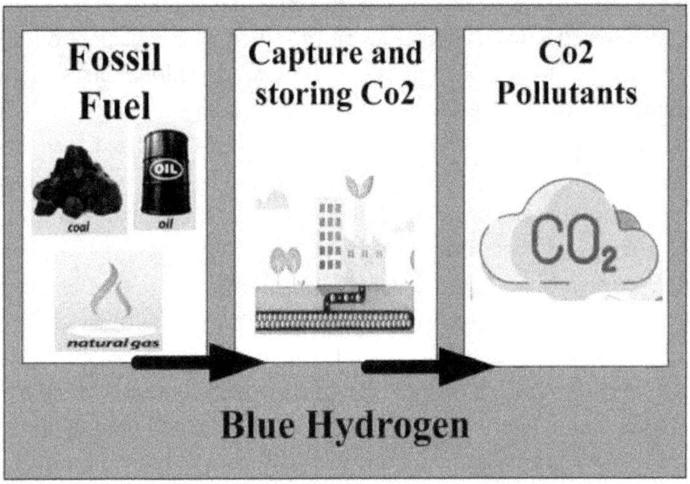

Figure 3.7 Schematic of grey hydrogen production [22]

By leveraging its existing infrastructure and expertise in natural gas extraction and distribution, the U.S. has positioned itself to play a pivotal role in the global hydrogen economy. This transition from fossil fuel-based natural gas production to hydrogen generation has the potential to reshape energy policies, contribute to decarbonization efforts, and strengthen energy security [23]. Similarly, countries like Russia, a significant natural gas exporter, have the potential to pivot from a hydrocarbon-based economy to becoming a major hydrogen supplier by capitalizing on SMR technology. This strategic shift aligns with global trends toward cleaner energy sources and offers economic diversification opportunities. Furthermore, the adoption of SMR is not limited to energy production; it also holds implications for policy frameworks. Countries are recognizing the role of hydrogen as a key enabler of clean energy transitions and are integrating its production into their long-term energy strategies. The affordability of the SMR process enhances its attractiveness for governments aiming to scale up hydrogen utilization for various applications, from powering vehicles to integrating renewable energy systems.

In the realm of hydrogen production, the SMR market undergoes segmentation based on system type, primarily categorized as "Merchant" and "Captive". The distinction between these system types carries profound implications for the overall hydrogen ecosystem and the strategies adopted by various stakeholders. In the Merchant segment, hydrogen producers operate as independent entities, specializing in the efficient production of hydrogen through SMR. These producers generate hydrogen as a product and supply it to various end-users, spanning industries such as transportation, energy, and manufacturing. The Merchant system type often involves economies of scale, where centralized production facilities optimize efficiency and cost-effectiveness. On the other hand, the Captive segment pertains to hydrogen production within industries for their own consumption. This category is characterized by on-site hydrogen generation tailored to specific industrial requirements, such as refineries or chemical plants that utilize hydrogen as a feedstock. Captive systems offer benefits in terms of localized production, reduced transportation costs, and increased operational control. Understanding the differentiation between Merchant and Captive SMR systems is crucial for shaping energy policies, infrastructure development, and investment strategies. While the Merchant system may enhance supply chain resilience and offer the potential for market expansion, the Captive approach can promote self-sufficiency and tighter integration of hydrogen utilization within industrial processes. These segmentation distinctions ultimately influence the deployment strategies pursued by various countries and companies as they seek to harness the potential of SMR to advance their energy security, economic growth, and sustainability objectives. The SMR market within the transportation segment is predicted to achieve a valuation of USD 9.5 billion by the year 2032. This projection indicates a significant growth trajectory for the SMR market, specifically in the transportation sector. It suggests that the adoption and utilization of SMR technology for various transportation applications, such as fueling hydrogen-powered vehicles or other mobility solutions, will lead to substantial economic value in terms of market size. This underscores the increasing recognition of SMR's potential benefits and its role in advancing sustainable transportation solutions.

The power generation segment is expected to experience a compound annual growth rate (CAGR) of 8% until the year 2032. This projection indicates a steady and substantial expansion of the use of hydrogen-based fuels for generating power. This growth is primarily driven by two key factors: the increasing demand for hydrogen-based fuels in the power generation sector and the growing recognition of hydrogen's utility in commercial buildings. As the world shifts towards cleaner and more sustainable energy sources, the SMR process is gaining prominence as an efficient option for producing electricity. This trend is likely to result in a significant expansion of the power generation segment, both in terms of its capacity and its contribution to the broader energy landscape. The SMR technology had an assessed value of 662.9 BN dollars in the year 2022. It is expected to experience a CAGR of 6.3% from the year 2023 to the year 2029. This projection suggests that the SMR technology's market value will continue to expand at a relatively steady pace over the specified period, reflecting its growing importance and adoption within the energy sector (Figure 3.8) [24]. The Asia Pacific region emerges as a pivotal player in the global hydrogen production landscape, propelled by its significant market presence and large population base. The region's embrace of hydrogen production, particularly through the SMR process, reflects a multifaceted approach to meeting its energy demands and sustainable development goals across diverse sectors. The sheer magnitude of the population in the Asia Pacific has spurred an escalating need for energy, fostering an environment where hydrogen fuel gains substantial traction. Countries within this region recognize the potential of reforming methane, notably through SMR, to address their energy security, economic growth, and environmental imperatives.

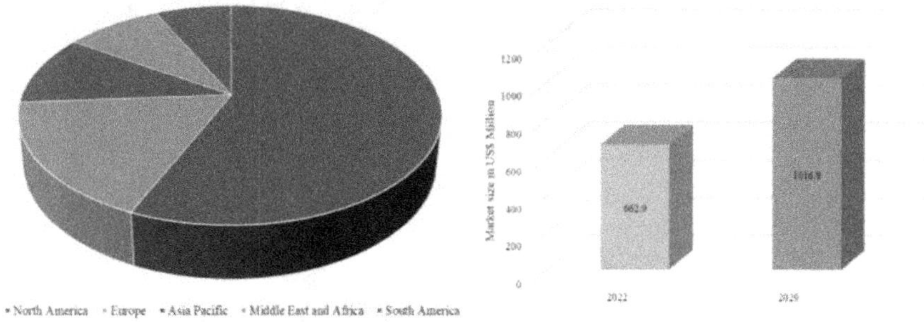

Figure 3.8 Steam methane reforming hydrogen production market

The industrial and economic diversity across the Asia Pacific further amplifies the appeal of hydrogen production from methane reforming. Industries ranging from petrochemicals, manufacturing, and transportation to power generation stand to benefit from hydrogen's versatility as a fuel and feedstock (Figure 3.9) [25]. This strategic shift aligns with the evolving global energy transition, where reducing carbon emissions and achieving sustainability are paramount objectives. The favorable policy frameworks and investments in research and development further bolster the momentum toward hydrogen production, mainly through SMR. As the largest market, the Asia Pacific region not only consumes substantial hydrogen but also significantly contributes to its production. This momentum is driven by collaborations between governments, industries, and research institutions, aiming to create a robust hydrogen ecosystem. With a commitment to harnessing the potential of methane reforming, Asia Pacific exemplifies how large and diverse populations can be harnessed as catalysts for innovative energy solutions, ultimately shaping a more sustainable and cleaner energy future.

The Europe region is poised for substantial growth in the field of SMR as it deploys hydrogen technology as a key enabler in its transition towards a low-carbon and sustainable energy future. Out of the 536 hydrogen production facilities located across Europe, a significant portion of 270 plants employ reforming processes to generate hydrogen. The region's well-established industrial infrastructure, coupled with its commitment to ambitious climate goals outlined in initiatives like the European Green Deal, positions SMR as a critical component of its energy strategy [26]. As the region works towards decarbonizing sectors such as transport, industry, and heating, hydrogen produced through SMR emerges as a versatile solution to address the challenges associated with emissions reduction and energy transition. The projected growth of SMR in Europe is not only expected to cater to domestic energy needs but also to foster international partnerships and collaborations, driving innovation and expertise exchange on a global scale.

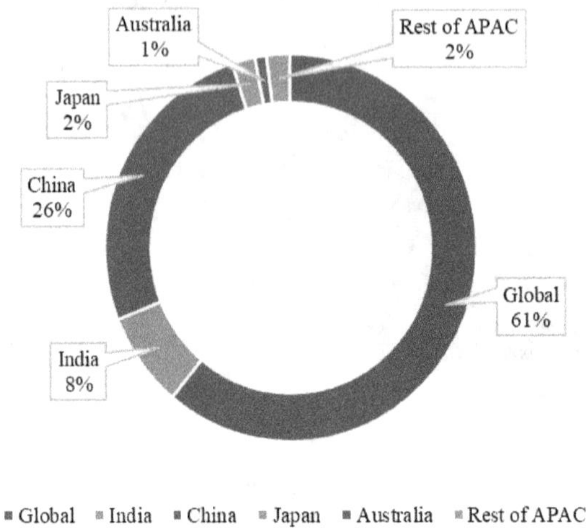

Figure 3.9 Hydrogen demand from Asia Pacific region

As the region aligns its economic, environmental, and policy objectives, SMR stands as a cornerstone in Europe's journey towards achieving energy sustainability and playing a leading role in shaping the global hydrogen landscape. Methane reforming has consistently stood as a mature and well-established market within North America. The region's long-standing history with hydrocarbon extraction and production, particularly natural gas, has laid a strong foundation for the adoption of SMR processes. With abundant reserves of natural gas, North America possesses a valuable resource base that can be harnessed to produce hydrogen through SMR. This inherent advantage has allowed the region to develop a robust hydrogen production ecosystem that capitalizes on the efficient conversion of methane into hydrogen-rich streams. The maturity of the methane reforming market in North America is emblematic of the region's pragmatic approach to energy solutions. Leveraging its rich energy heritage, North America recognizes the potential of methane reforming not only for hydrogen production but also for addressing energy transition challenges. The well-established infrastructure, technical expertise, and familiarity with hydrocarbon-based processes have contributed to the continued growth of the methane reforming market in the region. Moreover, as North America pursues energy diversification and emission reduction goals, the maturity of the methane reforming market aligns with the need for scalable and efficient hydrogen production methods. The region's proactive stance towards innovation and sustainable practices positions methane reforming as a key pillar in the transition to a cleaner energy landscape. By leveraging its mature methane reforming market,

North America showcases its capacity to balance energy demands, economic growth, and environmental stewardship in the pursuit of a resilient and sustainable energy future.

In the foreseeable future, the production of clean hydrogen is poised to emerge as a substantial source of sustainable energy. This burgeoning market exhibits vast potential for development and expansion. Presently, various sustainable and green technologies are employed for hydrogen generation, such as electricity, solar-powered systems, photovoltaic technology, and offshore wind. Notably, these methods do not entail the environmental cost of carbon emissions. However, it is worth mentioning that these technologies are relatively more expensive compared to traditional sources like natural gas or coal for hydrogen production. Nevertheless, ongoing technological advancements and innovation are likely to facilitate the commercial viability of these green technologies by the year 2029. The United States has positioned itself as a global leader in sustainable energy solutions across diverse sectors like power generation, manufacturing, and transportation. The increasing growth observed in these market segments, coupled with the global trend of industrialization, paints a promising future for the steam methane reform market. Notably, the rapid growth of fuel cells in recent years has contributed significantly to boosting sales opportunities. As industries gravitate towards smaller, more portable products, it is anticipated that the demand for SMR technology will continue to surge in the years to come. When considering System Type, the SMR Market is categorized into Merchant and Captive.

In 2022, the captive segment held the largest market share, accounting for 64%. This can be attributed to the efficiency of on-site hydrogen production, which often outperforms central production methods. In contrast, the merchant segment provides hydrogen to consumers from central production facilities and conveys it to end-users via cylinder trucks, bulk tanks, or pipelines. In an SMR facility, greenhouse gas emissions arise from the combustion of fossil fuels to generate heat and steam, as well as from the reforming reaction itself (Table 3.2). Contemporary SMR-based hydrogen production plants have achieved efficiencies that could potentially bring down CO_2 emissions to approximately 10% above the theoretical minimum. To achieve even greater reductions in CO_2 emissions from hydrogen production, the integration of CCS would be necessary. The prevailing industry practice for CO_2 capture in an SMR based hydrogen plant is employing Methyldiethanolamine (MDEA) solvent to capture CO_2 from the shifted syngas. Four alternative CO_2 capture methods are under consideration: utilizing an H_2-rich burner alongside capturing CO_2 from the shifted syngas using MDEA; capturing CO_2 from PSA's tail gas with MDEA; employing Cryogenic and Membrane Separation; and capturing CO_2 from flue gas using Monoethanolamine (MEA). These alternatives encompass CO_2 capture rates ranging from 56% to 90% [27].

Table 3.2 Greenhouse gas emissions from SMR/CCS [27]

Process unit/stage	Key emissions sources	Other emissions sources
Natural gas processing	• Electricity and/or liquid fuel combustion for NG extraction and movement • Fugitive methane and/or CO_2 from NG extraction	
Natural gas transport	• Electricity and/or liquid fuel combustion for materials movement • Methane leakage	
Heat recovery and electricity generation	• No significant emissions other than those covered under common emissions sources	
Hydrogen enrichment	• Water gas shift reactions occurring as part of hydrogen enrichment	
Syngas purification	• Electricity and/or heat for operation of the relevant purification units	• Exhaust CO_2 due to sulphur removal of exhaust gases (where applicable)
CO_2 capture and separation	• Electricity and/or heat for relevant separation units • Capture rate	
Compression and transportation of CO_2	• Electricity for compression of CO_2 • Electricity and/or gaseous fuel combustion for pipeline transport • Liquid and/or gaseous fuel combustion for motive transport • Fugitive carbon dioxide from CO_2 compression and transportation	
Storage of CO_2	• Electricity for injection or transformation	• Fugitive CO_2 from permanent storage location
Hydrogen compression and storage	• Electricity for compression and storage maintenance	• Fugitive hydrogen emissions

3.2.2 Coal gasification

In terms of Technology Type, the SMR Market is divided into Coal Gasification and Steam Methane Reforming. The coal gasification segment is projected to exhibit the highest growth rate by 2029. Coal gasification technology ranks among the most widely employed methods in the hydrogen generation market. This is partly due to the abundant availability of coal in several nations, such as India, which has driven its extensive utilization in the market. Coal stands as the most abundant finite resource globally, with current assessments projecting recoverable reserves of 216 to 500 years at present consumption levels [28]. By the close of 2021, the collective worldwide recoverable coal reserves were evaluated to be approximately 1,161 billion short tons, with five nations (the United States 22% (Figure 3.10), Russia 15%, Australia 14%, China 14%, and India 11%) accounting for roughly 75% of the globe's verified coal reserves. The generation of electricity through existing coal utilization methods, such as combustion systems, is associated with substantial greenhouse gas emissions and adverse environmental effects. An emerging alternative technology gaining momentum is coal gasification for hydrogen production [29].

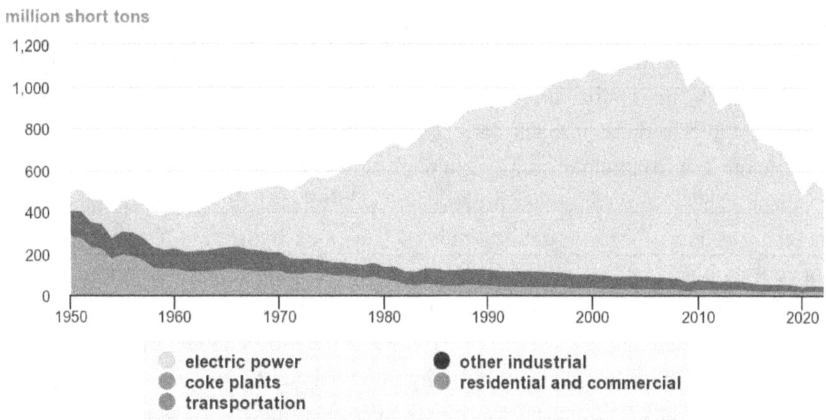

Figure 3.10 US coal consumption by major end users, 1950-2022 [30]

Coal gasification is a prominent method within the spectrum of hydrogen production, representing a transitional approach from fossil fuel-based technologies towards more sustainable alternatives. This process involves converting coal, a carbon-rich fossil fuel, into hydrogen-rich gas through a series of chemical reactions. Despite its association with fossil fuels, coal gasification offers the potential to mitigate certain environmental concerns associated with direct coal combustion. In coal gasification, coal is subjected to high temperatures and controlled amounts of oxygen and steam. This thermochemical process results in the breakdown of coal into its constituent elements, primarily carbon, hydrogen, and some trace elements. The produced gases, known as synthesis gas or syngas, consist mainly of CO and hydrogen. This syngas can then be further refined to separate hydrogen from other components. Coal gasification offers several advantages that make it an intriguing pathway for hydrogen production. One key advantage is the relative abundance of coal reserves in various regions, providing a reliable and accessible feedstock for gasification. Additionally, coal gasification has the potential to capture and utilize CO_2 emissions, making it a candidate for carbon capture and utilization (CCU) strategies. By capturing and storing or utilizing CO_2 emissions, coal gasification can contribute to the reduction of greenhouse gas emissions and promote a more sustainable energy landscape.

However, coal gasification also presents certain challenges and considerations. The process requires careful management of the waste products and byproducts generated during gasification, including various pollutants that must be effectively captured and treated to minimize environmental impacts. Moreover, the technical complexities of coal gasification demand advanced engineering and operational expertise, which could affect the overall cost-effectiveness of the method. The term "black hydrogen" is employed to characterize hydrogen generated through the utilization of coal as a primary feedstock. This production method involves coal gasification, a thermochemical process in which coal is subjected to high temperatures and controlled reactions to release a mixture of gases, including hydrogen. While this approach does produce hydrogen, it is often associated with significant environmental and sustainability concerns due to its high carbon footprint. The coal gasification process not only releases hydrogen but also results in the emission of substantial amounts of CO_2 as a byproduct. Consequently, the concept of black hydrogen highlights the environmental trade-offs and challenges posed by certain hydrogen production pathways, emphasizing the need for cleaner and more sustainable alternatives to ensure a transition to a low-carbon energy future. Brown hydrogen is a type of hydrogen that is produced using a process known as coal gasification, where brown coal, also referred to as lignite, serves as the primary source material. This production method involves subjecting brown coal to high temperatures and controlled reactions to release a mixture of gases, including hydrogen. However, the term brown hydrogen is often used to emphasize the environmental and climate-related challenges associated with this particular production pathway. Brown coal, being a lower-grade form of coal, contains higher levels of impurities and moisture compared to higher-grade coals, which can result in higher emissions of CO_2 and other pollutants during the gasification process. As a result, brown hydrogen is viewed as a less sustainable option compared to cleaner sources of hydrogen, such as renewable hydrogen. The concept of "brown hydrogen" underscores the importance of transitioning towards more environmentally friendly and carbon-neutral hydrogen production methods to mitigate the impacts of climate change and ensure a sustainable energy future.

At present, coal gasification contributes approximately 18% of the global hydrogen supply and stands as the second largest and most economically efficient technique for hydrogen generation. However, the landscape differs notably in China, where coal gasification dominates hydrogen production, accounting for 62% of the country's hydrogen output. It's noteworthy that coal gasification-based hydrogen production exhibits a more pronounced environmental impact. This stems from the process yielding 18-20 kg of CO_2 emissions per 1 kg of H_2, a higher rate compared to the 8-12 kg emitted per kilogram of H_2 produced via SMR [31]. The United States is another prominent player, accounting for a notable portion of coal gasification-based hydrogen production. The U.S. has been utilizing coal gasification as a means to extract hydrogen and other useful products from its coal resources, aligning with its focus on domestic energy production and technological innovation. Additionally, Germany, India, and Australia have been actively engaged in coal gasification for hydrogen generation. These countries possess significant coal reserves and have explored coal gasification as a strategy for both energy and industrial feedstock purposes.

From a policy perspective, governments play a pivotal role in steering the direction of coal gasification's integration into the energy landscape. Policymakers need to address challenges related to emissions reduction, energy security, and economic diversification. Environmental regulations, emission targets, and incentives for clean energy sources can influence the viability and attractiveness of coal gasification projects. Incentivizing research and development, fostering public-private partnerships, and creating a conducive regulatory environment can accelerate the technology's advancement. Economic implications are equally significant. Coal gasification projects entail substantial upfront investments in infrastructure and technology development. Policymakers must consider how to manage these costs and ensure a favorable risk-reward profile for investors. The cost competitiveness of hydrogen produced through coal gasification needs to be evaluated against other hydrogen production methods. This involves assessing the cost of carbon capture and storage, which can considerably impact the economic viability of coal gasification. Furthermore, economic diversification is a key driver. Regions heavily reliant on traditional coal industries can leverage coal gasification to transition their economies while ensuring a just transition for affected communities. Job creation potential and technology spillover effects should be weighed against any potential negative impacts. Balancing these policies and economic considerations is a complex endeavor. The success of coal gasification hinges on a harmonious alignment of policies that incentivize sustainable practices, address environmental concerns, and promote economic growth. By carefully navigating these intricacies, governments, and stakeholders can effectively harness the potential of coal gasification for hydrogen production, fostering a cleaner energy future while minimizing adverse economic consequences.

3.3 Hydrogen production from renewables

Hydrogen production from renewables (green hydrogen) stands as a beacon of hope in the quest for sustainable energy solutions. It represents a pivotal shift away from conventional fossil fuel-based methods, offering a pathway to cleaner, greener, and more environmentally responsible energy generation. The utilization of renewable sources such as wind, solar, hydropower, and biomass as the driving force behind hydrogen production not only addresses the pressing concern of reducing greenhouse gas emissions but also fosters energy independence and resilience. The renewable hydrogen paradigm revolves around harnessing the boundless power of nature to initiate electrolysis or other innovative processes that extract hydrogen from water or organic matter. By capitalizing on the virtually unlimited supply of renewable resources, the production of hydrogen gains a transformative edge, aligning with global commitments to mitigate climate change, enhance energy security, and usher in a future where sustainable development is at the forefront. The intricacies of hydrogen production from renewables reveal the intricate synergies between science, technology, policy, and international cooperation that shape this dynamic facet of the global energy landscape (Figure 3.11). The production of green hydrogen is still at a relatively early stage, and its global production capacity is not as significant as other forms of hydrogen production. Green hydrogen is produced through electrolysis, using renewable energy sources to split water into hydrogen and oxygen. The capacity for renewable energy-based hydrogen production is growing but is not as widespread as hydrogen production from fossil fuels. At the end of 2021, only merely around 4% of the world's total hydrogen production originated from the process of electrolysis. During the same period, electricity production globally maintained an average renewable energy contribution of approximately 33%, indicating that only about 1% of the total global hydrogen supply was generated using renewable energy sources [32]. During the period from 2022 to 2027, the primary scenario predicts the allocation of approximately 50 gigawatts (GW) of renewable capacity for the purpose of hydrogen production. This accounts for approximately 2% of the overall growth in renewable capacity during this timeframe. China is expected to spearhead this expansion, followed by Australia, Chile, and the United States.

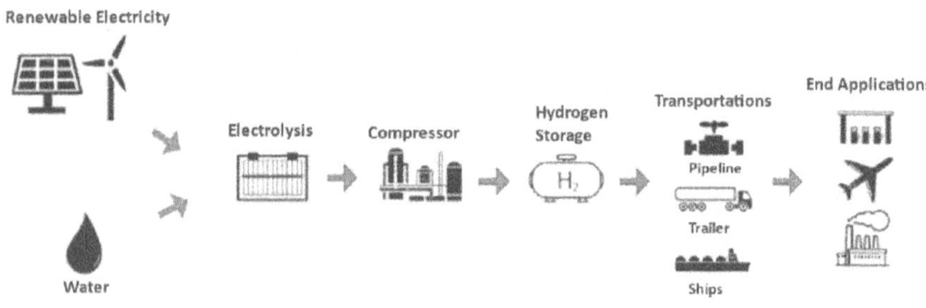

Figure 3.11 Green hydrogen production [33]

Collectively, these four markets contribute approximately two-thirds of the dedicated renewable capacity specifically earmarked for hydrogen production. On a global scale, the addition of fresh energy capacity is distributed equally between photovoltaic (PV) and onshore wind installations, although the distribution varies across regions based on the accessibility of resources. For example, the majority of the expansion in the Middle East and North Africa (MENA) region comes from solar PV installations, whereas in Latin America, the pipeline for electrolyzer projects is projected to be predominantly populated by onshore wind initiatives in Chile. Due to their lengthier development timelines, offshore wind ventures constitute less than 1% of the new renewable capacity intended for electrolyzer plants during the period spanning from 2022 to 2027 [34]. Figure 3.12 illustrates green hydrogen production on various continents [35], and Figure 3.13 depicts the top green hydrogen producers in the world in 2022. The most recent evaluations focusing on disclosed green hydrogen initiatives and objectives worldwide indicate that by the end of the decade, Australia, the United States, and Spain are poised to take the lead in the market. Following closely behind are Canada, Chile, Egypt, Germany, India, Brazil, and Oman (Figure 3.14).

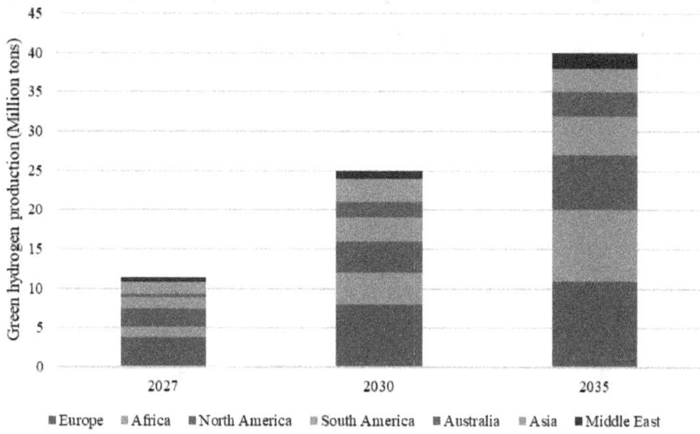

Figure 3.12 Green hydrogen production on various continents

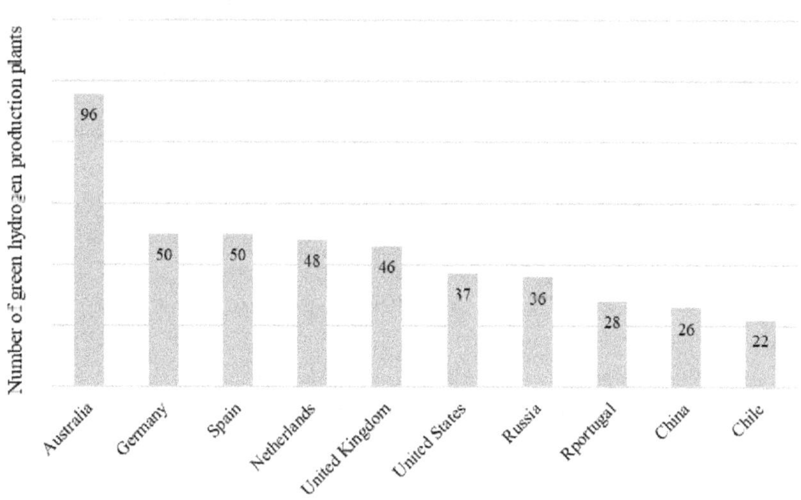

Figure 3.13 Top green hydrogen producers in 2022

The production of green hydrogen requires a multifaceted strategy that involves various aspects. Firstly, scaling up renewable energy capacity is crucial to provide the necessary electricity for electrolysis. This necessitates the deployment of additional wind, solar, and hydro facilities, coupled with advancements in energy storage technologies to ensure a stable supply of power even during periods of low renewable energy generation. Secondly, optimizing the electrolysis process itself is vital to enhance efficiency and cost-effectiveness. This involves research and development efforts to improve the efficiency of electrolysis cells, reduce the use of rare and expensive materials, and increase the durability and lifespan of the equipment. Furthermore, policies and incentives play a pivotal role in accelerating the adoption of green hydrogen.

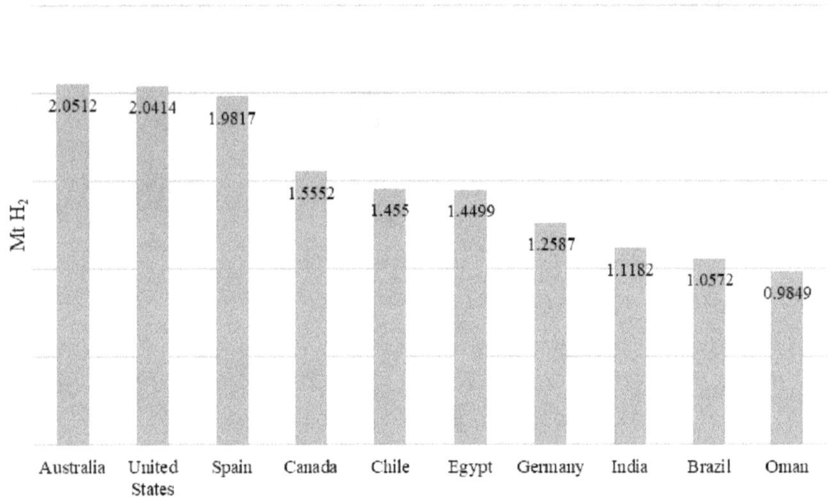

Figure 3.14 Top green hydrogen producers by 2030 [36]

Governments need to implement supportive regulatory frameworks, financial incentives, and subsidies to drive investment in renewable energy projects and electrolysis facilities. Collaboration between public and private sectors can foster innovation and drive down the costs of technology, making green hydrogen economically competitive with fossil fuels over time. Another critical aspect is international cooperation. As the global energy landscape is interconnected, collaboration among countries can facilitate the exchange of best practices, technology transfer, and the creation of a global market for green hydrogen. This can drive economies of scale, further reducing production costs and enhancing the attractiveness of green hydrogen as an energy solution.

The capability of green hydrogen lies in its potential to revolutionize energy systems, reduce carbon emissions, and enable sectors that are otherwise challenging to decarbonize. However, realizing this potential requires a holistic approach that involves ramping up renewable energy capacity, refining electrolysis technologies, implementing supportive policies, building infrastructure, and fostering international collaboration. By addressing these strategies comprehensively, green hydrogen can play a transformative role in achieving a more sustainable and low-carbon energy future [37]. The trajectory of dedicated renewable capacity expansion is intricately linked to the uncertain variable of demand for renewable hydrogen. The global demand for hydrogen, particularly in its sustainable form, will play a pivotal role in dictating the speed and scale of renewable capacity growth. Numerous countries have strategically incorporated hydrogen into their industrial development plans, setting ambitious production targets and signaling their commitment to a cleaner energy landscape. Moreover, demand-side policies, such as targets for fuel-cell vehicles, have been established with the intention of promoting hydrogen adoption and reducing emissions. However, these policies might not inherently prioritize renewable hydrogen over its non-renewable counterparts, especially if the former comes with a higher production cost.

The economic dynamics of renewable hydrogen production versus that of conventional methods pose a challenge, as the cost differential might impact the viability of policies solely focused on demand creation. This intricate interplay between demand, policy frameworks, and economic considerations underscores the complex nature of the renewable hydrogen landscape. As nations strive to align their strategies with sustainable energy goals, finding a balance between incentivizing renewable hydrogen production, fostering demand, and addressing cost implications becomes crucial in ensuring a successful transition towards a greener and more resilient energy future. Nations have unveiled ambitious goals for green hydrogen in alignment with their decarbonization strategies. For instance, the European Union aims to achieve an annual production of 10 million megatons of green hydrogen by 2030, with several individual member states also establishing their distinct targets. Likewise, the Chilean government has articulated its ambition to become the global leader in producing the most cost-efficient green hydrogen by the year 2030. Australia possesses the potential to emerge as a global hub for hydrogen production, owing to its abundant hydrogen resources that extend far and wide. Covering approximately three percent of the nation's vast land area, equivalent to an expansive 262,000 square kilometers, Australia possesses ideal conditions conducive to green hydrogen generation. These favorable conditions stem from ready access to renewable sources of electricity and ample water supplies. This unique combination of factors positions Australia to capitalize on its hydrogen potential, paving the way for the production of quantities exceeding the projected global demand anticipated for the year 2050. In essence, Australia's geographical advantages and resource availability set the stage for it to not only meet but potentially surpass the world's hydrogen requirements, showcasing the nation's pivotal role in shaping the future hydrogen landscape. Australia has established an ambitious goal known as the "H_2 under 2" target, which seeks to achieve the production of hydrogen at a cost below AU$2, approximately equivalent to US$1.40 per kilogram.

At present, the production of green hydrogen comes at a significant expense, exceeding US$7 per kilogram at its upper range. However, there is a promising trend of decreasing costs as technology advances, benefits from economies of scale, and witnesses a continuous decline in solar and wind energy prices (Figure 3.15) [1]. The initial investment required for electrolyzers, the devices employed to extract hydrogen from water, has the potential to plummet by up to 80 percent by the year 2030. This projected reduction in capital costs underscores the ongoing evolution of the hydrogen production landscape, indicating a more cost-effective and accessible future for green hydrogen. As these cost barriers continue to diminish, green hydrogen stands to become a more viable and competitive solution, paving the way for broader adoption and integration into global energy systems. According to the projections by the IEA, in their net-zero emissions scenario, green hydrogen is expected to contribute to approximately 60 percent of the total global hydrogen production by the year 2050.

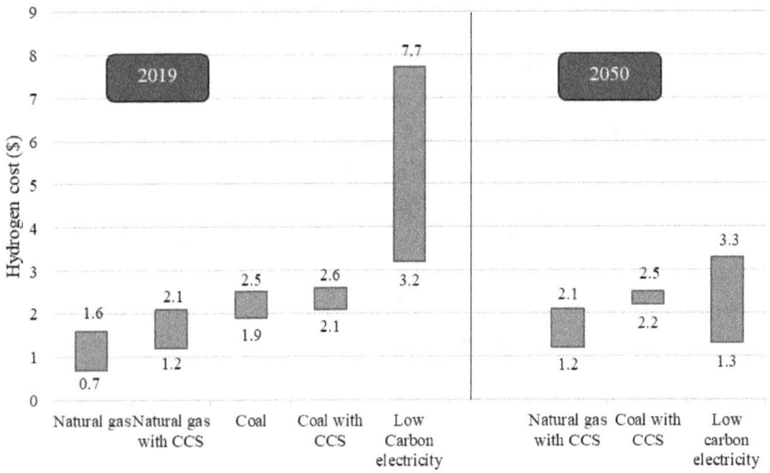

Figure 3.15 Global average levelized cost of hydrogen production by energy source and technology, 2019 and 2050

Fostering the growth of the green hydrogen economy involves establishing a comprehensive policy framework that provides both short-term and long-term support mechanisms. These mechanisms include waivers and tax reductions to incentivize investments, regulations requiring a certain portion of hydrogen to be sourced from renewables, incentives tied to actual production levels to encourage scaling, and financial backing to bridge the affordability gap between green hydrogen and natural gas-based hydrogen production. The ultimate aim is to drive down the cost of green hydrogen production, enabling it to effectively compete with its fossil fuel-based counterpart. By doing so, policymakers strive to foster a thriving green hydrogen economy that can play a pivotal role in reducing carbon emissions and promoting sustainable energy practices on a global scale.

Broadly, there exist two modes of coupling between power and hydrogen: coupling at the power source side and coupling at the power load side. In the context of power-hydrogen coupling at the load side, illustrated in Figure 3.16, centralized renewable power generation facilities are established in regions abundant in renewable energy resources, such as China's northern and western areas. The electricity generated is then transmitted to the central and eastern regions via ultra-high-voltage (UHV) transmission lines. Subsequently, on the load side, hydrogen production takes place, either for local consumption or to power hydrogen-fueled turbines. These turbines can serve various purposes, including voltage support and capacity reserve for the power grid located at the load center. This approach offers several advantages [38].

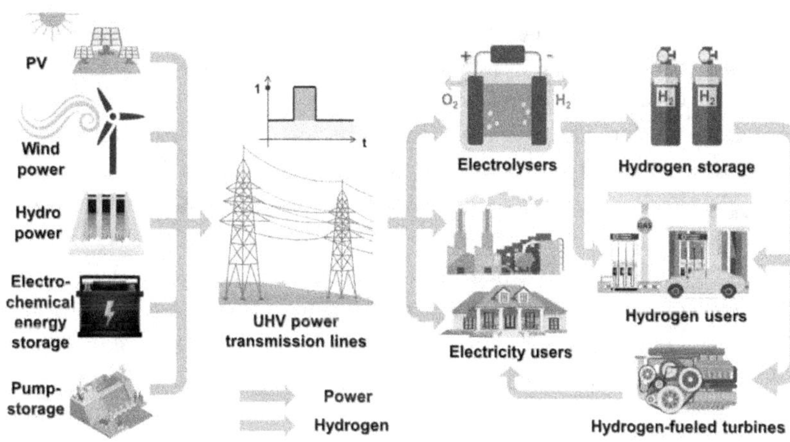

Figure 3.16 Power–hydrogen coupling at the load side [38]

Firstly, hydrogen production facilities have the flexibility to choose their locations. Unlike the power-hydrogen coupling at the source side, where hydrogen production facilities often need to install their own PV or wind power stations, typically situated in remote areas, the load-side approach utilizes electricity from the grid. By situating factories closer to regions with significant hydrogen demand, expenses related to conversion, storage, and long-distance hydrogen transportation can be reduced. Additionally, this approach mitigates potential safety risks associated with the extensive transport of hydrogen over long distances.

Secondly, it optimizes the utilization of UHV power transmission lines for large-scale allocation of renewable energy. Previous research has shown that as the transport distance increases from 400 km to 1000 km, the cost of hydrogen transport through pipelines rises significantly compared to the cost of electricity transmission via UHV lines. The unit cost of power transmission, when converted to caloric value, is considerably lower than that of hydrogen transport, making UHV lines more economically competitive for energy transmission [39].

Thirdly, it capitalizes on larger peak-valley price differences for electricity near the load center. This economic compensation helps offset losses in the energy conversion processes involved in power-hydrogen-power systems. However, this approach also presents potential drawbacks: A comparatively lower capital investment in UHV lines doesn't necessarily translate to lower total energy transmission costs. To accommodate the high volatility of renewable energy, it must be combined with power from other sources like coal, hydropower, electrochemical storage, and pumped storage, forming a smoother power curve for UHV lines. This additional investment in adjustable power sources and energy storage cells may result in a higher system cost. The approach underutilizes the dynamic response capabilities of electrolyzers because the high volatility of renewable energy is already smoothed out by adjustable power sources and energy storage cells at the source side. Consequently, the power transmitted to the load side is relatively stable, limiting the capacity of electrolyzers to provide grid-balancing services and absorb fluctuations in variable renewable energy at the source side. The conversion processes in power-hydrogen-power systems can lead to low efficiency and substantial energy losses. Alkaline and proton exchange membrane electrolyzers, for instance, have efficiencies no greater than 70%. Hydrogen-fueled gas turbines typically operate with an efficiency of around 50%. This results in an overall efficiency of only 35% for power-hydrogen-power systems, meaning that 65% of the energy is lost during energy conversion processes. In power–hydrogen coupling at the source side, the source side deployment consists of electrolyzers, hydrogen storage tanks, and hydrogen-fueled gas turbines, with the primary goal of mitigating curtailment of variable renewable energy and supporting the stable transmission of power over long distances via UHV lines, as illustrated in Figure 3.17. The inherent volatility of renewable power resources can be effectively harnessed by electrolyzers due to their exceptional dynamic responsiveness. The hydrogen generated can serve multiple purposes, including local consumption or utilization in gas turbines for electricity generation. It plays a pivotal role as a supplementary component to electrochemical energy storage, pumped storage, and regulatory power sources, offering grid-balancing capabilities and ensuring a relatively consistent power flow through UHV lines. The primary advantage of this mode lies in its ability to ensure a relatively high utilization rate of both renewable energy resources and the power transmission infrastructure. As coal and gas power plants are gradually phased out in the future, electrolyzers, hydrogen storage, and hydrogen-fueled gas turbines can become critical tools for peak regulation and providing grid-balancing services. When renewable power output is abundant, electrolyzers operate at maximum capacity, storing excess hydrogen in tanks. Conversely, when renewable power output is diminished, hydrogen-fueled gas turbines can be activated.

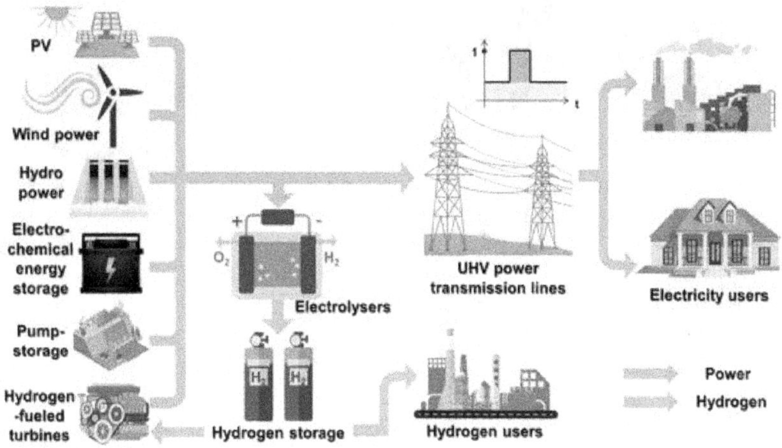

Figure 3.17 Power hydrogen coupling at the source side [38]

This approach guarantees a relatively stable power transmission through UHV lines while maximizing the utilization of renewable energy resources. However, a potential drawback of this approach is the relatively low efficiency and substantial energy losses incurred during the power–hydrogen–power conversion processes, mirroring the challenges seen in the mode of power–hydrogen coupling at the load side. The schematic diagram representing power–hydrogen coupling at the source side for hydrogen transport is depicted in Figure 3.18. In this mode, hydrogen generated from renewable energy sources at the source side is transported over extended distances through pipelines to reach the demand center. This mode offers several advantages, notably in terms of alleviating pressure on the power grid by transmitting renewable energy as hydrogen across long distances. This, in turn, can facilitate the extensive deployment and utilization of renewable energy resources. Unlike electric power, hydrogen exhibits greater tolerance for fluctuations because there's no immediate requirement for an instant supply-demand balance. Moreover, pipelines can also double as a means of storage. However, there are potential challenges associated with this approach. Hydrogen pipelines are relatively costly and are still in the early stages of development, which limits their near-term applicability. Presently, the most common method for hydrogen transport involves trucks, which are suitable primarily for small-scale transportation over short distances. Long-distance hydrogen pipelines are currently in the design phase, with costs soaring as high as $1,000,000 per kilometer. One approach to mitigate costs is to blend hydrogen with natural gas in pipelines, which can reduce expenses. However, it's important to note that the maximum permissible hydrogen content in such pipelines should not exceed 10%, placing constraints on the scale of hydrogen transport.

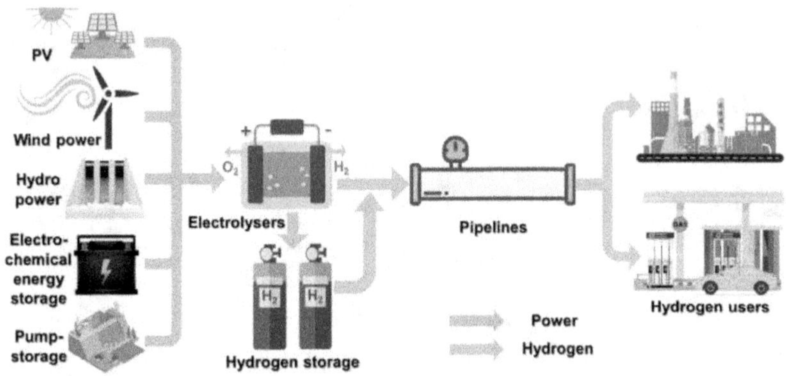

Figure 3.18 Power–hydrogen coupling at the source side for hydrogen transport

3.3.1 Hydrogen production from solar energy

Benefiting from its abundant availability, solar energy emerges as a highly promising energy source with the potential to yield substantial energy savings and a reduction in carbon emissions. In this context, various solar technologies, such as PV, solar thermal collector (STC), and hybrid photovoltaic/thermal (PV/T) systems, are currently being extensively employed [40,41]. Figure 3.19 shows how hydrogen is produced using PV systems. The generation of solar power is significantly influenced by geographical and climatic factors, resulting in certain regions of the world being particularly well-suited for harnessing solar energy. The best places to generate solar power exhibit characteristics that optimize the efficiency and output of PV systems. These areas are typically located near the equator, where they receive abundant and consistent sunlight throughout the year. Countries within the "solar belt," including regions in North Africa, the Middle East, parts of Southeast Asia, and portions of Central and South America, experience prolonged periods of high solar irradiance due to their proximity to the equator (Figure 3.20). This positioning ensures that solar panels receive direct sunlight for extended periods, maximizing energy production. Additionally, regions with minimal cloud cover and atmospheric pollutants are ideal for solar power generation. Deserts, such as the Sahara Desert in Africa, the Atacama Desert in South America, and parts of the Arabian Desert, often possess these conditions. The arid climate limits cloud cover and minimizes the obstruction of sunlight, resulting in consistently high levels of solar irradiance. In recent years, technological advancements in PV systems have also expanded the potential for solar power generation in less traditionally sun-rich areas.

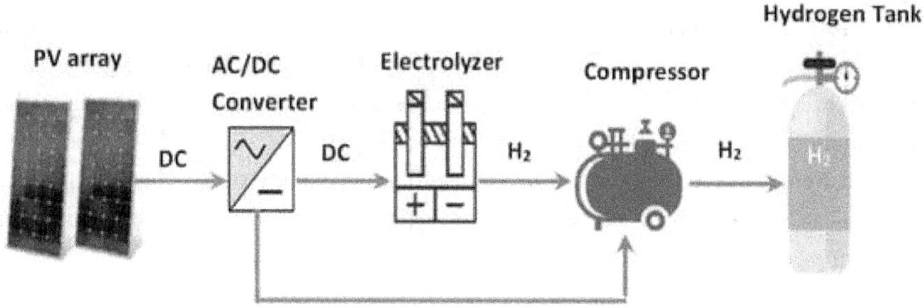

Figure 3.19 Solar-to-hydrogen using PV [42]

Figure 3.20 World solar PV power potential

Countries with high latitudes, such as Germany and China, have managed to become major solar power producers due to their adoption of solar technologies and supportive policies. Solar hydrogen represents an environmentally friendly alternative fuel with the potential to supplant fossil fuels, thereby mitigating the reliance on oil and gas resources. Additionally, it contributes to the reduction of carbon emissions, serving as a crucial tool in the fight against global climate change. The utilization of solar energy, which is abundantly available for the electrolysis process, offers a sustainable avenue. Earth's surface annually receives an estimated 3.93×1024 megajoules of incident solar energy, a quantity roughly 10,000 times greater than the current global energy demand. This immense potential underscores the possibility of fulfilling humanity's entire energy requirements using just a fraction of 1% of this remarkable energy reservoir. Figure 3.21 depicts the potential for producing hydrogen fuel through diverse pathways utilizing solar energy. These approaches encompass distinct methodologies, including harnessing solar energy for thermal processes like thermochemical cycles and thermolysis, utilizing photovoltaic (PV) systems to convert solar energy into electrical power for electrolysis, and leveraging photonic energy for hydrogen production through processes like photobiological mechanisms or photocatalysis. Given the anticipated pivotal role of solar energy in meeting the impending energy requirements, the development of cost-effective mechanisms for storing solar energy has emerged as a critical imperative. This becomes especially pertinent due to the intermittent inherent in solar power, necessitating a dependable strategy for sustaining an uninterrupted electricity supply in alignment with demand patterns.

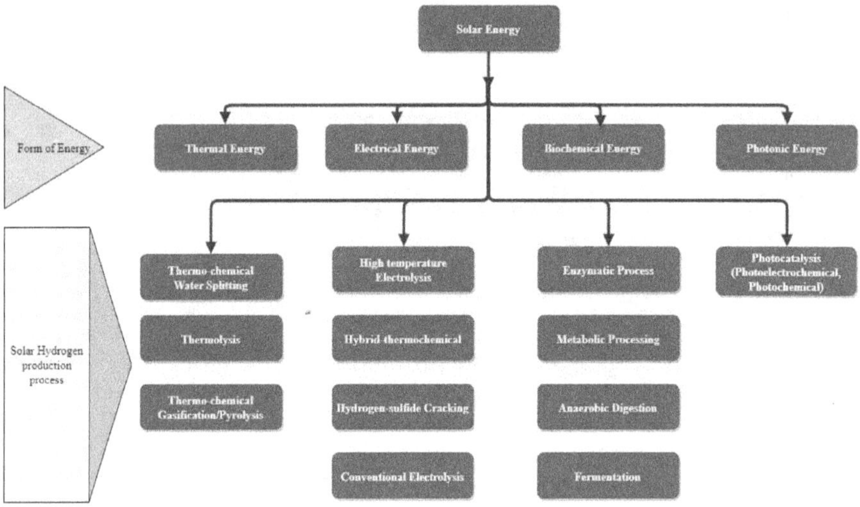

Figure 3.21 Various methods of hydrogen production from solar energy

Given the capricious nature of solar energy availability, integrating substantial storage capacity becomes indispensable for maintaining grid stability. Amidst this challenge, a viable solution emerges in the form of harnessing solar hydrogen as a storage medium. The fusion of solar systems with the electrolysis process holds the potential to generate solar hydrogen, which in turn finds versatile applications spanning industries, households, transportation, and power generation. Notably, this process mitigates the intermittency issue by enabling the storage of excess electricity as hydrogen, effectively transforming it into a reservoir for power generation when solar energy is not readily accessible. An additional advantage materializes in the context of fuel cell utilization, where the prices of these technologies become pivotal. As the expense of fuel cells diminishes, the prospect of utilizing surplus electricity for hydrogen production and subsequent storage gains viability, positioning it as an economic strategy for ensuring a consistent energy supply during periods of solar energy scarcity [43].

Among the various types of solar-to-hydrogen technologies, electrolysis is considered one of the most mature and commercially viable methods that is ready for use in the market. Electrolysis involves using electricity to split water molecules into hydrogen and oxygen, and it can be powered by electricity generated from solar energy. There are different electrolysis technologies, including proton exchange membrane (PEM) electrolysis and alkaline electrolysis, which have been developed and refined over the years. While other solar-to-hydrogen methods like thermolysis, biochemical, and photonic processes hold promise and are being researched, they might be at various stages of development, facing challenges related to scalability, efficiency, and cost-effectiveness. These methods might require further advancements and technological innovations before they become commercially viable on a large scale.

Solar-to-hydrogen conversion encounters ongoing hurdles in terms of economic viability and reliability, compounded by the modest efficiency levels exhibited by solar energy systems. Currently, the cost of hydrogen derived from electrolysis falls within the range of approximately \$3.26 to \$6.62 per gallon of gasoline-equivalent. Nonetheless, a paradigm shift looms on the horizon as the ascendancy of decentralized electricity generation emerges poised to supplant large-scale centralized fossil fuel power plants and conventional national electric grids. In this transformative context, the efficiency of solar-to-hydrogen systems emerges as a pivotal performance metric. The research underscores that the efficacy of solar hydrogen production can be markedly enhanced by up to 12% within a solar panel-based infrastructure. Furthermore, within the realm of concentrated photovoltaics (CPV)-based hydrogen production, reported efficiencies span a spectrum ranging from 18% to 24.8%, contingent upon the specific contextual parameters that govern the process [44]. The U.S. Department of Energy (DOE) has set an ambitious target price range of \$2 to \$3 per kilogram of hydrogen, which serves as a pivotal benchmark in advancing the feasibility and economic viability of hydrogen production through solar energy conversion. This target encapsulates the DOE's commitment to fostering sustainable and competitive hydrogen-based energy solutions that can significantly contribute to the transition towards cleaner energy sources. One of the most significant challenges in realizing this target is intricately tied to the reduction of costs associated with PV systems, which are at the heart of converting solar energy into hydrogen. These PV systems, responsible for capturing sunlight and initiating the water electrolysis process to generate hydrogen, encompass a complex network of technologies, components, and infrastructure. The cost-effectiveness of the entire process hinges on achieving substantial advancements in PV technology, manufacturing efficiency, and deployment practices. Achieving the DOE's target entails innovations across multiple fronts, including enhanced PV cell efficiency, streamlined manufacturing processes, and optimized integration of PV systems into hydrogen production setups. Successfully reducing the cost of the PV system is an essential step towards realizing the DOE's vision of cost-competitive hydrogen production and facilitating the broader integration of hydrogen as a clean energy source in various sectors of the economy [45].

The adoption of solar-to-hydrogen technology has the potential to catalyze a transformative shift in global energy policy regimes, challenging the dominance of traditional oil-based policies. This transition can be driven by several factors that highlight the advantages of solar-to-hydrogen and the imperative to address the challenges posed by the current oil policy framework. Solar-to-hydrogen technologies can enhance energy security and diversification. Unlike oil, which often involves geopolitical complexities and market volatility, solar power is abundant and domestically producible in many regions. By investing in solar-to-hydrogen infrastructure, countries can reduce their dependence on oil imports and insulate themselves from the uncertainties of global oil markets, thereby strengthening energy security. Moreover, solar-to-hydrogen introduces opportunities for economic growth and job creation. Establishing new policy frameworks that prioritize renewable hydrogen production can stimulate innovation, research, and development, leading to the growth of a new green energy industry. This can translate into job opportunities across the entire value chain, from solar technology manufacturing to hydrogen production, distribution, and application.

As an example, in July 2023, Chinese-based Sinopec commenced operations at the largest solar-to-hydrogen project globally and the first of its kind in China. Situated in the Xinjiang region, the facility comprises a photovoltaic (PV) generation complex, power transmission lines, a water electrolysis plant for hydrogen production, hydrogen storage, and transport infrastructure. Its objective is to generate 20,000 tons of green hydrogen annually using solar energy for electrolysis. The facility has the capacity to store 210,000 cubic meters of hydrogen and transport 28,000 cubic meters per hour. In Europe, the Spanish government has revised its 2030 hydrogen targets, significantly increasing them in the latest version of its national energy and climate plan. The government now plans to deploy approximately 11 GW of electrolyzers by 2030 for the production of renewable hydrogen, primarily for industrial applications. This target represents a substantial increase compared to the initial estimates in the hydrogen roadmap, which aimed for 4 GW of electrolysis capacity by 2030. It is anticipated that by 2030, the cheapest option for Germany to supply its hydrogen demand would be importing green hydrogen from Spain via pipeline. The projected cost for producing green hydrogen in Germany by 2030, generated through electrolyzers directly linked to renewable power facilities, is estimated to fall within the range of €3.90 ($4.25) to €5 per kilogram, according to the analyst's findings [46]. In contrast, the cost of producing renewable hydrogen in sunlit Spain would be €3.10/kg, with an additional €0.36/kg required for pipeline transportation to Germany. Alternatively, if transported as liquid hydrogen, the cost would be €1.25/kg, or €1.47/kg if shipped as ammonia. It's worth noting that the recently announced €2.5 billion H2Med hydrogen pipeline, originally planned to connect Spain and France and set to be finalized by 2030, will now be extended to Germany.

The cost of renewable hydrogen would also be lower in Australia (€3.10/kg), Chile (€3.10/kg), Morocco (€3.20), and the United Arab Emirates (€3.60) when considering the levelized cost, primarily due to their superior solar resources. When factoring in transportation and conversion expenses, it is anticipated that importing liquefied green hydrogen from Morocco would be the most economically attractive option from non-EU sources in 2030, with an estimated cost of €4.58/kg. If this hydrogen from Morocco were imported in the form of ammonia or liquid organic hydrogen carrier (LOHC), the costs would slightly increase to €4.72/kg and €4.68/kg, respectively. Imports from Australia and Chile would only be cost-competitive when transported as ammonia, with expected costs of €4.84/kg and €4.86/kg, respectively. In the scenario where a hydrogen pipeline is established between Morocco and Germany by 2030, green hydrogen from Morocco could be delivered at a cost of €3.72/kg.

As the world seeks to transition towards sustainable and renewable energy sources, policies governing the production of green hydrogen are of paramount significance. Governments and international bodies need to establish clear and supportive frameworks to incentivize and regulate the production of solar hydrogen. These policies should encompass a range of aspects, from financial incentives and subsidies to the development of infrastructure, research funding, and international collaborations. Financial incentives can significantly lower the production costs of solar hydrogen, making it more competitive in the energy market. Moreover, policies that facilitate the construction of hydrogen pipelines and transportation networks are essential for efficiently moving hydrogen from regions with abundant solar resources to areas with high demand. Additionally, government-supported research initiatives can drive innovation in solar hydrogen production technologies, ultimately leading to more efficient and cost-effective processes. Finally, international cooperation and agreements on hydrogen production and trade are crucial for creating a global hydrogen economy. Countries can come together to establish global agreements and protocols that prioritize renewable hydrogen over oil, fostering cooperation in research, technology transfer, and investment. New alliances and diplomatic efforts can be directed toward promoting renewable hydrogen as a shared solution to address energy security, environmental concerns, and economic growth.

3.3.2 Hydrogen production from wind power

Wind power generates more than 5% of the world's electricity. Wind power is anticipated to play a significant role as an energy source in future scenarios; however, its intermittent nature poses a challenge. To counter this challenge, the deployment of storage solutions and the implementation of super grids have been suggested. The necessity for storage in wind power revolves around managing peak demand and redistributing loads to accommodate continuous fluctuations in energy consumption. Among the most viable approaches for storing electrical energy are energy storage systems, and they can be classified into four main categories: chemical, thermal, mechanical, and electrical storage [47]. The utilization of hydrogen as an energy storage method has a historical background, reaching as far back as 1891 when the generation of hydrogen was initiated using wind energy. This process involved a water electrolysis system driven by an electric generator linked to a windmill [48].

Offshore and onshore wind power generation are two distinct methods of harnessing wind energy to produce electricity. Onshore wind power generation involves the installation of wind turbines on land, typically in areas with favorable wind conditions, such as ridges, open plains, or coastal regions. These turbines have varying capacities and heights, with the kinetic energy of the wind turning the turbine blades, which then generate mechanical power that is converted into electrical power through a generator. Onshore wind farms have been a significant contributor to renewable energy generation due to their established technology and relatively lower installation costs. They can be integrated into existing power grids, employing hydrogen production processes to reduce greenhouse gas emissions and reliance on fossil fuels. Offshore wind power generation, on the other hand, involves the placement of wind turbines in bodies of water, usually in coastal or marine environments. Offshore turbines are typically larger and more powerful than their onshore counterparts due to the stronger and more consistent wind speeds available at sea. The operation of offshore wind farms requires more advanced engineering techniques to anchor the turbines securely to the seabed and transmit electricity back to the shore. While the initial investment for offshore wind farms tends to be higher, they can take advantage of larger wind resources and less visual impact on densely populated areas. Offshore wind has the potential to generate substantial amounts of electricity, contributing significantly to the transition to cleaner energy sources and addressing the challenges of climate change.

The technical potential of offshore wind energy can be categorized into two types: shallow water (less than 60 meters) and deep water (60–2000 meters), with deep water offering a higher energy resource and presenting potential opportunities with reduced concerns from coastal communities. Another way to categorize offshore wind potential is by the distance from the shore, where near-shore sites are within 60 kilometers, and far-shore sites are up to 300 kilometers away from the coast. Certain countries, like Japan, possess significant offshore wind potential, exceeding 9000 TWh annually in deep water areas, necessitating the use of floating platform technology for effective harnessing [49].

Both onshore and offshore wind power have their advantages and challenges. Onshore wind farms benefit from relatively lower installation costs, easier maintenance access, and established infrastructure connections. They can contribute to local economies, create jobs, and provide a reliable source of renewable energy. Offshore wind farms, although more expensive to set up, offer the potential for higher energy yields, less visual and noise impact on communities, and more consistent wind conditions. However, they require more complex engineering and logistics for installation and maintenance, and there are environmental considerations related to marine ecosystems and navigation. Considering the projected increase in the need for clean energy and considering that approximately 80% of the worldwide offshore wind potential exists in waters exceeding 60 meters in depth, where only floating technologies are feasible, it is probable that extensive deployment of floating wind power will be combined with hydrogen production in the forthcoming years [50]. Offshore wind initiatives designed for the purpose of generating green hydrogen present the potential for substantial cost benefits compared to projects relying on grid-supplied electricity. This is attributed in part to the potential for cost reductions when transmission expenses are minimized or eliminated.

Numerous researchers have been investigating the environmental advantages of utilizing electrolytic hydrogen in conjunction with renewable energy sources as opposed to alternative hydrogen production techniques. Their findings indicate that when wind energy is employed to produce hydrogen through electrolysis, the overall CO_2 emissions are significantly lower, with approximately 1 kilogram of CO_2 produced per 1 kilogram of hydrogen over the entire life cycle. This stands in stark contrast to other methods like coal gasification and natural gas steam reforming, which result in substantially higher CO_2 emissions, approximately 21.7 and 9 kilograms of CO_2 per 1 kilogram of hydrogen, respectively [51,52]. A hydrogen ESS generally comprises components for hydrogen generation, like a water electrolysis setup, a means of hydrogen storage, and additional elements, including compressors, pipelines, and control mechanisms. It also encompasses safety mechanisms and tools for the safe utilization of hydrogen. Figure 3.22 illustrates an instance of a wind-to-hydrogen setup capable of producing electricity through a wind turbine and converting power into fuel using a PEM electrolyzer. The expense associated with hydrogen production displays significant disparities across various geographical locations. This divergence primarily stems from the differing wind resource potentials in each region. Nonetheless, the general estimate for the cost of hydrogen generation stands at approximately 6 h/kgH_2. This assessment finds support in a publication by the National Renewable Energy Laboratory (NREL), which outlines that the expenses for producing hydrogen in grid-connected setups span from 2.6 to 8.8 h/kgH_2. Employing hydrogen technology alongside wind power presents both benefits and disadvantages. The integration of hydrogen technology can enhance the grid's wind energy incorporation and reduce instances of power curtailment. Research also highlights notable hydrogen production efficiencies, typically ranging from 70% to 87%. However, it's worth noting that efficiency levels decrease to approximately 50% to 65% when hydrogen storage is introduced into the equation. This remarkable efficiency observed in the electrolysis process can be attributed to its electrochemical effectiveness, which aligns with principles from thermodynamics. However, in practical application, electrolyzers tend to function with cell voltages higher than the thermoneutral voltage, resulting in decreased efficiency due to heat losses. Additionally, the inclusion of extra components in hydrogen storage systems, such as compressors, has the potential to diminish overall system efficiency. Concerns also arise regarding the possibility of hydrogen leaks within the piping system. However, a wind-to-hydrogen system can also encompass a fuel cell for re-electrification within the network. Nonetheless, this incorporation leads to a reduction in the system's round-trip efficiency by at least 83%, which aligns with the maximum theoretical efficiency of fuel cells.

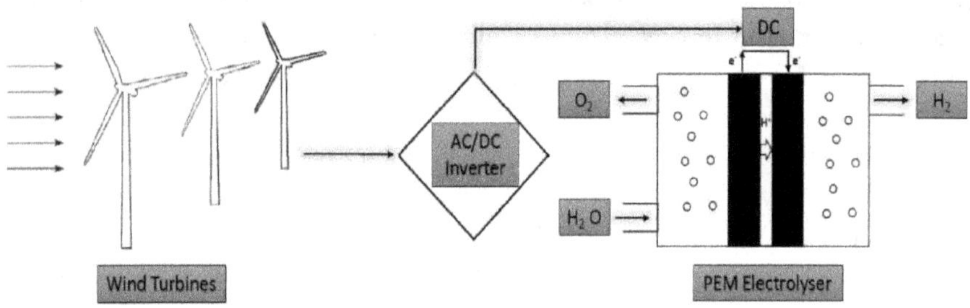

Figure 3.22 A schematic of hydrogen production from wind energy [53]

From a business perspective, hydrogen systems coupled with wind energy could prove financially advantageous as long as the cost of hydrogen fuel remains below 9 h/kgH_2, thereby generating adequate revenue and shortening the payback period. Despite these potential benefits, wind-to-hydrogen systems designed to enhance wind power contribution and support the grid or produce hydrogen fuel do face several drawbacks that limit their widespread adoption as a primary renewable energy storage solution. Although the constituent technologies within these systems have reached a state of maturity, their integration introduces complexity and escalates the frequency of operational failures. In comparison to alternative energy storage methods, the overall efficiency of a hydrogen system encompassing hydrogen production, storage, and utilization is relatively low. The round-trip efficiency spans from 30% for residential applications to 66%. Moreover, these systems are burdened by substantial capital and operational expenses, driven by the cost of electricity employed in hydrogen generation. Notably, the efficiency aspect holds notable uncertainty, with discrepancies of up to 15% for stand-alone electrolysis and up to 13% when incorporating storage and fuel cells. Furthermore, the variance between the highest and lowest mean values varies between 73% and 83% for electrolysis, 40% to 62% for electrolysis and storage, and 35% to 60% for electrolysis, storage, and fuel cells. These uncertainties necessitate careful consideration when assessing and strategizing for feasible hydrogen storage systems [54]. Since in offshore wind power generation, transmitting electricity over long distances underwater can be technically complex and expensive due to issues like transmission losses, cable installation, and maintenance costs, generating hydrogen locally at the offshore wind site allows for energy storage and transportation in a more convenient form [55]. In contrast, the transportation of gas via pipelines incurs significantly smaller losses (less than 0.1%) than electricity carried through offshore cables. Based on Miao B et al. [56] investigation, despite the higher initial installation costs of hydrogen pipelines, their ability to transport larger amounts of energy makes them a more efficient option for long-distance energy transmission. This is especially advantageous when renewable energy generation is located far from where the energy is needed. The greater energy transmission capacity of hydrogen pipelines compensates for their higher upfront costs, making them economically viable in certain scenarios [57].

From an economic standpoint, the cost per unit length of an offshore pipeline exceeds that of an offshore cable. However, the pipeline possesses a higher energy transmission capacity than the cable, resulting in reduced normalized capital costs for the pipeline when compared to an equivalent offshore electrical cable tasked with transmitting the same amount of energy. Figure 3.23 illustrates two different methods of offshore wind hydrogen production. The initial configuration comprises an offshore wind farm, an offshore electrolyzer, and onshore hydrogen storage. Conversely, the second system involves the placement of the electrolyzer onshore. In both setups, the option to integrate a fuel cell exists, serving the dual purpose of supplying electricity during high-demand periods and regulating grid frequency. In the first system, electricity generated by the offshore wind turbines is transmitted over a short distance to the electrolyzer platform. Here, hydrogen is generated, compressed, and subsequently conveyed to the shore through a pipeline. In contrast, in the second system, electricity is conveyed to the shore via a conventional cable. Upon reaching the shore, a decision can be made either feed the electricity directly into the grid or initiate hydrogen production. This configuration is commonly referred to as a hybrid system, providing the operator with the flexibility to control the quantity of power sold to the grid or directed into the electrolyzer. It even allows for the purchase of electricity from the grid for hydrogen production during periods of exceptionally low electricity prices, thereby enhancing load flexibility for the grid operator. Importantly, because the electricity powering the electrolyzer originates from wind farms, the hydrogen production process is entirely carbon-free [58]. The anticipated global offshore wind capacity is set to experience a significant increase in the coming decade, surging from approximately 50 GW in 2021 to 225 GW by 2030.

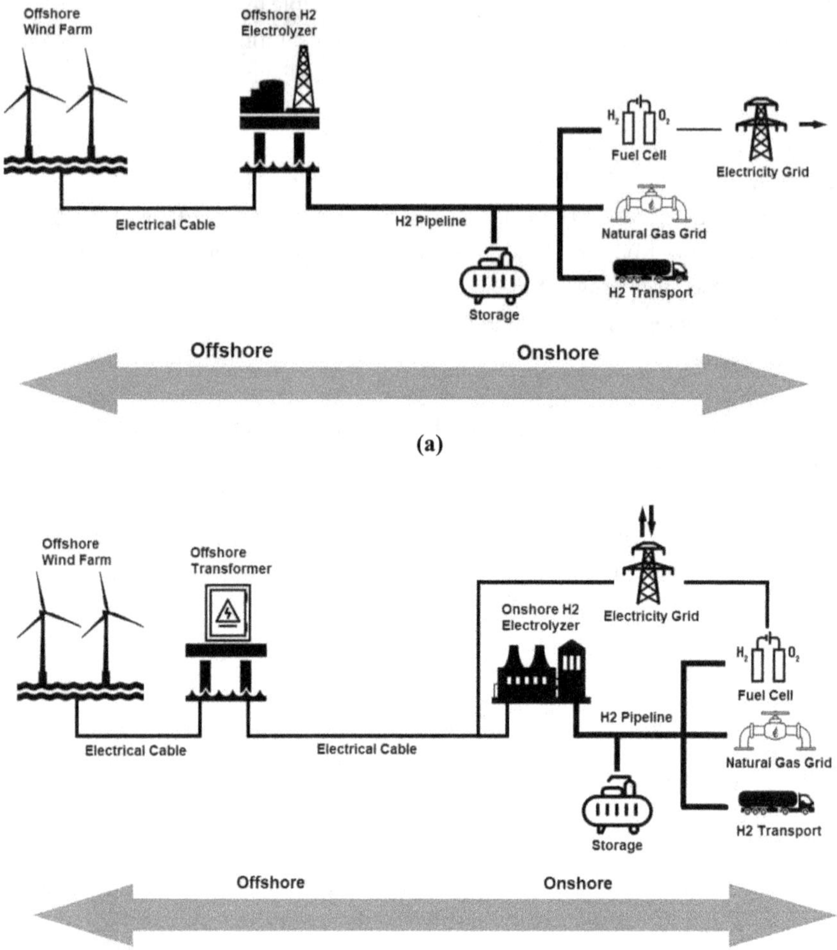

Figure 3.23 Offshore wind hydrogen production (a) Offshore electrolyzer system (b) Onshore electrolyzer system [58]

By the year 2030, Europe is expected to contribute to nearly half of the global offshore wind capacity; the installed offshore wind capacity within the Baltic Sea is projected to make up approximately 22% of the total capacity commissioned in Europe, Asia accounting for roughly 40% of the worldwide installed capacity, and the United States making up the remaining 10%. This growth in offshore wind is seen as a strategic transition for numerous oil and gas companies, leveraging their extensive offshore technical expertise as they increasingly pivot from fossil fuels to renewable energy sources. Offshore wind offers the potential to supply energy to offshore platforms, curbing carbon emissions associated with oil and gas operations and facilitating hydrogen production [59]. Utilizing hydrogen technology alongside wind power presents both advantages and drawbacks. Integration of hydrogen technology can enhance wind power integration into the grid, reducing power curtailments. Research also suggests that hydrogen production efficiencies are generally high, ranging from 70-87%. However, efficiency drops to 50-65% when hydrogen storage is introduced. The strong efficiency in electrolysis results from the electrochemical efficiency supported by thermodynamic theory. In practice, however, electrolyzers operate at higher cell voltages than the thermoneutral voltage, leading to efficiency reduction due to heat losses. Additionally, potential hydrogen leaks through the piping system and the use of additional components like compressors in a hydrogen storage system can decrease overall efficiency. On the positive side, a wind-to-hydrogen system can be employed for re-electrification through a fuel cell in the network, but this lowers the round-trip efficiency by a minimum of 83%, which is the maximum theoretical efficiency of fuel cells.

From a financial perspective, hydrogen systems combined with wind energy can be economically viable as long as the cost of hydrogen fuel remains above 9€/kgH_2 to generate adequate revenue and shorten the payback period. Nevertheless, the practical use of wind-to-hydrogen systems for improving wind power contribution and grid support or hydrogen fuel production has several disadvantages that hinder their widespread adoption as a primary renewable energy storage method. Despite the maturity of the technologies involved, these systems introduce complexity when integrated, leading to a higher frequency of operational failures. Moreover, the overall efficiency of a hydrogen system, encompassing hydrogen production, storage, and utilization, is comparatively low when compared to other energy storage technologies. Round-trip efficiency varies from 30% for residential use to 66%. These systems also entail substantial capital and operational expenses, mainly driven by the cost of electricity required for hydrogen production. Furthermore, there is notable uncertainty in the efficiency figures, with variations of up to 15% for standalone electrolysis and up to 13% when including storage and fuel cells. Additionally, the discrepancy between the highest and lowest mean values ranges from 73% to 83% for electrolysis, 40% to 62% for electrolysis and storage, and 35% to 60% for electrolysis, storage, and fuel cells. These uncertainties must be considered when evaluating and planning viable hydrogen storage systems [54]. The adoption of wind-to-hydrogen systems, whether integrated into the grid or functioning independently, offers substantial environmental benefits in terms of curbing greenhouse gas emissions. These systems can avert the release of several thousand tons of CO_2 annually and, contingent on their size can make a noteworthy contribution to cleaner energy generation and utilization in both transportation and power generation. Figure 3.24 illustrates hydrogen costs from onshore wind systems and hybrid solar PV in the long term [60].

Figure 3.24 Hydrogen costs from onshore wind system and hybrid solar PV in the long term

3.3.3 Hydrogen production from biomass

Biomass, a source of renewable and sustainable energy derived from a variety of natural materials, including animal remains and various plants such as grass, wood from forests, crop residues, municipal waste, industrial byproducts, and aquatic vegetation like algae, plays a vital role in the global energy landscape (Figure 3.25). This organic matter, primarily derived through the miraculous process of photosynthesis, serves as a reservoir for the energy captured from sunlight in the form of chemical bonds.

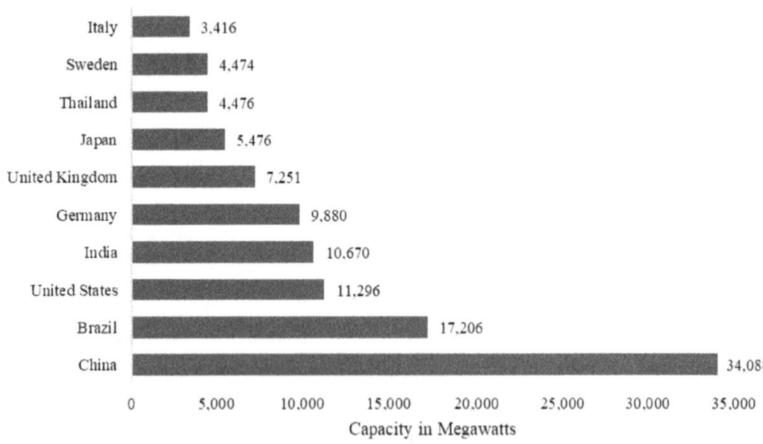

Figure 3.25 Leading countries in bioenergy (megawatts)

Traditionally, biomass has been harnessed for its energy potential through direct combustion, a method widely employed across the globe. While the combustion of biomass for energy generation does release CO_2 into the atmosphere, it's important to note that this emission is balanced by the amount of CO_2 absorbed by the plants during their growth phase when they are still alive [61]. In light of the growing concerns regarding the depletion of non-renewable fossil fuels, extensive research efforts have been directed toward finding sustainable ways to harness energy from biomass and replace these finite resources. The generation of energy from biomass, often referred to as bioenergy, has garnered significant attention in recent years, primarily due to the dwindling reserves of fossil fuels, necessitating a greater reliance on renewable alternatives. Among the various sources of renewable energy, bioenergy holds a distinctive place for several compelling reasons:

(i) Poverty Mitigation: It contributes significantly to alleviating poverty in developing countries by providing opportunities for employment and income generation through activities related to biomass production and processing.

(ii) Reliable Energy Supply: Bioenergy ensures a consistent energy supply without the need for expensive conversion devices. It can cater to energy demands in various forms, including heat, electricity, and liquid and gaseous fuels.

(iii) Versatility: Biomass-based energy can be readily transformed into different forms to meet diverse energy requirements.

(iv) Carbon Neutrality: Bioenergy is carbon-neutral, meaning it doesn't contribute to a net increase in atmospheric CO_2 levels. In fact, it has the potential to act as a carbon sink by sequestering carbon in vegetation and soil.

(v) Environmental Benefits: Bioenergy initiatives can contribute to the restoration of degraded and unproductive lands, enhance biodiversity, improve soil fertility, and increase water retention.

Considering its vast availability and sustainability, biomass stands out as one of the most significant renewable energy sources, holding tremendous potential to meet future energy demands and secure fuel supplies. The world possesses an annual biomass production capacity of approximately 220 billion oven dry tons, equivalent to about 4500 EJ [62]. Remarkably, a mere 5% of this enormous energy potential, amounting to 225 EJ, can already cover nearly half of the world's total primary energy requirements. Looking ahead, projections indicate that by 2050, forest and urban residues, alongside the cultivation of perennial energy crops, could generate approximately 150-400 EJ/year, accounting for approximately 25% of the world's energy needs [63].

Biomass can be efficiently converted into various energy carriers, including liquid and solid fuels, electricity, and heat, employing a range of energy conversion methods such as chemical, thermal, and biochemical processes. Presently, direct combustion remains the predominant approach, constituting over 97% of global bioenergy production, with thermochemical biomass conversion methods playing a relatively minor role. In recent years, the utilization of renewable biomass as a feedstock for hydrogen production has emerged as a topic of significant interest and exploration. This shift towards biomass-based hydrogen production is driven by several compelling advantages:

(i) Energy Independence: Biomass-derived hydrogen offers a path towards energy independence, reducing reliance on fossil fuel imports.
(ii) Domestic Energy Production: Producing hydrogen from biomass allows nations to keep the product within their borders, promoting self-sufficiency.
(iii) Stability in Pricing: The use of biomass for hydrogen production can contribute to more stable pricing of this crucial energy carrier.
(iv) Environmental Benefits: Harnessing hydrogen from biomass can enhance carbon dioxide balances, potentially improving them by up to 30% compared to conventional methods.

To produce hydrogen from biomass, two primary categories of feedstocks are readily available:
(i) Dedicated Bioenergy Crops: These are crops specifically grown for the purpose of energy production, offering a dedicated and controlled source of biomass.
(ii) Residues and Organic Waste: Biomass residues, such as organic waste from wood processing or regular agricultural farming, provide a more cost-effective and sustainable source of biomass for hydrogen production.

In 2021, the global biomass power market achieved a substantial size, reaching a valuation of USD 121,340.76 million. Projections indicate a CAGR of 6.0% from 2022 through 2030. This market expansion has been largely fueled by the growing concern for the environment, prompting many nations to augment the presence of renewable energy within their energy portfolios. Countries including India, China, Germany, the U.K., and France have set ambitious renewable energy targets with a vision of achieving carbon neutrality in the foreseeable future. Additionally, favorable policies and regulatory support have contributed to the increased adoption of biomass power. Moreover, European Union countries are actively seeking to phase out coal-based energy generation, further bolstering the demand for biomass power. Furthermore, regions like India, the U.S., and China have implemented co-firing strategies, combining biomass feedstock with coal-based power plants to reduce carbon emissions. These factors collectively pave the way for robust market growth in the upcoming forecast period.

Within the U.S., there has been a substantial increase in biomass power capacity, with 177 MW added in 2019 through the completion of 14 projects. These new projects were undertaken in various locations, including Florida, Georgia, California, and Hawaii. Furthermore, the country has over 384 MW of biomass power projects in different stages of development. These factors are poised to stimulate significant market growth in the U.S. in the coming years [64]. In North America, the U.S. and Canada are at the forefront of the regional biomass power market. Historically reliant on coal for power generation, the discovery of shale gas reserves in this region has led to a notable shift toward gas-based power generation, outpacing coal-based alternatives over the past decade. The trend of using installers and system providers to maintain major equipment in inventory has been disrupted by manufacturing bottlenecks and shortages, primarily attributed to limited production during the COVID-19 pandemic. Additionally, the development of biomass power projects faced delays in commissioning due to supply chain disruptions and lockdown-induced halts in on-site construction activities in heavily affected areas. Technology-wise, the global biomass power market has been categorized into combustion, gasification, and anaerobic digestion. In 2021, the combustion segment assumed a dominant position in the market, contributing to over 88.0% of the total revenue. This trend is expected to persist, with the combustion technology segment projected to maintain a steady growth rate during the forecast period. This technology involves the direct combustion of biomass feedstock in a furnace with air to produce steam, subsequently driving a steam turbine for electricity generation. Combustion technology distinguishes itself through relatively straightforward operations and lower costs compared to more advanced biomass power technologies, a characteristic likely to fuel its demand in the market. Biomass power has versatile applications, ranging from power generation to lighting, heating, and cooking gas. These multifaceted applications are anticipated to fuel the growth of the anaerobic digestion technology segment throughout the forecast period. However, the gasification technology segment is poised to exhibit the fastest CAGR among the technologies considered in the forecast period.

Green hydrogen is derived from biomass through either thermochemical or biological means. Biological approaches are more eco-friendly and require less energy but tend to generate lower hydrogen yields depending on the specific raw materials used [65]. Conversely, biomass thermochemical methods are considerably faster and result in higher hydrogen production rates, with gasification standing out as a promising technology supported by environmental and economic evaluations [66]. Various thermochemical biomass conversion techniques, including liquefaction [67], gasification [68], pyrolysis [69], and supercritical fluid extraction [70], are employed to maximize liquid output. Biological-based hydrogen generation encompasses processes like photo-fermentation, dark-fermentation, biophotolysis of water employing green algae, blue-green algae (cyanobacteria), and hybrid reactor systems. Comparatively, the thermochemical process boasts superior overall efficiency (thermal to hydrogen) with an efficiency of 52%, and it is more cost-effective. However, hydrogen yields from biomass tend to be relatively low, typically ranging from 16% to 18%, based on the dry weight of the biomass [71]. The production of hydrogen from biomass using the thermochemical approach encompasses pyrolysis and gasification processes (Figure 3.26) [72]. These methods yield various gaseous byproducts, including CH_4 and CO, which can be harnessed for additional hydrogen generation via processes like the water gas shift (WGS) reaction or steam reforming. However, it's worth noting that two alternative methods, liquefaction and combustion, are considered less economical due to their relatively low hydrogen production rates. Additionally, liquefaction's operational conditions are notably challenging to achieve, requiring pressures of 5 to 20 MPa in an air-free environment, while combustion tends to release pollutants [73]. Within the pyrolysis process, biomass undergoes a transformation into liquid oils, solid charcoal, and gaseous compounds. This conversion occurs at temperatures ranging from 650 to 800 K and pressures of 0.1 to 0.5 MPa, all in the absence of air [74]. Pyrolysis is categorized into fast and slow pyrolysis, with slow pyrolysis often disregarded for hydrogen production as its primary product is predominantly charcoal. On the other hand, high-temperature fast pyrolysis involves the rapid heating of biomass feedstock without the presence of air. This leads to the formation of vapors, subsequently condensed into a dark brown, mobile bio-liquid.

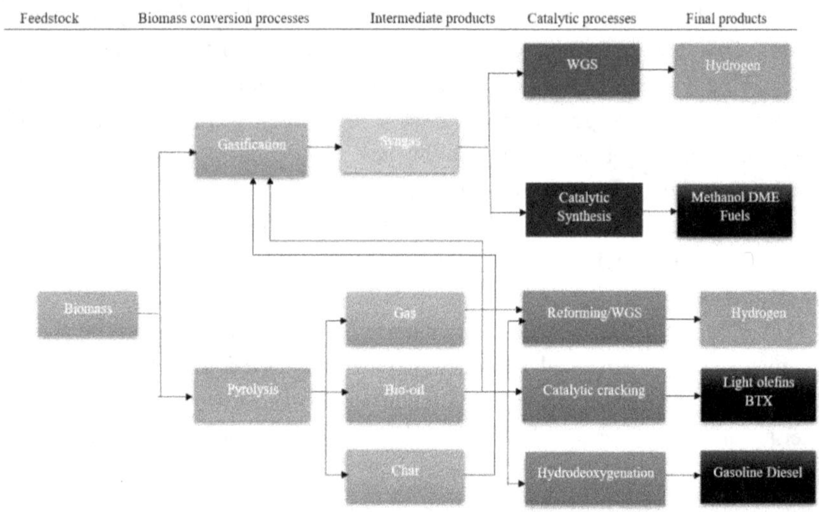

Figure 3.26 The production of hydrogen from biomass using the thermochemical approach

Consequently, fast pyrolysis results can be found in all three phases:

(i) Solid products: This category includes char, almost pure carbon, and other inert materials.

(ii) Liquid products: These comprise oils and tar, remaining in liquid form even at room temperature. Examples of these liquid products include acetic acid and acetone.

(iii) Gaseous products: Depending on the organic nature of the biomass used, gaseous products may encompass H_2, CO_2, CO, CH_4, and various other gases.

Pyrolysis: Pyrolysis methods initially designed for biofuel production, have the potential to yield hydrogen directly through rapid or flash pyrolysis, provided that an adequate volatile phase residence time is allowed at elevated temperatures:

$$Biomass + heat \rightarrow H_2 + CO + CH_4 + Other\ products \quad (3.3)$$

The methane and other hydrocarbon vapors produced in this process can be further converted into hydrogen by steam reforming:

$$C_nH_m + nH_2O \rightarrow nCO + (n + \frac{1}{2}m)H_2 \tag{3.4}$$

Furthermore, an additional source of hydrogen is generated through the application of a water-gas shift reaction:

$$CO + H_2O \rightarrow CO_2 + H_2 \tag{3.5}$$

In the conversion of CO into H_2 and CO_2, the resulting purified hydrogen can be obtained through a process known as PSA [75]. The incorporation of steam reforming and the water-gas shift reaction results in a substantial increase in the rate of hydrogen production. By employing a Ni-based catalyst, the maximum hydrogen yield attainable through pyrolysis can be significantly improved, reaching up to 90%. Crucial factors that govern the pyrolysis process include the heating rate, temperature, catalyst type, and residence time. These parameters are carefully chosen within the context of different reactor types and heat transfer mechanisms, including gas-solid convective heat transfer and solid-solid conductive heat transfer. A variety of biomass resources, including agricultural leftovers [76], peanut shells [77], rapeseed [78], and post-consumer materials like plastics, trap grease, mixed biomass, and synthetic polymers [79], have been utilized to assess hydrogen generation through pyrolysis. An important concern is the deposition of coke and char in the reactor or on the catalyst's surface, which can detrimentally impact reforming efficiency. To enhance hydrogen production rates during biomass pyrolysis, fluidized catalyst beds are often employed. Depending on factors such as the type of biomass and the scale of the facility, the cost of hydrogen produced via pyrolysis typically falls within the range of \$1.25 per kilogram to \$2.20 per kilogram

Gasification: Biomass gasification is conducted at elevated temperatures, surpassing 1000 K, causing biomass particles to undergo partial oxidation, resulting in the production of charcoal and gas. Essentially, biomass gasification can be viewed as a variation of pyrolysis occurring at higher temperatures, yielding a gas mixture containing approximately 6 to 6.5% hydrogen content [80]. This charcoal can subsequently be transformed into hydrogen (H_2), methane (CH_4), carbon monoxide (CO), carbon dioxide (CO_2), and light and heavy hydrocarbons:

$$\begin{aligned} &Biomass + heat + steam \rightarrow H_2 + CO + CO_2 + CH_4 + \\ &light\ and\ heavy\ hydrocarbons + char \end{aligned} \tag{3.6}$$

Distinguishing it from pyrolysis, biomass gasification occurs in the presence of oxygen. Its primary objective is the generation of gaseous products, whereas pyrolysis aims at producing bio-oils and charcoal. The gases produced in biomass gasification can be further processed through steam reforming to yield hydrogen, with potential enhancements achieved through WGS reactions, as previously discussed. It's essential that the moisture content of the biomass feedstock remains below 35% for effective use in the gasification process [81]. An issue that often arises in biomass gasification is the formation of tar, which can create tar aerosols and polymerize into more complex structures during the hydrogen production process. To mitigate tar formation, three approaches have been employed: (i) gasifier design improvement, (ii) precise control and operation, and (iii) the utilization of additives or catalysts. In addition, operating parameters such as residence time, process temperature, and the gasifying agent play pivotal roles in tar formation and decomposition. Elevating the temperature above 1273 K is one method to thermally crack tar compounds. The use of additives like olivine, dolomite, and char within the gasifier has been effective in tar reduction, with dolomite, in particular, being credited with eliminating 100% of tar [82].

Ni-based catalysts, dolomite, and alkali metal oxides have been widely employed as gasification catalysts. These catalysts not only aid in reducing tar but also improve the quality of gas production and conversion efficiency. An additional method to reduce tar and enhance gasification quality is the modification of the process through two-stage gasification and secondary air injection in the gasifier. In biogas gasification, ash formation leading to issues like deposition, slagging, sintering, fouling, and agglomeration is a significant concern [83]. To address these ash-related problems, fractionation and leaching have been used to mitigate ash formation within the reactor [84]. While employing fractionation may reduce ash, it could also impact the quality of the remaining ash. Conversely, leaching can eliminate inorganic biomass components, thereby improving the quality of the remaining ash. Research on hydrogen production via the gasification of leached olive oil waste in a circulating fluidized bed reactor has shown that leaching can serve as a pretreatment technique to enhance the gasification process and increase gas production rates [85].

Given that gasification predominantly yields gases as its primary products, this hydrogen production method holds advantages over pyrolysis. Through the utilization of various biomass feedstocks such as sawdust, wood, almond shells, switchgrass, sewage sludge, and cotton stems in conjunction with fluidized bed gasifiers and suitable catalysts, it's possible to boost hydrogen production rates by approximately 60%. This impressive conversion efficiency renders biomass gasification a highly attractive technology for hydrogen production. Notably, the cost of hydrogen produced via biomass gasification is competitive with natural gas reforming. Additionally, the environmental impact of biomass gasification-based hydrogen production is markedly lower compared to natural gas reforming, making this technology highly promising from both economic and environmental perspectives.

The biomass gasification market size, valued at $95.6 billion in 2022, is poised for substantial growth, with projections indicating an increase from $103.7 billion in 2023 to a substantial USD 199.2 billion by 2032 (Figure 3.27). This growth trajectory underscores an impressive CAGR of 8.50% during the forecast period from 2023 to 2032. Several key factors are propelling this upward momentum in the biomass gasification market. Primary among these drivers is the mounting environmental consciousness among the global populace. As concerns about the environment grow, adopting eco-friendly energy sources is on the rise. Moreover, the escalating levels of greenhouse gas emissions are pressing the need for sustainable alternatives, further stimulating the market's expansion [86].

The CAGR for biomass gasification owes its momentum to a multitude of factors. One significant contributor is municipal solid waste (MSW), encompassing materials like paper, wood chippings, cardboard, leaves, and leather products, as well as non-biomass elements like plastic, glass, and metal. The utilization of municipal waste, wood processing, and wood residues stand as primary sources of biomass gasification. Wood waste arising from pellets, firewood, sawdust, lumber, and furniture remnants is incinerated via gasification to produce energy, becoming another key driver propelling the growth of the biomass gasification industry. Moreover, crops and food processing residues, along with sources like corn, soybeans, sugar cane, switchgrass, algae, and woody plants, play crucial roles in biofuel production through biomass gasification. Being sourced from wood or forest waste, biomass is renowned for its eco-friendliness, comprising materials such as logs, sawmill waste, and dried wood strands.

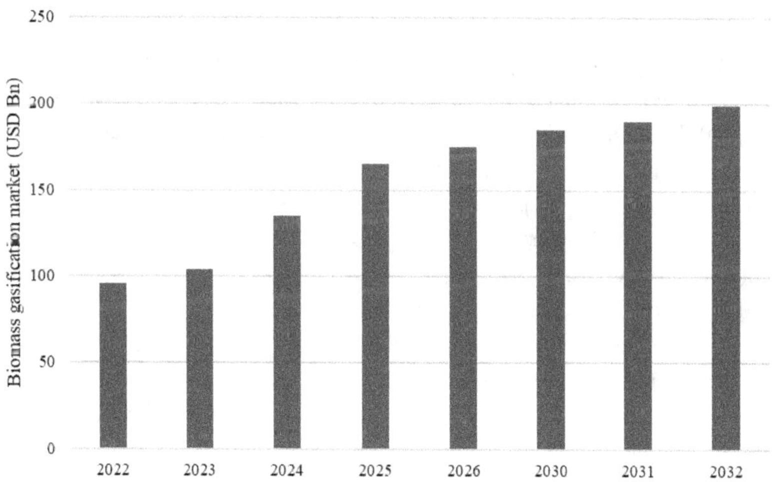

Figure 3.27 Biomass gasification market, 2018-2032 (USD Bn)

Notably, biomass gasification plants were once limited in their applications, primarily serving residential or small-scale industrial setups. However, increased awareness and investments, along with recognition of the benefits and incentives associated with this technology, have led to a surge in demand for biomass gasification. Consequently, this has fueled its commercialization and driven the expansion of the market. The transition from finite fossil fuel sources, which will inevitably deplete over time, plays a pivotal role in this narrative. Coal-based nuclear power plants and other coal applications have spurred intensified coal extraction, contributing to elevated carbon emissions throughout the coal extraction value chain, from mining to combustion. Simultaneously, oil and gas resources are dwindling, necessitating the exploration and extraction of alternate energy sources. In this context, renewable sources, with biomass being a prominent example, are emerging as favored alternatives. Hence, the biomass gasification market is positioned to flourish in response to the accelerating depletion of fossil fuels. This trend is further fortified by the abundant availability of biomass, accompanied by governmental initiatives across many nations that promote the adoption of renewable energy sources. Nevertheless, it's essential to note that the gasification process might encounter cost-related hurdles. Furthermore, while biomass gasification has been harnessed for several decades to produce heat and energy across a spectrum of residential and industrial applications, current gasification technologies face limitations in terms of efficiency and their claimed carbon-neutral status. Therefore, the imperative need for research and development is paramount to enhance these technologies, maximize their output, and foster cleaner energy production. Advancements in gasification technologies hold immense potential to propel the growth of the biomass gasification industry and thereby bolster revenues within the Biomass Gasification market.

In the realm of biomass gasification market segmentation by application, three primary categories emerge: chemicals, liquid fuels, and power & gas fuels (hydrogen). Notably, the power & gas fuels segment stands as a dominant contributor to the revenue. This prominence can be attributed to the active endeavors within the power sector to undergo an energy transition, shifting away from coal-based resources and steering toward more cost-effective and environmentally sustainable alternatives. It is anticipated that biomass's role in this transition will expand as gasification technology continues to advance [86].

In terms of geographical coverage, the analysis of the market can be extended to four major regions: North America, Europe, Asia-Pacific, and the rest of the world. Within this scope, the North American biomass gasification market is positioned as the leader, exerting a dominant influence. The primary factors contributing to this dominance are the rapid pace of globalization and the burgeoning global population. These dynamics are anticipated to drive the increased adoption of biomass gasification across various sectors, particularly in transportation, as well as in the production of ethanol, biodiesel, and biogas. The biomass gasification market in Europe commands the second-largest share, driven by substantial demand for biomass across various regions within this continent. Notably, the German biomass gasification market has emerged as the market leader in terms of share, while the United Kingdom's biomass gasification market has exhibited remarkable growth, positioning it as the fastest-growing market in Europe. In the Asia-Pacific region, the biomass gasification market is poised to experience the swiftest CAGR between 2023 and 2032. This growth is attributed to the increasing electrification initiatives in both urban and rural areas of various countries. Additionally, China's biomass gasification market has secured the largest market share, with the Indian biomass gasification market emerging as the fastest-growing in the Asia-Pacific region. These developments are bolstered by favorable government incentives in India, such as customs duty reductions or exemptions on parts of biomass-operated electricity generators, excise duty exemptions on these components, and relief from central sales tax. These incentives are instrumental in propelling market expansion for biomass gasification systems. Key market players in the biomass gasification industry are making substantial investments in research and development to diversify their product portfolios, thereby facilitating further growth in the market. These industry participants are also engaged in a variety of strategic initiatives to extend their market reach. These include the introduction of new products, contractual agreements, mergers and acquisitions, increased investments, and collaborations with other organizations. Adapting the business strategy of lean manufacturing to reduce operational costs is a vital approach employed by manufacturers in the biomass gasification industry.

The advancement of biomass-based hydrogen production has confronted notable hurdles. Primarily, biomass yields relatively low hydrogen content compared to other sources, starting at approximately 6%, as opposed to around 25% for CH_4. Additionally, biomass contains a significant proportion of oxygen, roughly 40% by weight, which limits its energy content. Over half of the hydrogen derived from biomass is produced by splitting water in a steam-reforming reaction. This intrinsic characteristic of biomass feedstock, with lower energy content, inherently restricts the process [87]. The cost of hydrogen produced through supercritical water gasification is estimated at approximately 0.35 $/kg. This suggests that although hydrogen production through supercritical water gasification is still in its nascent stages, this method has demonstrated its economic competitiveness when compared to other hydrogen production techniques. Nevertheless, since the hydrogen produced via supercritical water gasification is accompanied by other gases, the necessity for hydrogen separation and purification is inevitable. Biomass has emerged as a significant renewable resource globally, poised to replace fossil fuel reserves. It is expected to assume a pivotal role in the future's worldwide energy landscape, serving purposes in heat and power generation and facilitating the production of chemicals and fuels. Consequently, there is an anticipation that biomass thermochemical conversion technologies, particularly innovative approaches such as supercritical water gasification, will soon become available for large-scale hydrogen production [88].

3.3.4 Hydrogen production from geothermal energy

Geothermal energy is regarded as one of the most reliable and consistent forms of renewable energy sources, mainly because it remains unaffected by weather conditions. The reliability of a geothermal system hinges on factors like depth, soil characteristics, and the materials used for grouting. Geothermal energy finds applications across a spectrum of purposes, encompassing power generation, energy storage, heating, and cooling. Geothermal energy systems fall into two primary categories: shallow and deep systems, each involving distinct installation methods. Shallow systems employ ground heat exchangers, while deep systems utilize production injection wells [89]. The utilization of geothermal energy for heating and cooling typically entails the deployment of a ground-source heat pump, an advanced iteration of an air-source heat pump. These systems feature various heat exchanger configurations, including vertical [90], horizontal [91], or coiled [92], all of which necessitate the encasement of grout material for safeguarding and facilitating heat transfer. Modern variants of grouting materials, such as phase change materials, are being incorporated to enhance capacity and stability. The installation of both grout and ground heat exchangers collectively constitutes a borehole heat exchanger. An alternative, though less frequently employed, method for providing heating and cooling is an earth-air heat exchanger, which circulates air through an underground duct using a fan or blower [93]. Geothermal energy application represents a carbon-neutral energy source, and in contrast to wind and solar power, it has the capability to provide a relatively consistent and steady power output throughout the day. According to data, 82 nations make use of geothermal energy directly, and 26 countries employ it to generate electricity. The United States, Indonesia, the Philippines, Turkey, and New Zealand are the leading countries in geothermal power generation [94]. Figure 3.28 demonstrates the prediction of the geothermal energy installation capacity for the top 10 countries utilizing geothermal energy by 2030 [95].

One of the most commonly employed geothermal technologies today is the geothermal power plant (GPP). This system primarily produces electricity by extracting hot geothermal fluid from deep underground rocks and subsequently returning it to the earth. There are three primary types of GPPs: flash-steam [96], dry-steam [97], and binary cycles [98]. The first two cycles typically utilize the geothermal fluid directly, depending on the specific conditions. The dry-steam cycle is implemented when the geothermal fluid exists solely in the form of steam. In contrast, the flash cycle takes hot geothermal water and transforms it into steam by employing a flash separator, which separates the steam from the water. This steam then powers a turbine for electricity generation. Flash cycles can be categorized as single flash cycles or double flash cycles, depending on the number of separators used. More recently, a novel cycle known as the multi-flash cycle has been developed. It employs more than two separators to extract the maximum available power from the geothermal fluid before reinjection.

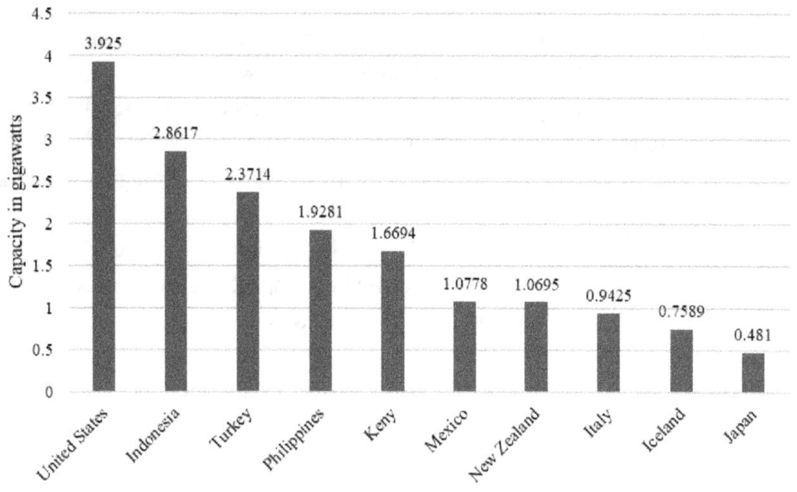

Figure 3.28 The prediction of the geothermal energy installation capacity for the top 10 countries by 2030 [95]

On the other hand, the binary GPP does not utilize geothermal fluid directly. Instead, it initiates an organic Rankine cycle (ORC) when the geothermal source is of low grade [99]. In this scenario, geothermal fluid is employed to heat a working fluid via a heat exchanger. The choice of working fluid is a critical factor in such GPP cycles.

In hydrogen production systems driven by geothermal energy, PEM electrolyzers are predominantly employed to separate water into hydrogen and oxygen molecules. Yilmaz et al. [100] investigated the impact of temperature on the electrolysis process within a geothermal-based hydrogen production system using an artificial neural network. The findings revealed that as the temperature increased from 25°C to 70°C, the energy required for hydrogen production decreased from 43.51 kW/kg to 42.2 kW/kg, representing a roughly 3% reduction. To accomplish this, a preheater is commonly utilized to elevate the water's temperature before it enters the electrolyzer. In systems where geothermal power plants are integrated with hydrogen production, the geothermal fluid serves a dual role: generating electricity within the power plant and heating water. Initially, the geothermal fluid is extracted from the production well and routed through the power plant. Subsequently, residual heat is employed to preheat water. The air preheater is responsible for capturing the remaining heat from the geothermal fluid before it is reinjected into the ground [101]. When operating with a geothermal resource temperature of 160°C and a flow rate of 100 kg/s, it was documented that the hydrogen production rate reached 0.0253 kg/s. This implies that roughly 0.253

grams of hydrogen are generated for each kilogram of geothermal water [102]. The production of hydrogen through geothermal systems is not solely contingent on the prevailing geothermal conditions. The cooling system, which is responsible for the highest exergy destruction, plays a pivotal role. There has been a proposal to enhance the cooling system with liquefied natural gas (LNG) due to its effective heat dissipation capabilities and the extra power it can generate through a secondary power generator [103]. To minimize the overall cost of products, thermo-economic optimizations are routinely carried out, often employing genetic algorithm methodologies. This analysis takes into account the impact of all system components on exergetic performance and product cost. Factors considered include product exergy and cost, fuel exergy and cost, exergy destruction and cost, exergetic unit cost of electricity, and the exergetic cost of hydrogen production. Cogeneration plants, also referred to as combined heat and power (CHP) facilities, are frequently the subject of investigation for their capacity to simultaneously generate hydrogen and electricity using geothermal energy sources. These plants typically encompass a dual-cycle setup, consisting of a topping cycle responsible for electricity generation and a bottoming cycle dedicated to converting the residual thermal energy into hydrogen. These cycles are typically propelled by ORCs [104]. In certain scenarios, the entirety of the electricity generated is transformed into hydrogen fuel. Yuksel et al. [105] introduced a geothermal-based system that integrates district cooling with hydrogen production. This system makes effective use of the surplus heat present in the geothermal fluid before it is reinjected by passing it through an absorption cooling system. Another strategy to extract the maximum heat from the geothermal fluid involves the application of a multi-flash GPP, which enhances the system's overall energy and exergy efficiencies. The multi-flash cycle incorporates a flash separation process before each turbine stage to segregate water and steam [106]. Choosing the right working medium for geothermal-driven hydrogen production systems is a matter of significant importance. Geothermal binary cycles commonly employ ORCs due to the typically low temperature of geothermal fluids. ORCs prove to be efficient in harnessing low-quality energy sources since they utilize fluids with lower boiling points in contrast to conventional Rankine cycles. Numerous investigations have been carried out with the aim of elevating the geothermal fluid's temperature prior to its application in hydrogen production, especially in cases where geothermal energy is harnessed as a low-quality heat source. This enhancement is accomplished by introducing an extra energy source, ideally one from renewable origins like solar energy, to uphold the system's eco-friendliness. The inclusion of this supplementary source not only yields additional power but also serves to enhance the facility's efficiency by warming up the geothermal fluid or working fluid through preheating or reheating processes [107,108].

3.3.5 Hydropower-based hydrogen production

Hydropower stands as a well-established technology for generating clean energy and holds a prominent position among various renewable energy sources (Figure 3.29) [109]. As of 2021, hydropower played a significant role in global electricity production, contributing to approximately 16% of the total output. It is also the cheapest source of renewable energy, averaging \$0.05 per kilowatt-hour. This positioned it as the third-largest energy source, trailing behind only coal and natural gas, surpassing nuclear power and all other renewable energy sources combined (Figure 3.30) [22].

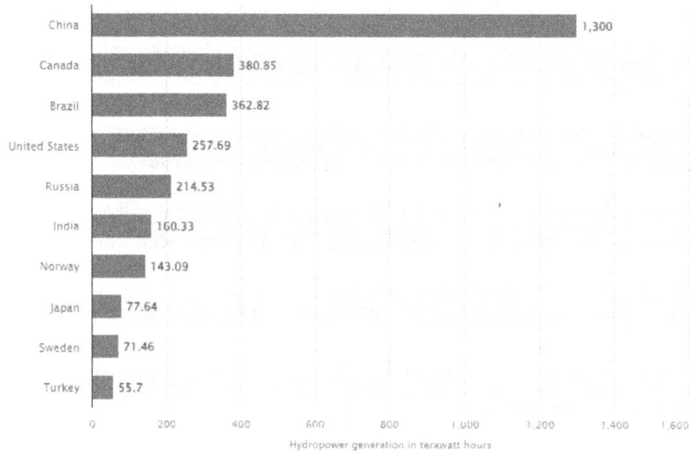

Figure 3.29 Largest hydroelectric power generating countries worldwide in 2021

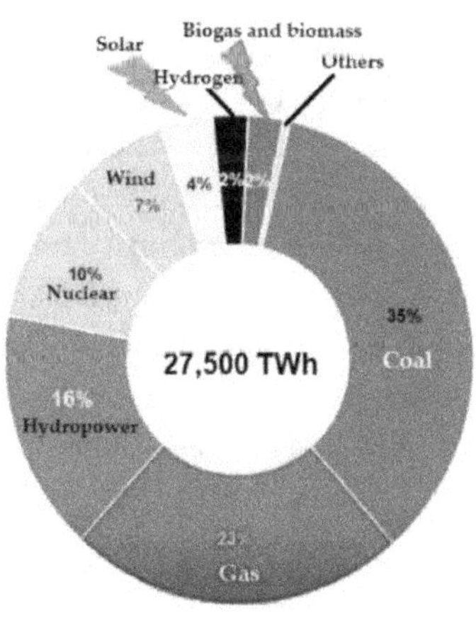

Figure 3.30 Global share of energy sources 2021 and the role of hydropower in the energy mix of the world [22]

One of the key advantages of hydropower is its resilience to weather conditions, setting it apart from wind and solar technologies. Consequently, hydropower remains vital in low-carbon electricity generation, currently accounting for nearly half of it. Hydropower capacity varies widely, ranging from small-scale systems with just a few kilowatts to massive installations like the Three Gorges Dam in Yichang, China, boasting a net electric power generation capacity of 22.5 GW. A notable application of hydropower lies in energy storage for the electrical grid, achieved through pumped hydro-storage and river hydro-storage. These methods entail storing excess hydropower energy during periods of low electricity costs and subsequently releasing it into the grid during peak-demand times. Hydropower systems are typically designed with reservoirs to store water upstream, and many hydro turbine types can operate efficiently across different load levels, from partial to full capacity [110]. Hydropower has the potential to have a transformative impact on supporting the growth of green hydrogen. In this context, it could provide a substantial portion, at least 1,000 TWh (equivalent to 13%), of the additional emission-free electricity demand required, as outlined in IRENA's 2050 decarbonization scenario [111]. Moreover, as the understanding of hydrogen's potential grows, new analyses project even more extensive utilization of hydrogen electrolysis. This, in turn, implies that hydropower's contribution could be significantly larger within a significantly increased demand for low-carbon electricity. For instance, IRENA's 1.5°C scenario, as highlighted in their World Energy Transitions Outlook, envisions the possibility of deploying as much as 5,000 GW of hydrogen electrolyzers by the year 2050 [112,113].

New initiatives are also taking shape in areas endowed with hydropower resources. For instance, in 2020, a 2 MW electrolysis facility was inaugurated in Switzerland at the Gösgen hydropower plant as part of the 'Hydrospider' joint venture. Additionally, Southeast Asia saw the launch of its first integrated hydrogen production and vehicle refueling station in Sarawak, Malaysia, which incorporates a 500 kW electrolysis plant powered by hydropower [114,115]. More projects involving green hydrogen are in progress in regions like Iceland, Tasmania, and Canada, where hydropower assumes a primary role. Examples of green hydrogen and hydropower initiatives in various countries are detailed in Table 3.3 [116]. Choosing to establish projects in these nations offers access to dependable, cost-effective, and environmentally friendly electricity that is available throughout the year. This not only ensures a consistent supply of renewable green power for hydrogen producers but also reduces production costs. Presently, green hydrogen is, on average, two to three times more expensive than grey or brown hydrogen generated from fossil fuels. This cost differential remains a significant challenge to address. The primary cost driver for green hydrogen is the price of electricity, with other significant factors being the initial capital investment in the electrolyzer plant and the utilization rate, which measures how frequently the electrolyzer operates annually. By targeting countries with affordable and easily accessible hydropower, it becomes evident that this can significantly enhance the economics of green hydrogen producers, particularly in regions where there is substantial government backing for expanding green hydrogen production. Furthermore, as a reliable and adaptable source of electricity, hydropower can complement wind and solar energy to provide continuous capacity, which further boosts the efficient use of hydrogen electrolysis plants compared to intermittent operation. Additionally, run-of-river hydropower can achieve high utilization rates in areas with a continuous year-round water supply, making it especially well-suited for hydrogen production [117].

Table 3.3 Hydropower-based hydrogen production projects

Country	Projects	Reference
Norway	Back in 1927, Norsk Hydro established a 167 MW electrolysis facility to provide hydrogen for ammonia production in Rjukan, utilizing hydropower from plants located along the Måna river. Subsequently, in 1953, another hydro-powered electrolysis plant with a capacity of 135 MW was inaugurated in Glomfjord, in proximity to a hydropower facility of the same name, presently owned by Statkraft. This collective experience laid the groundwork for NEL Hydrogen, which has evolved into a prominent electrolyser supplier in the present day.	[118]
Switzerland	The 'Hydrospider' initiative is based at the Gösgen run-of-river hydro plant and is a collaborative project involving Alpiq, H2 Energy, Linde Gas, Hyundai Hydrogen Mobility, and the H2 Mobility Switzerland Association. This endeavor, which commenced in 2020, established a 2 MW electrolysis facility at the 51 MW Gösgen hydro plant. This installation has the capacity to generate around 300 tons of green hydrogen annually, adequate to fuel approximately 40-50 heavy-duty trucks or 1700 passenger cars.	[119]
Iceland	In 2020, Landsvirkjun revealed intentions to initiate a 10 MW electrolysis project located at the Ljósifoss hydropower plant. This endeavor has the potential to serve as an alternative to the consumption of fossil fuels, particularly in sectors like heavy transportation.	[120]
Sarawak, Malaysia	In 2019, Sarawak Energy introduced the inaugural integrated hydrogen production facility and vehicle refueling station in Southeast Asia, serving H_2-powered vehicles in Kuching. The electrolysis system generates environmentally friendly hydrogen using hydropower as its source. In 2020, Sarawak also entered into Memorandums of Understanding (MOUs) with Japanese firms, aiming to expand and export their green hydrogen production.	[121,122]
Canada	Air Liquide Canada has recently set up a 20 MW electrolysis unit for producing environmentally friendly hydrogen at its Bécancour, Québec facility. This development capitalizes on the ample and cost-effective hydropower resources accessible in the area. Furthermore, Hydro Québec, in its 2020-2024 Strategic Plan, has identified green hydrogen as a significant growth opportunity. It has also unveiled intentions to establish a 90 MW electrolysis plant for generating green hydrogen dedicated to biofuel production. Similarly, various proposals in other Canadian provinces are in the works, including the Renewable Hydrogen Canada (RH2C) project in British Columbia, which intends to leverage wind resources and BC Hydro's grid stabilization capabilities.	[123–126]

Tasmania, Australia	Hydro Tasmania published a Whitepaper in 2019, which was subsequently followed by the Tasmanian State Government's release of the Tasmanian Renewable Hydrogen Action Plan. This strategic move identified green hydrogen as a competitive advantage for the state. Notably, this advantage stems from the availability of renewable energy sources, such as wind power, and the flexibility of hydropower operations. These factors facilitate the production of high-capacity and cost-effective green hydrogen. To further promote this initiative, the Tasmanian Government is allocating $2.6 million to support three feasibility studies examining large-scale renewable hydrogen projects in Tasmania, with project capacities of 500 MW, 100 MW, and 90-100 MW. Additionally, the government has pledged $50 million over a decade to foster the growth of the Tasmanian hydrogen industry.	[127]
Russia	In the eastern region of Russia, RusHydro is in the process of establishing a test facility for the production of liquefied hydrogen at the 310 MW Ust-Srednekanskaya hydropower station. This project is being undertaken in collaboration with the Japanese company Kawasaki Heavy Industries.	[116]

Regarding initial expenses, even though they remain somewhat high, the costs associated with both Alkaline electrolyzers and PEM technologies have decreased over the past decade. These costs are anticipated to keep decreasing as production scales up. In general, cost projections indicate that green hydrogen could become cost-competitive, reaching approximately US$ 2 per kilogram by 2030 in the most favorable locations. With the expansion of the hydrogen market, hydroelectric power producers have the opportunity to incorporate electrolyzers either directly at hydropower facilities or in their vicinity. This integration allows for the production of green hydrogen, which is entirely carbon-free, while also offering a revenue-generating avenue. Green hydrogen production can efficiently utilize excess hydropower capacity, particularly when it can contribute to achieving high utilization rates for electrolysis plants. Green hydrogen production serves two key purposes: (a) it offers a stable market for hydropower during periods of low electricity demand, allowing for the full utilization of existing hydropower resources, and (b) it creates a new market for a zero-carbon product, potentially encouraging the development of new hydropower projects in various parts of the world. In a broader context, the push for green hydrogen hinges on the decarbonization of the power sector and a substantial surge in demand for clean electricity. As the primary source of renewable and dependable power, this strongly underscores the importance of hydropower capacity and its pivotal role, in conjunction with other renewables, in the transition to a net-zero emissions future. In North America, hydropower facilities exhibit a utilization rate ranging from 30% to 50%, indicating the potential for generating surplus power that could be redirected for hydrogen production. Through strategic deployment, hydropower can maintain its role in electricity generation while simultaneously contributing to electrolysis for hydrogen manufacturing. The adaptability of electrolyzers adds to this operational versatility. Hydropower stands as a crucial player in assisting countries in their transition to achieving net-zero greenhouse gas emissions by facilitating hydrogen production. Nations endowed with abundant hydropower resources, like Ethiopia and Nepal, currently reliant on oil and gas imports, could leverage hydropower for hydrogen generation across various sectors, including transportation. Similarly, island economies such as New Zealand and Puerto Rico stand to gain by decreasing their reliance on fossil fuel imports through the utilization of hydropower for hydrogen production. Moreover, the substitution of diesel-powered urban transportation with hydrogen vehicles promises multifaceted economic, health, and societal advantages [128].

For the past century, hydropower has played a pivotal role in the generation of green hydrogen. Noteworthy historical instances include the Rjukan and Glomfjord electrolysis facilities in Norway, boasting capacities of 167 MW and 135 MW, respectively. These facilities utilized locally sourced hydropower to produce hydrogen for ammonia used in fertilizer production. Other significant electrolysis sites from the 20th century include the Trial facility in British Columbia, Canada (commissioned in 1939), as well as those in Nangal, India, and Aswan, Egypt, both established in the 1950s. All these sites are strategically located near low-cost hydroelectric generation and abundant water supply. Presently, the integration of hydropower with other renewable sources for hydrogen production is recognized as a cost-effective and sustainable approach [129]. Hydropower resources are actively contributing to the development of new hydrogen production initiatives globally. In 2020, a 2 MW electrolysis plant was introduced in Switzerland through the "Hydrospider" joint venture at the Gösgen hydropower plant. Furthermore, the commissioning of Southeast Asia's inaugural integrated hydrogen production and vehicle refueling station, featuring a 500 kW electrolysis plant powered by hydropower, took place in Sarawak, Malaysia. Ongoing endeavors are evident in regions like Iceland, Tasmania, and Canada, where hydropower stands as a significant energy source [114,115].

In the Pacific Northwest region of the United States, there exists an opportunity to explore innovative approaches to manage the multi-use hydropower system of the Columbia River for the production of renewable hydrogen. In this context, Andrus et al. [130] conducted a study to assess the feasibility of generating green hydrogen using excess energy derived from the Columbia River System. Similar to European systems, this river system exhibits substantial variability in water runoff but has limited storage capacity. The research aimed to estimate the surplus energy and potential hydrogen production from 11 hydroelectric projects along the Columbia River System, focusing on the period from March to July. The findings indicated that the average monthly hydrogen production potential of the system ranges from 2.22×10^6 to 8.96×10^6 kg H_2 over an 80-year period (1928-2008). The study proposed that utilizing surplus energy for hydrogen production, specifically for transportation purposes, represents a viable pathway toward transitioning to a hydrogen-based economy. The expenses associated with Alkaline electrolysis and PEM electrolysis technologies, although still relatively elevated, have witnessed a decline over the past ten years, and further reductions are anticipated as production scales up.

3.3.6 Hydrogen from tidal power

Tidal energy is produced by capturing the power of the tides, a phenomenon driven by the gravitational influence of the sun and moon on Earth. This energy can be transformed into electricity and is classified as a renewable resource because it originates from natural processes that renew more rapidly than they are consumed. Additionally, tidal energy generates fewer greenhouse gas emissions in comparison to fossil fuels. The potential of tidal energy is considerable, with projections indicating it could reach 500 gigawatts by 2020, constituting approximately a quarter of the global coal power capacity at that time. Tidal energy technology can be segmented into three main categories: tidal range, tidal current, and hybrid forms technologies. Tidal range technology involves tapping into the potential energy created by the variance in water level between high and low tides. This is often implemented through a tidal barrage (Figure 3.31), a structure resembling a dam or barrier constructed to enclose a substantial body of water. The disparity in water levels between the enclosed area and the external environment propels water through turbines embedded in the barrage, thereby generating electricity [131].

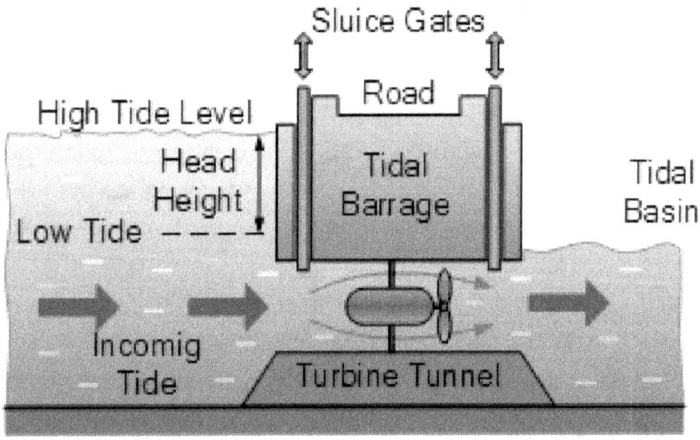

Figure 3.31 Tidal barrage power generation [20]

An example of a power plant employing tidal barrage technology is the Annapolis Royal Generating Station located in Canada. Tidal lagoons share similarities with tidal barrages but are not necessarily connected to the shore, enabling placement in open ocean waters. Tidal lagoons are preferred due to their lower environmental impact compared to tidal barrages. Emerging tidal range technologies encompass tidal reefs, tidal fences, and low-head tidal barrages. Tidal current technologies, also referred to as tidal stream technologies, leverage the kinetic energy of moving water to drive turbines, akin to the mechanism behind wind turbines harnessing air currents. This approach has gained popularity for its cost-effectiveness and minimal environmental footprint. Examples include horizontal and vertical-axis turbines, where tidal currents rotate the rotors, and tidal kites, particularly effective in areas with slower tidal flows. Tidal kites are anchored to the seabed, traversing through the water with an attached turbine to generate power from the movement. The MeyGen Tidal Energy project in Scotland is anticipated to be the largest tidal power station of its kind, boasting a potential generation capacity of up to 398 megawatts. Hybrid forms technology integrates both tidal current and tidal range approaches to electricity generation. A recent advancement in this domain is Dynamic Tidal Power (DTP). DTP involves the construction of a dam perpendicular to the coast, featuring a barrier at its end. This dam harnesses the variance in water level to generate potential energy and employs turbines to produce electricity. Tidal energy is regarded as more dependable and predictable when compared to wind and solar energy. The cycles of low tide and high tide are easily forecasted and experience minimal unforeseen variations. Tidal currents can be accurately predicted for hundreds of years in advance, and the tidal range remains largely unaffected by weather conditions. Unlike wind and solar energy, which can be influenced by atmospheric factors, tidal currents exhibit low and consistent fluctuations. This predictability reduces the likelihood of sudden drops in generation, as observed in wind energy in the UK [132].

Tidal energy boasts greater efficiency than wind energy due to water's approximately 830 times higher density than air, enabling the capture of more energy in a smaller area. The Sihwa Lake Tidal Power Plant in South Korea, the world's largest, exemplifies this efficiency, requiring only a 12.7-kilometer seawall. In contrast, wind and solar projects typically occupy much larger spaces, such as the Roscoe wind farm in Texas, covering 400 square kilometers, and the Bhadla Industrial Solar Park in India, spanning 45 square kilometers. According to a 2019 US Department of Energy study, the average cost for a commercial tidal energy project is approximately $280 per megawatt-hour, significantly higher than the cost of wind energy at around $20 per megawatt-hour. The elevated prices are primarily attributed to the substantial upfront costs of plant construction and the expenses associated with maintaining machinery resistant to corrosive seawater. Additionally, as tidal energy is a less mature and less widely adopted renewable energy source, costs remain relatively high compared to more established sources like wind and solar. Without current plans for supply chain development and technology expansion, tidal energy costs are expected to persist at higher levels, necessitating further research to identify ways to reduce costs.

Tidal energy systems are constrained in their potential installation locations, restricted to coastlines, and unsuitable for shallow waters with turbulent waves or deep waters lacking sufficient currents. Tidal turbines require specific conditions, with tidal range technologies demanding a minimum tidal range of 3.05 meters and tidal current technologies necessitating stream speeds of 1.5 to 2 m/s. Countries with notable tidal ranges suitable for tidal range technologies include Australia, Canada, the UK, the USA, France, and certain parts of East Africa [133]. Meanwhile, countries with potential for tidal current technologies include Australia, Spain, Africa, and Norway, although limited studies have been conducted in this area. The environmental impact of tidal energy varies based on the power plant site, with both positive and negative effects on the ecosystem. The construction of tidal power plants can alter water flow and quality, potentially harming marine habitats. Turbine blades and underwater noise can adversely affect sea creatures, as exemplified by the closure of the Annapolis Royal Generating Station in Canada due to its impact on fish. Conversely, tidal power plants can offer environmental benefits, such as altering water gradients in ways that benefit aquatic ecology and improve water quality. The Sihwa Lake Tidal Power Station and La Rance Tidal Power Station exemplify both advantages and disadvantages. Several proposed tidal power plant projects are in progress, including the Morlais project in Wales, aiming to install turbines at one of the world's largest tidal stream energy sites with the potential to power 180,000 homes. Other upcoming projects include the Incheon Tidal Power Station in South Korea and the Penzhinskaya Tidal Power Plant in Russia. However, overcoming challenges like high costs, geographical limitations, and environmental impacts requires further research and innovation, emphasizing the importance of funding for the transition to a society powered by renewable energy [133].

H2Tec, a Scotland-based provider of hydrogen energy solutions, has been chosen to implement significant enhancements to the hydrogen production plant located at the European Marine Energy Centre (EMEC) in Orkney. The facility at EMEC serves as a focal point for renewable energy development and research into green hydrogen. The planned upgrades are part of the Integrating Tidal Energy into the European Grid (ITEG) initiative, funded by the EU's Interreg North-West Europe regional development body. The improvements involve enhancing system efficiency through the procurement, installation, and commissioning of a control panel for piping to facilitate the transfer of newly generated hydrogen to on-site storage tanks. This EU-funded project aims to devise a solution that integrates tidal energy and hydrogen production to tackle grid export limitations in remote communities. The upgrades specifically enhance the efficiency of existing equipment and the connection between the hydrogen production plant and storage facilities on Eday Island. Furthermore, Logan Energy has entered into an agreement with TORC, a collaborative research and incubation center, to explore joint projects tailored to the Chinese market. This collaboration aligns with a £1 million investment from Kelvin Capital, intended to support the advancement of the company's hydrogen technology and its expansion initiatives in China [134].

3.4 Nuclear hydrogen production

The utilization of nuclear power has consistently been regarded as an environmentally friendly and sustainable means of generating electricity (Figure 3.32), particularly in conjunction with the electrolysis process for large-scale hydrogen production [135]. Currently, stakeholders in the energy sector and decision-makers are looking to nuclear hydrogen as a strategy for leveraging nuclear power plants and advancing low-carbon hydrogen production systems. This strategic approach aligns with the stance of the International Atomic Energy Agency (IAEA), emphasizing the necessity of expanding the use of low-carbon nuclear power to facilitate the global transition to clean energy. Towards the end of 2019, the initiation of three nuclear hydrogen projects by different companies in the United States was announced. These projects received funding from the US Department of Transportation (DOT) in collaboration with various national laboratories. The majority of presently operational nuclear power generation systems adhere to the Pressurized Water Reactor (PWR) technique, where the maximum process temperature does not surpass 350°C.

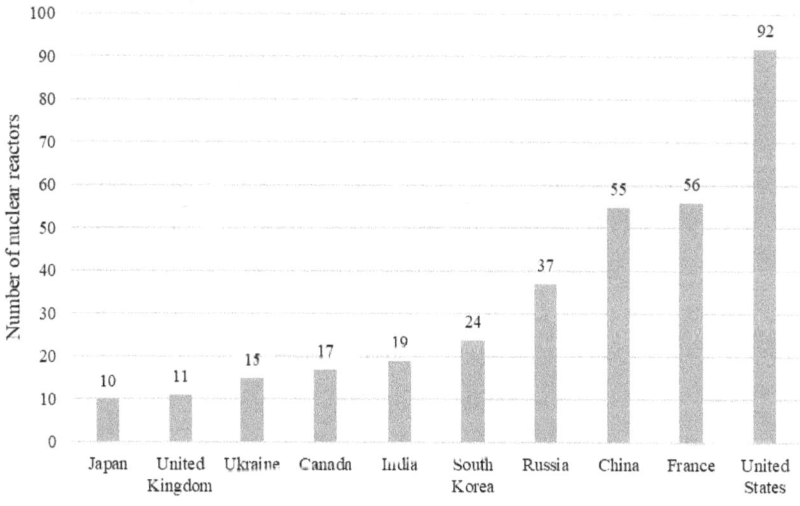

Figure 3.32 Top countries with the most nuclear reactors

Incorporating low-temperature hydrogen production technologies, such as conventional electrolysis, into a nuclear power plant represents a means of revitalizing the nuclear industry by enabling the co-generation of electricity alongside clean hydrogen [136]. In recent times, numerous advanced nuclear reactors have been proposed to enhance system thermal efficiency by elevating the operating temperature of the reactor. These advanced reactors can serve as a viable energy source for driving thermochemical hydrogen production processes.

Figure 3.33 illustrates the potential for integrating a nuclear reactor with hydrogen production. Broadly, the thermal energy generated by nuclear reactors serves as the primary heat source for producing hydrogen through various thermochemical processes such as steam-methane reforming (SMR), thermochemical water splitting, and electrochemical processes like water electrolysis and high-temperature steam electrolysis. For direct water electrolysis, a minimum temperature of 2500°C is necessary for hydrogen production. However, thermochemical processes like the water-splitting cycle, involving a series of chemical reactions, require a lower temperature to achieve similar overall results. The overall efficiency of the nuclear-hydrogen system hinges on factors such as operating conditions, choice of coolant, conversion efficiency, and reactor type. These variables play a crucial role in determining the economic and technological feasibility of the system [137].

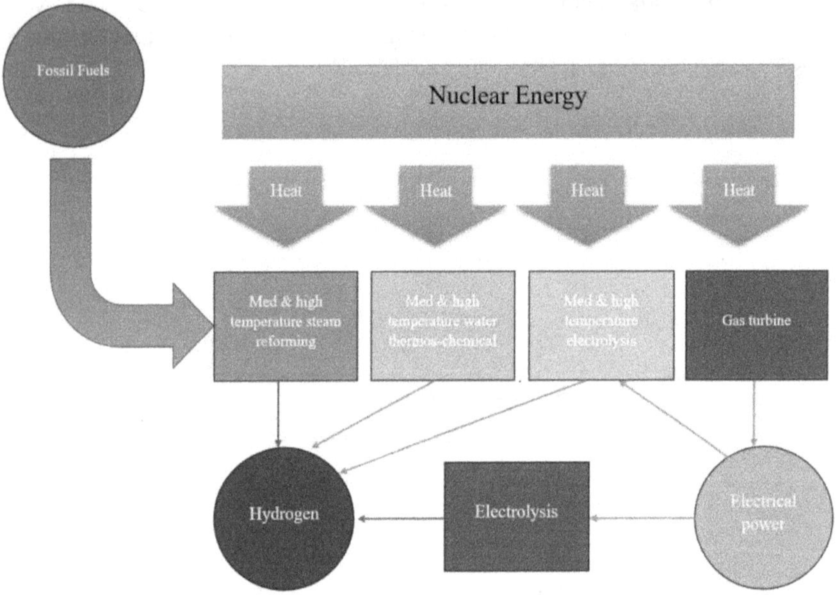

Figure 3.33 The potential for integrating a nuclear reactor with hydrogen production

Despite the maturity achieved by nuclear technology, the implementation of this power generation concept remains largely theoretical. Specifically, the conversion efficiency from thermal to electrical energy stands out as the main technical challenge, directly influencing the total cost of the system. Nuclear hydrogen (or so-called pink hydrogen) emerges as a viable option for environmentally friendly hydrogen production owing to its abundance, transportability, and stability. However, a multitude of factors must be considered, spanning from economic considerations to stakeholder involvement. According to El-Emam and Ozcan [138], conventional electrolysis stands out as the potentially most cost-effective method when the necessary electrical power is generated through geothermal or nuclear energy systems. Despite the development of thermochemical processes for efficient and economical hydrogen production, their high-temperature requirements and challenges related to material compatibility contribute to the overall cost and safety considerations of this technology. The cost of pink hydrogen is contingent on factors such as capacity, size, and specific plant configurations.

The details of various methods of nuclear hydrogen production, including the S-I cycle, HyS cycle, Cu-Cl cycle, Mg-Cl cycle, Ca-Br cycle, and other available methods, were discussed in the book titled "Fundamentals of Hydrogen Production and Utilization in Fuel Cell Systems" [139]. The HyS-based nuclear hydrogen has introduced a wide range of cost variations, ranging from 2.27 $/kg to 6.27 $/kg. The cost of pink hydrogen generated through Cu-Cl and Mg-Cl cycles falls within specific ranges, with the former ranging from 2.36 $/kg to 3.6 $/kg, and the latter approximately between 3.09 $/kg to 3.87 $/kg. In contrast, the high-temperature steam electrolysis (HTSE) process, which does not require a controlled cycle process or a large reactor, exhibits lower costs compared to thermochemical cycles. The primary contributor to the HTSE cycle's cost is the fuel, although the HTSE electrical power consumption is more than 20% lower than that of low-temperature electrolysis. Table 3.4 summarizes nuclear hybrid energy systems for hydrogen production [140].

Table 3.4 Nuclear hybrid energy systems for hydrogen production

	Alkaline electrolysis	**PEM electrolysis**	**Solid oxide electrolysis**	**Steam methane reforming**	**Thermochemical S-I**
Technology Readiness [141]	9	6-8	5	9	4
Temperature (°C)	60	60	800	870	910
Pressure (atm)	1	1	1.57	4.1	3.851
Efficiency (HHV, %)	30	27	36	79	25
Electricity (MJ)	180	200	146	1.4	75
Heat (MJ)	26	26	30	0	375
Water (kg)	11.5	11.5	83	10.3	9
Natural gas (kg)	0	0	0	2.9	0
CO_2 out (kg)	0	0	0	5-11	0
Production cost	$5.92	$3.56-5.46	$2.24-3.73	$1.54-2.30	$2.18-5.65

In February 2010, the Japan Atomic Energy Agency (JAEA) established the high-temperature gas-cooled reactor (HTGR) Hydrogen and Heat Application Research Centre in Oarai to advance operational technology for an Iodine-Sulfur (I-S) plant designed for thermochemical hydrogen production. Laboratory-scale and bench-scale demonstrations of hydrogen production using the I-S process, reaching up to 30 liters/h, have been conducted. Concurrently, a pilot plant test project utilizing helium heated with 400 kW assessed the engineering feasibility of the I-S process by producing hydrogen at a rate of 30 m^3/h. An I-S plant with a capacity to produce 1000 m^3/h (90 kg/h, 2 t/day) of hydrogen was planned to be connected to the high-temperature engineering test reactor (HTTR) to validate the performance of an integrated production system, slated for the 2020s. Demonstrations in 2014 showcased hydrogen production at up to 20 L/h, and in January 2019, the HTTR was utilized for over 150 hours of continuous operation to produce hydrogen using the iodine-sulfur process. JAEA aims to achieve hydrogen production at less than \$3/kg by approximately 2030 with very high-temperature reactors. In the United States, the Nuclear Energy Research Initiative (NERI), initiated in 1999, was redirected in 2004 to include the Nuclear Hydrogen Initiative (NHI) aligned with the Next Generation Nuclear Plant (NGNP) program established in 2005. NGNP envisioned the construction and operation of a prototype HTGR and associated facilities for electricity or hydrogen production by 2021. However, funding reductions occurred under the Obama administration, and prelicensing activities were suspended in 2013. Under an international NERI agreement, Sandia National Laboratories in the USA, along with the French CEA and General Atomics, were also exploring the I-S process, focusing on utilizing high-temperature reactors. They successfully built and operated a laboratory-scale loop for thermochemical water-splitting. In February 2022, the U.S. DOE launched a \$6 billion program aimed at supporting America's clean nuclear energy infrastructure. Prior to this, in the fourth quarter of 2021, the DOE allocated \$20 million in funding for technologies enabling the production of clean hydrogen from nuclear power. This strategic approach underscores the production of hydrogen from zero-carbon electricity as a pivotal economic product for nuclear power plants, extending beyond electricity generation. The DOE's plan involves reducing the cost of hydrogen to \$1 per kilogram within a decade, thereby opening up new markets for the nuclear industry [142].

South Korea, supported by General Atomics, demonstrated thermochemical water-splitting at the laboratory scale. In December 2008, the Korea Atomic Energy Commission formally endorsed nuclear hydrogen development as a national program, aiming to develop key and basic technologies by 2017. The goal was to demonstrate nuclear hydrogen production using the I-S process and a very high-temperature reactor (VHTR) by 2026.

The economic viability of hydrogen production is contingent on the efficiency of the chosen method. The I-S cycle coupled with a modular high-temperature reactor is anticipated to yield hydrogen at around \$2.00/kg. Additionally, the oxygen byproduct is considered valuable. General Atomics had earlier projected costs of \$1.53/kg based on a 2400 MWt HTR operating at 850 °C with 42% overall efficiency and \$1.42/kg at 950 °C with 52% efficiency, both at a 10.5% discount rate. Such a plant could potentially produce 800 tons of hydrogen per day.

Nuclear energy exhibits the capacity to supply low-cost hydrogen and energy to industrial hubs. Its stability and high power density enable the provision of a substantial, continuous stream of low-carbon hydrogen and heat. Leveraging nuclear energy opens avenues to optimize infrastructure costs associated with hydrogen delivery and enables strategic co-location with industrial activities that are otherwise challenging to decarbonize. In terms of grid and system-level expenses, nuclear energy proves advantageous with low costs. Achieving net-zero decarbonization objectives, particularly with an increased focus on hydrogen production solely through variable renewables, would necessitate unprecedented levels of generation capacity. The report from the Nuclear Energy Agency (NEA) demonstrates that incorporating nuclear energy into the generation mix diminishes the overall capacity requirements at the system level and enhances the cost-effectiveness of the global power system at the grid level (Figure 3.34) [143].

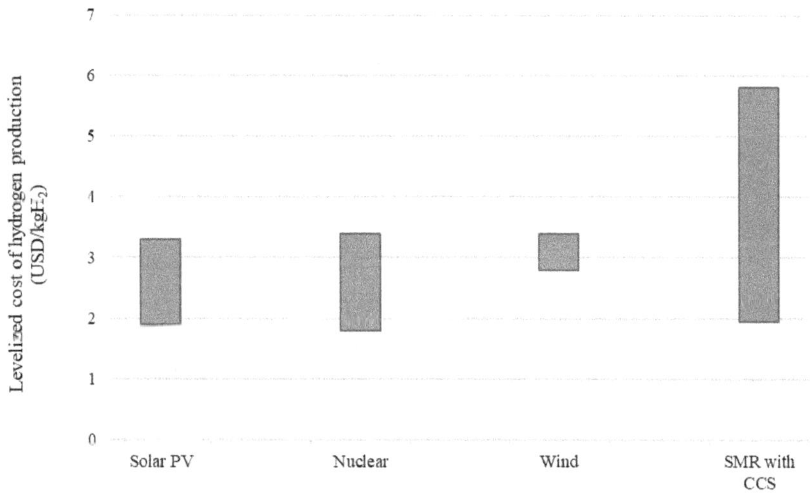

Figure 3.34 Hydrogen cost by 2035

3.5 Natural hydrogen production

In 2012, coinciding with humanity's quest for sustainable energy sources, an inadvertent revelation of naturally occurring hydrogen took place. Originally, this discovery was traced back to 1987 in Mali's Bourakebougou village during drilling operations for a water well. When the drill reached a depth of 108 meters, a driller casually discarded his cigarette into the well collar, triggering an unexpected eruption of a massive flame. Astonishingly, this fire burned for several weeks without producing any carbon dioxide or other byproducts. Eventually, the drillers successfully capped the well and abandoned it in pursuit of water in different locations. This abandonment persisted until 2012, when a new investor secured state permission to explore the region. The exploration team sampled the gas emanating from the dormant well, revealing that it consisted of almost 98% molecular hydrogen. Subsequently, an engine was installed to convert the gas into electricity, providing uninterrupted power to the residents of Bourakebougou village from that moment onward [144].

Before the Mali incident, occurrences of hydrogen gas flowing from various mines, faults, seeps, and hydrocarbon wells were frequently documented worldwide. Despite this, the concept of natural hydrogen as a viable renewable energy source was either overlooked or underestimated for three primary reasons. Firstly, most deep wells were originally drilled for oil and gas exploration in sedimentary rocks, where the presence of free hydrogen is less likely due to its rapid conversion to hydrocarbons. Secondly, hydrogen's highly lightweight and diffusive nature results in its escape from rock pores or potential underground traps to the ground surface. The third reason stems from the widespread belief that sources of free hydrogen are scarce, leading to a general lack of interest in sampling natural hydrogen concentrations in gases released from underground during geoscience studies. Even if someone were inclined to do so, the absence of appropriate equipment would hinder the measurement of hydrogen content in gases released from subterranean geological structures. However, in the present era, the majority of gas detectors used in geoscience lack an H_2 detection sensor.

The commercial utilization of natural hydrogen in Mali has motivated numerous researchers and companies to investigate subsurface natural hydrogen resources worldwide. Additionally, the United States drilled its inaugural hydrogen borehole in 2019. In September 2022, the US government disclosed an investment of approximately USD 700 million for the exploration of natural hydrogen. In 2021, the Australian government granted permission for investors to explore potential natural hydrogen sources in Australia [145]. Evidently, Australia possesses noteworthy hydrogen-bearing formations, including rocks containing iron and uranium. As will be detailed in subsequent sections, the iron-containing rocks undergo a reaction with hot groundwater, releasing hydrogen through the serpentinization process. Furthermore, uranium-containing rocks emit radioactive rays, inducing the splitting of groundwater and generating natural hydrogen through the electrolysis process. It's important to note that the Australian continent is surrounded by water, resulting in crustal rocks that are partially or entirely saturated. Various researchers have proposed multiple explanations for the existence of geologic hydrogen [146]. The divergence in these hypotheses might stem from the inadequate equipment and knowledge available for characterizing natural hydrogen.

The literature outlines four primary origins suggested for natural hydrogen, which are:

- The natural radiolysis of deep groundwater is caused by the radioactive elements present in high-radioactivity rocks.
- The serpentinization process, wherein hot groundwater reacts with iron-rich formations like ophiolites, results in the release of natural hydrogen.
- The deep-seated origin, wherein natural hydrogen emanates from the Earth's core or mantle.
- Changes in the in situ stress regime can induce fault reactivation, earthquakes, and activity in regional tectonics.

These four origins are illustrated in Figure 3.35, and the subsequent paragraphs delve into detailed discussions on each of them.

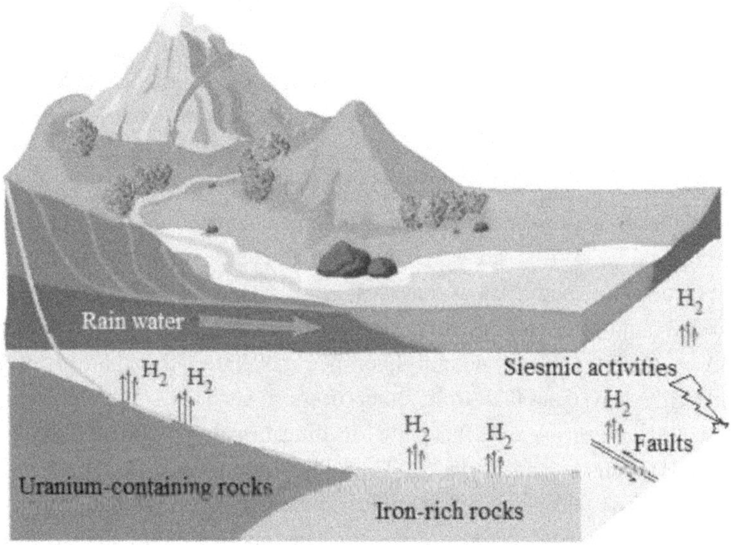

Figure 3.35 Four main origins of geological hydrogen [147]

Vernadsky [148] initiated an investigation into the concept of radiolysis of groundwater through the energy released from radioactive decay. This process involves the easy splitting of groundwater molecules into oxygen and hydrogen due to the energy of radioactive decay, and it has been considered the most likely mechanism for significant geologic hydrogen generation [149]. Approximately 1% of the total energy released from radioactive decay is dedicated to decomposing groundwater molecules, with the remaining 99% converted to heat within the solid rock structure. Consequently, higher rock porosity is correlated with a greater volume of generated geologic hydrogen [150].

Another source of natural hydrogen is the serpentinization process, wherein ultramafic formations are exposed to high-temperature groundwater, resulting in an oxidization reaction and the release of geologic hydrogen [151]. The Earth's mantle primarily consists of ultramafic rocks like peridotite, with olivine as the dominant mineral. When olivine reacts with hot groundwater, it produces hematite, silica, and molecular hydrogen. This process is not limited to the mantle, as crustal basaltic formations can also react with hot groundwater to generate natural hydrogen, a fact confirmed by investigations into crustal basaltic rocks, revealing high hydrogen concentrations [152].

The third source of natural hydrogen is the deep-seated origin, linked to the deep rocks in the Earth's core and mantle. Studies report an increase in hydrogen concentration with depth in subsurface rocks, leading to the designation of this type of natural hydrogen as the deep-seated source [153,154]. However, given the current inability to drill to the Earth's core or mantle, characterizing deep-seated sources remains challenging. Some researchers associate these sources with mantle rocks [155–157], while others attribute them to core rocks. For instance, natural hydrogen detected in the Kansas region was linked to mantle rocks, while some suggest the core may contain tens of times more natural hydrogen than oceanic hydrogen [158,159].

The fourth origin of natural hydrogen is the change in the in-situ stress regime in tectonically active areas. Experimental tests indicate high hydrogen concentrations on active fault surfaces [160,161], suggesting that the reaction between groundwater and Fe-rich rock surfaces of faults generates hydrogen [162]. However, the amount of natural hydrogen generated through this process is considerably less than that from other origins, such as serpentinization [163]. Earthquakes can activate crustal faults, leading to natural hydrogen generation in subsurface fractures and faults [164]. Additionally, studies suggest that newly propagated earthquake-induced fractures contain a larger volume of H_2 content compared to old, non-activated fractures, leading some researchers to propose hydrogen concentration as an indicator for predicting earthquake occurrences [165,166].

Irrespective of its origin, natural hydrogen in the subsurface is characterized by its lightweight and diffusive nature, allowing it to easily migrate toward the Earth's surface. During its travel, subsurface natural hydrogen may undergo reactions with rock minerals, be consumed by subsurface microbes, or become trapped by impermeable rocks like salt layers. Upon reaching the Earth's surface, it may give rise to circular formations known as fairy circles. The cause of this phenomenon is still not fully understood, although Carolina bays, a variety of fairy circles along the eastern seaboard of the United States, have been extensively studied. In many instances, a consistent release of hydrogen was observed in these formations, with researchers noting an increase in hydrogen concentration with depth. The formation of these fairy circles could potentially be explained by weak H_2 seepages [145].

Large fairy circles in Brazil result from hydrogen leakage from underground to the surface, which is an example of this phenomenon. In these instances, hydrogen has caused damage to local vegetation. Similar phenomena have been observed in Namibia and Australia. Some researchers speculate that the degradation of vegetation due to hydrogen seepage from fairy circles may be related to hydrogen-loving microbes consuming other nutrients, but further investigations are necessary to determine the primary cause of this phenomenon. Another manifestation of natural hydrogen is through seeps, where continuous, long-term fires burn on the Earth's surface. Ref. [167] presents an image of a hydrogen seep in the Chimarea region near Antalya, Turkey, where the hydrogen concentration was reported to exceed 11%. The burning of hydrogen in the Chimaera region is attributed to the presence of ophiolite rocks. These local ophiolites contain iron-rich minerals like olivine, and the continuous reaction of subsurface ophiolites with hot groundwater leads to the constant generation of natural hydrogen, which travels through crustal fractures and reaches the Earth's surface [168]. There are three distinct technologies employed for the retrieval of natural hydrogen from subsurface reservoirs (Figure 3.36) [147]:

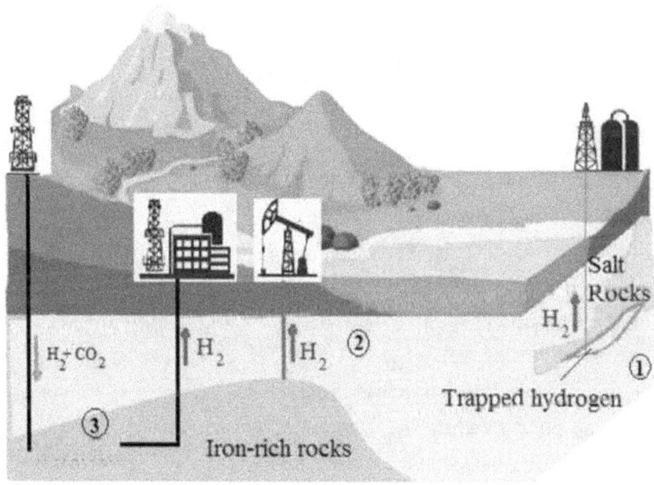

Figure 3.36 Three different extraction methods for geologic hydrogen from underground reserves

1. Single wellbore extraction from trapped hydrogen: If natural hydrogen is confined in an underground space, a single wellbore can be drilled to extract it. Impermeable formations, like salt rocks, can serve as cap rocks, effectively trapping natural hydrogen beneath the surface.

2. Drilling a single wellbore for iron-rich formations: If shallow formations rich in iron are consistently exposed to hot groundwater, the resulting natural hydrogen can be gathered and extracted by drilling a single wellbore from the ground surface to the depth of the iron-rich formations.
3. Enhanced hydrogen recovery system in water-limited conditions: In situations where there is insufficient groundwater for the serpentinization of shallow iron-rich formations, an enhanced hydrogen recovery system can be deployed. This involves injecting water into the iron-rich formation through an injection well to generate economically viable concentrations of natural hydrogen. Subsequently, the produced hydrogen is extracted from the production well. This method also allows for the injection of carbon dioxide into the ground for sequestration purposes.

The extraction of geologic hydrogen has been developed in some countries such as Oman [169], Turkey [170], Philippines [171], Brazil [172], New Caledonia [173], Russia [174] and the United States [175].

Abbreviations

CAGR: Compound annual growth rate
CCS: Carbon capture and storage
CCU: Carbon capture and utilization
CHP: Combined heat and power
CPV: Concentrated photovoltaic
DOE: US Department of Energy
DOT: US Department of Transportation
DTP: Dynamic Tidal Power
EJ: Exajoules
EMEC: European Marine Energy Centre
GPP: Geothermal power plant
GW: Gigawatts
HTGR: High-temperature gas-cooled reactor
HTS: High-temperature shift
HTSE: High-temperature steam electrolysis
IEA: International Energy Agency
IAEA: International Atomic Energy Agency
ITEG: Integrating Tidal Energy into the European Grid
JAEA: Japan Atomic Energy Agency
LCOH: Levelized cost of hydrogen
LNG: Liquefied natural gas
LTS: Low-temperature shift
MENA: Middle East and North Africa
MDEA: Methyldiethanolamine
MEA: Monoethanolamine

MSW: Municipal solid waste
NEA: Nuclear Energy Agency
NERI: Nuclear Energy Research Initiative
NHI: Nuclear Hydrogen Initiative
NGD: Natural gas decomposition
NGNP: Next Generation Nuclear Plant
NREL: National Renewable Energy Laboratory
ORC: Organic Rankine cycle
PEM: Proton exchange membrane
PSA: Pressure swing adsorption
PV: photovoltaic
PV/T: Hybrid photovoltaic/thermal
PWR: Pressurized Water Reactor
RTC: Round-the-clock
SMR: Steam methane reforming
STC: solar thermal collector
UHV: Ultra-high voltage
VHTR: Very high-temperature reactor
WGS: Water gas shift

References

[1] Explainer: the many shades of hydrogen | News | Eco-Business | Asia Pacific n.d. https://www.eco-business.com/news/explainer-the-many-shades-of-hydrogen/ (accessed August 19, 2023).

[2] Hossein Ali YR, Shin D. Green Hydrogen Production Technologies from Ammonia Cracking. Energies 2022, Vol 15, Page 8246 2022;15:8246. https://doi.org/10.3390/EN15218246.

[3] Liu X, Elgowainy A, Wang M. Life cycle energy use and greenhouse gas emissions of ammonia production from renewable resources and industrial by-products. Green Chemistry 2020;22:5751–61. https://doi.org/10.1039/D0GC02301A

[4] Demirbas A. Future hydrogen economy and policy. Http://DxDoiOrg/101080/155672492014950394 2017;12:172–81. https://doi.org/10.1080/15567249.2014.950394.

[5] Levin DB, Chahine R. Challenges for renewable hydrogen production from biomass. Int J Hydrogen Energy 2010;35:4962–9. https://doi.org/10.1016/J.IJHYDENE.2009.08.067.

[6] Blue Hydrogen: A key player in the future of energy transition n.d. https://sweetcrudereports.com/blue-hydrogen-a-key-player-in-the-future-of-energy-transition/ (accessed February 4, 2024).

[7] Li J, Cheng W. Comparative life cycle energy consumption, carbon emissions and economic costs of hydrogen production from coke oven gas and coal gasification. Int J Hydrogen Energy 2020;45:27979–93. https://doi.org/10.1016/J.IJHYDENE.2020.07.079.

[8] Li G, Cui P, Wang Y, Liu Z, Zhu Z, Yang S. Life cycle energy consumption and GHG emissions of biomass-to-hydrogen process in comparison with coal-to-hydrogen process. Energy 2020;191:116588. https://doi.org/10.1016/J.ENERGY.2019.116588.

[9] Howarth RW, Jacobson MZ. How green is blue hydrogen? Energy Sci Eng 2021;9:1676–87. https://doi.org/10.1002/ESE3.956.

[10] Postels S, Abánades A, von der Assen N, Rathnam RK, Stückrad S, Bardow A. Life cycle assessment of hydrogen production by thermal cracking of methane based on liquid-metal technology. Int J Hydrogen Energy 2016;41:23204–12. https://doi.org/10.1016/J.IJHYDENE.2016.09.167.

[11] Oni AO, Anaya K, Giwa T, Di Lullo G, Kumar A. Comparative assessment of blue hydrogen from steam methane reforming, autothermal reforming, and natural gas decomposition technologies for natural gas-producing regions. Energy Convers Manag 2022;254:115245. https://doi.org/10.1016/J.ENCONMAN.2022.115245.

[12] Leal Pérez BJ, Medrano Jiménez JA, Bhardwaj R, Goetheer E, van Sint Annaland M, Gallucci F. Methane pyrolysis in a molten gallium bubble column reactor for sustainable hydrogen production: Proof of concept & techno-economic assessment. Int J Hydrogen Energy 2021;46:4917–35. https://doi.org/10.1016/J.IJHYDENE.2020.11.079.

[13] Valente A, Iribarren D, Dufour J. Harmonising methodological choices in life cycle assessment of hydrogen: A focus on acidification and renewable hydrogen. Int J Hydrogen Energy 2019;44:19426–33. https://doi.org/10.1016/J.IJHYDENE.2018.03.101.

[14] Ghandehariun S, Kumar A. Life cycle assessment of wind-based hydrogen production in Western Canada. Int J Hydrogen Energy 2016;41:9696–704. https://doi.org/10.1016/J.IJHYDENE.2016.04.077.

[15] Siddiqui O, Dincer I. A well to pump life cycle environmental impact assessment of some hydrogen production routes. Int J Hydrogen Energy 2019;44:5773–86. https://doi.org/10.1016/J.IJHYDENE.2019.01.118.

[16] Wulf C, Kaltschmitt M. Life cycle assessment of hydrogen supply chain with special attention on hydrogen refuelling stations. Int J Hydrogen Energy 2012;37:16711–21. https://doi.org/10.1016/J.IJHYDENE.2012.03.028.

[17] Valente A, Iribarren D, Dufour J. Harmonising methodological choices in life cycle assessment of hydrogen: A focus on acidification and renewable hydrogen. Int J Hydrogen Energy 2019;44:19426–33. https://doi.org/10.1016/J.IJHYDENE.2018.03.101.

[18] Yu M, Wang K, Vredenburg H. Insights into low-carbon hydrogen production methods: Green, blue and aqua hydrogen. Int J Hydrogen Energy 2021;46:21261–73. https://doi.org/10.1016/J.IJHYDENE.2021.04.016.

[19] Boretti A. White is the color of hydrogen from concentrated solar energy and thermochemical water splitting cycles. Int J Hydrogen Energy 2021;46:20790–1. https://doi.org/10.1016/J.IJHYDENE.2021.03.178.

[20] Barbarelli S, Nastasi B, Rusu E, Bergillos RJ. Tides and Tidal Currents—Guidelines for Site and Energy Resource Assessment. Energies 2021, Vol 14, Page 6123 2021;14:6123. https://doi.org/10.3390/EN14196123.

[21] Suleman F, Dincer I, Agelin-Chaab M. Environmental impact assessment and comparison of some hydrogen production options. Int J Hydrogen Energy 2015;40:6976–87. https://doi.org/10.1016/J.IJHYDENE.2015.03.123.

[22] Dash SK, Chakraborty S, Elangovan D. A Brief Review of Hydrogen Production Methods and Their Challenges. Energies 2023, Vol 16, Page 1141 2023;16:1141. https://doi.org/10.3390/EN16031141.

[23] What is Steam Methane Reforming (SMR)? n.d. https://www.hydrogennewsletter.com/what-is-steam-methane-reforming-smr/ (accessed August 18, 2023).

[24] Steam Methane Reforming Market - Global Analysis and Forecasts 2029 n.d. https://www.maximizemarketresearch.com/market-report/steam-methane-reforming-market/72593/ (accessed October 20, 2023).

[25] The Future of Energy Transition: A Look at the Asia-Pacific Region's Hydrogen Strategies - PTR Inc. n.d. https://ptr.inc/asia-pacific-emerging-as-the-hydrogen-powerhouse/ (accessed August 18, 2023).

[26] The EU's Hydrogen Market Landscape - PTR Inc. n.d. https://ptr.inc/the-eus-hydrogen-market-landscape/ (accessed August 18, 2023).

[27] Methodology for Determining the Greenhouse Gas Emissions Associated With the Production of Hydrogen V E R S I O N 1-O C T O B E R 2 0 2 1 n.d. www.iphe.net/_files/ugd/45185a_ef588ba32fc54e0eb57b0b7444cfa5f9.pdf (accessed January 4, 2024).

[28] Shoko E, McLellan B, Dicks AL, da Costa JCD. Hydrogen from coal: Production and utilisation technologies. Int J Coal Geol 2006;65:213–22. https://doi.org/10.1016/j.coal.2005.05.004.

[29] How much coal is left - U.S. Energy Information Administration (EIA) n.d. https://www.eia.gov/energyexplained/coal/how-much-coal-is-left.php (accessed December 24, 2022).

[30] Use of coal - U.S. Energy Information Administration (EIA) n.d. https://www.eia.gov/energyexplained/coal/use-of-coal.php (accessed October 20, 2023).

[31] Hydrogen Production from Coal Gasification n.d. https://www.hydrogennewsletter.com/hydrogen-production-from-coal-gasification/ (accessed August 18, 2023).

[32] IRENA – International Renewable Energy Agency n.d. https://www.irena.org/ (accessed August 19, 2023).

[33] Noyan OF, Hasan MM, Pala N. A Global Review of the Hydrogen Energy Eco-System. Energies 2023, Vol 16, Page 1484 2023;16:1484. https://doi.org/10.3390/EN16031484.

[34] IEA – International Energy Agency n.d. https://www.iea.org/ (accessed March 11, 2022).

[35] Africa and Europe: green hydrogen economy n.d. https://www.rystadenergy.com/news/africa-and-europe-set-to-be-the-dynamos-for-the-global-green-hydrogen-economy (accessed February 16, 2024).

[36] Marzouk OA. 2030 Ambitions for Hydrogen, Clean Hydrogen, and Green Hydrogen. Engineering Proceedings 2023, Vol 56, Page 14 2023;56:14. https://doi.org/10.3390/ASEC2023-15497.

[37] Hosseini SE, Wahid MA. Hydrogen production from renewable and sustainable energy resources: Promising green energy carrier for clean development. Renewable and Sustainable Energy Reviews 2016;57. https://doi.org/10.1016/j.rser.2015.12.112.

[38] Zhang S, Zhang N, Dai H, Liu L, Zhou Z, Shi Q, et al. Comparison of Different Coupling Modes between the Power System and the Hydrogen System Based on a Power–Hydrogen Coordinated Planning Optimization Model. Energies 2023, Vol 16, Page 5374 2023;16:5374. https://doi.org/10.3390/EN16145374.

[39] Demir ME, Dincer I. Cost assessment and evaluation of various hydrogen delivery scenarios. Int J Hydrogen Energy 2018;43:10420–30. https://doi.org/10.1016/J.IJHYDENE.2017.08.002.

[40] Nazri NS, Fudholi A, Bakhtyar B, Yen CH, Ibrahim A, Ruslan MH, et al. Energy economic analysis of photovoltaic–thermal-thermoelectric (PVT-TE) air collectors. Renewable and Sustainable Energy Reviews 2018;92:187–97. https://doi.org/10.1016/J.RSER.2018.04.061.

[41] Tripanagnostopoulos Y. Aspects and improvements of hybrid photovoltaic/thermal solar energy systems. Solar Energy 2007;81:1117–31. https://doi.org/10.1016/J.SOLENER.2007.04.002.

[42] Hassan Q, Abdulrahman IS, Salman HM, Olapade OT, Jaszczur M. Techno-Economic Assessment of Green Hydrogen Production by an Off-Grid Photovoltaic Energy System. Energies 2023, Vol 16, Page 744 2023;16:744. https://doi.org/10.3390/EN16020744.

[43] Götz M, Lefebvre J, Mörs F, McDaniel Koch A, Graf F, Bajohr S, et al. Renewable Power-to-Gas: A technological and economic review. Renew Energy 2016;85:1371–90. https://doi.org/10.1016/J.RENENE.2015.07.066.

[44] Burhan M, Oh SJ, Chua KJE, Ng KC. Solar to hydrogen: Compact and cost effective CPV field for rooftop operation and hydrogen production. Appl Energy 2017;194:255–66. https://doi.org/10.1016/J.APENERGY.2016.11.062.

[45] Hosseini SE. The US hydrogen fuel industry today and future. Future Energy 2022;1:1. https://doi.org/10.55670/fpll.fuen.1.1.7.

[46] Green hydrogen imported to Europe would be cost-competitive with locally produced H2 by 2030: analyst | Hydrogen news and intelligence n.d. https://www.hydrogeninsight.com/production/green-hydrogen-imported-to-europe-would-be-cost-competitive-with-locally-produced-h2-by-2030-analyst/2-1-1393655 (accessed October 3, 2023).

[47] Luo X, Wang J, Dooner M, Clarke J. Overview of current development in electrical energy storage technologies and the application potential in power system operation. Appl Energy 2015;137:511–36. https://doi.org/10.1016/J.APENERGY.2014.09.081.

[48] Carton JG, Olabi AG. Wind/hydrogen hybrid systems: Opportunity for Ireland's wind resource to provide consistent sustainable energy supply. Energy 2010;35:4536–44. https://doi.org/10.1016/J.ENERGY.2010.09.010.
[49] Offshore Wind Outlook 2019 – Analysis - IEA n.d. https://www.iea.org/reports/offshore-wind-outlook-2019 (accessed August 22, 2023).
[50] Eurek K, Sullivan P, Gleason M, Hettinger D, Heimiller D, Lopez A. An improved global wind resource estimate for integrated assessment models. Energy Econ 2017;64:552–67. https://doi.org/10.1016/J.ENECO.2016.11.015.
[51] Burmistrz P, Chmielniak T, Czepirski L, Gazda-Grzywacz M. Carbon footprint of the hydrogen production process utilizing subbituminous coal and lignite gasification. J Clean Prod 2016;139:858–65. https://doi.org/10.1016/J.JCLEPRO.2016.08.112.
[52] Bhandari R, Trudewind CA, Zapp P. Life cycle assessment of hydrogen production via electrolysis – a review. J Clean Prod 2014;85:151–63. https://doi.org/10.1016/J.JCLEPRO.2013.07.048.
[53] Javaid A, Javaid U, Sajid M, Rashid M, Uddin E, Ayaz Y, et al. Forecasting Hydrogen Production from Wind Energy in a Suburban Environment Using Machine Learning. Energies 2022, Vol 15, Page 8901 2022;15:8901. https://doi.org/10.3390/EN15238901.
[54] Bhogilla SS, Ito H, Segawa T, Kato A, Nakano A. Experimental study on laboratory scale Totalized Hydrogen Energy Utilization System using wind power data. Int J Hydrogen Energy 2017;42:13827–38. https://doi.org/10.1016/J.IJHYDENE.2016.12.125.
[55] d'Amore-Domenech R, Leo TJ, Pollet BG. Bulk power transmission at sea: Life cycle cost comparison of electricity and hydrogen as energy vectors. Appl Energy 2021;288:116625. https://doi.org/10.1016/J.APENERGY.2021.116625.
[56] Miao B, Giordano L, Chan SH. Long-distance renewable hydrogen transmission via cables and pipelines. Int J Hydrogen Energy 2021;46:18699–718. https://doi.org/10.1016/J.IJHYDENE.2021.03.067.
[57] Ibrahim OS, Singlitico A, Proskovics R, McDonagh S, Desmond C, Murphy JD. Dedicated large-scale floating offshore wind to hydrogen: Assessing design variables in proposed typologies. Renewable and Sustainable Energy Reviews 2022;160:112310. https://doi.org/10.1016/J.RSER.2022.112310.
[58] Calado G, Castro R. Hydrogen Production from Offshore Wind Parks: Current Situation and Future Perspectives. Applied Sciences 2021, Vol 11, Page 5561 2021;11:5561. https://doi.org/10.3390/APP11125561.
[59] Ligęza K, Łaciak M, Ligęza B. Centralized Offshore Hydrogen Production from Wind Farms in the Baltic Sea Area—A Study Case for Poland. Energies 2023, Vol 16, Page 6301 2023;16:6301. https://doi.org/10.3390/EN16176301.
[60] Hydrogen – Analysis - IEA n.d. https://www.iea.org/reports/hydrogen (accessed November 10, 2022).
[61] Flamos A, Georgallis PG, Doukas H, Psarras J. Using Biomass to Achieve European Union Energy Targets—A Review of Biomass Status, Potential, and Supporting Policies. Http://DxDoiOrg/101080/154350752011576292 2011;8:411–28. https://doi.org/10.1080/15435075.2011.576292.
[62] Jekayinfa SO, Scholz V. Potential Availability of Energetically Usable Crop Residues in Nigeria. Https://DoiOrg/101080/15567030701750549 2009;31:687–97. https://doi.org/10.1080/15567030701750549.
[63] Charles MB, Ryan R, Oloruntoba R, Heidt T von der, Ryan N. The EU–Africa Energy Partnership: Towards a mutually beneficial renewable transport energy alliance? Energy Policy 2009;37:5546–56. https://doi.org/10.1016/j.enpol.2009.08.016.
[64] Global Biomass Power Market Size Report, 2022-2030 n.d. https://www.grandviewresearch.com/industry-analysis/biomass-power-market (accessed October 20, 2023).
[65] Balat H, Kırtay E. Hydrogen from biomass – Present scenario and future prospects. Int J Hydrogen Energy 2010;35:7416–26. https://doi.org/10.1016/j.ijhydene.2010.04.137.
[66] Parthasarathy P, Narayanan KS. Hydrogen production from steam gasification of biomass: Influence of process parameters on hydrogen yield – A review. Renew Energy 2014;66:570–9. https://doi.org/10.1016/J.RENENE.2013.12.025.

[67] Demirbas A. Liquefaction of Biomass Using Glycerol. Http://DxDoiOrg/101080/15567030601100654 2008;30:1120–6. https://doi.org/10.1080/15567030601100654.
[68] Bhuiyan MNA, Ota M, Murakami K, Yoshida H. Pyrolysis Kinetics of Newspaper and Its Gasification. Http://DxDoiOrg/101080/15567030802464347 2009;32:108–18. https://doi.org/10.1080/15567030802464347.
[69] Caglar A. Valorization of tea wastes by pyrolysis. Energy Education Science and Technology Part A: Energy Science and Research 2009:135–44.
[70] Şener Ü, Genel Y, Saka C, Kiliçel F, Küçük MM. Supercritical Fluid Extraction of Cotton Stalks. Http://DxDoiOrg/101080/15567030903060424 2009;32:20–5. https://doi.org/10.1080/15567030903060424.
[71] Demirbaş A. Yields of hydrogen-rich gaseous products via pyrolysis from selected biomass samples. Fuel 2001;80:1885–91. https://doi.org/10.1016/S0016-2361(01)00070-9.
[72] Arregi A, Amutio M, Lopez G, Bilbao J, Olazar M. Evaluation of thermochemical routes for hydrogen production from biomass: A review. Energy Convers Manag 2018;165:696–719. https://doi.org/10.1016/J.ENCONMAN.2018.03.089.
[73] Ni M, Leung DYC, Leung MKH, Sumathy K. An overview of hydrogen production from biomass. Fuel Processing Technology 2006;87:461–72. https://doi.org/10.1016/j.fuproc.2005.11.003.
[74] Demirbaş A. Biomass resource facilities and biomass conversion processing for fuels and chemicals. Energy Convers Manag 2001;42:1357–78.
[75] Iribarren D, Susmozas A, Petrakopoulou F, Dufour J. Environmental and exergetic evaluation of hydrogen production via lignocellulosic biomass gasification. J Clean Prod 2014;69:165–75. https://doi.org/10.1016/J.JCLEPRO.2014.01.068.
[76] Fahmy TYA, Fahmy Y, Mobarak F, El-Sakhawy M, Abou-Zeid RE. Biomass pyrolysis: past, present, and future. Environ Dev Sustain 2020;22:17–32. https://doi.org/10.1007/S10668-018-0200-5/FIGURES/2.
[77] Zhang L, Zhang B, Yang Z, Yan Y. Pyrolysis behavior of biomass with different Ca-based additives. RSC Adv 2014;4:39145–55. https://doi.org/10.1039/C4RA04865B.
[78] Nowicki L, Siuta D, Markowski M. Pyrolysis of Rapeseed Oil Press Cake and Steam Gasification of Solid Residues. Energies 2020, Vol 13, Page 4472 2020;13:4472. https://doi.org/10.3390/EN13174472.
[79] Czernik S, French R, Evans R, Chornet E. Hydrogen from post-consumer residues. US DOE Hydrogen and Fuel Cells Merit Review Meet Ing, Berkeley, CA 2003.
[80] Demirbas A. Hydrogen Production from Carbonaceous Solid Wastes by Steam Reforming. Energy Sources, Part A: Recovery, Utilization, and Environmental Effects 2008;30:924–31. https://doi.org/10.1080/10826070601082658.
[81] Demirbaş A. Hydrogen Production from Biomass by the Gasification Process. Http://DxDoiOrg/101080/00908310252712307 2010;24:59–68. https://doi.org/10.1080/00908310252712307.
[82] Sutton D, Kelleher B, Ross JRH. Review of literature on catalysts for biomass gasification. Fuel Processing Technology 2001;73:155–73. https://doi.org/10.1016/S0378-3820(01)00208-9.
[83] Arvelakis S, Koukios EG. Physicochemical upgrading of agroresidues as feedstocks for energy production via thermochemical conversion methods. Biomass Bioenergy 2002;22:331–48. https://doi.org/10.1016/S0961-9534(01)00056-3.
[84] Arvelakis S, Gehrmann H, Beckmann M, Koukios EG. Effect of leaching on the ash behavior of olive residue during fluidized bed gasification. Biomass Bioenergy 2002;22:55–69. https://doi.org/10.1016/S0961-9534(01)00059-9.
[85] García-Ibañez P, Cabanillas A, Sánchez JM. Gasification of leached orujillo (olive oil waste) in a pilot plant circulating fluidised bed reactor. Preliminary results. Biomass Bioenergy 2004;27:183–94. https://doi.org/10.1016/J.BIOMBIOE.2003.11.007.
[86] Biomass Gasification Market Size, Share, Trends Report 2032 - Industry Growth Analysis n.d. https://www.marketresearchfuture.com/reports/biomass-gasification-market-7313 (accessed October 21, 2023).

[87] Kalinci Y, Hepbasli A, Dincer I. Biomass-based hydrogen production: A review and analysis. Int J Hydrogen Energy 2009;34:8799–817. https://doi.org/10.1016/j.ijhydene.2009.08.078.
[88] Unal H, Alibas K. Agricultural residues as biomass energy. Energy Sources, Part B: Economics, Planning and Policy 2007;2:123–40. https://doi.org/10.1080/15567240600629401.
[89] Mahmoud M, Ramadan M, Naher S, Pullen K, Ali Abdelkareem M, Olabi AG. A review of geothermal energy-driven hydrogen production systems. Thermal Science and Engineering Progress 2021;22:100854. https://doi.org/10.1016/J.TSEP.2021.100854.
[90] Ahmadfard M, Bernier M. A review of vertical ground heat exchanger sizing tools including an inter-model comparison. Renewable and Sustainable Energy Reviews 2019;110:247–65. https://doi.org/10.1016/J.RSER.2019.04.045.
[91] Gan G. A numerical methodology for comprehensive assessment of the dynamic thermal performance of horizontal ground heat exchangers. Thermal Science and Engineering Progress 2019;11:365–79. https://doi.org/10.1016/J.TSEP.2019.04.013.
[92] Huang G, Liu Y, Zhuang C, Zhang H, Lu J, Zhu L. Experimental study on heat transfer characteristics of helix ground heat exchanger coil under dynamic load. Thermal Science and Engineering Progress 2020;18:100546. https://doi.org/10.1016/J.TSEP.2020.100546.
[93] Agrawal KK, Misra R, Agrawal G Das, Bhardwaj M, Jamuwa DK. The state of art on the applications, technology integration, and latest research trends of earth-air-heat exchanger system. Geothermics 2019;82:34–50. https://doi.org/10.1016/J.GEOTHERMICS.2019.05.011.
[94] Top Five Geothermal Power Generation Markets in 2021 - GlobalData n.d. https://www.globaldata.com/data-insights/power-and-utilities/the-top-five-geothermal-power-generation-markets-in-2021/ (accessed November 4, 2023).
[95] Salhein K, Kobus CJ, Zohdy M. Forecasting Installation Capacity for the Top 10 Countries Utilizing Geothermal Energy by 2030. Thermo 2022, Vol 2, Pages 334-351 2022;2:334–51. https://doi.org/10.3390/THERMO2040023.
[96] Hernández Martínez E, Avitia Carlos MCP, Cisneros Solís JI, Prieto Avalos MCM del C. Thermodynamic simulation and mathematical model for single and double flash cycles of Cerro Prieto geothermal power plants. Geothermics 2020;83:101713. https://doi.org/10.1016/J.GEOTHERMICS.2019.101713.
[97] Prananto LA, Juangsa FB, Iqbal RM, Aziz M, Soelaiman TAF. Dry steam cycle application for excess steam utilization: Kamojang geothermal power plant case study. Renew Energy 2018;117:157–65. https://doi.org/10.1016/J.RENENE.2017.10.029.
[98] Nasruddin N, Dwi Saputra I, Mentari T, Bardow A, Marcelina O, Berlin S. Exergy, exergoeconomic, and exergoenvironmental optimization of the geothermal binary cycle power plant at Ampallas, West Sulawesi, Indonesia. Thermal Science and Engineering Progress 2020;19:100625. https://doi.org/10.1016/J.TSEP.2020.100625.
[99] Zare V. A comparative exergoeconomic analysis of different ORC configurations for binary geothermal power plants. Energy Convers Manag 2015;105:127–38. https://doi.org/10.1016/J.ENCONMAN.2015.07.073.
[100] Yilmaz C, Koyuncu I, Alcin M, Tuna M. Artificial Neural Networks based thermodynamic and economic analysis of a hydrogen production system assisted by geothermal energy on Field Programmable Gate Array. Int J Hydrogen Energy 2019;44:17443–59. https://doi.org/10.1016/J.IJHYDENE.2019.05.049.
[101] Kanoglu M, Bolatturk A, Yilmaz C. Thermodynamic analysis of models used in hydrogen production by geothermal energy. Int J Hydrogen Energy 2010;35:8783–91. https://doi.org/10.1016/J.IJHYDENE.2010.05.128.
[102] Yilmaz C, Kanoglu M, Abusoglu A. Thermoeconomic cost evaluation of hydrogen production driven by binary geothermal power plant. Geothermics 2015;57:18–25. https://doi.org/10.1016/J.GEOTHERMICS.2015.05.005.
[103] Ebadollahi M, Rostamzadeh H, Pedram MZ, Ghaebi H, Amidpour M. Proposal and assessment of a new geothermal-based multigeneration system for cooling, heating, power, and hydrogen production, using LNG cold energy recovery. Renew Energy 2019;135:66–87. https://doi.org/10.1016/J.RENENE.2018.11.108.

[104] Kianfard H, Khalilarya S, Jafarmadar S. Exergy and exergoeconomic evaluation of hydrogen and distilled water production via combination of PEM electrolyzer, RO desalination unit and geothermal driven dual fluid ORC. Energy Convers Manag 2018;177:339–49. https://doi.org/10.1016/J.ENCONMAN.2018.09.057.

[105] Yuksel YE, Ozturk M, Dincer I. Thermodynamic analysis and assessment of a novel integrated geothermal energy-based system for hydrogen production and storage. Int J Hydrogen Energy 2018;43:4233–43. https://doi.org/10.1016/J.IJHYDENE.2017.08.137.

[106] Ratlamwala TAH, Dincer I. Comparative efficiency assessment of novel multi-flash integrated geothermal systems for power and hydrogen production. Appl Therm Eng 2012;48:359–66. https://doi.org/10.1016/J.APPLTHERMALENG.2012.04.039.

[107] Siddiqui O, Ishaq H, Dincer I. A novel solar and geothermal-based trigeneration system for electricity generation, hydrogen production and cooling. Energy Convers Manag 2019;198:111812. https://doi.org/10.1016/J.ENCONMAN.2019.111812.

[108] Yuksel YE, Ozturk M. Thermodynamic and thermoeconomic analyses of a geothermal energy based integrated system for hydrogen production. Int J Hydrogen Energy 2017;42:2530–46. https://doi.org/10.1016/J.IJHYDENE.2016.04.172.

[109] Hydropower generation by country 2021 | Statista n.d. https://www.statista.com/statistics/474799/global-hydropower-generation-by-major-country/ (accessed November 7, 2023).

[110] Jovan DJ, Dolanc G, Pregelj B. Utilization of excess water accumulation for green hydrogen production in a run-of-river hydropower plant. Renew Energy 2022;195:780–94. https://doi.org/10.1016/J.RENENE.2022.06.079.

[111] Renewable energy roadmaps n.d. https://www.irena.org/Energy-Transition/Outlook/Renewable-energy-roadmaps (accessed November 7, 2023).

[112] World Energy Transitions Outlook n.d. https://www.irena.org/publications/2021/March/World-Energy-Transitions-Outlook (accessed November 7, 2023).

[113] Hydrogen Decarbonization Pathways | Hydrogen Council n.d. https://hydrogencouncil.com/en/hydrogen-decarbonization-pathways/ (accessed November 7, 2023).

[114] The green hydrogen ecosystem has been launched: Hydrospider delivers first container to St. Gallen | Hydrospider AG n.d. https://hydrospider.ch/en/the-green-hydrogen-ecosystem-has-been-launched-hydrospider-delivers-first-container-to-st-gallen/ (accessed November 7, 2023).

[115] Sarawak Launches South East Asia's First Integrated Hydrogen Production Plant and Refueling Station, Unveils Hydrogen Buses - Sarawak Energy n.d. https://www.sarawakenergy.com/media-info/media-releases/2019/sarawak-launches-south-east-asias-first-integrated-hydrogen-production-plant-and-refueling-station-unveils-hydrogen-buses (accessed November 7, 2023).

[116] The green hydrogen revolution: hydropower's transformative role n.d. https://www.hydropower.org/publications/the-green-hydrogen-revolution-hydropowers-transformative-role (accessed November 7, 2023).

[117] Hydrogen n.d. https://www.hydro.com.au/clean-energy/hydrogen (accessed November 7, 2023).

[118] About | Nel Hydrogen n.d. https://nelhydrogen.com/about/#timeline (accessed November 7, 2023).

[119] Gigastack phase 2 | Ørsted n.d. https://orsted.co.uk/media/newsroom/news/2020/02/gigastack-phase-2 (accessed November 7, 2023).

[120] Green hydrogen is an environmentally friendly energy carrier - The National Power Company of Iceland n.d. https://www.landsvirkjun.com/news/green-hydrogen-is-an-environmentally-friendly-energy-carrier (accessed November 7, 2023).

[121] Shell unveils world's largest offshore wind plan to power green hydrogen | Recharge n.d. https://www.rechargenews.com/wind/shell-unveils-worlds-largest-offshore-wind-plan-to-power-green-hydrogen/2-1-763610 (accessed November 7, 2023).

[122] Japan's ENEOS, Sumitomo, Malaysian SEDC Energy mull green hydrogen supply network | S&P Global Commodity Insights n.d.

https://www.spglobal.com/commodityinsights/en/market-insights/latest-news/electric-power/102320-japans-eneos-sumitomo-malaysian-sedc-energy-mull-green-hydrogen-supply-network (accessed November 7, 2023).
[123] RH2C - Home n.d. https://www.rh2ca.com/ (accessed November 7, 2023).
[124] Hydro-Québec to operate one of the world's most powerful electrolyzers to produce ... | Hydro-Québec n.d. http://news.hydroquebec.com/en/press-releases/1667/hydro-quebec-to-operate-one-of-the-worlds-most-po/ (accessed November 7, 2023).
[125] Hydro-Québec's Strategic Plan 2022–2026 n.d. https://www.hydroquebec.com/about/publications-reports/strategic-plan.html (accessed November 7, 2023).
[126] World's largest green-hydrogen plant inaugurated in Canada by Air Liquide | Recharge n.d. https://www.rechargenews.com/transition/worlds-largest-green-hydrogen-plant-inaugurated-in-canada-by-air-liquide/2-1-952085 (accessed November 7, 2023).
[127] Hydro-Québec's Strategic Plan 2022–2026 n.d. https://www.hydroquebec.com/about/publications-reports/strategic-plan.html (accessed November 7, 2023).
[128] Hydropower: A Cost-Effective Source of Energy for Hydrogen Production n.d. https://www.powermag.com/hydropower-a-cost-effective-source-of-energy-for-hydrogen-production/ (accessed January 25, 2023).
[129] Largest Hydrogen Electrolyser Plant in the World (135/167 MW) | DeepResource n.d. https://deepresource.wordpress.com/2019/01/14/largest-hydrogen-electrolyser-plant-in-the-world-135-167-mw/ (accessed January 25, 2023).
[130] Andrus SR, Diffely RJ, Alford TL. Theoretical analysis of green hydrogen from hydropower: A case study of the Northwest Columbia River system. Int J Hydrogen Energy 2022. https://doi.org/10.1016/J.IJHYDENE.2022.11.027.
[131] Tidal Barrage and Tidal Barrage Energy Devices n.d. https://www.alternative-energy-tutorials.com/tidal-energy/tidal-barrage.html (accessed November 8, 2023).
[132] El Tawil T, Charpentier JF, Benbouzid M. Tidal energy site characterization for marine turbine optimal installation: Case of the Ouessant Island in France. International Journal of Marine Energy 2017;18:57–64. https://doi.org/10.1016/j.ijome.2017.03.004.
[133] What is Tidal Energy? Advantages, Disadvantages, and Future Trends | Earth.Org n.d. https://earth.org/what-is-tidal-energy/ (accessed January 27, 2023).
[134] H2Tec to upgrade EMEC's Orkney tidal energy hydrogen plant. Fuel Cells Bulletin 2020;2020:10–1. https://doi.org/10.1016/S1464-2859(20)30300-X.
[135] World Nuclear Industry Status Report n.d. https://www.worldnuclearreport.org/ (accessed November 10, 2023).
[136] El-Emam RS, Ozcan H, Zamfirescu C. Updates on promising thermochemical cycles for clean hydrogen production using nuclear energy. J Clean Prod 2020;262:121424. https://doi.org/10.1016/J.JCLEPRO.2020.121424.
[137] Hydrogen from Nuclear Reactors n.d. http://large.stanford.edu/courses/2010/ph240/chen2/ (accessed November 10, 2023).
[138] El-Emam RS, Özcan H. Comprehensive review on the techno-economics of sustainable large-scale clean hydrogen production. J Clean Prod 2019;220:593–609. https://doi.org/10.1016/J.JCLEPRO.2019.01.309.
[139] Hosseini SE. Fundamentals of Hydrogen Production and Utilization in Fuel Cell Systems. Elsevier; 2023. https://doi.org/10.1016/C2020-0-03183-X.
[140] Hydrogen Production and Uses - World Nuclear Association n.d. https://world-nuclear.org/information-library/energy-and-the-environment/hydrogen-production-and-uses.aspx (accessed November 11, 2023).
[141] Technology readiness level - Wikipedia n.d. https://en.wikipedia.org/wiki/Technology_readiness_level (accessed November 11, 2023).
[142] Hosseini SE. Retrieving nuclear power plants by producing hydrogen. Future Energy 2022;1:1–2. https://doi.org/10.55670/fpll.fuen.1.2.1.
[143] Nuclear Energy Agency (NEA) - The role of nuclear energy in clean hydrogen production n.d. https://www.oecd-nea.org/jcms/pl_73201/the-role-of-nuclear-energy-in-clean-hydrogen-production (accessed November 11, 2023).

[144] Prinzhofer A, Tahara Cissé CS, Diallo AB. Discovery of a large accumulation of natural hydrogen in Bourakebougou (Mali). Int J Hydrogen Energy 2018;43:19315–26. https://doi.org/10.1016/J.IJHYDENE.2018.08.193.

[145] Hand E. HIDDEN HYDROGEN. Science (1979) 2023;379:630–6. https://doi.org/10.1126/SCIENCE.ADH1477.

[146] Vacquand C, Deville E, Beaumont V, Guyot F, Sissmann O, Pillot D, et al. Reduced gas seepages in ophiolitic complexes: Evidences for multiple origins of the H2-CH4-N2 gas mixtures. Geochim Cosmochim Acta 2018;223:437–61. https://doi.org/10.1016/J.GCA.2017.12.018.

[147] Knez D, Ahmad O, Zamani M. Up-to-Date Status of Geoscience in the Field of Natural Hydrogen with Consideration of Petroleum Issues. Energies 2023, Vol 16, Page 6580 2023;16:6580. https://doi.org/10.3390/EN16186580.

[148] VOVK IF. Radiolysis of underground waters as the mechanism of geochemical transformation of the energy of radioactive decay in sedimentary rocks. Lithology and Mineral Resources 1982;16:328–34.

[149] Lin LH, Hall J, Lippmann-Pipke J, Ward JA, Lollar BS, DeFlaun M, et al. Radiolytic H2 in continental crust: Nuclear power for deep subsurface microbial communities. Geochemistry, Geophysics, Geosystems 2005;6. https://doi.org/10.1029/2004GC000907.

[150] Lin LH, Slater GF, Sherwood Lollar B, Lacrampe-Couloume G, Onstott TC. The yield and isotopic composition of radiolytic H2, a potential energy source for the deep subsurface biosphere. Geochim Cosmochim Acta 2005;69:893–903. https://doi.org/10.1016/J.GCA.2004.07.032.

[151] Suzuki N, Saito H, Hoshino T. Hydrogen gas of organic origin in shales and metapelites. Int J Coal Geol 2017;173:227–36. https://doi.org/10.1016/J.COAL.2017.02.014.

[152] Lin HT, Cowen JP, Olson EJ, Lilley MD, Jungbluth SP, Wilson ST, et al. Dissolved hydrogen and methane in the oceanic basaltic biosphere. Earth Planet Sci Lett 2014;405:62–73. https://doi.org/10.1016/J.EPSL.2014.07.037.

[153] Nivin VA. Diffusively disseminated hydrogen-hydrocarbon gases in rocks of nepheline syenite complexes. Geochemistry International 2009;47:672–91. https://doi.org/10.1134/S0016702909070039/METRICS.

[154] Jr. JHM, Cunningham KI, Roberts AA, Dietrich JA. Soil gas studies around hydrogen-rich natural gas wells in northern Kansas. Open-File Report 1986. https://doi.org/10.3133/OFR86461.

[155] Rohrbach A, Ballhaus C, Golla-Schindler U, Ulmer P, Kamenetsky VS, Kuzmin D V. Metal saturation in the upper mantle. Nature 2007 449:7161 2007;449:456–8. https://doi.org/10.1038/nature06183.

[156] Smith EM, Shirey SB, Nestola F, Bullock ES, Wang J, Richardson SH, et al. Large gem diamonds from metallic liquid in Earth's deep mantle. Science (1979) 2016;354:1403–5. https://doi.org/10.1126/SCIENCE.AAL1303/SUPPL_FILE/AAL1303-SMITH-SM.PDF.

[157] Yang X, Keppler H, Li Y. Molecular hydrogen in mantle minerals. Geochem Perspect Lett 2016;2:160–8. https://doi.org/10.7185/GEOCHEMLET.1616.

[158] Coveney RM, Goebel ED, Zeller EJ, Dreschhoff GAM, Angino EE. Serpentinization and the Origin of Hydrogen Gas in Kansas. Am Assoc Pet Geol Bull 1987;71:39–48. https://doi.org/10.1306/94886D3F-1704-11D7-8645000102C1865D.

[159] Ikuta D, Ohtani E, Sano-Furukawa A, Shibazaki Y, Terasaki H, Yuan L, et al. Interstitial hydrogen atoms in face-centered cubic iron in the Earth's core. Scientific Reports 2019 9:1 2019;9:1–8. https://doi.org/10.1038/s41598-019-43601-z.

[160] Hirose T, Kawagucci S, Suzuki K. Mechanoradical H2 generation during simulated faulting: Implications for an earthquake-driven subsurface biosphere. Geophys Res Lett 2011;38:17303. https://doi.org/10.1029/2011GL048850.

[161] Ware RH, Roecken C, Wyss M. The detection and interpretation of hydrogen in fault gases. Pure and Applied Geophysics PAGEOPH 1984;122:392–402. https://doi.org/10.1007/BF00874607/METRICS.

[162] Wakita H, Nakamura Y, Kita I, Fujii N, Notsu K. Hydrogen Release: New Indicator of Fault Activity. Science (1979) 1980;210:188–90. https://doi.org/10.1126/SCIENCE.210.4466.188.

[163] Freund F, Dickinson JT, Cash M. Hydrogen in Rocks: An Energy Source for Deep Microbial Communities. Https://HomeLiebertpubCom/Ast 2004;2:83–92. https://doi.org/10.1089/153110702753621367.

[164] Su Q, Zeller E, Angino E. Inducing action of hydrogen migrating along faults on earthquakes. Acta Seismologica Sinica 1992;5:841–7. https://doi.org/10.1007/BF02651032/METRICS.

[165] Fang Z, Liu Y, Yang D, Guo L, Zhang L. Real-time hydrogen mud logging during the Wenchuan earthquake fault scientific drilling project (WFSD), holes 2 and 3 in SW China. Geosciences Journal 2018;22:453–64. https://doi.org/10.1007/S12303-017-0068-7/METRICS.

[166] Sugisaki R, Ido M, Takeda H, Isobe Y, Hayashi Y, Nakamura N, et al. Origin of Hydrogen and Carbon Dioxide in Fault Gases and Its Relation to Fault Activity. Https://DoiOrg/101086/628769 1983;91:239–58. https://doi.org/10.1086/628769.

[167] Gas That Makes a Mountain Breathe Fire Is Turning Up Around the World - The New York Times n.d. https://www.nytimes.com/2019/05/01/science/flames-chimaera-turkey-methane.html (accessed November 17, 2023).

[168] Hosgörmez H. Origin of the natural gas seep of Çirali (Chimera), Turkey: Site of the first Olympic fire. J Asian Earth Sci 2007;30:131–41. https://doi.org/10.1016/J.JSEAES.2006.08.002.

[169] Deville E, Prinzhofer A, Pillot D, Vacquand C. Natural Flows Of H2 And Associated Diagenetic Processes Of Atmospheric CO2 Capture And Sequestration: A Study In The Ophiolites Of Oman 2011.

[170] Hosgörmez H. Origin of the natural gas seep of Çirali (Chimera), Turkey: Site of the first Olympic fire. J Asian Earth Sci 2007;30:131–41. https://doi.org/10.1016/J.JSEAES.2006.08.002.

[171] Vacquand C, Deville E, Beaumont V, Guyot F, Sissmann O, Pillot D, et al. Reduced gas seepages in ophiolitic complexes: Evidences for multiple origins of the H2-CH4-N2 gas mixtures. Geochim Cosmochim Acta 2018;223:437–61. https://doi.org/10.1016/J.GCA.2017.12.018.

[172] Prinzhofer A, Moretti I, Françolin J, Pacheco C, D'Agostino A, Werly J, et al. Natural hydrogen continuous emission from sedimentary basins: The example of a Brazilian H2-emitting structure. Int J Hydrogen Energy 2019;44:5676–85. https://doi.org/10.1016/J.IJHYDENE.2019.01.119.

[173] Deville E, Prinzhofer A. The origin of N2-H2-CH4-rich natural gas seepages in ophiolitic context: A major and noble gases study of fluid seepages in New Caledonia. Chem Geol 2016;440:139–47. https://doi.org/10.1016/J.CHEMGEO.2016.06.011.

[174] Larin N, Zgonnik V, Rodina S, Deville E, Prinzhofer A, Larin VN. Natural Molecular Hydrogen Seepage Associated with Surficial, Rounded Depressions on the European Craton in Russia. Natural Resources Research 2015;24:369–83. https://doi.org/10.1007/S11053-014-9257-5/TABLES/3.

[175] Zgonnik V, Beaumont V, Deville E, Larin N, Pillot D, Farrell KM. Evidence for natural molecular hydrogen seepage associated with Carolina bays (surficial, ovoid depressions on the Atlantic Coastal Plain, Province of the USA). Prog Earth Planet Sci 2015;2:1–15. https://doi.org/10.1186/S40645-015-0062-5/FIGURES/7.

CHAPTER 4

Global developments in hydrogen storage and delivery

4.1 Introduction

As the world seeks cleaner and more efficient energy sources, hydrogen stands out for its abundance, versatility, and zero-emission potential. However, realizing the full potential of hydrogen as a viable energy carrier requires overcoming significant challenges, particularly in storage and delivery. Because hydrogen is both lightweight and gaseous, storing it poses significant challenges, with practical hydrogen storage representing a major obstacle to the widespread adoption of the hydrogen economy. Various physical and chemical methods exist for storing hydrogen, each with its own set of strengths and limitations. Hydrogen storage stands as a pivotal technology facilitating the progress of hydrogen and fuel cell technologies across various applications, spanning from stationary power and portable power to transportation. While hydrogen boasts the highest energy per mass among fuels, its low density at ambient temperature translates to low energy per unit volume. Consequently, there arises a need for the exploration and development of advanced storage techniques capable of achieving higher energy density.

Hydrogen boasts the highest energy content per unit mass at 120 MJ/kg H_2, yet its volumetric energy density remains notably low due to its extremely sparse density under standard temperature and pressure conditions. Under ideal gas conditions at standard atmospheric pressure and 25 °C, hydrogen's density is merely 0.0824 kg/m^3, whereas air density under the same circumstances stands at 1.184 kg/m^3. Consequently, hydrogen's minimal density results in an exceptionally low volumetric energy content, measuring at 0.01 MJ/L H_2 under ambient conditions and 8.5 MJ/L H_2 for liquefied hydrogen (LH2) [1]. In contrast, methane and gasoline offer volumetric energy contents of 0.04 MJ/L and 32 MJ/L, respectively. The notably low volumetric energy density of hydrogen poses a significant challenge in economically and efficiently storing hydrogen, ultimately impacting the success of the hydrogen economy. In an advanced hydrogen economy, hydrogen is anticipated to serve both stationary and onboard purposes. Storage for stationary applications is comparatively less demanding than for onboard use. For stationary applications, the

DOI: https://doi.org/10.55670/fpll.book/1

weight of the storage system, known as gravimetric hydrogen density, is less critical than the system's volume or volumetric hydrogen density. Conversely, both gravimetric and volumetric densities are crucial for onboard applications, although volumetric energy density is relatively less significant for larger vehicles. Onboard storage necessitates a sufficient amount of hydrogen to enable reasonable travel distances without the need for frequent refueling. Therefore, an ideal storage method must offer both high gravimetric energy density and high volumetric energy density alongside moderate operating conditions, minimal enthalpy change, and rapid kinetics for hydrogen storage and release. Safety, affordability, and public acceptance are also pivotal considerations. Various physical and chemical hydrogen storage techniques exist, each with distinct features and storage capacities that may prove advantageous in shaping the future hydrogen economy. A robust infrastructure for hydrogen transportation and storage is essential for establishing global value chains for hydrogen and connecting regions with low-cost production to potential users. While considering total supply costs, which encompass conversion and transport expenses, it's important to note that the production costs of hydrogen are just one element. Efficient and cost-effective hydrogen delivery and storage systems are indispensable for the widespread adoption of hydrogen as an energy carrier. Ensuring a reliable supply of hydrogen to end-users, particularly in applications like fuel cell vehicles, heavily relies on effective storage and transportation methods. The expense associated with hydrogen delivery and storage depends on several factors, including production techniques, transportation modes, delivery distances, and operational scales. Various storage methods, such as compressed gas, liquid hydrogen, metal hydride, and underground storage, come with their own capital and operational costs. Similarly, different transportation modes, including tube trailers, liquid tankers, pipelines, trucks, and rail transport, incur varying expenses. As per a National Renewable Energy Laboratory (NREL) report, underground storage emerged as the most economical option, while pipeline delivery proved to be the most cost-effective for large hydrogen quantities. For smaller amounts of hydrogen and longer delivery distances, liquid hydrogen demonstrated advantages over other storage methods [2].

4.2 Hydrogen storage

Hydrogen can be physically stored either in its gaseous form or as a liquid. Storing hydrogen as a gas usually entails high-pressure tanks, with pressure ranging from 350 to 700 bar (equivalent to 5,000 to 10,000 psi). Conversely, storing hydrogen as a liquid necessitates cryogenic temperatures due to its boiling point at one-atmosphere pressure, which is -252.8°C. Additionally, hydrogen can be stored on solid surfaces through adsorption or within solids through absorption (Figure 4.1) [3].

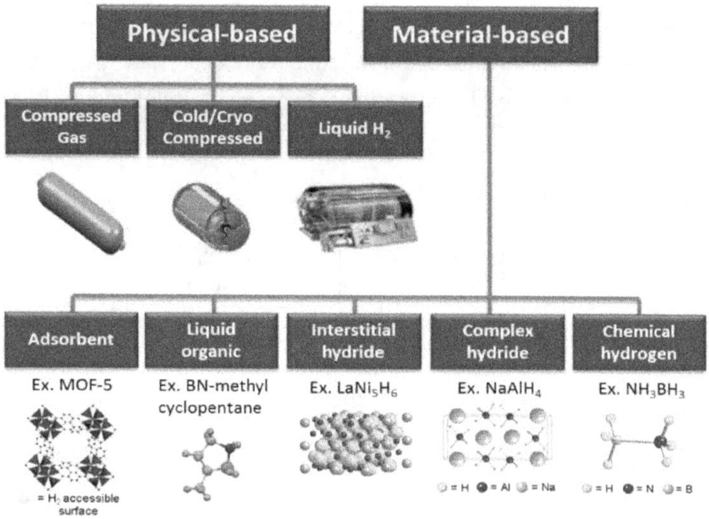

Figure 4.1 Hydrogen storage methods [3]

Hydrogen storage methods are usually categorized into chemical and physical hydrogen storage [4]. Chemical hydrogen storage involves the utilization of hydrogen compounds formed through chemical reactions or transformations, which encompass metal hydrides, coordination hydrides, and organic liquid hydrides. On the other hand, physical hydrogen storage represents a conventional approach to hydrogen storage, encompassing both liquid and high-pressure storage methods. Liquid hydrogen storage entails storing hydrogen in its liquefied state at low temperatures (around 20 K) within a tank, often employing a high-vacuum adiabatic low-pressure tank to minimize vaporization. High-pressure hydrogen storage, on the other hand, involves directly storing gaseous hydrogen at high pressures in a tank, typically requiring pressures in the range of tens of MPa to achieve high volume density for hydrogen storage [5].

Comparing the aforementioned hydrogen storage methods, chemical hydrogen storage falls short of meeting current requirements due to factors such as material costs, challenges in controlling actual reaction temperatures, the pressure needed for hydrogen absorption and release, and the weight of materials involved. Liquid hydrogen storage, despite its high-density quality, is hindered by drawbacks, including high energy consumption and the complexity of equipment, as it necessitates more than 30% consumption of liquid hydrogen for hydrogen storage energy, with the adiabatic layer further increasing tank weight. Consequently, it is not a favorable option, especially for mobile applications. Conversely, high-pressure storage emerges as a prevalent method for vehicle-mounted hydrogen storage presently due to its rapid hydrogen absorption and release rates, straightforward equipment structure, and lower energy consumption for compression compared to other hydrogen storage methods [6]. Various methods exist for storing hydrogen fuel in different forms, as detailed in Table 4.1, and more details are given in Figure 4.2. Moreover, Table 4.2 compares the available hydrogen technologies [7].

Table 4.1 Storing hydrogen in different forms

Classification	Type	Features	Comments
High-pressure storage	Compressed hydrogen	High efficiency, low energy consumption, simple equipment structure	High material cost
Liquid hydrogen storage	Liquified low-temperature hydrogen	High-quality hydrogen storage density	High energy consumption and complex equipment
Chemical hydrogen storage	Metal hybrid	High safety, high purity, and high hydrogen storage density	The high cost of hydrogen storage and the chemical reaction process is difficult.
Physical hydrogen storage	Porous coordination network	High hydrogen storage density	Ultra-low temperature, Complex equipment structure

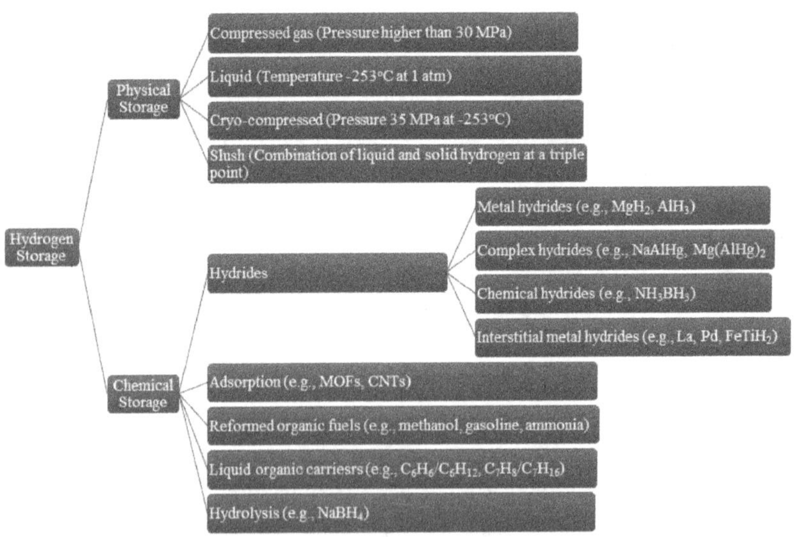

Figure 4.2 Physical and chemical hydrogen storage technologies

Table 4.2 Comparison of hydrogen storage technologies

Properties	**Compressed hydrogen**	**Metal Hydrid (MgH_2-10wt%Ni)**	**Liquid organic (C_7H_8/C_7H_{14})**	**Liquid hydrogen**	**Liquid ammonia**
Density (kg/m^3)	39 (69 MPa, 25 °C)	1,450	769 (1 atm, 20 °C)	70.9 (1 atm, −253 °C)	682 (1 atm, −33.33 °C)
Boiling point (°C)	-253	-	101	-253	-33.33
Gravimetric hydrogen density (wt%)	100	7.10	6.16	100	17.8
Volumetric hydrogen density ($kg\text{-}H_2/m^3$)	42.2		47.1	70.9	120.3
Hydrogen release temp. (°C)	-	250	200-400	-253	350-900
Regeneration temp. (°C)	-	-	100-200	-	400-600
Enthalpy change to release hydrogen (kJ/mol-H_2)	-	118	67.5	0.899	30.6

Compressed hydrogen storage techniques are anticipated to dominate the landscape by 2035, constituting approximately 67% of all hydrogen storage methods. This percentage is projected to increase to 56% by 2050. The cost of storing 1 kg of hydrogen in a tank utilizing the compressed gas method ranges between USD 400 and USD 700, as outlined in Table 4.3.

Table 4.3 The cost of various hydrogen storage methods

Storage Technologies	Compressed gas (H_2)	Cryogenic (H_2)	Cryo-compressed (H_2)	High pressure-solid	Sorbents (H_2)	Metal hydrides (H)	Complex hydrides (H)	Chemical hydrides (H)
Cost (USD/kg H_2)	400-700	200-270	400	-	-	>500	300-450	160-270

4.2.1 Compressed hydrogen gas storage

Compression represents the conventional and most straightforward method for hydrogen storage. The density of hydrogen stored in compressed form is heavily reliant on the storage pressure. Typically, at 10 MPa pressure, the volumetric density of stored hydrogen stands at 7.8 kg-H_2/m^3 (at a temperature of 20°C). This density increases to 39 kg-H_2/m^3 with pressure raised to around 70 MPa. Compressed hydrogen finds application across various sectors, including vehicle propulsion, hydrogen refueling infrastructure, and industrial processes. Achieving high density requires the use of advanced vessel materials like carbon fiber and glass fiber-reinforced plastics. However, manufacturing constraints impose limitations on vessel size. Furthermore, the elevated pressure increases the risk of hydrogen gas permeation into the vessel wall, leading to heightened concerns regarding accelerated embrittlement [8].

Compressed gas hydrogen (CGH_2) storage involves storing hydrogen gas at high pressure within a tank, enabling a greater amount of hydrogen to be stored in a relatively small space. This approach is widely employed in industrial and transportation settings. The storage tanks are engineered to endure high pressure and are often constructed from durable materials like carbon fiber or metal alloys. Upon release, the stored hydrogen can be utilized as a fuel in a range of applications, including powering fuel cell vehicles and facilitating industrial processes. Physical storage represents the most advanced technology for storing hydrogen. Currently, the primary technology for onboard automotive physical hydrogen storage involves compressed gas vessels operating at pressures of 350 and 700 bar (5,000 and 10,000 psi). Although low-pressure liquid hydrogen, stored near its normal boiling point of 20 K, is commonly utilized for bulk storage and transport of hydrogen, there is limited ongoing development focused on adapting it for onboard automotive applications. While compressed hydrogen is typically stored at temperatures close to ambient, there is ongoing research into cold (sub-ambient but above 150 K) and cryogenic (150 K and below) compressed hydrogen storage due to the potential for achieving higher hydrogen densities at lower temperatures (Figure 4.3).

During the cyclic testing of hydrogen gas, there is a notable temperature change, particularly as the temperature increases sharply during the filling phase due to hydrogen compression, conversion of kinetic energy into internal energy, and the negative Joule-Thomson effect [9]. In the process of high-pressure filling, the tank's internal pressure rises, potentially leading to thermal stress due to the increase in hydrogen temperature. Hence, it is essential to investigate both the rise in temperature and the amount of hydrogen charged to ensure hydrogen safety. The market for hydrogen pressure vessels is experiencing significant growth, with a notable compound annual growth rate (CAGR) of 18.1%, reaching a substantial value of £900 billion by 2027. This expansion is attributed to the growing focus on highly efficient hydrogen energy storage systems, alongside the escalating utilization of hydrogen energy across various sectors, including automotive, aerospace, and marine industries. Consequently, manufacturers are actively seeking innovative methods to enhance the bonding and protection of composite materials utilized in hydrogen pressure vessels. These efforts aim to improve efficiencies while simultaneously reducing maintenance costs [10].

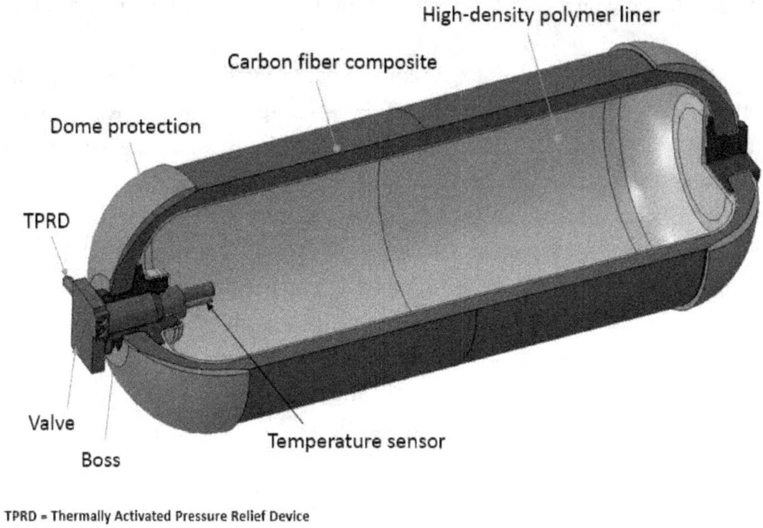

Figure 4.3 A type of compressed hydrogen tank [11]

Currently, hydrogen energy storage is more expensive than fossil fuel alternatives. Most of the hydrogen storage technologies are still in their early developmental phases. Additionally, the amount of energy that fuel cells can produce from hydrogen and subsequently utilize to fulfill the energy demands of commercial and residential buildings is relatively limited. Moreover, the market for storing hydrogen energy in liquid form entails substantial capital expenditures, mainly due to the high insulation costs required to prevent vaporization. Consequently, both the time and cost associated with charging and discharging hydrogen in these systems, as well as the operational expenses, are significant [12].

There are five categories of pressure vessels designed for hydrogen storage [13,14]:

Type I: These vessels are entirely metallic, typically constructed from aluminum or steel. They represent the most traditional method of storing hydrogen, being the least costly option but also the heaviest, weighing around 3 pounds per liter. They can endure pressures of up to 50 MPa.

Type II: These vessels are constructed from steel and feature a glass fiber composite overwrap. Although they are approximately 50% pricier than Type I vessels, they offer a significant reduction in weight (around 30% to 40% lighter) and can withstand the highest pressures.

Type III: These vessels utilize a full composite wrap with a metal liner, with the structural load primarily borne by the composite material while the liner serves sealing purposes. They are reliable up to 45 MPa but face challenges with pressures of 70 MPa. They weigh half as much as Type II vessels but come with double the cost.

Type IV: These vessels are entirely composite, typically employing high-density polyethylene as a liner and carbon fiber or carbon-glass composites for structure. While they are the lightest option, they are also the most expensive, capable of withstanding pressures of up to 100 MPa.

Type V: A linerless pressure vessel constructed entirely from composite material has been developed for pre-commercial purposes. Introduced in 2010 by Composites Technology Development Inc., the initial design boasted a weight reduction of 20% compared to Type IV vessels and had the capacity to withstand operational pressures of 1.37 MPa. However, the current operational pressure remains too low for practical hydrogen storage beyond laboratory applications [15].

4.2.2 Liquid/cryogenic hydrogen storage

Cryogenic hydrogen storage, which includes cooling hydrogen to extremely low temperatures (approximately -253°C), faces the primary challenge of maintaining this temperature. The process of liquefying hydrogen is both time and energy-intensive, resulting in energy losses of up to 40% compared to the 10% energy loss in compressed hydrogen storage [16]. Losses can occur in various stages of the process, including heat exchangers, ortho-para conversion, hydrogen gas purification, compression equipment, expansion equipment, nitrogen liquefaction, and insulation [17]. The extent of these losses also depends on the specific configuration of the process employed. An essential process involves several steps: initially, hydrogen is compressed to a pressure exceeding the critical point (13 bar) up to 20-30 bar. Subsequently, the hydrogen feed is precooled, typically utilizing liquid nitrogen or other refrigerants, to attain temperatures around -190°C. Further cooling is achieved through a cryogenic refrigeration cycle, commonly employing hydrogen or helium, followed by expansion to near atmospheric pressure within the liquid hydrogen storage tank. The liquefaction process encompasses various modifications, including differing pressure levels, refrigerant selections, utilization of expansion turbines or throttling valves, open or closed cycles, employment of cascade systems (wherein one cycle provides cooling for the subsequent cycle), and adjustment of mass flow through the expander, among others. These modifications have given rise to distinct cycles such as Linde, Linde-Hampson, Claude, Brayton, and mixed-refrigerant.

Cryogenic storage is mainly utilized for medium to large-scale storage and transportation purposes, such as truck deliveries and intercontinental hydrogen shipping. A cryogenic tanker typically has a capacity of holding 5000 kg of hydrogen, which is five times greater than the capacity of compressed hydrogen gas tube trailers. The safety of cryogenic vessels is ensured by their additional protective layer (vacuum jacket) in case of accidents, and hydrogen exhibits low energy for adiabatic expansion at cryogenic temperatures. In the event of a leak or tanker rupture, there is unlikely to be a significant explosion unless the gas is ignited. However, leaked hydrogen gas at low temperatures can cause malfunctions and damage to nearby valves or pressure relief devices that are not designed to withstand such conditions. An incident in a cryogenic hydrogen lab in 2016 serves as an example, where a pressure relief valve failed to function due to exposure to cryogenic temperatures in an area that was not rated for such conditions [18].

Liquid hydrogen is stored within specialized cryogenic tanks that are meticulously insulated and pressurized, featuring mechanisms for cooling, heating, and venting. These tanks are designed with vacuum insulation to minimize evaporation losses and incorporate redundant pressure relief devices as a safety measure against over-pressurization. Operational pressures typically range up to 850 kPa (~123 psi), and these tanks adhere to established regulations such as those outlined by the United States Department of Transportation and Transport Canada. While cylindrical tanks are common, spherical tanks may be employed for larger volumes (Figure 4.4) [19].

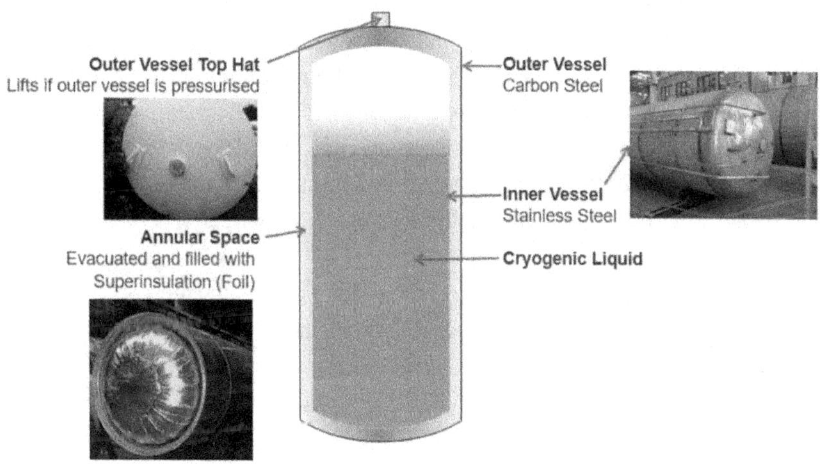

Figure 4.4 Typical cryogenic tank [19]

These tanks are outfitted with top-tier insulation, employing vacuum insulation comprising dual walls with a high-vacuum layer in between. This configuration offers exceptional insulation properties, while the double-wall design serves as a dual-containment system. Vacuum insulation is also utilized for the transfer lines connected to these tanks [20]. Primary applications for cryogenic hydrogen storage systems encompass refueling stations, aerospace facilities, and automotive uses.

Liquid hydrogen finds extensive application in space flight, serving as fuel for space shuttles, as well as in fuel cell electric vehicles, including cars, buses, and trucks. International standards and certifications governing hydrogen storage systems include compliance with the ASME Boiler and Pressure Vessel Code section VIII and specifications and regulations set forth by the US Department of Transportation (DOT).

The energy required for liquefying hydrogen ranges from approximately 12.5 to 15.0 kWh/kg, whereas compression to 70 MPa consumes about 6.0 kWh/kg [21]. Cryogenic hydrogen storage vessels contribute significantly to CO_2 emissions, with 98% of total emissions attributed to the storage process due to the higher energy demand to maintain hydrogen in its liquid state at lower temperatures. Moreover, these vessels necessitate additional energy to sustain operational conditions, with an energy consumption of 724 kWh/kgH_2. The liquefaction of hydrogen is notably energy-intensive, with a specific energy consumption of 10.85 kWh/kgLH_2 for a daily production capacity of 250 tons of liquid hydrogen.

Several strategies have been identified to enhance the efficiency of the hydrogen liquefaction cycle [22]:

- Harnessing power from the shafts of the hydrogen turbines (reducing exergy loss by 0.65 kWh/kg).
- Implementing an optimized mixed-refrigerant cycle with a pre-cooling temperature of 80 K (altering the chemical composition of the refrigerant and adjusting high- and low-pressure levels).
- Substituting the throttling valve with an expander (decreasing the enthalpy of the stream rather than utilizing isenthalpic expansion).
- Broadening the temperature range of the mixed-refrigerant cycle by lowering the pre-cooling temperature.
- Minimizing pressure drop in the ejector that regulates the pre-cooled hydrogen.
- Utilizing high-temperature chillers to complement ambient cooling.
- Recapturing waste heat from the intercoolers.
- Employing a combination of helium, neon, and hydrogen in the refrigerant cycle, resulting in a higher molecular mass conducive to employing centrifugal compressors (instead of reciprocating piston compressors), which offer superior economies of scale.

Two additional factors can significantly impact the specific energy consumption of the liquefaction process: plant size and partial load. It is anticipated that enlarging the plant capacity from the current typical scale (5-10 t/d) to a larger scale of 150-200 t/d will result in a reduction of energy consumption by 30-38%. Efficiency at a larger scale follows an asymptotic trend, where further increases in size yield only marginal efficiency gains. Valenti and Macchi [23] proposed an even larger plant with a capacity of 860 t/d, achieving a specific energy consumption of 5.29 kWh/kg. Regarding partial load, the liquefaction plant can operate down to 25% of its design load. Liquefaction additionally involves a catalytic conversion phase. Hydrogen, being a diatomic molecule, can exist in two forms based on the spin of its nucleus: ortho and para. The proportion between them is primarily determined by temperature. At room temperature, the equilibrium concentration is approximately 75% ortho and 25% para, whereas at the boiling point, it becomes 0.2% ortho and 99.8% para. The conversion from ortho to para releases heat. Therefore, if this conversion occurs after the hydrogen has turned into liquid, it can cause some of the liquid hydrogen to boil, undoing the process for which a considerable amount of energy has been invested. The heat generated during the ortho-para conversion is 0.388 kcal/mol, while the latent heat of vaporization is 0.213 kcal/mol. Consequently, this conversion should ideally be encouraged at higher temperatures. However, since this reaction is slow, catalysts are necessary. Modern liquefiers typically incorporate a continuous ortho-para conversion below the precooling step, usually below approximately -190°C [17].

Ensuring the cost-effectiveness of maintaining a leak-proof cryogenic hydrogen storage system is crucial to prevent potential issues, given that a leak from such a tank can diminish its capacity and pose an explosion risk. The expenses associated with converting gas into liquid form and the energy expenditure for active cooling are also considerable considerations. In the realm of cryogenics, the process of transforming gas into liquid form can consume up to 30% of the energy content inherent in the stored hydrogen. Furthermore, the intricacy and expense involved in maintaining the requisite low temperatures for storing liquid hydrogen have thus far constrained its widespread adoption. Conversely, high-pressure gas storage remains a preferred option for routine hydrogen storage needs, particularly in mobile applications, owing to its reduced volume, rapid delivery capabilities, and simplified operational aspects [24,25].

Several nations utilize liquid/cryogenic hydrogen storage technology. Countries such as the United States, Germany, and the Netherlands actively engage in advancing and implementing this technology. Recent pioneers in this field include entities such as H2FLY and Airbus, which spearheaded the utilization of liquid hydrogen for emissions-free aviation and superconducting power systems [26,27]. Furthermore, NASA remains at the forefront of innovative liquid hydrogen storage solutions to support space exploration endeavors, continually conducting research and development in this domain. The emergence of cryogenic fuels, notably liquid hydrogen, is increasingly recognized as a groundbreaking force within the energy sector [28]. Storage of liquid hydrogen has already been implemented extensively, with notable examples including the largest sphere currently in operation, utilized by NASA, boasting a volume of 3,200 m^3 (227 tH_2). The spherical shape of storage is used to decrease the surface area and heat transfer. Construction of a new spherical tank with a capacity of 4,700 m^3 (334 tH_2) commenced in 2019 and operated in 2022. Kawasaki Heavy Industries, in December 2020, finalized the basic design for an 11,200 m^3 spherical tank with a working capacity of 10,000 m^3, featuring a boil-off ratio of less than 0.1% per day. Additionally, the company has pioneered advanced ship design, envisioning vessels with four tanks, each with a capacity of 40,000 m^3 [29,30]. This marks a necessary fourfold scale-up from current designs. Consequently, akin to the challenges facing liquefaction, storage infrastructure would require considerable expansion to meet the demands of global trade, with commercial-scale projects anticipated to utilize tanks of 50,000 m^3. Presently, there are no ships engaged in the transportation or utilization of liquid hydrogen, with such endeavors remaining confined to demonstration projects, indicating that large-scale implementation in this domain is approximately a decade away. Hydrogen regasification, a relatively straightforward process, involves heating liquid hydrogen using seawater or air. While materials capable of withstanding cryogenic conditions are necessary, this aspect is not anticipated to be a focal point for significant research and development efforts [31].

To meet the demands of global trade, liquefaction plant capacities must undergo substantial increases from their current levels. This presents an opportunity for potential cost reductions by simply expanding the capacity of a single train. However, such scaling is constrained by equipment size limitations. There comes a point where the equipment becomes too large to manufacture or transport efficiently, considering factors like road width or bridge height. For instance, in LNG liquefaction, the size of a single train is restricted by the main cryogenic heat exchanger, which achieves the lowest temperatures in the process. A similar practical limitation is anticipated for hydrogen liquefaction. The maximum feasible dimensions for the coldbox are constrained by the size that can be transported by road or ship, which typically range from 4 to 5.5 meters in diameter and 20 to 40 meters in length [32]. To overcome these constraints and enable larger sizes, one approach is to utilize transport by ship rather than road. Another option is to deliver pre-packaged transportable coldboxes or systems for assembly on-site, similar to practices seen in air separation units. However, this approach involves additional on-site work, such as pressure testing, compared to completing the work in the workshop. Notably, a single cryogenic column, a key component of the plant, can weigh up to 170 tons and measure up to 7 meters in diameter and 45 meters in length, while preassembled coldboxes may be sized at 5 by 7 meters and 40 meters in length. Finally, situating workshops near the harbor or construction site allows for some welding and final manufacturing to occur on-site.

A plant exceeding 1,000 tons per day could introduce opportunities for implementing innovative cryogenic refrigeration technologies utilizing centrifugal compressors, potentially leading to reduced costs through enhanced economies of scale [33]. Larger-capacity designs are inclined towards employing a mixed-refrigerant Joule-Thomson Cycle for pre-cooling and a dual-cycle system involving hydrogen and neon for cryogenic refrigeration [34]. As the capacity surpasses 200 tons per day for a single train, the coldbox would approach its maximum practical dimensions for transportation. Similarly, beyond the 500 tons per day mark, the electric motor drives of the compressors would reach their current maximum sizes. However, these limitations could be circumvented by utilizing equipment in parallel configurations.

4.2.3 Cryo-compressed hydrogen storage

Cryo-compressed hydrogen (CcH_2) storage, initially introduced by Aceves et al. [35], involves compressing supercritical cryogenic gas at approximately 233°C rather than liquefying it. This approach shows promise in terms of storage density and safety, offering an 80 g/L storage density, which is 10 g/L higher than traditional cryogenic storage, alongside rapid and efficient refueling and enhanced safety due to the vacuum enclosure [36]. Ahluwalia et al. [37] conducted a comprehensive technical assessment of this storage method and determined its potential to fulfill the Department of Energy's targets for both gravimetric and volumetric capacity, as well as hydrogen loss during dormancy. Given its potential for widespread application, it is crucial to include it in ongoing research and development endeavors. Nevertheless, challenges such as infrastructure availability and cost persist.

CcH_2 storage involves storing hydrogen at cryogenic temperatures in a pressurized vessel, typically ranging from 250 to 350 atm. This approach provides a higher storage density compared to LH_2 while retaining the advantages of a gaseous state. Commercialization efforts are underway for various applications, including trucks, commercial vehicles, small aircraft, and small ships, aiming to enable long-range transportation with the support of refueling stations. CcH_2 storage systems consist of insulated pressure vessels that can be filled through cryo-compression of LH_2 or cryo-cooling of hydrogen gas. This technology eliminates boil-off concerns and enables refueling irrespective of the tank's temperature [38]. Technical evaluations and research have focused on CcH_2 storage's performance, safety, and cost-effectiveness for vehicular applications. It is anticipated that this technology could meet targets for system gravimetric and volumetric capacity, durability, and operability. Safety assessments have highlighted its high safety level due to the low adiabatic expansion energy of hydrogen at cryogenic temperatures. The design of a cryo-compressed storage tank resembles that of a liquid hydrogen tank, comprising an inner vessel, vacuum super-insulation, and a lightweight outer vessel. The cryo-compressed storage system possesses both inherent and additional safety attributes. Inherent safety features include those offered by the vacuum enclosure, such as protection against external mechanical impacts, intrinsic leakage detection, and controlled hydrogen discharge via safety lines in case of leakage. Moreover, the low adiabatic expansion energy of hydrogen at cryogenic temperatures enhances safety levels in the event of system failure. Overall, cryo-compressed hydrogen storage offers a promising solution for high-density automotive hydrogen storage, featuring enhanced safety and potential cost-effectiveness. Figure 4.5 compares the adiabatic expansion energies of various hydrogen storage systems. The data illustrates that the temperature of the hydrogen being stored notably influences the energy released in the event of sudden vessel failure. Therefore, the low adiabatic expansion energy exhibited by hydrogen at cryogenic temperatures suggests a potentially elevated level of safety [39].

CcH_2 tanks provide a combined approach, bridging the gap between LH_2 and CGH_2 storage methods. By employing low temperatures (for instance, 40-80K or -233°C to -193°C) and moderate pressures (such as 350 bar), BMW has successfully overcome the boil-off challenges associated with LH_2- which begins to boil at temperatures above -253°C- and realized storage densities significantly exceeding those of both CGH_2 and LH_2 (Figure 4.6) [25].

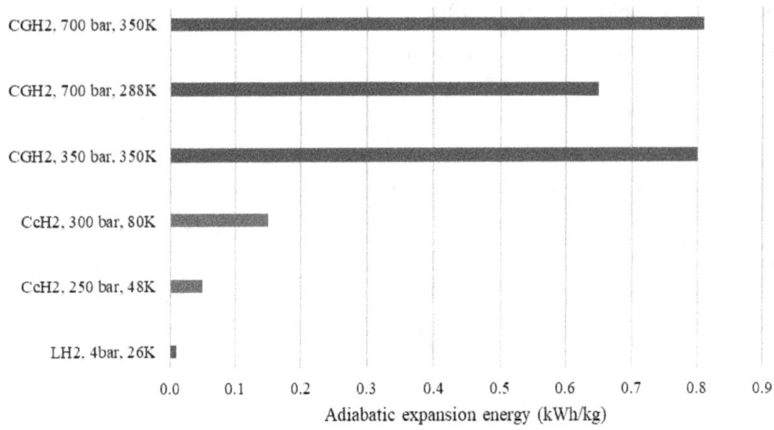

Figure 4.5 Comparison of adiabatic expansion energies per kilogram of hydrogen stored across various hydrogen storage methods

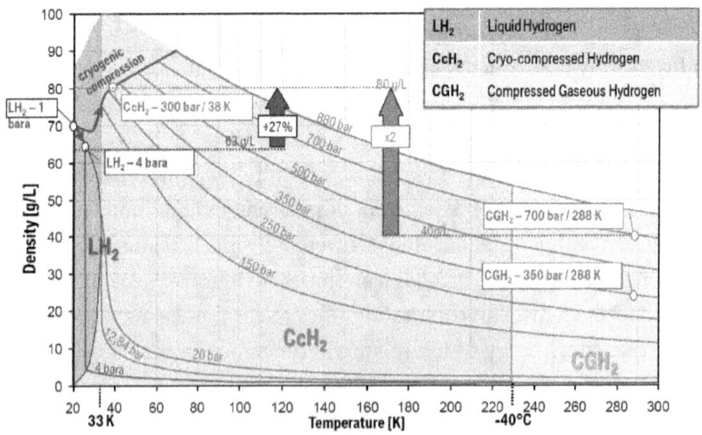

Figure 4.6 BMW cryo-compressed hydrogen is denser than liquid hydrogen [25]

It was found that when LH_2 is compressed into a cryogenic gas at 30 or 40 megapascals (300 or 400 bar), its density can essentially be increased. The initial hypothesis was proven by constructing a pump with Linde, demonstrating the production of high-pressure CcH_2 gas at 30 megapascals with a density of 80 grams/liter, compared to between 65 and 70 grams/liter for LH_2. In other words, CcH_2 offers a higher density than LH_2 while maintaining the advantages of CGH_2 [40].

Recently, ZeroAvia and Verne have joined forces to investigate the utilization of CcH_2 in airport and aircraft settings. Together, they will examine the advantages of expanding CcH_2 storage and refueling facilities at airports as hydrogen-electric propulsion systems grow to accommodate larger aircraft. This partnership seeks to analyze the feasibility of integrating CcH_2 into aircraft and establish a blueprint for selecting initial airport sites in California. This collaboration holds importance in driving forward zero-emission aviation and promoting hydrogen as a vital solution for addressing the aviation industry's environmental footprint [41]. Moreover, H2Fly, a German company, and Universal Hydrogen, a U.S.-based firm, have teamed up to create a powertrain for regional aircraft based on CcH_2. Their collaboration seeks to showcase the feasibility of using CcH_2 as a reliable and effective energy solution for aviation, emphasizing the reduction of emissions and operational expenses [42]. Airbus, a prominent aircraft manufacturer, has been collaborating with Air Liquide, a French multinational company, to investigate the potential of CcH_2 for future hydrogen-powered aircraft. This partnership involves research and development initiatives aimed at enhancing the storage and distribution of cryogenic hydrogen for aviation purposes, with the ultimate objective of enabling long-distance emissions-free flights [43]. In addition, HyPoint, a company based in the United States, has forged a partnership with ZeroAvia, a British/American hydrogen-electric aircraft developer, to integrate CcH_2 fuel cells into regional aircraft. This joint effort aims to capitalize on the high energy density and rapid refueling capabilities offered by CcH_2, facilitating efficient and environmentally sustainable air travel while addressing the ecological concerns associated with conventional aviation fuels [44].

Cryomotive has developed prototype tanks for use in trucks that, upon reaching the market, will measure between 600 to 700 millimeters across and 2,350 to 2,650 millimeters long, capable of storing 75 to 115 kilograms of CcH_2 gas. These will be available in configurations ranging from two to four tanks. Designed with tanks consisting of an aluminum liner wrapped in a carbon fiber/epoxy composite material (Type III), all housed within an aluminum external shell. This shell is distanced from the inner tank by vacuum-sealed multi-layer insulation (MLI), which uses several layers of aluminum foil and glass fiber fleece to block heat transfer via radiation. The inner tank is held in place by nonconductive composite materials that act as suspension/supports (Figure 4.7). Cryomotive aims to roll out its first commercial offerings in heavy-duty trucks by 2025, with plans to ramp up production to hundreds of CcH_2 tanks by 2026 and thousands by 2027.

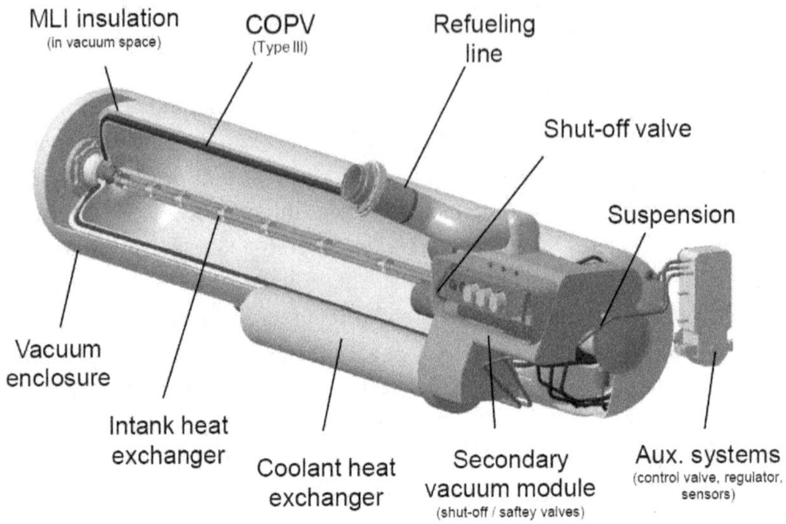

Figure 4.7 The structure of a cryo-compressed hydrogen storage tank [45]

The company has engineered advanced cryogenic hydrogen gas storage systems equipped with rapid refueling capabilities tailored for long-distance hydrogen trucks, with the objective of attaining a range of 1,000 kilometers per tank fill and a refueling duration of approximately 10 minutes [46]. Various advancements and initiatives are anticipated to drive the scaling up of hydrogen storage in the future. Opportunities highlighted by the International Energy Agency (IEA) include leveraging industrial ports as central hubs for expanding the utilization of clean hydrogen and integrating hydrogen into long-term energy strategies. Additionally, research conducted by the U.S. Department of Energy's NREL has pinpointed crucial opportunities for hydrogen within an integrated energy system, projecting a potential increase in U.S. hydrogen demand to 22-41 million metric tons annually by 2050 [47,48]. Furthermore, progress in technologies such as electrolysis, along with the advancement of storage solutions like metal-organic frameworks and chemical carriers, contributes to the scaling up of hydrogen storage capabilities. These collective endeavors, coupled with ongoing research, development, and deployment initiatives, are poised to play a pivotal role in unlocking the full potential of hydrogen storage on a broader scale in the years ahead.

4.3 Hydrogen delivery

Hydrogen, being the smallest and lightest element on Earth, contrasts notably with the mass of other substances like gasoline. For instance, while a gallon of gasoline weighs approximately 2.75 kilograms, a gallon of hydrogen only weighs 0.00075 kilograms (at 1 atm pressure and 0°C). To transport significant quantities of hydrogen, it needs to be either pressurized and conveyed as a compressed gas or liquefied. The location of hydrogen production can significantly influence both the cost and the optimal delivery method. For instance, a large-scale hydrogen production facility situated centrally can yield hydrogen at a lower cost due to higher production volumes. However, transportation expenses are higher as the point of use is farther away. Conversely, decentralized production facilities generate hydrogen on-site, resulting in relatively lower delivery costs. Nonetheless, the production costs are likely to be higher due to lower production volumes [49]. The transportation of hydrogen is crucial in determining the expenses, energy usage, and environmental impacts associated with hydrogen routes. In the case of centralized hydrogen production, delivering it to end users encompasses two phases: transmission from production facilities to urban entry points and distribution from these entry points to refueling stations or consumers. Different methods of hydrogen delivery, including gaseous, liquid, and material-based carriers, are selected considering various factors such as the demographic targeted, consumer habits, population density, size of refueling stations, and the adoption rate of fuel cell vehicles and hydrogen-utilizing units in the market.

The challenges associated with hydrogen delivery encompass cost reduction, enhanced energy efficiency, preservation of hydrogen purity, and leakage mitigation. Ongoing research is concentrated on developing more cost-effective compression technology, novel materials for pipelines, and liquefaction processes that are more energy-efficient. To address the hurdles of long-distance transportation, some advocate for distributed production and efficient delivery through localized ecosystems and the utilization of high-pressure bulk hydrogen transport trailers. These endeavors are aimed at cutting costs, enhancing reliability, and minimizing emissions, ultimately hastening the widespread adoption of hydrogen as a clean energy source.

Presently, hydrogen is conveyed from its production site to its destination through various means, including pipelines, cryogenic liquid tanker trucks, and gaseous tube trailers. Pipelines are typically installed in regions with significant and stable demand, often amounting to hundreds of tons per day, projected to endure over the long term. Conversely, liquefaction facilities, liquid tanker trucks, and tube trailers are deployed in areas with lower or emerging demand. Additionally, there are ongoing demonstrations of hydrogen transportation via chemical carriers, such as barges, particularly in large-scale applications like export markets. Upon reaching the point of use, additional infrastructure components are typically installed, including compression units, storage facilities, dispensers, meters, and technologies for detecting and purifying contaminants. For instance, stations designed to dispense hydrogen for medium- and heavy-duty fuel cell vehicles are anticipated to compress hydrogen to pressures ranging from 350 to 700 bar and dispense at rates of up to 10 kg per minute. Technologies capable of achieving these performance benchmarks are currently in the developmental stage.

4.3.1 Gaseous hydrogen delivery

Compressed hydrogen gas can be conveyed either via gas pipelines or in tube trailers equipped with compressed hydrogen pressure vessels. Gaseous hydrogen transportation via pipelines mirrors the current infrastructure used for natural gas distribution. Pipelines represent the most efficient and economical means of hydrogen transportation for distances spanning between 2,500 and 3,000 kilometers, especially suited for capacities of around 200,000 tons per year. The United States currently boasts approximately 2,600 kilometers of hydrogen pipelines in operation, while Europe maintains around 2,000 kilometers, predominantly under the ownership of private entities and serving to interconnect industrial consumers. Various nations are formulating strategies for enhancing hydrogen infrastructure, with Europe taking a leading role. The European Hydrogen Backbone initiative, formed in 2020, unites 32 gas infrastructure operators with the objective of establishing an extensive hydrogen network across Europe. In June 2022, the Dutch government unveiled a EUR 750 million investment plan aimed at developing a national hydrogen transmission grid spanning 1,400 kilometers. To align with the Net Zero Emissions (NZE) Scenario, it is estimated that approximately 15,000 kilometers of hydrogen pipelines (including both new installations and repurposed pipelines) will be required by 2030 [50]. The hydrogen infrastructure could expand into a comprehensive pan-European network, spanning nearly 53,000 kilometers by 2040, primarily utilizing repurposed existing natural gas infrastructure [51].

In the United States, the pipelines are typically owned by merchant hydrogen producers. They are strategically positioned in regions with a high concentration of major hydrogen consumers, such as petroleum refineries and chemical plants, notably in the Gulf Coast region. Utilizing existing pipelines for gaseous hydrogen transport presents an economically feasible option for conveying large quantities of hydrogen. However, the significant upfront investment required for constructing new pipelines poses a substantial obstacle to expanding hydrogen pipeline infrastructure. Consequently, ongoing research endeavors aim to address technical challenges associated with pipeline transmission, including mitigating the potential for hydrogen-induced embrittlement of pipeline materials and welds, developing effective strategies for controlling hydrogen permeation and minimizing leaks, and advancing the development of more affordable, reliable, and durable hydrogen compression technologies [52]. One potential solution involves hydrogen distribution using fiber-reinforced polymer (FRP) pipelines. The setup expenses for FRP pipelines are approximately 20% lower compared to steel pipelines, primarily because FRP sections can be obtained in longer lengths than steel, thereby reducing the need for welding [53].

The primary cost elements of the gaseous hydrogen pipeline delivery include the pipeline infrastructure, compressors, energy consumption associated with compression, and expenses related to repurposing or replacing components like seals and meters, where applicable. The capital investment required for a new pipeline is contingent upon its diameter and operational pressure. Scaling up the diameter results in a nonlinear escalation in steel usage, which is the primary determinant of costs, as well as a nonlinear increase in capacity. Consequently, it is generally more advantageous to enlarge the pipeline and configure it to accommodate future capacity requirements, leading to the deployment of a larger pipeline instead of multiple smaller ones. Cost estimates for hydrogen pipelines exhibit a considerable range, influenced by three principal factors. Firstly, the designated pressure level, higher pressures necessitate thicker materials, resulting in greater steel consumption and consequently higher costs. However, the benefit of higher pressure, namely increased energy transport capacity for a given pipeline, creates a trade-off between capital expenditure and energy transported. Secondly, the scope of costs varies, ranging from material expenses only to encompassing laying procedures, labor, right-of-way acquisition, project management, and contingency funds. Thirdly, the implicit steel price utilized also plays a significant role. Over the past decade, the average ratio between minimum and maximum steel prices has been approximately four. Given that steel constitutes the primary material cost of the pipeline, the timing of the estimate can impact the outcome. Considering these factors, a new hydrogen pipeline may incur costs 10-50% higher than those of a new natural gas pipeline [54]. Regarding offshore pipelines, the investment outlay surpasses that of onshore counterparts, with the cost differential contingent upon factors such as water depth and the specific pipeline diameter. Niermann et al. [55] suggested a cost disparity of 57%, while the Hydrogen Council [56] projected a range of 1.3-2.3 times higher. Meanwhile, d'Amore-Domenech et al. [57] proposed a factor of 2.3, and Miao et al. [58] indicated a factor approaching threefold.

An alternative and the most cost-effective strategy for rapidly expanding the hydrogen delivery infrastructure entails repurposing a segment of the existing natural gas distribution network to accommodate hydrogen. Adapting natural gas pipelines to transport a blend of natural gas and hydrogen (up to around 15% hydrogen) may necessitate only minor adjustments to the pipeline configuration. However, converting current natural gas pipelines to convey pure hydrogen may entail more extensive modifications. When repurposing natural gas pipelines, several factors must be taken into account, including material compatibility, compression requirements, and pipeline capacity. Material suitability is particularly crucial, with a significant challenge being hydrogen embrittlement. This phenomenon occurs when atomic hydrogen permeates the steel structure, diminishing its ductility, toughness, and tensile strength, thereby increasing the likelihood of pipeline failure. The susceptibility to hydrogen embrittlement varies depending on the specific steel composition, and factors such as existing cracks, cyclic loading (e.g., pressure fluctuations), and sulfur content can exacerbate this issue. Therefore, determining the suitability of a pipeline for repurposing necessitates a thorough, case-by-case assessment. Strategies to address the heightened vulnerability introduced by cracks include enhancing pipeline monitoring systems, albeit at potentially higher operational costs. Other repurposing methods include internal coating, the use of inhibitors, or employing a secondary pipe within the pipeline [59]. To manage hydrogen embrittlement, continuous monitoring to detect and track crack growth, along with minimizing operations involving pressure fluctuations, are viable options.

Hydrogen emissions can take place in three primary ways: permeation through steel, leakage, and operational emissions. Permeation rates, albeit minimal compared to other methods, exist. Hydrogen tends to leak more readily than methane due to its smaller molecular size, necessitating the replacement of components such as pressure regulators, valves, and seals when repurposing natural gas pipelines for hydrogen. Additionally, different gas meters are required to accommodate hydrogen's distinct gas properties, along with adjustments to pipeline operation methods, including pressure fluctuation control. Depending on the required hydrogen purity, a comprehensive purging and cleaning procedure is essential to eliminate any remaining impurities within the pipeline [60]. Although these requirements contribute a relatively small fraction to the total costs, if anticipated sufficiently in advance, these components could be gradually replaced with hydrogen-ready devices as part of routine maintenance schedules. In terms of energy content, hydrogen's volumetric energy content (MJ/m^3) is approximately one-third that of natural gas. Consequently, transporting the same energy amount necessitates approximately three times more volume. However, hydrogen's lower molecular weight and gas properties yield a mass density roughly nine times lower than natural gas. These factors largely offset each other, resulting in hydrogen's energy transport capacity being between 80-98% of that of a natural gas pipeline, given an existing pipeline with the same diameter and a fixed pressure drop [61].

The power demand for compressing hydrogen is significantly higher, typically three to four times greater than that for natural gas. This discrepancy arises from hydrogen's lower volumetric energy density and the larger volumes involved in handling it. The power requirement exhibits non-linear behavior concerning the pressure differential, suggesting that it might be more efficient to compress hydrogen more frequently rather than using larger compressors placed farther apart. Consequently, the optimal distance between compressors is likely to differ from existing natural gas compression stations. In cases where the pressure profile deviates from the optimal, alternatives include increasing compression power or delivering hydrogen at lower pressures. Various factors, such as pipeline utilization, terrain, and injection and withdrawal points, can influence compressor placement.

Due to the larger volume of hydrogen, investment costs for hydrogen compressors can be 40-80% higher than those for natural gas, with energy requirements approximately three times higher. Nonetheless, this increased energy consumption typically represents less than 1.5-2% of the energy transported over every 1,000 km in most scenarios [51].

An alternative approach to leverage existing infrastructure involves blending hydrogen with natural gas within the current network. This strategy offers several advantages, particularly in regions where the gas network is widespread, such as Europe, North America, Russia, and eastern China. It enables production scaling without the necessity of coordinating transport and end-use simultaneously, simplifying financing and execution due to a narrower project scope. Moreover, blending provides a guaranteed market, reducing project risks. Some residential heating equipment can already tolerate blending of up to 20% hydrogen by volume [62]. However, there are several drawbacks that limit the viability of this option:

- Limited CO_2 reduction: Different components within the gas network have varying tolerances to hydrogen. The network's tolerance is dictated by its least tolerant component. Sensitive components such as gas turbines, compressors, compressed natural gas (CNG) tanks, metering equipment, and industrial users set the limits. For instance, the tolerance of CNG tanks can be as low as 0.1% hydrogen by volume [63]. While blending up to 20% hydrogen without significant investment may be feasible where downstream users can accommodate it, exceeding this threshold requires further research and investment. Given the energy content disparity due to molecular weight differences, a 20% hydrogen fraction only equates to approximately 7% in energy terms, resulting in a limited potential for CO_2 emissions reduction.
- Higher energy costs: Hydrogen is relatively costly compared to methane, which is among the most affordable options. Even at a relatively low cost of \$2/GJ for natural gas (in the United States), it remains four times cheaper than hydrogen, priced at \$1/kg. Consequently, blending hydrogen into the grid would progressively raise gas prices, necessitating government subsidies (which are not sustainable) or passing the costs to end consumers, potentially leading to increased energy prices and equity concerns [64].
- User tolerance: The tolerance level of the gas network is determined by its most sensitive component. While many residential appliances have been tested for blending ratios up to 23% [65], and some projects have successfully demonstrated blending levels up to 30% without issues [66], the majority of downstream users can only tolerate a few percentage points of hydrogen. Sensitive components such as existing gas turbines, CNG vehicles, and certain feedstock applications are among the most susceptible [67]. Consequently, if any of these applications are present on the network, the blending threshold will be low.

- High mitigation costs: Blending hydrogen with natural gas results in limited CO_2 reduction benefits and a significant increase in energy costs. This leads to a considerable expense in mitigating the greenhouse gas emissions associated with natural gas usage. Considering the current production costs for renewable hydrogen, the mitigation cost could exceed \$500 per ton of CO_2 for most gas prices.
- Purity degradation: When hydrogen is introduced into the gas network, it cannot be directly utilized as pure hydrogen for fuel cells. Various technologies exist for separating hydrogen from natural gas, including pressure swing adsorption, membrane separation, and cryogenic methods. However, these separation processes entail considerable costs, rendering them economically unappealing. The cost ranges from at least \$1.5 per kilogram in the best-case scenario to approximately \$7-8 per kilogram in a more realistic scenario [68]. Furthermore, even after separation, the purity of the hydrogen obtained is inadequate for fuel cell applications, necessitating an additional expense for a polishing step.
- Gas variability: One aspect contributing to variability is the seasonal fluctuation in gas demand observed in many countries, particularly driven by residential heating requirements. Consequently, while gas demand varies, hydrogen supply remains constant, resulting in fluctuating hydrogen content throughout the year. Another significant factor is that if hydrogen is produced via electrolysis, hydrogen injection will be subject to fluctuations in renewable electricity input, leading to temporal variations in hydrogen content that may not be tolerable for all gas users. This challenge can be mitigated by implementing gas storage near the injection point, albeit at the expense of incurring additional costs.
- Regulatory considerations: Given the temporal variability of hydrogen content, regulatory measures would need to be implemented over short intervals to monitor compliance and safeguard gas users from adverse impacts. This could increase the complexity and costs of the regulatory framework or pose challenges for compliance, as many systems lack the acceptable temporal resolution required for such regulation.
- Disparate standards across countries: Blending limits exhibit significant variations across countries, with hydrogen often not even defined in gas quality standards in many regions. This poses an additional barrier to global trade, as all countries involved in a particular route would need to adhere to uniform quality standards.
- Technological maturity: Blending is currently in the pilot phase, with numerous ongoing demonstration projects aimed at validating its feasibility. However, it has yet to be proven at scale, necessitating further research and development before the option can be fully operationalized.

Repurposing existing pipelines can yield investment savings ranging from 65% to 94% compared to constructing entirely new hydrogen pipelines [60,69]. In smaller-diameter pipelines, the capital cost is predominantly attributed to the pipeline itself, accounting for nearly 98% of the total, whereas compressor expenses become more significant in larger-diameter pipelines, constituting approximately half of the total cost if a repurposed line is considered. Conversely, when new pipelines are required, compressor costs represent a minor fraction of the overall expenditure. Fixed costs, including maintenance, typically comprise around 13-15% of the total investment. The electricity expenses are contingent upon factors like pipeline diameter, distance between compression stations, and pressure profile, with smaller pipelines and flows incurring approximately 0.5% of the total transport cost. In comparison, larger pipelines with higher velocities may allocate up to 15% of costs to electricity. Additionally, when repurposing pipelines, significant costs may arise from cleaning and preparing the pipeline for use, amounting to as much as 35% of the total investment for reuse. Nevertheless, this cost remains only about 10% of the expense associated with constructing an entirely new pipeline [60]. Factoring in all expenses, it was determined that cost savings from repurposing natural gas pipelines range from 75% in scenarios where the network relies entirely on repurposed pipelines (with compression costs being the primary expenditure) to approximately 20% when 70% of the pipeline is new (with costs showing a linear relationship for proportions in between) [69]. A drawback of repurposing is the potential need to operate the pipeline at lower pressures to reduce the risk of crack propagation. This results in a larger volume and reduced energy transport, consequently increasing the cost per unit of hydrogen and compression expenses (Figure 4.8).

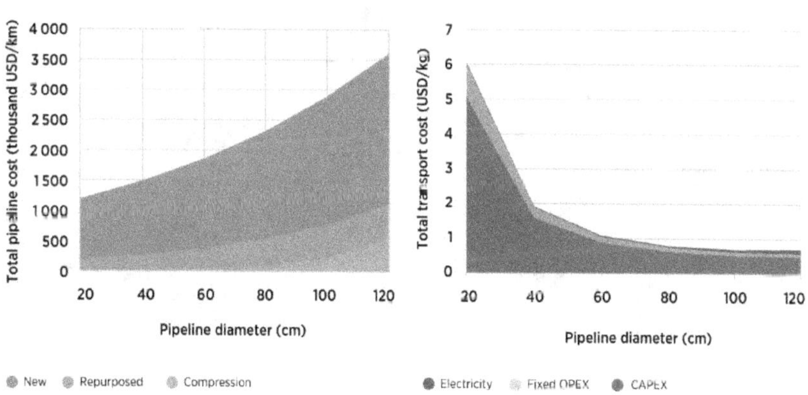

Figure 4.8 Capital expenditure for a hydrogen pipeline (left), and the complete transportation expenses (right) categorized by cost components (OPEX: Operational expenditure, CAPEX: Capital expenditure) [70]

Trucks designated for transporting gaseous hydrogen are referred to as tube trailers (Figure 4.9). Gaseous hydrogen is compressed to pressures of 180 bar (~2,600 psig) or greater and stored in elongated cylinders arranged on a trailer pulled by the truck. The elongated shape of the cylinders resembles tubes, thus earning the designation tube trailer. Tube trailers are presently constrained by U.S. DOT regulations to pressures of up to 250 bar, although exemptions have been granted to allow operation at higher pressures (e.g., 500 bar or higher). Steel tube trailers are the most commonly used and typically carry around 380 kg of hydrogen onboard; their capacity is limited by the weight of the steel tubes. Recently, composite storage vessels have been developed, boasting capacities of 560-900 kg of hydrogen per trailer. These tube trailers are currently utilized to deliver compressed natural gas in various other countries [71].

A HEXAGON Lincoln report outlines the development of the high-pressure tube trailer named "TITAN." These tanks operate at a pressure of 250 bar and have a capacity of 616 kg of hydrogen, approximately 7% of the tank weight. Tube trailers are an appealing option for hydrogen transportation due to their minimal infrastructure requirements and proven track record in transporting other gases. The cost of compression at fueling stations is low and can potentially be reduced by 60% compared to the transport of liquid hydrogen. Safety tests, including hydrostatic burst, penetration test, leak before burst, and pressure/temperature cycle tests, were conducted to ensure the safety of tube trailers. However, there are several challenges to address, such as liner blistering from saturation/decompression, the high cost associated with producing composite overwrapped pressure vessels, limited storage capacity, and regulations regarding tank dimensions and maximum pressure [72].

Figure 4.9 Tube trailer for gaseous hydrogen delivery [71]

The utilization of tube trailers for delivering gaseous hydrogen is widespread across various countries, particularly within the transportation and industrial sectors. These trailers play a crucial role in transporting compressed hydrogen, which is driven by demand in industrial and automotive applications. There's a significant projected growth in the market for hydrogen tube trailers, emphasizing the production of lightweight composite variants to propel market expansion. Many industry participants are directing their efforts toward innovating lightweight materials and composites for crafting highly resilient, dependable, and efficient hydrogen tube trailers. The global market size for hydrogen tube trailers is anticipated to increase from $340.6 million in 2023 to $557.5 million by 2030, exhibiting a CAGR of 7.3% [73].

4.3.2 Liquid hydrogen delivery

Gaseous hydrogen undergoes liquefaction through a cooling process to temperatures below −253°C. Following liquefaction, hydrogen can be stored in sizeable, insulated tanks at the liquefaction facility. Liquefying hydrogen necessitates energy expenditure, with current technology consuming over 30% of the hydrogen's energy content, rendering the process financially burdensome. Moreover, a portion of the stored hydrogen may escape through evaporation, commonly referred to as boil-off, which is particularly evident when utilizing small tanks with high surface-to-volume ratios. Research endeavors aimed at enhancing liquefaction technology, coupled with advancements in economies of scale, hold promise for reducing energy requirements and costs associated with the process [74].

Liquid hydrogen is commonly transported using liquid semitrailers, which typically have a capacity ranging from 12,000 to 17,000 gallons. The process involves initial precooling and a catalytic ortho-para isotope conversion from the pressurized hydrogen feed gas, followed by the liquefaction process to produce liquid hydrogen. Utilizing liquid hydrogen for delivery offers the benefit of reduced necessity for frequent delivery runs compared to compressed hydrogen, which is particularly advantageous for heavy-duty applications such as trucks, ships, airplanes, and trains. The transportation of liquid hydrogen plays a crucial role in facilitating the widespread adoption of hydrogen as a fuel, particularly in scenarios where long-distance transit is required and pipelines are not a viable option. Liquid hydrogen delivery, although resulting in increased energy loss, is deemed economically viable for scenarios of substantial demand (exceeding 500 kg/day) and intermediate distances. According to a report by the California Air Resources Board, hydrogen refueling stations are anticipated to opt for liquid hydrogen due to its superior storage capacity. The process of cryogenic hydrogen delivery encompasses three primary stages: liquefaction, storage, and transportation to end-users. Presently, there are eight liquefaction facilities across North America, each capable of producing 5 to 10 metric tons daily. However, for liquid hydrogen to become the preferred delivery method in the future, establishing additional plants boasting higher production capacities, reduced energy consumption, lower costs, and heightened efficiency is imperative. Ongoing efforts are directed towards optimizing the liquefaction process and developing more efficient methodologies, including a proposed novel approach with an energy consumption rate of 7.69 kWh/(kgLH_2), contrasting with the current range of 12.5 to 15 kWh/(kgLH_2) [75].

The delivery of liquid hydrogen presents numerous advantages that render it appealing across diverse sectors. Liquid hydrogen serves as an efficient means of storing energy over extended durations, thus serving as a valuable energy reservoir [76]. Moreover, the transportation of liquid hydrogen proves more economical than gaseous hydrogen owing to its higher energy density in liquid form, enabling the transportation of larger volumes at competitive costs. During hydrogen transportation and storage, a portion of the stored hydrogen evaporates or boils off, which is particularly notable when employing small tanks with significant surface-to-volume ratios. This phenomenon results in energy dissipation and potential safety risks. Efforts are underway to mitigate boil-off occurrences during transportation and storage through the advancement of tank insulation technologies and the exploration of strategies to diminish evaporative losses, particularly in smaller tanks characterized by substantial surface-to-volume ratios. Liquid hydrogen boasts a superior energy density relative to its gaseous counterpart, allowing for the storage and transportation of greater energy quantities within a smaller volume [77]. Liquid hydrogen poses various risks, including physical (such as embrittlement and storage tank failure), physiological (like cold burns and hypothermia), and chemical (involving ignition and fire hazards). Exposure to liquid hydrogen can result in severe burns and frostbite upon direct contact, while inhaling its cold vapor may cause respiratory problems [78]. While the handling of liquid hydrogen carries inherent risks, adherence to appropriate safety protocols can effectively mitigate these risks. Specially designed insulated storage tanks and transportation systems ensure the safe management of liquid hydrogen. It should be noted that the construction of storage and transportation infrastructure for liquid hydrogen necessitates the use of costly alloy systems to guarantee safety and mitigate embrittlement concerns. This requirement introduces complexity and additional expenses to the overall infrastructure [79].

Liquid hydrogen finds application across a spectrum of industries, including aerospace for rocket propulsion, maritime for sustainable energy via fuel cells, and potentially in heavy-duty transportation and superconductivity ventures. Liquid hydrogen plays a pivotal role in advancing the transition towards a decarbonized future, serving as a clean energy carrier capable of contributing to the reduction of greenhouse gas emissions.

4.3.3 The cost of hydrogen delivery

The hydrogen transportation cost primarily depends on two factors: the volume being transported, and the distance covered (Figure 4.10) [70]. Hydrogen can be transported via trucks, pipelines, or ships. Trucks are suitable for small volumes and short distances, but as the volume increases, the use of trucks carrying liquid hydrogen becomes necessary. With further increases in volume or distance, pipelines become the most economical option, starting from distribution pipelines for smaller volumes and progressing to transmission pipelines for larger volumes. For long distances or routes across water bodies, ships become the most viable choice. Consequently, for most cross-border trade scenarios, trucks are not considered suitable as they are more suitable for smaller volumes. Rail transportation is also a viable option with energy efficiency comparable to ships; however, it has a capacity limitation compared to pipelines. It has been indicated that transporting ammonia by rail is more cost-effective per hydrogen unit [80].

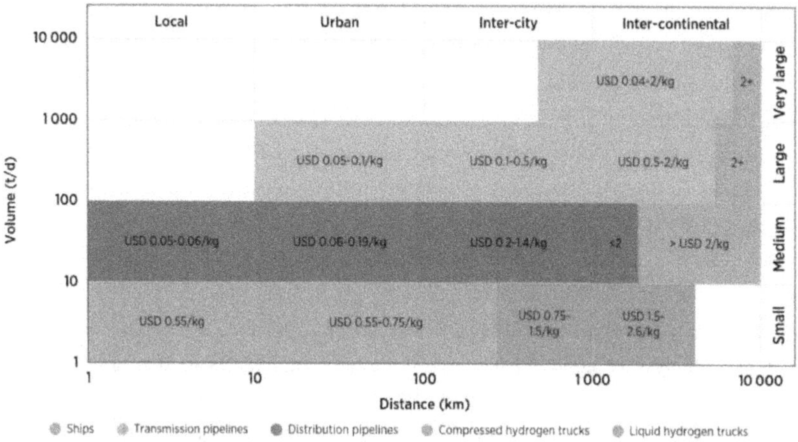

Figure 4.10 Hydrogen transport cost based on distance and volume [70]

Transporting hydrogen faces a significant challenge due to its low volumetric energy density. Compared to hydrogen, batteries have approximately 220 times higher energy density, while jet fuel boasts nearly 3,000 times higher energy density [81]. Therefore, hydrogen needs to be compressed, liquefied, or converted into other forms for economical transportation. Compression typically involves pressures ranging from 70 to 100 bar for pipeline transport and 350 to 700 bar for road transport, with lower pressures for buses and trucks (350 bar) and higher pressures for passenger vehicles (700 bar). This compression process can reduce the volume of hydrogen by 95, 290, and 500 times, respectively, compared to hydrogen at atmospheric pressure. Liquefaction can further reduce the volume by nearly 800 times, but the most significant volume reduction is achieved by converting hydrogen into other carriers. Ammonia, for example, is approximately 1,800 times denser than atmospheric hydrogen, methanol is about 2,500 times denser, and jet fuel is approximately 3,000 times denser [82]. The dilemma lies in balancing the convenience of transport and storage against the associated costs and energy requirements. Ideally, it would be preferable to avoid the inefficiency of converting hydrogen solely for transportation purposes. However, the low energy density of hydrogen makes its transportation economically unfeasible. As a result, an energy penalty is imposed, leading to increased energy input to offset lower transportation costs. This energy penalty varies depending on the transportation method: it may be only a few percentage points for pipelines, around 30-36% for hydrogen liquefaction at present, and nearly 50% for the round-trip efficiency of ammonia, although this can improve if cracking is not required. Therefore, the energy density (and the consequent energy loss) is contingent upon factors such as the volume to be transported, the distance, and the form of the final energy demand.

4.3.4 International maritime hydrogen transport

For maritime delivery, hydrogen must undergo a conversion process to increase its energy density before being reconverted at the destination terminal. The most energy-efficient option would involve compression. Global Energy Ventures is currently investigating this option and has patented a ship designed to transport compressed hydrogen, along with a ship designed for CNG. The company suggests that distances of up to 8,500 km can be competitively covered using a ship carrying 2,000 tons of H_2 (at 250 bar) powered by a fuel cell and the cargo [83]. This design received approval from a maritime classification society in March 2021. To gradually scale up, a 430-ton design was proposed and received preliminary approval from the American Bureau of Shipping (ABS) in October 2021, with full-class approval by late 2022. However, this approach results in the lowest energy density, meaning it contains the least mass of hydrogen [84]. If the hydrogen cargo were used as fuel for the ship, it would be quickly depleted, leaving little for delivery at the destination. Compressing hydrogen to 275 bar (a pressure similar to CNG ships) allows for a maximum shipping distance of 2,600 km (assuming a ship size of 2,700 m^3 or approximately 50 tons of H_2) [57]. Other studies suggest that even journeys of 1,000 km may not be feasible. As a result, the cost per unit of hydrogen delivered increases significantly with distance, rendering compressed hydrogen ships unattractive for large-scale, long-distance voyages [85].

Oxygenated products offer an alternative for maritime hydrogen transport, including methanol, ethanol, and dimethyl ether (DME). However, a common drawback is that these products leave behind carbon, which is released as CO_2 when the hydrogen is extracted. This necessitates either carbon capture and storage (CCS) at the destination terminal or sourcing carbon from the atmosphere (either via direct air capture or biogenic) to prevent a net increase in CO_2 emissions. For instance, to produce 100 million tons of transported methanol annually, approximately 0.73 gigatons of air-captured or biogenic CO_2 would be required. By 2050, a 1.5°C scenario requires 4.7 gigatons of negative emissions, mainly achieved through bioenergy with carbon capture and storage (BECCS) or potentially direct air capture with CCS (DACCS), although these technologies are not yet widely deployed. Therefore, any additional CO_2 capture needed for methanol production would further strain the ambitious BECCS and DACCS targets and incur significant cost penalties. In the long term, the levelized cost of DAC could reach USD 100 per ton of CO_2, translating to approximately \$140 per ton of methanol, equivalent to about

\$0.75-\$1.25/kg H_2. Comparatively, current fossil-based methanol production costs range from \$100-\$250 per ton of CO_2 [86].

Another alternative is methane, but it encounters similar challenges to the aforementioned compounds. Moreover, any methane leakage occurring during transfer or engine slip would contribute to increased greenhouse gas (GHG) emissions along the pathway. This option is particularly sensitive to methane loss rates due to methane's significantly higher global warming potential compared to CO_2, which is 29.8 times greater over a 100-year time horizon (increasing to 82.5 over a 20-year horizon [87]. Additionally, using methane as a hydrogen carrier would result in significant inefficiencies if the initial hydrogen is derived from natural gas. Instead of the double hydrogen production process, the initial methane could be transported directly as liquefied natural gas (LNG). However, this approach would entail the release of fossil CO_2 upon reconversion at the importing terminal, necessitating permanent sequestration. It's worth noting that the importing country may face challenges related to limited CO_2 storage capacity or social opposition to the technology, as observed in countries like Japan and South Korea [88].

Another option involves exporting fossil natural gas and subsequently importing the CO_2-generated post-hydrogen production. This concept has been proposed, for instance, from Saudi Arabia to South Korea, with the intention of utilizing CO_2 for enhanced oil recovery [89]. However, a limitation of this approach is that the volume of CO_2 produced exceeds the volume of the exported LNG, posing logistical challenges. Additionally, LNG carriers are typically designed for atmospheric pressure, whereas liquid CO_2 necessitates pressurized tanks reaching up to 70 bar [90]. Transporting CO_2 in its gaseous or solid state at atmospheric pressure is not economically viable [91]. Moreover, the relatively low technology readiness level of this approach suggests that it may become viable just as renewable hydrogen becomes more cost-effective than fossil-based alternatives Alternatively, importing methane and employing pyrolysis, where the carbon is converted into a solid carbon black, is another option. However, this technology is still in its early stages of development and has yet to reach commercial scale.

Ammonia, liquid organic hydrogen carriers (LOHC), and liquid hydrogen represent the three remaining alternatives for transporting hydrogen via ship. Each of these methods entails three distinct phases: converting gaseous hydrogen into a suitable form and state for transportation, the transportation process itself, and subsequently converting the carrier back into pure gaseous hydrogen under the necessary conditions. Ammonia is presently manufactured, stored, and traded extensively. Its primary drawback lies in the process of reconverting ammonia back to hydrogen, known as cracking, which can consume between 13% to 34% of the energy contained in the hydrogen.

LOHC possesses characteristics akin to diesel and can be transported in liquid form, resulting in minimal losses during transportation. However, the principal limitation lies in the energy consumption during reconversion, termed dehydrogenation, which can consume approximately 25% to 35% of the energy contained in hydrogen.

Liquid hydrogen faces constraints related to the energy consumed during liquefaction, accounting for at least 25% of the energy in the hydrogen. Other challenges include the necessity to significantly scale up liquefaction plants from current levels, the absence of commercially available large-scale ships, the high cost of adapting equipment for cryogenic conditions (-253°C), and some technological hurdles in the liquefaction process that must be addressed to enhance scalability (Table 4.4).

Table 4.4 Pros and cons of the potential hydrogen carriers

Carrier	Pros	Cons
Ammonia	• Produced extensively on a large scale • Traded globally on a wide scale • Minimal losses during transportation • Boasts high energy density and hydrogen content • Serves as a carbon-free carrier • Applicable directly in certain uses (e.g., fertilizers, power generation, maritime fuel) • Can be readily liquefied (at 20°C under 7.5 bar or -33°C at 1 bar)	• Ammonia synthesis involves significant energy consumption (12-26%) • Reconverting ammonia during importation involves high energy consumption (13-34%) and requires elevated temperatures (up to 900°C, but typically in the 500-550°C range) • Demonstrating ship engines using ammonia as fuel is necessary • Further purification of the produced hydrogen may be necessary • Hydrogen compression is required for the majority of applications • Increased production of nitrogen oxides (NO_x) during shipping necessitates flue gas treatment • Ammonia is both toxic and corrosive • The effectiveness of ammonia synthesis and cracking flexibility is yet to be fully demonstrated
LOHC	• Can be conveyed similarly to oil using current infrastructure, allowing for versatile transport options • Minimal initial investment required across all stages • Simple storage capabilities available	• The dehydrogenation process demands significant energy consumption (25-35%) in the importing region • Dehydrogenation necessitates high-temperature conditions (150-400°C) • Additional purification of the resultant hydrogen is necessary • Hydrogen generation occurs at 1 bar pressure, mandating compression • Hydrogen content in carriers is only 4-7% of their total weight • Lack of a definitive preferred chemical compound • Existing carriers incur high expenses • Losses in carriers occur with each cycle (0.1% per cycle) • Probable presence of fossil CO_2 in carriers

		• Majority of potential carriers require substantial scale-up from current global production levels
Liquid hydrogen	• The regasification process entails minimal energy usage (with the bulk of energy consumption occurring in the exporting region, where renewable energy costs are anticipated to be low) • Elimination of purification system requirement at the destination • Simplified transport logistics at the importing terminal • Low energy expenditure for boosting hydrogen pressure upon delivery • Established commercial viability of liquefaction technology • Carrier characterized by its carbon-free nature	• Significant energy dissipation during liquefaction processes (currently at 30-36%), necessitating expanded energy provisions • Evaporative losses (0.05-0.25% per day) encountered during transportation and storage • Cryogenic conditions contribute to elevated equipment expenses • Presently limited to small-scale implementation

The HySTRA project, initiated in Japan in 2015, stands as one of the earliest endeavors aimed at addressing the transportation of liquid hydrogen. Its primary objective was to establish a pilot scale by 2020, featuring a 1,250 m^3 ship, with plans for subsequent scaling to commercial levels by the mid-2020s, culminating in a full commercial scale (2 x 160,000 m^3) by 2030. The energy source for hydrogen production stems from brown coal sourced from the Latrobe Valley in Victoria, Australia, which undergoes gasification to yield hydrogen. Subsequently, the hydrogen is conveyed over 150 km to the port site in Hastings, where it undergoes liquefaction and is subsequently loaded onto the ship. Embarking on a 9,000 km voyage by sea, as depicted in Figure 4.11, the ship reaches Kobe Airport Island in Japan upon completion of the journey. At Kobe Airport Island, storage tanks with a capacity of 2,500 m^3 are available, facilitating the utilization of hydrogen for the airport's operations and for demonstrating power generation, with modifications solely to the combustors. Key partners in this initiative encompass Kawasaki Heavy Industries, J-POWER, Iwatani, Marubeni, AGL, Sumitomo, Shell, ENEOS, and K-Line.

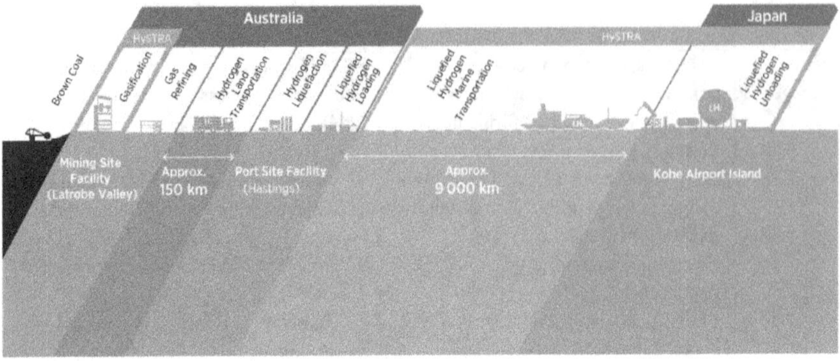

Figure 4.11 Hydrogen transport from Australia to Japan [70]

4.3.5 Ammonia for hydrogen delivery

Ammonia production is already widespread, with global production reaching approximately 183 Mt, requiring nearly 32.4 million metric tons of hydrogen. China accounted for a quarter of this production, with an additional 30% coming from Russia, the United States, and India [92]. The majority of ammonia production, about 72%, relies on natural gas, while approximately 22% is derived from coal, 5% from oil, and less than 1% from renewable sources [93]. Compared to hydrogen, ammonia offers distinct advantages, primarily in its ease of liquefaction and higher hydrogen content per unit volume, approximately 1.7 times that of liquid hydrogen. To achieve liquefaction, liquid ammonia necessitates pressurization to approximately 8 bar at ambient temperature or cooling to -33°C under ambient pressure. Additionally, there exists a semi-refrigerated option with a pressure below 8 bar. While pressurized storage is typically favored for smaller volumes, refrigerated storage becomes preferable as the scale increases, as these tanks require less steel and consequently offer cost savings. These characteristics bear similarities to liquefied petroleum gas (LPG), enabling the utilization of LPG carriers for ammonia transportation [94,95].

The production process is associated with significant CO_2 emissions, totaling 300 Mt from fossil fuel energy use and 270 Mt from feedstock sources. Of the feedstock emissions, around 140 million metric tons are captured and utilized for urea production, only to be released into the atmosphere when the urea is used in agricultural applications. Currently, approximately 10% of ammonia production is traded globally, facilitated by pipelines and over 70 LPG tankers ranging in cargo capacity from 2,500t to 40,000t [96]. The significant ammonia trade flows included shipments from the Middle East to Asia, intra-regional trade within Asia, and Eurasia to Europe. Eurasia stands out as the largest exporting region, while Europe leads as the largest importing region. Despite this established trade, certain regions anticipated to become significant players in future green hydrogen exports currently have minimal roles in ammonia exports (such as Australia and African countries). Present ammonia trade predominantly occurs on an interregional basis, with the most substantial flows occurring between neighboring regions rather than across continents (Eurasia to Europe, Middle East to Asia, and Latin America to North America). With the extensive global trade in ammonia, substantial infrastructure exists that could serve as a foundation for expanded ammonia trade as a hydrogen carrier. Currently, ammonia terminals are operational at over 38 ports for ammonia export and 88 ports for ammonia import, with several of them serving both roles (Figure 4.12). These terminals are equipped with loading and unloading systems and storage facilities, some shared with nearby ammonia or fertilizer plants. Storage tanks primarily come in two forms: isothermal tanks, capable of holding up to 45,000 metric tons, and spherical pressure tanks, with capacities ranging from 1,000 to 2,000 metric tons [97]. Ports with high cargo throughput and existing ammonia infrastructure are poised to lead in handling ammonia as a hydrogen carrier, driven not only by potential cost savings (albeit relatively small within the entire value chain) but also by their familiarity and expertise with ammonia handling. For instance, the Port of Rotterdam has entered into a memorandum of understanding (MoU) with the Transhydrogen Alliance to facilitate the import of green ammonia (and green hydrogen) into Europe via Rotterdam [98]. The initiative aims for operational readiness by 2024, with long-term import capacity reaching 0.5 million metric tons of hydrogen per year (equivalent to 3.8 million metric tons per year of ammonia). Similarly, the Port of Onahama in Japan is evaluating necessary adaptations to potentially import up to 1.6 million metric tons of ammonia per year [99]. Furthermore, ports like Singapore are exploring ammonia utilization as a marine fuel, presenting potential synergies with its role as a hydrogen carrier.

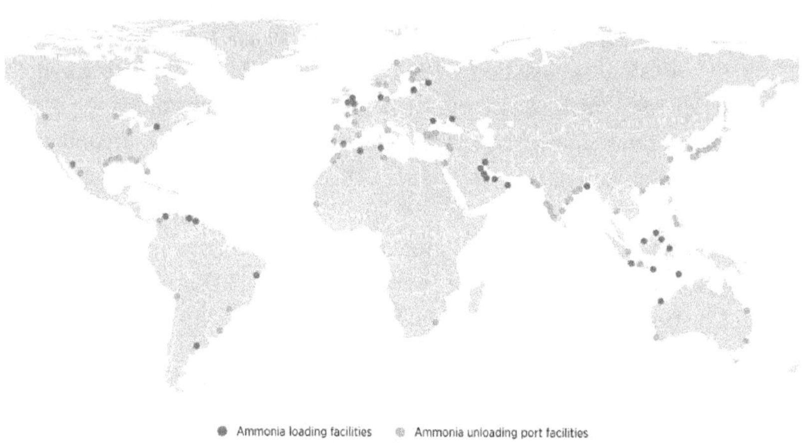

Figure 4.12 Ports equipped with facilities for loading and unloading ammonia [70]

Inland, ammonia facilities are also prominently situated. Notably, the United States boasts a 3,060-kilometer pipeline network stretching from Texas to Iowa, while a pipeline spanning 2,420 kilometers connects Russia to Ukraine, primarily utilized for transporting ammonia to fertilizer and chemical plants [100]. These pipelines are characterized by their smaller diameter, typically less than 35 centimeters, in contrast to the larger dimensions of oil and gas pipelines, which can exceed 122 centimeters. Additionally, numerous ammonia storage facilities are scattered across the globe, with the United States alone hosting over 10,000 such storage facilities [101]. Consequently, the conversion and transportation processes are already executed on a substantial scale, with existing infrastructure poised to support the integration of ammonia as a hydrogen carrier. Further advancements are required primarily in two key areas: the utilization of ammonia as a fuel for maritime vessels (currently transported solely as cargo with vessels reliant on fossil fuels) and the reconversion from ammonia back to hydrogen (via cracking, if necessary).

Ammonia production and trade are global phenomena. China is the foremost ammonia producer, contributing to 30% of global production (and 45% of associated CO_2 emissions), while the United States, the European Union, India, Russia, and the Middle East each contribute 8-10% of the production. Ammonia is traded internationally, with global exports amounting to approximately 10% of total production. Its primary derivative, urea, is traded even more extensively, comprising nearly 30% of its production. The availability of feedstock and process energy plays a pivotal role in determining the locations and methods of ammonia production. The prevalence of low-cost natural gas in regions such as the United States, the Middle East, and Russia elucidates their significant roles and reliance on natural gas-based plant fleets. Conversely, China's abundant coal reserves account for its heavy dependence on coal, constituting around 85% of its ammonia production. European and Asian nations have initiated strategies for importing hydrogen and ammonia. In efforts to reduce dependence on Russian fossil fuels, the European Union's REPowerEU initiative aims to domestically produce 10 Mt of renewable hydrogen and import an additional 10 Mt by 2030. The import goal comprises 6 Mt of renewable hydrogen and 4 Mt of hydrogen in the form of ammonia (equivalent to 24 Mt NH_3) or other derivatives [102]. Japan's Roadmap for Fuel Ammonia outlines plans to import 3 Mt of clean ammonia by 2030, with demand projected to escalate to 30 Mt by 2050. While the specific proportion of low-emission hydrogen imports is not specified, Japan intends to consume 3 Mt H_2/yr by 2030 (including imports), with 0.42 Mt H_2/yr designated as low-emission hydrogen [103].

Ammonia has an established marine transport infrastructure, utilizing chemical and semi-refrigerated LPG tankers. Despite the costs associated with converting hydrogen to ammonia, the overall sea transport cost is lower than liquid hydrogen, making ammonia the preferred carrier for long-distance hydrogen transport in the current decade (Figure 4.13) [104]. Since December 2021, the announced green ammonia initiatives collectively project a capacity of 34 million tons per year over the next decade, with half of this capacity concentrated in Latin America [105]. The overall project pipeline, encompassing ventures without set start-up dates, reaches nearly 100 million tons per year, with nearly half of these projects situated in Australia. Notably, the most extensive project is the Western Green Energy Hub in Australia, aiming to establish 50 gigawatts of renewable energy capacity, comprising 30 gigawatts of wind and 20 gigawatts of photovoltaic (PV) energy, to yield 3.5 million tons of hydrogen per year or 20 million tons of ammonia. CWP and Intercontinental Energy are the principal stakeholders in this venture, necessitating a total investment of AUD 70 billion. The final decision on investment is anticipated by 2028, with initial production slated to commence from the first of three phases in 2030 [106].

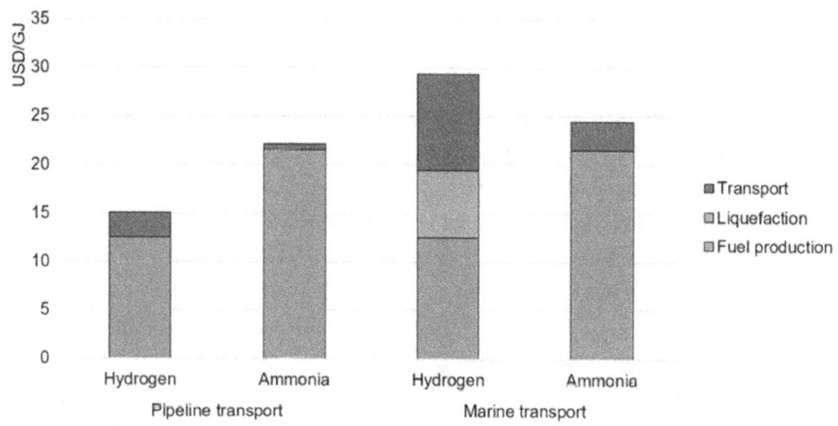

Figure 4.13 Cost comparison for supplying hydrogen and ammonia considering pipeline or marine transport

Another significant Australian project is the Asian Renewable Energy Hub, proposing a capacity of 10 million tons of ammonia per year for export to Japan or South Korea. This initiative comprises 26 gigawatts of renewable energy capacity and 14 gigawatts of electrolyzer capacity, entailing a USD 36 billion investment. While the final investment decision was scheduled for 2025, with exports set to begin in 2028, the project faced rejection by the federal government, prompting the Clean Energy Council to seek clarification on the grounds for this decision [107]. Additional large-scale projects in Australia include a 2.4 million tons per year initiative in Carnarvon, powered by onshore wind and PV, targeted for completion by 2030 or earlier, and a 1.9 million tons per year undertaking in Murchison, also leveraging onshore wind and PV, with a decision pending between ammonia and liquid hydrogen. Furthermore, a 1.8 million tons per year endeavor in Gladstone, driven by PV, is set to initiate its first phase in 2025, facilitated by Hydrogen Utility employing 3 gigawatts of electrolyzer capacity.

Beyond Australia, Green Energy Oman emerges as one of the prominent initiatives, aiming to generate 10 million tons per year of green ammonia from a renewable capacity of 25 gigawatts. This project is a collaboration between Oman's national oil company, Enertech (a renewables development firm), and Intercontinental Energy. The projected investment stands at USD 30 billion, with a planned final investment decision scheduled for 2026. Another significant endeavor is Aman in Mauritania, targeting a production of 10 million tons per year of green ammonia through 30 gigawatts of wind and solar PV energy. Feasibility studies for this project are anticipated to conclude in 2022, with operations slated to commence before 2030 [108]. The Helios project in Saudi Arabia envisions 4 gigawatts of renewable capacity powering 2 gigawatts of electrolyzers to produce 1.2 million tons per year of ammonia, with a total investment of USD 5 billion and operations expected to commence in 2025 [109]. The majority of the disclosed initiatives primarily center on ammonia production, with the inclusion or clarity regarding reconversion to hydrogen being either absent or vague in many cases. For a comprehensive inventory of green ammonia projects, please refer to the detailed list.

The energy consumption associated with ammonia synthesis constitutes a pivotal aspect of the process. Energy is predominantly utilized across multiple units, including the air separation unit, refrigeration compressor, feed compressor, recycle compressor, and heater. Under typical operating conditions, the ammonia reactor's conversion is not exhaustive, prompting the separation of hydrogen from the outlet for recycling to enhance overall conversion efficiency. This recycling process ensures that nearly all (97-99%) of the hydrogen is converted to ammonia, with only losses occurring through heat due to the exothermic nature of the reaction. These heat losses represent approximately 8% of the energy contained in the ammonia and can be harnessed to preheat the feed and achieve autothermal reactor operation [110,111]. Ammonia production holds significant sway over global energy consumption and emissions. Presently, it constitutes around 2% of total final energy consumption and contributes 1.3% of CO_2 emissions from the energy system. The production of ammonia is paramount for mineral nitrogen fertilizers, serving as a crucial link between atmospheric nitrogen and food production. Roughly 70% of the produced ammonia is earmarked for fertilizers, while the remainder is allocated across diverse industries such as plastics, explosives, and synthetic fibers. Ongoing efforts are aimed at bolstering the efficiency and sustainability of ammonia production. Future endeavors center on curtailing process energy demands through enhancements in faradaic efficiency and process configurations [112].

Ammonia finds application in turbines, fuel cells, and combustion engines. Solid oxide fuel cells (SOFCs) possess the capability to directly utilize NH_3 and exhibit high efficiency rates ranging between 40% to 60%. However, they suffer from limitations such as low power density, inadequate load response capability, high costs exceeding USD 1,650 per kilowatt (kW), sensitivity to contaminants, and are in the nascent stages of development. In the case of alternative fuel cell types like proton exchange membrane (PEM) fuel cells, pure hydrogen is necessitated. Consequently, ammonia must undergo cracking, potentially leveraging heat from the ammonia itself, thereby compromising overall process efficiency and bringing it closer to that of large two-stroke maritime diesel engines [113]. Fuel cells function by converting fuel into electricity, necessitating an electric motor to further convert the electricity into rotational kinetic energy. Conversely, engines offer a more economical option, priced at less than USD 550 per kW, exhibiting greater robustness, acceptable power density, and load response. Engine efficiencies with ammonia can surpass 50%. However, engines are associated with higher emissions of nitrogen oxide (NO_x) pollutants. This issue could be addressed through the implementation of a selective catalytic reduction system utilizing ammonia as a reducing agent [114]. Ammonia's low flammability necessitates the use of a pilot fuel to ensure complete combustion. Initially, diesel may serve as the pilot fuel, with the potential transition to hydrogen (derived from onboard ammonia cracking) or biofuel at a later stage. Partial cracking suffices for this scenario, reducing energy requirements and enabling the utilization of high pressures (to obviate compression), albeit at the expense of conversion efficiency. In this context of partial conversion, exhaust gases, and the engine's cooling system can be harnessed to supply heat to the cracker, evaporator, and tank heater, with recycled heat proving adequate given the small percentage of hydrogen needed as pilot fuel.

Ammonia dissociation, splitting, or cracking represents the reverse reaction of synthesis, transforming ammonia into pure hydrogen and nitrogen. This process thrives under high temperatures and low pressures. Since the synthesis of ammonia releases heat, cracking necessitates additional heat to disintegrate the bonds within the ammonia molecules. Regarding temperature, two options exist: higher temperatures ranging from 950 to 1,050°C, eliminating the need for a finely tuned catalyst, or lower temperatures between 500 and 550°C, utilizing a catalyst to expedite the reaction toward equilibrium. Traditional catalysts typically rely on nickel and operate at temperatures between 600 and 900°C, whereas alternative catalysts leverage critical minerals like iridium and ruthenium, enabling operation at lower temperatures of 350 to 600°C. The next wave of catalyst development may incorporate sodium or lithium, functioning at even lower temperatures of 250°C [115]. From a thermodynamic viewpoint, pressure should ideally be minimized, preferably at atmospheric levels, to enhance ammonia conversion and mitigate slippage. However, lower pressures correspond to heightened downstream demands for hydrogen compression, striking a balance with a typical operating pressure of 20 to 40 bar for the cracker [116].

Despite the pressure increase, the energy penalty remains modest since it occurs during the liquid phase using pumps, which consume less energy than compressors. Additionally, hydrogen separation post-ammonia cracking can leverage a high-pressure to low-pressure gradient, necessitating elevated operating pressures. The optimal catalyst for ammonia cracking remains uncertain, but catalysts based on nickel, ruthenium, and cobalt appear to demonstrate superior performance. Considerations include the catalyst itself, the promoter, and the support material. Performance metrics include the ammonia conversion rate, the required temperature, and the catalyst quantity per unit of hydrogen generated. Many high-temperature ammonia cracking processes employ nickel-based catalysts, balancing activity and cost. These catalysts, classified as first-generation, boast compositions akin to those used in hydrocarbon reforming. Utilizing potassium as a promoter for ruthenium catalysts and employing carbon nanotubes as support achieves high conversion rates at lower temperatures with relatively high yields, as observed with second-generation catalysts [117]. However, these catalysts necessitate critical minerals, such as ruthenium. A cost-effective alternative that avoids platinum-group metals entails sodium amide-based catalysts, which can achieve notable conversion rates even at 500°C [118]. Classified as third-generation catalysts, these options are still in the early stages of technological development, with technology readiness levels ranging from 2 to 4 [119]. In cases of complete conversion, the resulting product stream comprises approximately 75% hydrogen (though this may vary based on the equilibrium state of the mixture), with the remaining components primarily consisting of nitrogen and unconverted ammonia. Separating hydrogen from this mixture may not be necessary, as the resulting gas can power the ammonia cracker's heater. However, additional pure hydrogen- approximately 7%- may be required to meet all heating needs [120].

In scenarios necessitating ammonia separation, one approach involves employing absorption/desorption columns with water. While this method can effectively separate most ammonia, it falls short when extremely low ammonia content is required, such as for use in PEMFCs (which may demand levels below 0.1 ppm of ammonia). To achieve such stringent requirements, a pressure swing adsorption (PSA) unit can be employed. In this process, the gas traverses through beds that selectively adsorb or release ammonia based on operating pressure, operating in a batch mode to undergo cycles of adsorption and release. The energy consumption for this option typically ranges from 1.5 to 2 kWh/kgH_2, including compression. Another alternative involves utilizing palladium-based membranes, which exhibit selectivity for hydrogen and could be integrated with the cracking process in a single step. This approach offers the advantage of reduced equipment count, along with a smaller footprint, and enhances conversion by removing equilibrium limitations, necessitating lower temperatures, and reducing energy consumption. However, the cost associated with palladium-based membranes is considerable. Separating nitrogen from hydrogen poses greater challenges compared to ammonia separation. One method involves employing a cryogenic cycle to liquefy the nitrogen, which entails compression to 240 bar using a multi-stage compressor, followed by successive expansion to cool the stream to -230°C [118]. While commercial-scale implementation of ammonia cracking remains nascent, there are analogous technologies that could aid in the de-risking process. Conceptually, the process could mirror the approach used in steam methane reforming (SMR) today. Both processes require elevated temperatures, typically achieved using a firebox containing parallel, vertical tubes filled with catalysts through which the gas flows. Given the endothermic nature of both reactions, a minimum temperature is essential to achieve adequate conversion. This enables a rough comparison between the costs of an ammonia cracker and an SMR unit [121]. Currently available crackers cater to small-scale applications, such as the annealing of high- or low-carbon steels, stainless steel, and various nickel or copper alloys, as well as brazing, sintering, and deoxidation. Commercial designs are typically limited to sizes up to 1,000 normal cubic meters per hour (Nm^3/hr), equivalent to less than 700 tH_2 per year or the output of a 5 MW electrolyzer operating for 4,000 hours annually. These units operate at pressures of up to 7 bar and can achieve an ammonia content of 20 ppm.

More giant ammonia crackers are utilized in a process aimed at producing heavy water for use as a moderator in nuclear reactors. This process encompasses both ammonia synthesis and cracking stages. Ammonia serves as an intermediary to generate deuterium in the gaseous phase, with the cracking step facilitating the production of pure deuterium from the enriched ammonia liquid phase. Notably, a facility in Argentina (Arroyito) boasts a cracking capacity exceeding 1,400 t/d, while a plant in India (Thal) possesses a cracking capacity of 800 t/d. This translates to hydrogen production equivalent to 0.3 GW and 0.2 GW, respectively. Some ammonia crackers have been operational for over three decades, with designs available across a spectrum ranging from 100 to 2,400 t/d [122].

Abbreviations

ABS: American Bureau of Shipping
BECCS: Bioenergy with carbon capture and storage
CAGR: Compound annual growth rate
CAPEX: Capital expenditure
CCS: Carbon capture and storage
CcH_2: Cryo-compressed hydrogen
CGH_2: Compressed gas hydrogen
CNG: Compressed natural gas
DACCS: Direct air capture with carbon capture and storage
DME: Dimethyl ether
DOE: Department of Energy
DOT: Department of Transportation
FRP: Fiber-reinforced polymer
GHG: Greenhouse gas
IEA: International Energy Agency
LH2: Liquefied hydrogen

LOHC: Liquid organic hydrogen carriers
LPG: Liquefied petroleum gas
MLI: Multi-layer insulation
NREL: National Renewable Energy Laboratory
NZE: Net Zero Emissions
OPEX: Operational expenditure
PEM: Proton exchange membrane
PSA: Pressure swing adsorption
PV: Photovoltaic
SMR: Steam methane reforming

Reference

[1] Usman MR. Hydrogen storage methods: Review and current status. Renewable and Sustainable Energy Reviews 2022;167:112743. https://doi.org/10.1016/J.RSER.2022.112743.

[2] Wade A. Amos. Costs of Storing and Transporting Hydrogen. National Renewable Energy Laboratory 1998. https://www.nrel.gov/docs/fy99osti/25106.pdf (accessed February 17, 2024).

[3] Hydrogen Storage | Department of Energy n.d. https://www.energy.gov/eere/fuelcells/hydrogen-storage (accessed February 17, 2024).

[4] Zarezadeh Mehrizi M, Abdi J, Rezakazemi M, Salehi E. A review on recent advances in hollow spheres for hydrogen storage. Int J Hydrogen Energy 2020;45:17583–604. https://doi.org/10.1016/J.IJHYDENE.2020.04.201.

[5] Weisberg AH, Aceves SM, Espinosa-Loza F, Ledesma-Orozco E, Myers B, Spencer B. Cold hydrogen delivery in glass fiber composite pressure vessels: Analysis, manufacture and testing. Int J Hydrogen Energy 2013;38:9271–84. https://doi.org/10.1016/J.IJHYDENE.2013.05.068.

[6] Wang Y, Dai X, You H, Gao M. Research on the design of hydrogen supply system of 70 MPa hydrogen storage cylinder for vehicles. Int J Hydrogen Energy 2018;43:19189–95. https://doi.org/10.1016/J.IJHYDENE.2018.08.138.

[7] Li JQ, Li JC, Park K, Jang SJ, Kwon JT. An Analysis on the Compressed Hydrogen Storage System for the Fast-Filling Process of Hydrogen Gas at the Pressure of 82 MPa. Energies 2021, Vol 14, Page 2635 2021;14:2635. https://doi.org/10.3390/EN14092635.

[8] Fujiwara H, Ono H, Onoue K, Nishimura S. High-pressure gaseous hydrogen permeation test method -property of polymeric materials for high-pressure hydrogen devices (1)-. Int J Hydrogen Energy 2020;45:29082–94. https://doi.org/10.1016/J.IJHYDENE.2020.07.215.

[9] Liu YL, Zhao YZ, Zhao L, Li X, Chen H gang, Zhang LF, et al. Experimental studies on temperature rise within a hydrogen cylinder during refueling. Int J Hydrogen Energy 2010;35:2627–32. https://doi.org/10.1016/J.IJHYDENE.2009.04.042.

[10] What is a Hydrogen Tank & Tank-Types n.d. https://www.addcomposites.com/post/what-is-a-hydrogen-tank-tank-types (accessed February 18, 2024).

[11] Rivard E, Trudeau M, Zaghib K. Hydrogen Storage for Mobility: A Review. Materials 2019, Vol 12, Page 1973 2019;12:1973. https://doi.org/10.3390/MA12121973.

[12] Hydrogen Energy Storage Companies and Startups - GreyB n.d. https://www.greyb.com/blog/hydrogen-energy-storage-companies/ (accessed February 18, 2024).

[13] Hydrogen Storage Tanks: The Types, The Pitfalls, & the Solutions n.d. https://didionvessel.com/hydrogen-storage-tanks/ (accessed February 18, 2024).

[14] A Guide to Hydrogen Pressure Vessels & Pipes - Antala Ltd. n.d. https://www.antala.uk/hydrogen-pressure-vessels/ (accessed February 18, 2024).

[15] The first commercial Type V composite pressure vessel | CompositesWorld n.d. https://www.compositesworld.com/articles/next-generation-pressure-vessels (accessed February 18, 2024).
[16] Ye L, Lu L. Environmental and economic evaluation of the high-pressured and cryogenic vessels for hydrogen storage on the sedan. International Journal of Low-Carbon Technologies 2023;18:144–9. https://doi.org/10.1093/IJLCT/CTAC126.
[17] Aasadnia M, Mehrpooya M. Large-scale liquid hydrogen production methods and approaches: A review. Appl Energy 2018;212:57–83. https://doi.org/10.1016/J.APENERGY.2017.12.033.
[18] Hosseini SE. Fundamentals of Hydrogen Production and Utilization in Fuel Cell Systems. Elsevier; 2023. https://doi.org/10.1016/C2020-0-03183-X.
[19] Liquid Storage Vessels | Hydrogen Tools n.d. https://h2tools.org/bestpractices/liquid-storage-vessels (accessed February 20, 2024).
[20] Liquid hydrogen storage: increasingly larger storage tanks n.d. https://demaco-cryogenics.com/blog/liquid-hydrogen-storage/ (accessed February 20, 2024).
[21] Marshall HA. Advanced transmission systems for new propulsion technologies. Elsevier; 2022. https://doi.org/10.1016/B978-0-323-90979-2.00016-0.
[22] Berstad D, Skaugen G, Wilhelmsen Ø. Dissecting the exergy balance of a hydrogen liquefier: Analysis of a scaled-up claude hydrogen liquefier with mixed refrigerant pre-cooling. Int J Hydrogen Energy 2021;46:8014–29. https://doi.org/10.1016/J.IJHYDENE.2020.09.188.
[23] Valenti G, Macchi E. Proposal of an innovative, high-efficiency, large-scale hydrogen liquefier. Int J Hydrogen Energy 2008;33:3116–21. https://doi.org/10.1016/J.IJHYDENE.2008.03.044.
[24] The storage performance of automotive cryo-compressed hydrogen vessels (Journal Article) | OSTI.GOV n.d. https://www.osti.gov/biblio/1558334 (accessed February 20, 2024).
[25] Shin HK, Ha SK. A Review on the Cost Analysis of Hydrogen Gas Storage Tanks for Fuel Cell Vehicles. Energies 2023, Vol 16, Page 5233 2023;16:5233. https://doi.org/10.3390/EN16135233.
[26] Airbus pioneers a superconducting powertrain cooled by liquid hydrogen n.d. https://newatlas.com/aircraft/airbus-liquid-hydrogen-cryogenic-superconductivity/ (accessed February 20, 2024).
[27] H2FLY Pioneers Emissions-Free Flight with Successful Liquid-Hydrogen-Powered Aircraft n.d. https://www.cryogenicsociety.org/index.php?day=07&id=272%3Ah2fly-pioneers-emissions-free-flight-with-successful-liquid-hydrogen-powered-aircraft&month=01&option=com_dailyplanetblog&view=entry&year=2024 (accessed February 20, 2024).
[28] Innovative Liquid Hydrogen Storage to Support Space Launch System - NASA n.d. https://www.nasa.gov/missions/artemis/innovative-liquid-hydrogen-storage-to-support-space-launch-system/ (accessed February 20, 2024).
[29] Kawasaki Completes Basic Design for World's Largest Class (11,200-cubic-meter) Spherical Liquefied Hydrogen Storage Tank | Kawasaki Heavy Industries, Ltd. n.d. https://global.kawasaki.com/en/corp/newsroom/news/detail/?f=20201224_8018 (accessed February 23, 2024).
[30] Kawasaki Develops Cargo Containment System for Large Liquefied Hydrogen Carrier with World's Highest Carrying Capacity—AiP Obtained from ClassNK | Kawasaki Heavy Industries, Ltd. n.d. https://global.kawasaki.com/en/corp/newsroom/news/detail/?f=20210506_9983 (accessed February 23, 2024).
[31] Kawasaki Heavy Industries Ltd. KHI's Activity for International Liquefied Hydrogen Supply Chain n.d. www.idealhy.eu/uploads/documents/IDEALHY_D3-16_Liquefaction_Report_web.pdf (accessed February 23, 2024).
[32] Integrated Design for Demonstration of Efficient Liquefaction of Hydrogen (IDEALHY) 2013.
[33] Berstad D, Gardarsdottir S, Roussanaly S, Voldsund M, Ishimoto Y, Nekså P. Liquid hydrogen as prospective energy carrier: A brief review and discussion of underlying assumptions

applied in value chain analysis. Renewable and Sustainable Energy Reviews 2022;154:111772. https://doi.org/10.1016/J.RSER.2021.111772.

[34] Cardella UF. Large-Scale Hydrogen Liquefaction under the Aspect of Economic Viability. Technical University of Munich n.d. https://mediatum.ub.tum.de/doc/1442078/1442078.pdf (accessed February 23, 2024).

[35] Aceves SM, Espinosa-Loza F, Ledesma-Orozco E, Ross TO, Weisberg AH, Brunner TC, et al. High-density automotive hydrogen storage with cryogenic capable pressure vessels. Int J Hydrogen Energy 2010;35:1219–26. https://doi.org/10.1016/J.IJHYDENE.2009.11.069.

[36] Stolten Detlef, Samsun RC, Garland Nancy. Fuel cells : data, facts, and figures. Wiley-VCH; 2016.

[37] Ahluwalia RK, Hua TQ, Peng JK, Lasher S, McKenney K, Sinha J, et al. Technical assessment of cryo-compressed hydrogen storage tank systems for automotive applications. Int J Hydrogen Energy 2010;35:4171–84. https://doi.org/10.1016/J.IJHYDENE.2010.02.074.

[38] Cryo-compressed hydrogen, the best solution for storage and refueling stations? | CompositesWorld n.d. https://www.compositesworld.com/articles/cryo-compressed-hydrogen-the-best-solution-for-storage-and-refueling-stations (accessed February 20, 2024).

[39] Kircher O, Greim G, Burtscher J, Brunner T. Validation of Cryo-Compressed Hydrogen Storage (CCH2) - A Probabilistic Approach | Hydrogen Tools n.d. https://h2tools.org/bibliography/validation-cryo-compressed-hydrogen-storage-cch2-probabilistic-approach (accessed February 20, 2024).

[40] Cryo-compressed hydrogen, the best solution for storage and refueling stations? | CompositesWorld n.d. https://www.compositesworld.com/articles/cryo-compressed-hydrogen-the-best-solution-for-storage-and-refueling-stations (accessed February 20, 2024).

[41] ZeroAvia and Verne Partner to Explore Cryo-Compressed Hydrogen for Airports and Aircraft - ZeroAvia n.d. https://zeroavia.com/zeroavia-and-verne-partner-to-explore-cryo-compressed-hydrogen-for-airports-and-aircraft/ (accessed February 20, 2024).

[42] H2FLY - Realizes hydrogen-powered zero-emission aviation. n.d. https://www.h2fly.de/ (accessed February 20, 2024).

[43] Airbus, Air Liquide and VINCI Airports announce a partnership to promote the use of hydrogen and accelerate the decarbonization of the aviation sector | Air Liquide n.d. https://www.airliquide.com/group/press-releases-news/2021-09-21/airbus-air-liquide-and-vinci-airports-announce-partnership-promote-use-hydrogen-and-accelerate (accessed February 20, 2024).

[44] ZeroAvia Acquires Leading Fuel Cell Stack Innovator HyPoint n.d. https://www.prnewswire.com/news-releases/zeroavia-acquires-leading-fuel-cell-stack-innovator-hypoint-301642136.html (accessed February 20, 2024).

[45] Xu Z, Zhao N, Hillmansen S, Roberts C, Yan Y. Techno-Economic Analysis of Hydrogen Storage Technologies for Railway Engineering: A Review. Energies 2022, Vol 15, Page 6467 2022;15:6467. https://doi.org/10.3390/EN15176467.

[46] Cryomotive - Clean Transportation Meets Clean Hydrogen n.d. https://www.cryomotive.com/ (accessed February 20, 2024).

[47] The Future of Hydrogen – Analysis - IEA n.d. https://www.iea.org/reports/the-future-of-hydrogen (accessed February 20, 2024).

[48] Study Shows Abundant Opportunities for Hydrogen in a Future Integrated Energy System | News | NREL n.d. https://www.nrel.gov/news/program/2020/study-shows-abundant-opportunities-for-hydrogen-in-a-future-integrated-energy-system.html (accessed February 20, 2024).

[49] Hydrogen Delivery | Department of Energy n.d. https://www.energy.gov/eere/fuelcells/hydrogen-delivery (accessed February 20, 2024).

[50] International Energy Agency I. Global Hydrogen Review 2022 n.d. www.iea.org/t&c/ (accessed February 17, 2023).

[51] European Hydrogen Backbone 2022. https://www.nrel.gov/docs/fy23osti/81704.pdf (accessed February 23, 2024).

[52] Hydrogen Pipelines | Department of Energy n.d. https://www.energy.gov/eere/fuelcells/hydrogen-pipelines (accessed February 20, 2024).

[53] Folga SM. Natural gas pipeline technology overview. 2007. https://doi.org/10.2172/925391.

[54] Cerniauskas S, Jose Chavez Junco A, Grube T, Robinius M, Stolten D. Options of natural gas pipeline reassignment for hydrogen: Cost assessment for a Germany case study. Int J Hydrogen Energy 2020;45:12095–107. https://doi.org/10.1016/J.IJHYDENE.2020.02.121.
[55] Niermann M, Timmerberg S, Drünert S, Kaltschmitt M. Liquid Organic Hydrogen Carriers and alternatives for international transport of renewable hydrogen. Renewable and Sustainable Energy Reviews 2021;135:110171. https://doi.org/10.1016/J.RSER.2020.110171.
[56] Hydrogen Insights 2021 | Hydrogen Council n.d. https://hydrogencouncil.com/en/hydrogen-insights-2021/ (accessed February 23, 2024).
[57] d'Amore-Domenech R, Leo TJ, Pollet BG. Bulk power transmission at sea: Life cycle cost comparison of electricity and hydrogen as energy vectors. Appl Energy 2021;288:116625. https://doi.org/10.1016/J.APENERGY.2021.116625.
[58] Miao B, Giordano L, Chan SH. Long-distance renewable hydrogen transmission via cables and pipelines. Int J Hydrogen Energy 2021;46:18699–718. https://doi.org/10.1016/J.IJHYDENE.2021.03.067.
[59] Cerniauskas S, Jose Chavez Junco A, Grube T, Robinius M, Stolten D. Options of natural gas pipeline reassignment for hydrogen: Cost assessment for a Germany case study. Int J Hydrogen Energy 2020;45:12095–107. https://doi.org/10.1016/J.IJHYDENE.2020.02.121.
[60] HyWay 27 › Gasunie n.d. https://www.gasunie.nl/en/expertise/hydrogen/hyway-27 (accessed February 23, 2024).
[61] Haeseldonckx D, D'haeseleer W. The use of the natural-gas pipeline infrastructure for hydrogen transport in a changing market structure. Int J Hydrogen Energy 2007;32:1381–6. https://doi.org/10.1016/J.IJHYDENE.2006.10.018.
[62] Kevin Topolski et al. Hydrogen Blending into Natural Gas Pipeline Infrastructure: Review of the State of Technology 2022. https://www.nrel.gov/docs/fy23osti/81704.pdf (accessed February 23, 2024).
[63] Agreement concerning the Adoption of Uniform Technical Prescriptions for Wheeled Vehicles, Equipment and Parts Which Can Be Fitted and/or Be Used on Wheeled Vehicles and the Conditions for Reciprocal Recognition of Approvals Granted on the Basis of These Prescriptions. n.d. https://digitallibrary.un.org/record/227567?ln=en (accessed February 23, 2024).
[64] IRENA. Green Hydrogen Supply: A Guide to Policy Making. International Renewable Energy Agency 2021:64. https://www.irena.org/publications/2021/May/Green-Hydrogen-Supply-A-Guide-To-Policy-Making (accessed February 23, 2024).
[65] Altfeld K, Pinchbeck D. Admissible hydrogen concentrations in natural gas systems. Gas Energy 2013:1–2.
[66] Kippers M, Laat JD, Hermkens R. Hydrogen Injection in Natural Gas on Island of Ameland in the Netherlands 2011. https://api.semanticscholar.org/CorpusID:59572559 (accessed February 23, 2024).
[67] Overview of test results & regulatory limits for hydrogen admission into existing natural gas infrastructure & end use | Marcogaz n.d. https://www.marcogaz.org/publications/overview-of-test-results-regulatory-limits-for-hydrogen-admission-into-existing-natural-gas-infrastructure-end-use/ (accessed February 23, 2024).
[68] Hydrogen Deblending in the GB Gas Network | ENA Innovation Portal n.d. https://smarter.energynetworks.org/projects/nia_nggt0156 (accessed February 23, 2024).
[69] Study on the reuse of oil and gas infrastructure for hydrogen and CCS in Europe 2021. www.concawe.eu/wp-content/uploads/Re-stream-final-report_Oct2021.pdf (accessed February 23, 2024).
[70] Global hydrogen trade to meet the 1.5°C climate goal: Part II – Technology review of hydrogen carriers. International Renewable Energy Agency 2022. www.irena.org/-/media/Files/IRENA/Agency/Publication/2022/Apr/IRENA_Global_Trade_Hydrogen_2022.pdf?rev=3d707c37462842ac89246f48add670ba (accessed February 22, 2024).
[71] Hydrogen Tube Trailers | Department of Energy n.d. https://www.energy.gov/eere/fuelcells/hydrogen-tube-trailers (accessed February 20, 2024).
[72] Elgowainy A, Reddi K, Sutherland E, Joseck F. Tube-trailer consolidation strategy for reducing hydrogen refueling station costs. Int J Hydrogen Energy 2014;39:20197–206. https://doi.org/10.1016/J.IJHYDENE.2014.10.030.

[73] Hydrogen Tube Trailer Market Size, Growth & Share [2030] n.d. https://www.fortunebusinessinsights.com/hydrogen-tube-trailer-market-106975 (accessed February 20, 2024).

[74] Liquid Hydrogen Delivery | Department of Energy n.d. https://www.energy.gov/eere/fuelcells/liquid-hydrogen-delivery (accessed February 20, 2024).

[75] Asadnia M, Mehrpooya M. A novel hydrogen liquefaction process configuration with combined mixed refrigerant systems. Int J Hydrogen Energy 2017;42:15564–85. https://doi.org/10.1016/J.IJHYDENE.2017.04.260.

[76] All about liquid hydrogen - Demaco Holland B.V. n.d. https://demaco-cryogenics.com/cryogenics/all-about-liquid-hydrogen/ (accessed February 22, 2024).

[77] Liquid hydrogen seen as "holy grail" for hydrogen uptake in mobility sector: Linde COO | S&P Global Commodity Insights n.d. https://www.spglobal.com/commodityinsights/en/market-insights/latest-news/energy-transition/111621-liquid-hydrogen-seen-as-holy-grail-for-hydrogen-uptake-in-mobility-sector-linde-coo (accessed February 22, 2024).

[78] Keeping your team safe through the decarbonisation journey n.d. https://www.ricardo.com/en/news-and-insights/insights/liquid-hydrogen-how-can-we-harness-its-benefits-safely (accessed February 22, 2024).

[79] Ahad MT, Bhuiyan MMH, Sakib AN, Becerril Corral A, Siddique Z. An Overview of Challenges for the Future of Hydrogen. Materials 2023;16. https://doi.org/10.3390/MA16206680.

[80] Papadias DD, Peng JK, Ahluwalia RK. Hydrogen carriers: Production, transmission, decomposition, and storage. Int J Hydrogen Energy 2021;46:24169–89. https://doi.org/10.1016/J.IJHYDENE.2021.05.002.

[81] Making the Hydrogen Economy Possible: Accelerating Clean Hydrogen in an Electrified Economy n.d. https://buyhydrogen.com.au/wp-content/uploads/2021/04/H2-Supply-Chain-Scoping-Study-Issued-GEV-210301.pdf (accessed February 22, 2024).

[82] Use ammonia for shipping, synthetic hydrocarbons for aviation | Carbon Commentary n.d. https://www.carboncommentary.com/blog/2021/12/1/use-ammonia-for-shipping-synthetic-hydrocarbons-for-aviation (accessed February 22, 2024).

[83] ASX Announcement: GEV Scoping Study Delivers Zero Emission Supply Chain for Green Hydrogen 2021.

[84] Global Energy Ventures, GEV, Developing A C-H2 Ship With Cargo Capacity of 430 Tonnes of Hydrogen - Hydrogen Central n.d. https://hydrogen-central.com/global-energy-venture-c-h2-ship-cargo-capacity-430-tonnes-hydrogen/ (accessed February 22, 2024).

[85] Decarbonizing bulk carriers with hydrogen fuel cells and wind-assisted propulsion: A modeled case study analysis - International Council on Clean Transportation n.d. https://theicct.org/publication/hydrogen-and-propulsion-ships-jan22/ (accessed February 22, 2024).

[86] Keith DW, Holmes G, St. Angelo D, Heidel K. A Process for Capturing CO2 from the Atmosphere. Joule 2018;2:1573–94. https://doi.org/10.1016/j.joule.2018.05.006.

[87] Chapter 7: The Earth's Energy Budget, Climate Feedbacks, and Climate Sensitivity | Climate Change 2021: The Physical Science Basis n.d. https://www.ipcc.ch/report/ar6/wg1/chapter/chapter-7/ (accessed February 22, 2024).

[88] Kearns J, Teletzke G, Palmer J, Thomann H, Kheshgi H, Chen YHH, et al. Developing a Consistent Database for Regional Geologic CO2 Storage Capacity Worldwide. Energy Procedia 2017;114:4697–709. https://doi.org/10.1016/J.EGYPRO.2017.03.1603.

[89] Saudi Arabia to Ship Gas to South Korea and Take CO2 Back - Bloomberg n.d. https://www.bloomberg.com/news/articles/2021-03-03/saudi-arabia-to-ship-gas-to-south-korea-and-take-back-the-co2?embedded-checkout=true (accessed February 22, 2024).

[90] Shipping carbon dioxide (CO2): UK cost estimation study - GOV.UK n.d. https://www.gov.uk/government/publications/shipping-carbon-dioxide-co2-uk-cost-estimation-study (accessed February 22, 2024).

[91] Geske J, Berghout N, van den Broek M. Cost-effective balance between CO2 vessel and pipeline transport. Part I – Impact of optimally sized vessels and fleets. International Journal of Greenhouse Gas Control 2015;36:175–88. https://doi.org/10.1016/J.IJGGC.2015.01.026.

[92] NITROGEN (FIXED)- AMMONIA n.d. https://pubs.usgs.gov/periodicals/mcs2021/mcs2021-nitrogen.pdf (accessed February 24, 2024).
[93] IRENA and AEA. Innovation Outlook: Renewable Ammonia. International Renewable Energy Agency 2022:144. https://www.topsoe.com/hubfs/DOWNLOADS/DOWNLOADS%20-%20White%20papers/Ammonfuel%20Report%20Version%2009.9%20August%203_update.pdf (accessed February 24, 2024).
[94] Ammonia: Fuel vs. Hydrogen Carrier | Black & Veatch n.d. https://www.bv.com/perspectives/ammonia-fuel-vs-hydrogen-carrier/ (accessed March 8, 2024).
[95] Why ammonia is the more efficient hydrogen carrier n.d. https://www.thyssenkrupp.com/en/stories/sustainability-and-climate-protection/why-ammonia-is-the-more-efficient-hydrogen-carrier (accessed March 8, 2024).
[96] Ammonfuel – an industrial view of ammonia as a marine fuel 2020.
[97] Rouwenhorst KHR, Van der Ham AGJ, Mul G, Kersten SRA. Islanded ammonia power systems: Technology review & conceptual process design. Renewable and Sustainable Energy Reviews 2019;114:109339. https://doi.org/10.1016/J.RSER.2019.109339.
[98] Alliance for green hydrogen production and import | Port of Rotterdam n.d. https://www.portofrotterdam.com/en/news-and-press-releases/alliance-green-hydrogen-production-and-import (accessed March 7, 2024).
[99] Japan Onahama port to examine ammonia, hydrogen imports | Latest Market News n.d. https://www.argusmedia.com/en/news-and-insights/latest-market-news/2202370-japan-onahama-port-to-examine-ammonia-hydrogen-imports (accessed March 7, 2024).
[100] Valera-Medina Agustin, Banares-Alcantara Rene. Techno-economic challenges of green ammonia as energy vector 2020. https://shop.elsevier.com/books/techno-economic-challenges-of-green-ammonia-as-an-energy-vector/valera-medina/978-0-12-820560-0 (accessed March 7, 2024).
[101] Ammonia Storage Tanks – AmmoniaKnowHow n.d. https://ammoniaknowhow.com/ammonia-storage-tanks/ (accessed March 7, 2024).
[102] REPowerEU n.d. https://ec.europa.eu/commission/presscorner/detail/en/IP_22_3131 (accessed February 4, 2024).
[103] Japan's Road Map for Fuel Ammonia – Ammonia Energy Association n.d. https://www.ammoniaenergy.org/articles/japans-road-map-for-fuel-ammonia/ (accessed February 4, 2024).
[104] International Energy Agency. Renewable Hydrogen from Oman A producer economy in transition n.d. https://iea.blob.core.windows.net/assets/338820b9-702a-48bd-b732-b0a43cda641b/RenewableHydrogenfromOman.pdf (accessed February 3, 2024).
[105] Atome announce biggest green hydrogen and ammonia project in Latin America n.d. https://www.renewableenergymagazine.com/hydrogen/atome-announce-biggest-green-hydrogen-and-ammonia-20221201 (accessed March 7, 2024).
[106] Western Green Energy Hub Infrastructure Pipeline n.d. https://infrastructurepipeline.org/project/western-green-energy-hub (accessed March 7, 2024).
[107] Asian Renewable Energy Hub - Infrastructure Pipeline n.d. https://infrastructurepipeline.org/project/asian-renewable-energy-hub (accessed March 7, 2024).
[108] The Glasgow Joint Declaration Islamic Republic of Mauritania & CWP – 4 November 2021 n.d. https://petrole-mines-energies.gov.mr/sites/default/files/2023-02/glascow_joint_declaration.pdf (accessed March 7, 2024).
[109] Helios Project in Saudia Arabia to start construction | Global Energy Infrastructure n.d. https://globalenergyinfrastructure.com/news/2022/03-march/helios-project-in-saudia-arabia-to-start-construction/ (accessed March 8, 2024).
[110] Allen J, Panquet S, Bastiani A. Electrochemical Ammonia: Power to Ammonia Ratio and Balance of Plant Requirements for Two Different Electrolysis Approaches. Frontiers in Chemical Engineering 2021;3:765457. https://doi.org/10.3389/FCENG.2021.765457/BIBTEX.

[111] Hollevoet L, De Ras M, Roeffaers M, Hofkens J, Martens JA. Energy-Efficient Ammonia Production from Air and Water Using Electrocatalysts with Limited Faradaic Efficiency. ACS Energy Lett 2020;5:1124–7. https://doi.org/10.1021/ACSENERGYLETT.0C00455/ASSET/IMAGES/LARGE/NZ0C00455_0003.JPEG.

[112] Executive Summary – Ammonia Technology Roadmap – Analysis - IEA n.d. https://www.iea.org/reports/ammonia-technology-roadmap/executive-summary (accessed March 8, 2024).

[113] Quach TQ, Giap VT, Keun Lee D, Pineda Israel T, Young Ahn K. High-efficiency ammonia-fed solid oxide fuel cell systems for distributed power generation. Appl Energy 2022;324:119718. https://doi.org/10.1016/J.APENERGY.2022.119718.

[114] Sekkesæter Ø. Evaluation of Concepts and Systems for Marine Transportation of Hydrogen 2019. https://ntnuopen.ntnu.no/ntnu-xmlui/bitstream/handle/11250/2623195/no.ntnu%3Ainspera%3A2525165.pdf?sequence=1&isAllowed=y (accessed March 8, 2024).

[115] Makhloufi C, Olivier P, Kezibri N, Liedes A. Ammonia As Hydrogen Carrier to Unlock the Full Potential of Green Renewables – Ammonia Energy Association n.d. https://www.ammoniaenergy.org/paper/ammonia-as-hydrogen-carrier-to-unlock-the-full-potential-of-green-renewables/ (accessed March 8, 2024).

[116] Hacker V, Kordesch K. Ammonia crackers / Ammoniak Cracker 2003:121–7. https://graz.elsevierpure.com/en/publications/ammonia-crackers-ammoniak-cracker (accessed March 8, 2024).

[117] René Bañares-Alcántara GDIMFPGJMLETAYLYSZ. Analysis of Islanded Ammonia-based Energy Storage Systems . University of Oxford 2015. https://eng.ox.ac.uk/media/11082/ammonia-based_ess.pdf (accessed March 8, 2024).

[118] Siemens. Ammonia to Green Hydrogen Project | Feasibility Study n.d. https://assets.publishing.service.gov.uk/government/uploads/system/uploads/attachment_data/file/880826/HS420_-_Ecuity_-_Ammonia_to_Green_Hydrogen.pdf (accessed March 8, 2024).

[119] The Royal Society. Ammonia: zero-carbon fertiliser, fuel and energy store . ISBN: 978-1-78252-448-9 2020. https://royalsociety.org/-/media/policy/projects/green-ammonia/green-ammonia-policy-briefing.pdf (accessed March 8, 2024).

[120] Ishimoto Y, Voldsund M, Nekså P, Roussanaly S, Berstad D, Gardarsdottir SO. Large-scale production and transport of hydrogen from Norway to Europe and Japan: Value chain analysis and comparison of liquid hydrogen and ammonia as energy carriers. Int J Hydrogen Energy 2020;45:32865–83. https://doi.org/10.1016/J.IJHYDENE.2020.09.017.

[121] Cesaro Z, Ives M, Nayak-Luke R, Mason M, Bañares-Alcántara R. Ammonia to power: Forecasting the levelized cost of electricity from green ammonia in large-scale power plants. Appl Energy 2021;282:116009. https://doi.org/10.1016/J.APENERGY.2020.116009.

[122] Rasmus Nielsen H. TopsoesAmmonia cracking technology –Delivering green Hydrogen 2021. https://www.ammoniaenergy.org/wp-content/uploads/2021/11/Rasmus-Topsoe-NH3-cracking-AEA-2021.pdf (accessed March 8, 2024).

CHAPTER 5

The hydrogen transport revolution: fuel cell vehicles and beyond

5.1 Introduction

Fossil fuel-based engines have dominated the transportation sector for the past century due to their well-established technology and affordability. However, environmental challenges, particularly global warming, have arisen as a result of the widespread use of fossil fuels in transportation systems. The depletion of fossil fuel reserves and the emissions of anthropogenic greenhouse gases (GHGs) are key factors driving the adoption of alternative fuels and cleaner propulsion technologies in transportation engines. Internal combustion engines (ICEs) fueled by fossil fuels account for about one-fifth of global CO_2 emissions and contribute to air pollution with substances like nitrogen oxides (NO_x), sulfur oxides (SO_x), and carbon monoxide (CO). Annually, this pollution leads to approximately 9 million premature deaths worldwide, surpassing the number of deaths related to tobacco smoking [1]. Developed countries have taken significant steps to improve air quality in cities by substituting petroleum with alternative fuels to reduce transportation-related emissions. In the United States, there is a goal to reduce petroleum use in the light-duty vehicle (LDV) fleet by 50% by 2030 and 80% by 2050 compared to 2005 levels, aiming for fossil-free road transportation [2]. Similarly, the European Union aims to achieve a 27% share of alternative fuels in the transportation sector by 2030, resulting in a 40% reduction in GHGs compared to 1990. There is also a nonbinding target to phase out fossil fuel vehicles in cities to achieve a 60% reduction in GHG emissions by 2050 [3].

In the quest for pollution reduction goals, hydrogen fuel cell (HFC) engines are poised to play a pivotal role. Extensive research spanning two decades and significant investments totaling billions of dollars have led to the development of various models of HFC vehicles (HFCVs) introduced by global automakers. Assessing the exergy efficiency of FC systems in comparison to other power generation systems like thermal solar power plants, hydroelectric plants, photovoltaic (PV) panels, wind turbines, diesel engines, Rankine cycles, gas turbines, combined Rankin-Brayton cycles, and FC systems, it has been determined that FCs boast one of the top exergy efficiencies [4]. When compared to battery electric vehicles (BEVs), HFCVs offer numerous advantages, including longer lifespan (over 200,000 km), quick refueling time (3-5 minutes), and a driving range of 600 km on a full tank. However, due to the inadequate development of hydrogen infrastructure relative to BEVs, the sales of HFCVs do not match up. Additionally, the capital and operational costs of HFCVs are roughly twice as much as BEVs due to low production volumes and the use of expensive platinum (Pt) catalysts [5]. A notable characteristic of HFCs is their ability to decouple energy and power ratings. In terms of power availability, HFCVs outperform BEVs because the power capacity is determined by the FC itself, while the duration of energy supply is managed by hydrogen storage, akin to ICEs. The essential elements found in an HFCV generally include the FC stack along with its cooling mechanism, an electric drive motor coupled with a transmission system, high-voltage batteries or ultracapacitors, the primary power control unit, and a hydrogen storage tank. By employing a Type 4 filament wound storage tank, HFCVs can elevate the pressure of compressed hydrogen to 700 bar for their operational needs [6].

5.2 An overview of fuel cell functions

A fuel cell (FC) is a device capable of producing electricity directly from hydrogen using an electrochemical process. In some ways, the functioning of an FC is similar to that of a conventional battery; however, unlike batteries, the reactants for FCs are stored externally to the device. Therefore, the performance of FCs not only relies on their design but also the accessibility of fuel and oxidant. This is why FCs are classified based on their power output (kW) rather than their capacity (kWh). The operation of an FC is essentially the reverse of the electrolysis process, where the cathode and anode of an electrolyzer switch roles to become the anode and cathode of an FC, respectively. Despite this change, the flow of current remains consistent, with cations and anions continuing to move in their respective directions. In FCs, the fuel undergoes oxidation at the negative electrode (anode), while the oxidant is reduced at the positive electrode (cathode).

FCs provide numerous benefits compared to conventional combustion-based power systems, such as superior efficiency, minimal emissions, and silent operation. Nevertheless, there are still obstacles to overcome, notably concerning costs, longevity, and the infrastructure required for hydrogen storage and distribution. In an FC system, electrical power is generated through the reaction between oxygen and hydrogen. To comprehend the release of electrons and their movement, it is crucial to analyze the reactions occurring at each electrode. In an acid FC system (Figure 5.1), which is the simplest type of FC system, the anode (negative electrode) facilitates the oxidation of hydrogen, releasing electrons, energy in the form of heat, and generating H^+ ions (Eq (5.1)). These electrons then flow through the electrical circuit to reach the cathode (positive electrode). Concurrently, the H^+ ions move through the acidic electrolyte solution (which contains free H^+ ions, serving this purpose effectively). At the cathode, oxygen reacts with the H^+ ions from the electrolyte, and electrons are transferred from the electrode to create water (Eq (5.2)). The overall reaction in an FC system is represented by Eq (5.3). When a load such as a light bulb is connected along the electron pathway, the electrons in motion transfer power to the load, causing the light bulb to illuminate [7].

$$2H_2 \rightarrow 4H^+ + 4e^- \tag{5.1}$$

$$O_2 + 4e^- + 4H^+ \rightarrow H_2O \tag{5.2}$$

$$2H_2 + O_2 \rightarrow 2H_2O + Heat \tag{5.3}$$

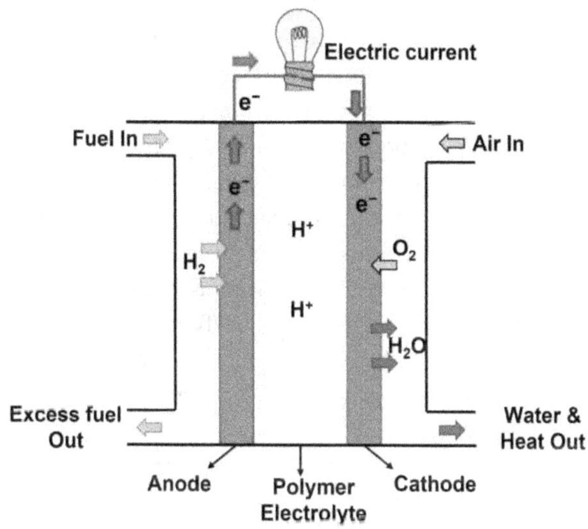

Figure 6.1 A schematic of an acidic fuel cell [8]

According to Eq (5.3), a balance in the FC system is maintained when two hydrogen molecules are consumed for each oxygen molecule. The process of removing electrons from a species is termed oxidation (Eq (5.1)), while the process of adding electrons to a species is known as reduction (Eq (5.2)). Thus, the corresponding electrochemical half reactions are identified as the hydrogen oxidation reaction (HOR) and the oxygen reduction reaction (ORR).

To produce electrical power in an FC system, a continuous supply of fuel and oxidants is necessary. While this process may seem straightforward, it can be complex, especially under high current and demand conditions. In such scenarios, if the reactants are not supplied adequately to the FC, it cannot function effectively. To address this issue, a combination of flow field plates and porous electrode structures is commonly utilized. The flow field plates are designed with grooves or fine channels to deliver gas flow to the FC. The specific characteristics of these flow field plates, including the arrangement, size, and shape of the channels, significantly impact the performance of the FC system. Additionally, the material properties of the flow field plate (such as mechanical strength, thermal conductivity, electrical properties, and corrosion resistance) are equally important as those of the electrodes. When hydrogen and oxygen are supplied to the electrodes, the electrochemical reaction begins. The current produced by the FC is directly related to the speed of the electrochemical reaction, with faster reactions resulting in higher current output from the FC. Given that a high current output is desirable in FC applications, catalysts are used to boost the speed of the electrochemical reaction, thereby enhancing the overall system efficiency. Hence, choosing the right catalyst and designing the reaction zone are pivotal factors that greatly impact the FC's performance. The transfer of ions from the anode to the cathode and their consumption at the cathode, along with the conduction of electrons through the external circuit, are essential processes in an FC system. While electrons can easily travel through a wire to move from one electrode to the other, ions present a more complex challenge due to their larger size and longer path compared to electrons. To facilitate the movement of ions, an electrolyte is used in FC systems. However, this ion transport process is less efficient than electron transport, leading to a noticeable decrease in the overall FC system performance. To improve ion transport efficiency, FC systems employ thin electrolytes to minimize the distance over which ionic conduction occurs. Apart from generating electrical power, all hydrogen fuel cells (HFCs) also produce water, which must be removed to prevent obstruction of the FC. Typically, removing the product is not complex because the introduction of reactants into the FC aids in the expulsion of product species from it.

The basic process of hydrogen oxidation in an alkaline fuel cell (AFC) mirrors that of an acidic fuel cell, albeit with different reactions occurring at the anode and cathode (Figure 5.2). In an AFC, mobile ions are represented by hydroxyl (OH^-) ions instead of H^+ ions. At the anode, OH^- ions interact with hydrogen to produce electrons, heat energy, and water (Eq (5.4)). Conversely, at the cathode, oxygen reacts with electrons flowing through the electrical circuit and water in the electrolyte to generate new OH^- ions (Eq (6.5)).

$$2H_2 + 4OH^- \rightarrow 4H_2O + 4e^- \tag{5.4}$$

$$O_2 + 4e^- + 2H_2O \rightarrow 4OH^- \tag{5.5}$$

Various types of FC systems are currently accessible in the market, categorized based on operating temperature, electrical efficiency, power output, and the type of electrolyte material used. Among these, Alkaline membrane FCs (AMFCs), also known as AFCs, and Proton exchange membrane FCs (PEMFCs) are prominently utilized. While both types function similarly, the key distinction lies in the membrane material, with AMFCs employing an alkaline membrane and PEMFCs utilizing an acid membrane. AMFCs often use non-precious metal catalysts at the electrodes, making them more cost-effective than PEMFCs. However, PEMFCs face challenges such as power density limitations, higher operating temperatures, membrane conductivity and durability issues, water management complexities, CO_2 tolerance, and anode electrocatalysis challenges [9]. Despite these challenges, PEMFCs are widely preferred in mobile power applications like automobiles. The core component of a PEMFC system is the membrane electrode assembly (MEA), consisting of a polymer membrane sandwiched between two electrodes and two gas diffusion layers (GDLs). However, a single unit cell's voltage potential is relatively low for practical use (between 0.5 and 0.8 V), leading to the combination of multiple unit cells in series to form an FC stack [10].

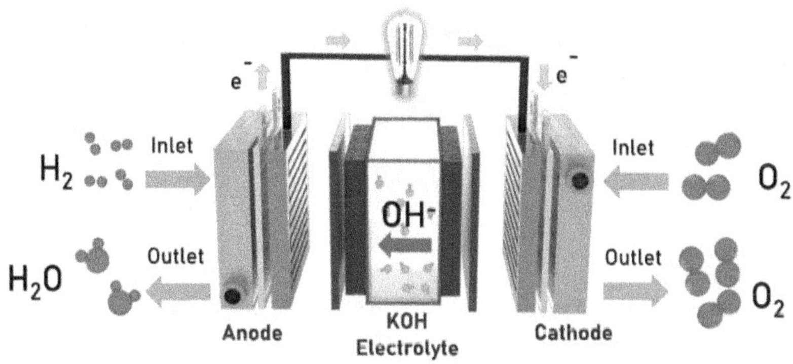

Figure 5.2 A schematic of an alkaline fuel cell [11]

For transportation purposes, PEMFCs offer advantages such as quicker start-up times than other FC types. The life cycle assessment (LCA) comparing a PEMFC engine, and an ICE revealed that the energy needed for hydrogen production for a PEMFC vehicle is approximately 3.5 times higher than the energy used for gasoline production for ICE vehicles. However, the overall life cycle energy consumption of PEMFC vehicles is about 2.3 times lower than that of ICE vehicles. While the GHG emissions of PEMFCs during the fuel cycle are roughly 8.5 times higher than ICEs due to greater GHG emissions in feedstock and fuel production/distribution, the overall life cycle GHG emissions of PEMFCs are approximately 2.6 times lower than ICE engines [12]. The primary challenge with PEMFCs is flooding, caused by increased gas humidity, which accelerates platinum dissolution-precipitation and carbon support corrosion [13].

5.2.1 Fuel cell efficiency

Closed systems like batteries, capacitors, and electric double-layer capacitors (EDLCs) are utilized for storing electrical energy. EDLCs employ high dielectric materials to swiftly store more energy, offering extended cyclic life and high-power densities compared to other storage devices [14]. Another energy storage system is the redox flow cell (RFC), combining a secondary battery and an FC, where energy is stored or supplied through two chemical components dissolved in liquids held within storage tanks. Figure 5.3 illustrates the specific energy density and power characteristics of FCs, batteries, capacitors, and electrochemical capacitors.

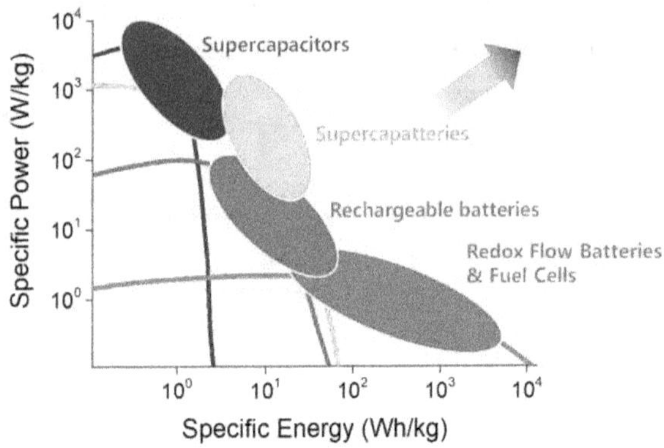

Figure 5.3 Specific power and energy density of various electrical energy devices [15]

Integrating energy storage systems and supercapacitors (SCAPs) with HFCs can enhance system affordability and performance. SCAPs offer higher power density due to low internal electrode resistance, quick charging, and improved longevity [16,17]. Combining a lithium-ion battery (LiB) with a PEMFC requires maintaining the battery state of charge (SOC) between 60% and 90% for backup power [18]. Although HFCs have lower power density and slower response than batteries or ICEs, their volumetric energy density exceeds chemical batteries. The efficiency of HFC engines ranges from 30% to 60%, considering that hydrogen production efficiency is crucial and varies depending on the production method [19]. This indicates room for improvement in hydrogen production and utilization systems to achieve higher efficiencies.

Research findings indicate that HFCVs exhibit lower well-to-wheel (WTW) fossil energy consumption and GHG emissions, ranging from 5% to 33% and 15% to 45% lower, respectively, compared to gasoline ICEs. This improvement is notable despite hydrogen production via steam methane reforming (SMR) from fossil fuels. Unlike conventional ICEs and hybrid electric vehicles (HEVs) that store energy in a fuel tank, BEVs store energy in batteries, eliminating the well-to-tank efficiency consideration. The overall efficiency of electrical power generation, transmission, and distribution is termed grid efficiency, which is around 40% for fossil fuel-based electrical power. BEVs achieve an efficiency of about 85% for battery charging and 80% for battery-to-wheel powertrain efficiency, resulting in a WTW efficiency of approximately 27% [20].

5.3 Economy of hydrogen fuel cell vehicles

Nearly half of the global transportation energy demand stems from LDVs, with projections indicating a rise in the number of car passengers from 1 billion to 2.5 billion by 2050 (Figure 5.4). The market for fuel cell technology extends significantly to light traction vehicles (LTVs) encompassing personal mobility aids like wheelchairs, recreational vehicles like golf carts and bicycles, as well as utility vehicles such as motorbikes, scooters, electric-assisted airport tugs, agricultural machinery like tractors, and material handling equipment including pallet trucks, tow trucks, and forklifts [21].

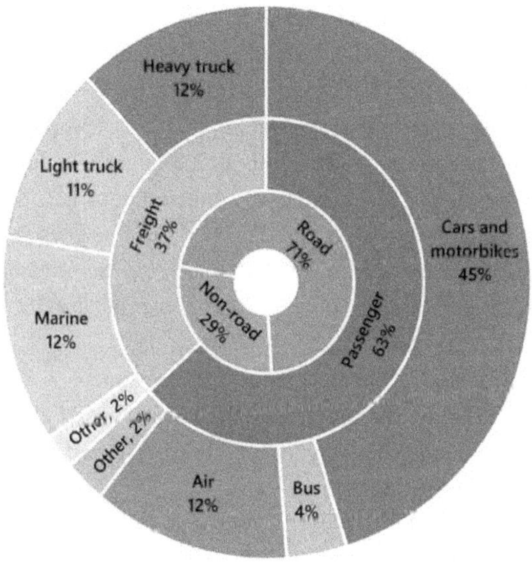

Figure 5.4 Worldwide energy utilization in transportation [5]

Motorbikes, particularly two-wheeled ones, are a prevalent mode of transportation in numerous regions. Notably, Intelligent Energy, in collaboration with Suzuki, has introduced a 4-kW HFC system for two-wheeled vehicles in London [22]. Forklifts, traditionally powered by combustion engines or rechargeable batteries, now benefit from HFC technology offering advantages such as rapid refueling in 2 to 5 minutes (versus 15 to 30 minutes for battery recharge), extended operational cycles of around 8 hours, resilience to temperature variations making them suitable for refrigerated environments, lack of self-degradation with charging cycles, emission-free operation unlike fossil fuel engines, versatility for indoor and outdoor use unlike fossil fuel counterparts unsuitable for indoor environments, high efficiency, and low maintenance needs. According to a report from the US Department of Energy (DOE), using hydrogen from wind energy in forklifts emits minimal GHG emissions when considering upstream emissions associated with converting energy sources into usable work in forklifts. Typically, HFC-powered forklifts rely on 5 to 20 kW PEMFCs for operation [7]. The future market for fuel cells also includes FC-powered electric bicycles and scooters capable of traveling distances up to 200 km with power requirements ranging from 4 to 6 kW. These vehicles are well-suited for short to medium-distance trips, providing an efficient alternative to mitigate traffic congestion issues [23–25].

The transportation sector contributes 17% of the annual GHG emissions [26]. Given that HFC engines offer nearly zero harmful emissions along with high engine efficiency, approximately twice that of ICEs, they have garnered significant attention from both automotive manufacturers and consumers [27]. In Europe, the typical engine power of an HFC car stands at 60 kW, surpassing that of residential FCs rated at 1 kW [28]. The cost of an HFC vehicle is calculated based on several factors, including the FC system, LCA, battery or traction motor, and hydrogen refueling infrastructure. Depending on the manufacturer and vehicle model, HFC vehicles are priced in the range of $55,000 to $65,000, with a target to reduce costs to $32,000 to $34,000 [29]. Since 2015, Hyundai, a Korean automaker, has exported 17 L-FCEVs to Denmark and Sweden at $50,000 per vehicle. These vehicles feature an FC system integrated with a lithium-ion battery for regenerative braking, capable of operating at low temperatures up to -25°C and achieving speeds of up to 160 km/h with a range of about 600 km [30].

Investigations indicate that no significant investment is required initially to establish the infrastructure for a fleet of HFCVs, estimated at around €24 to €25 million. It has been argued that hydrogen fuel is competitive with gasoline for vehicles despite the price difference between HFCVs and ICE vehicles. They suggest that a fuel availability rate of 20% can be economically viable with a hydrogen price of approximately €5 per kilogram [31]. Platinum is a crucial factor contributing to the capital cost of an HFCV [32]. Toyota has set a goal to reduce platinum content by 50% from current levels [33]. Early adopters of HFCVs typically exhibit high education levels, high household income, a strong interest in lifestyle changes, and similar attributes [34]. It is predicted that hydrogen will play a vital role in global vehicle power generation by 2050, with an estimated sale of around 150 million HFCVs [35]. In Japan, there are plans to deliver 800,000 HFCVs to customers by 2030 [36]. California has emerged as a leading market for HFCVs due to its extensive network of hydrogen refueling stations, with nearly 3000 vehicles delivered since June 2018 [37]. The latest HFCVs can be refueled within 4 minutes and offer maximum driving ranges of up to 580 kilometers (km) [38]. Some of the top-performing HFCVs currently available in the market include the Mercedes-Benz GLC F-CELL, Honda Clarity FCEV, Toyota Mirai FCEV, and Hyundai Tuscon/ix35 FCEV, Nexo [39–42]. These vehicles are equipped with PEMFC engines and composite compressed cylindrical fuel tanks. Honda has been leasing the high-performance CLARITY FCV since 2016, featuring a durable downsized FC system optimized for extended lifetime and improved water distribution control in the catalyst-coated membrane [43].

The increasing urban population has increased interest in public transportation, emphasizing the need for clean energy in this sector [44]. The integration of HFCVs into the taxi fleet has been proposed to make establishing hydrogen refueling infrastructure more affordable, considering taxis as a stable market for hydrogen consumption [45]. Replacing the entire taxi fleet with HFCVs in a large city could reduce CO_2 emissions by approximately 300 kilotons per year (kt/year), equivalent to around 3% of greenhouse gas emissions from urban transportation. The global distribution of HFC-powered LDVs by region is illustrated in Figure 5.5.

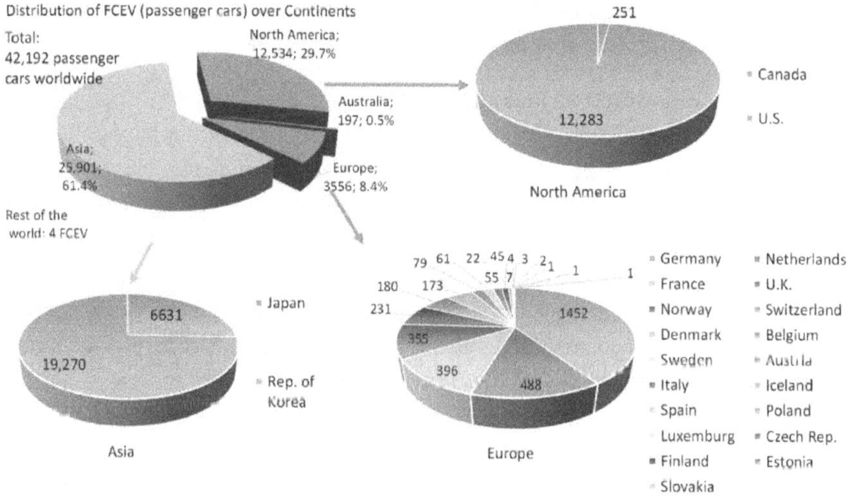

Figure 5.5 The global distribution of the HFC light-duty vehicles [46]

Efficient energy management is crucial for achieving cost-effective and efficient performance in a PEMFC engine [47]. Optimal sizing of the energy storage system and selecting a suitable power source based on the load requirements are key factors in maximizing the efficiency and cost-effectiveness of a PEMFC engine. Moreover, ensuring high purity of hydrogen is essential to enhance the performance and lifespan of an HFC stack, although this can increase fuel costs. ISO 14687-2 specifies that hydrogen purity should reach 99.97% in hydrogen PEMFC engines to maintain FC lifetimes [48]. While hydrogen from water electrolysis is typically pure enough for HFC engines, hydrogen produced from SMR requires additional cleanup processes, contributing to higher fuel costs [49]. Therefore, monitoring hydrogen purity should be integrated into refueling and production facilities. Manufacturers of HFCVs must see a return on their investment with a reasonable profit margin, and HFCs should be priced affordably to ensure market viability [50].

The hydrogen fueling station market size, valued at USD 0.32 Billion in 2022, is forecasted to increase from USD 0.38 Billion in 2023 to USD 1.35 Billion by 2032, with a compound annual growth rate (CAGR) of 17.30% during the forecast period. Government initiatives, including tax incentives, subsidies, and regulations favoring zero-emission vehicles, are driving the expansion of hydrogen filling station markets. Countries such as Japan, South Korea, Germany, and the United States are prioritizing the adoption of hydrogen fuel cell vehicles and investing significantly in hydrogen infrastructure development [51]. Numerous governments worldwide are planning to establish hundreds of hydrogen facilities, with major markets in Asia, Europe, and North America potentially hosting 1,000 hydrogen stations capable of accommodating 1 million to 2 million fuel cell vehicles. Given that over 15% of global greenhouse gas emissions originate from the transport sector, governments are actively seeking alternative energy sources for transportation.

5.4 Hydrogen-powered buses and trucks

HFC buses (HFCBs) offer numerous advantages over diesel buses, including improved fuel efficiency, reduced noise levels, and mitigated greenhouse gas emissions [52]. Consequently, the commercialization of HFCBs has begun, leading to a significant increase in the number of FC engines in public transportation systems in recent years. The primary drivers behind the rollout of HFCBs are concerns about urban noise pollution and air quality. This technology has garnered considerable attention, reaching a mature stage at the technology readiness level [53]. Public transportation, seen as a solution to urban traffic congestion and environmental pollution, has focused on integrating HFC engines into buses (Figure 5.6). The U.S. DOE aims to extend the lifetime of HFCBs to 25,000 hours, a significant improvement over the 8000-hour lifespan of LDVs [54]. HFCBs boast a more flexible design than mid-size HFC vehicles due to the fixed routes of buses and the unconstrained nature of hydrogen storage systems in terms of weight and size. Consequently, governments worldwide have funded HFCB programs. These initiatives have been implemented in various regions through programs such as the Zero Emission Bay Area (ZEBA) demonstration in California [52], the Brazilian Fuel Cell Bus Project in Brazil [55], the Urban-Route Buses Trial Project in China [56], the Hydrogen Fuel Cell Demonstration Project in Canada [57], the Sustainable Transport Energy Program in Australia [58], and the Clean Urban Transport for Europe in Europe [59]. Despite challenges such as hydrogen costs, availability, the initial expense of HFCBs, and the durability of FC power systems, these barriers can be overcome through the establishment of extensive refueling infrastructure and advancements in FC technology. In the United States, the National Fuel Cell Bus program, initiated by the Federal Transit Administration in 2006, aims to commercialize HFCBs and promote the adoption of eco-friendly fuel sources in the transit sector [61]. Vision Industries Corporation introduced the first commercialized HFC heavy-duty class-8 truck in the United States, operating on a hybrid HFC/Li-ion battery system. This truck, with a maximum weight of 36,000 kg, height of 2.9 m, and length of 8.4 m, can function in temperatures ranging from 26°C to 43°C, delivering a peak power of 400 kW and reaching speeds of around 105 km/h. Refueling the hydrogen tank takes 4 to 7 minutes, providing a driving range of 320 km.

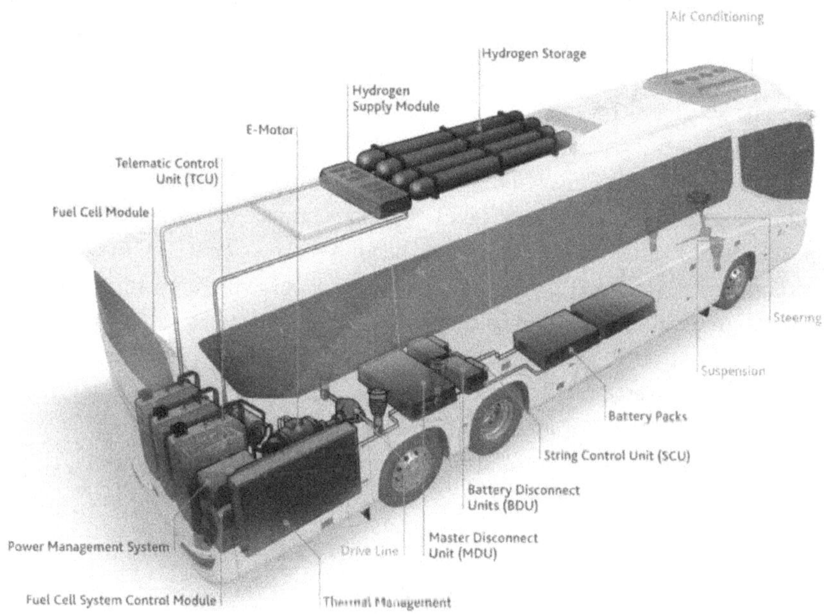

Figure 5.6 Components of hydrogen fuel cell bus [60]

In Asia, Toyota introduced more than 100 HFCBs in the Tokyo area ahead of the Tokyo 2020 Olympic and Paralympic Games [61]. Considering China's significant bus market, there is potential for widespread adoption of HFC technology in bus power systems in the country [62]. Foshan City in China has ordered 300 HFCBs, marking it as one of the largest hydrogen-powered bus fleets globally [63]. In the first phase of the clean hydrogen in European Cities project, 26 FCEBs were operational in various countries, including Italy, the United Kingdom, Norway, and Switzerland. The H2Bus consortium members have pledged their support to manufacture 1000 HFCBs along with the necessary infrastructure at competitive rates. Furthermore, the EU's Connecting Europe Facility has allocated 40 million euros to produce 600 FCEBs, of which 200 were deployed in Latvia, the UK, and Denmark in 2023 [64]. Additionally, the High V.LO City project, in collaboration with the Belgian bus company Van Hool, is developing 20 FCEBs. As part of the HyTransit program, Ballard and Van Hool will supply 6 HFCBs to the City of Scotland and Aberdeen, similar in design to the High V.LO City buses [65]. Currently, HFCBs have an availability of 90% in Europe, surpassing the 85% target, with refueling stations reaching an average availability of 95% [66]. In a particular Van Hool HFCB model, the electric motor and FC system are situated at the rear, while the hydrogen tanks are located on the bus's roof. Since buses do not face space constraints, hydrogen can be stored at lower pressures in the tanks, reducing compression costs. These tanks can hold approximately 40 kg of hydrogen stored at 350 bar [67]. The estimated total cost of HFCBs by 2030 is expected to be around 10% to 20% higher than diesel buses, a cost that could potentially decrease with large-scale development [68].

Trucks also demonstrate significant potential for HFC adoption. In Germany, Heliocentris developed a heavy-duty truck for waste disposal featuring a hybrid FC system alongside a diesel engine. The FC system is utilized for waste collection and disposal tasks, while the diesel engine handles propulsion [69]. In 2019, the California Air Resources Board partnered with Toyota and Kenworth to develop HFC T680 trucks for routes in and around Los Angeles, extending to San Bernardino [70,71]. In the United States, Nikola is deploying a long-distance heavy goods vehicle (HGV) powered by liquefied hydrogen [72]. FCs are additionally utilized as auxiliary power units (APUs) in HGVs, providing power for lighting, heating, cooling, refrigeration, and electronic devices [73]. Interest in HFC-powered HGVs may rise as cities implement bans on diesel trucks within city centers.

5.5 Zero-emission maritime with hydrogen fuel cells

The International Maritime Organization (IMO) foresees a significant rise in global shipping emissions over the next 35 years due to the growth of the global GDP and the subsequent increase in marine transportation. Even with improvements in maritime vessel efficiency, it is projected that GHG emissions from shipping in 2050 will be between 50% and 250% higher than the levels observed in 2012 [74]. Achieving a 50% reduction in GHG emissions from the maritime sector by 2050, in line with a 2°C pathway, necessitates the development of zero-emission vessels. The IMO has proposed a plan to achieve complete emissions neutrality from global marine transport at the earliest opportunity in this century. Additionally, the IMO aims to enhance emissions efficiency per unit of transportation by a minimum of 40% by 2030 compared to levels recorded in 2008. Traditionally, diesel generators are utilized on ships to convert fuel's chemical energy into electricity through thermal and mechanical means. However, HFCs offer a direct conversion of chemical energy into electrical power, eliminating combustion systems to enhance efficiency and reduce GHG emissions, noise, and vibrations. Researchers have been actively exploring the use of HFC technology in boats, ferries, ships, and submarines. HFC engines have shown significant reductions in GHG emissions, creating a quieter environment and improving overall system efficiency compared to conventional watercraft engines [75]. A theoretical comparison between a high-speed ferry powered by a hydrogen PEMFC (Figure 5.7) and a traditional diesel engine demonstrated that the hydrogen PEMFC significantly reduces NO_x and hydrocarbon emissions below Tier 4 criteria pollutant requirements [76]. Despite the urgent need to reduce emissions in the maritime sector, FC-powered maritime vessels are still in the prototyping phase. Most FC systems serve as APUs for energy needs beyond propulsion, and some ships use hybrid FC engines [77]. Cryogenic hydrogen storage is necessary for ferries consuming 2000 kg of hydrogen daily, with hydrogen cost being a critical factor, requiring fuel prices below $7 per kg [78].

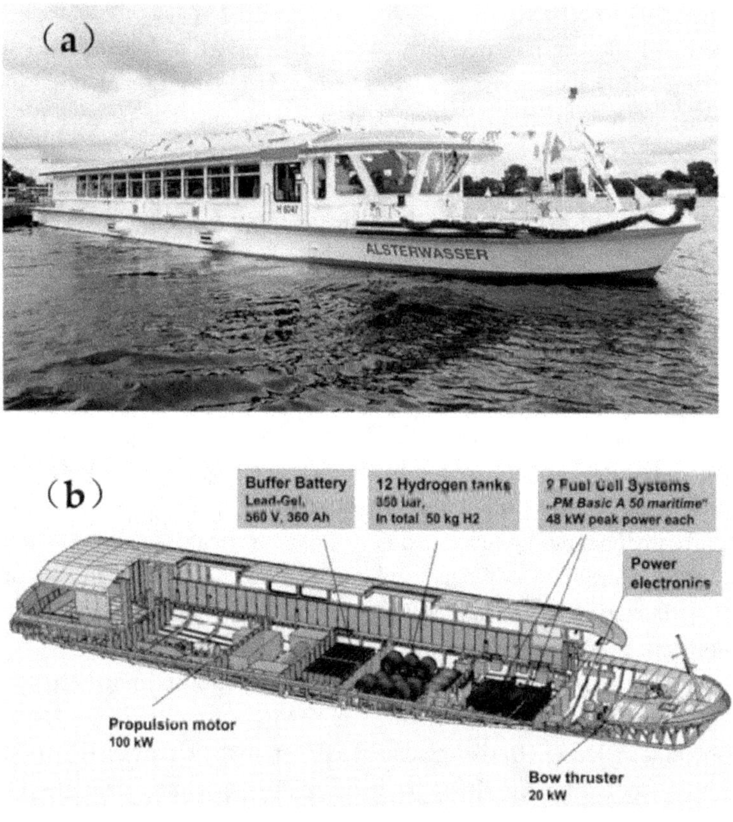

Figure 5.7 (a) The tourist ship PEMFCs-based Alsterwasser, (b) a schematic diagram of the electric propulsion system of "FCS Alsterwasser" [79]

The first commercial hydrogen-powered ferry was introduced by CMB Tech, the Hydroville, equipped with dual-fuel (H_2/diesel) internal combustion engines and supplied with 36 kg of usable hydrogen stored in 200 bar hydrogen tanks. This ferry could reach a top speed of 27 knots, cruising speed of 22 knots, and accommodate 16 passengers. Operating on the River Scheldt in the Port of Antwerp, Belgium, the Hydroville served as a commuter ferry. Presently, it serves as a testing and exhibition vessel and is also utilized for business meetings and excursions [80]. Following the Hydroville, CMB Tech developed a hydrogen-powered work boat called the Hydrocat 48 [81]. This vessel utilizes dual-fuel (H_2/diesel) engines and stores hydrogen as compressed hydrogen, holding 210 kg of compressed hydrogen and cruising at 30 knots. Recently, Norled introduced the MF Hydra, operating along the Hgelmeland-Nesvik route in Norway. The Hydra is equipped with PEM fuel cells fueled by hydrogen gas from a 4,000 kg LH2 tank, capable of carrying 299 passengers, 80 cars, 10 cargo trailers, and reaching a top speed of 9 knots [82].

The first hydrogen-powered ferry in the U.S., known as Water-Go-Round and later renamed Sea Change, was crafted by Golden Gate Zero Emission Marine, now Zero Emission Industries. Constructed on an aluminum catamaran hull, the Sea Change accommodates 78 passengers along with a crew of 2, boasting 360 kW of installed PEMFC power and reaching a maximum speed of 13 knots. Utilizing PEMFCs for propulsion, it stores approximately 250 kg of hydrogen in 250 bar hydrogen tanks [83]. The broader integration of hydrogen into maritime vessels may initially manifest in a more limited form, with HFC power supplementing primarily diesel-based powertrains in hybrid configurations. An important consideration arises: what role does HFC technology play as a component of hybrid power systems? This question was explored through the feasibility study of the H_2 Hybrid Research Vessel, intended as a replacement for the Scripps Institution of Oceanography's R/V Robert Gordon Sproul. This concept vessel, funded with $35 million from the State of California, is known as the Coastal Class Research Vessel and is currently undergoing functional design development [84].

In 2018, Hornblower Yachts explored the potential of hydrogen fuel-cell technology in maritime applications through Project Nautilus. This project involved designing and integrating a hydrogen PEMFc auxiliary power system onto an existing vessel, originally named the New York Hybrid and now known as the Discover Zero, with the aim of operating it in San Francisco Bay. Once Hornblower Yachts obtained a Design Basis Agreement letter from the US Coast Guard (USCG) and secured funding for retrofitting the hydrogen systems, the vessel was relocated to San Francisco for retrofitting. The design project received funding from the U.S. Department of Transportation's Maritime Administration and obtained a Design Basis Agreement Letter from the USCG [85]. In 2021, the HySeas III consortium secured funding from the EU to develop the world's first zero-emission ferry for marine use, powered by FC utilizing eco-friendly hydrogen derived from renewable sources. This vessel utilizes established Ballard PEMFC technology with a track record of success in land-based transportation. Yanmar, a Japanese firm, has been actively engaged in the development of a marine HFC system since 2015, leveraging its expertise in FC technology from the automotive sector. This initiative involves extensive research, safety assessments, and demonstration projects supported by funding from the Japanese government [86].

All evidence proves that the adoption of hydrogen-powered ships is already underway. Hydrogen diplomacy plays a crucial role in driving the maritime industry's transition towards sustainable solutions. Governments, international organizations, and industry stakeholders collaborate to establish the necessary infrastructure, regulations, and safety standards for widespread hydrogen adoption in maritime operations. Initiatives like the H2Ports project at the Port of Valencia, testing heavy-duty port equipment powered by HFCs, underscore global efforts towards emissions-free port operations [87]. Looking ahead, the potential for hydrogen in larger maritime vessels such as cargo ships, cruise liners, and offshore platforms presents substantial opportunities to further reduce emissions and operational costs in the global shipping industry. The successful transition to hydrogen-powered maritime transportation will depend on ongoing developments in hydrogen production facilities, refueling infrastructure, and the harmonization of policies and regulations across nations.

5.6 Hydrogen-powered trains

The current landscape of trains and locomotives worldwide is predominantly fueled by fossil fuels. However, stringent regulations are necessary to control emissions from petroleum-based locomotives. The introduction of hydrogen-powered trains marked a significant milestone, with the first unveiling taking place on August 22, 2003, during a presentation at the US Department of Transportation's Volpe Transportation Systems Center in Cambridge, MA, and the first hydrogen locomotive appearing on February 17, 2004 [88]. In 2016, Cummins Inc. from Columbus, OH, USA, partnered with Alstom, a French Railway Company based in La Rochelle, France, to develop, deliver, and incorporate a hydrogen fuel cell solution into its Coradia iLint trains. The world witnessed its first hydrogen-powered passenger train, the Coradia iLint (Figure 5.8), equipped with roof-mounted hydrogen tanks, entering service in Germany on September 19, 2018, to transport passengers within Germany's northern Lower Saxony state, covering a range of 500 miles [89]. The German Coradia iLint train achieved a remarkable feat by covering a distance of 1175 km without refueling. Subsequently, 14 FC trains were put into service in Bremervörde, Lower Saxony, in August 2022, following successful hydrogen-powered passenger train trials in Germany [90]. Similar trials were conducted outside Germany, including in the UK, Denmark, Norway, Sweden, Italy, the Netherlands, Canada, Austria, and France. In June 2019, the United Kingdom unveiled its first HFC train, named HydroFlex, to contribute to global sustainable transportation efforts. The aim is to establish a fleet of HFC trains in the UK to achieve the government's goal of phasing out diesel trains by 2040 [91]. Hydrogen-powered trains are equipped with FC stacks to generate the required power, employing a battery-super capacitors hybrid system and hydrogen storage tanks instead of traditional overhead or underground electric lines found in electric trains. Similar to hydrogen-fueled vehicles, hydrogen trains utilize compressed hydrogen gas to power an HFC engine. Excess electrical power is directed to lithium-ion batteries to drive the motors when insufficient electric circuitry is supplied from the HFC to the motor.

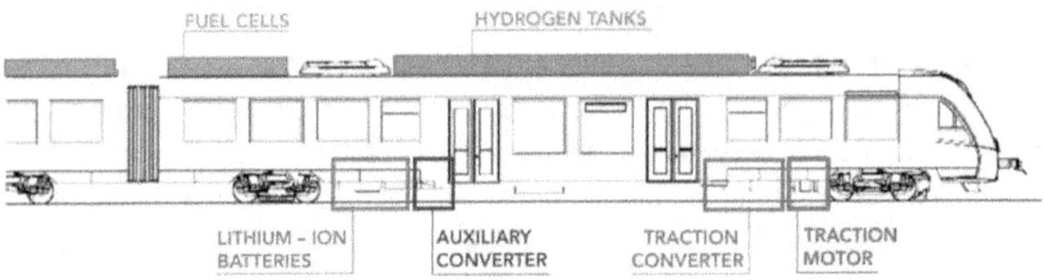

Figure 5.8 A schematic diagram of Coradia iLint train [92]

In Japan, Hitachi Ltd., in collaboration with Toyota Motor Corporation, developed, constructed, and tested the Hybari, a hydrogen-powered passenger train capable of reaching speeds of up to 100 km/h and traveling up to 140 km on a single filling [93]. In China, the China Railway Rolling Stock Corporation introduced a hydrogen fuel-cell-powered passenger train with a top speed of 160 km/h, covering a distance of 600 km before needing to refuel. This train also features advanced automatic driving technology and was manufactured at the CRRC Datong plant in Shanxi province. Notably, this represents China's first hybrid locomotive (Hyrail), marking a significant advancement in the country's adoption of hydrogen energy technology, enabling locomotives to achieve zero emissions of carbon and other pollutants. With a design speed of 80 km/h on flat, straight tracks and a maximum traction load exceeding 5000 metric tons, this locomotive can operate continuously for 24.5 hours on a full tank of hydrogen. China has become the first Asian nation and the second globally to develop hydrogen trains, signaling a growing trend in research and development across various countries [92].

A propulsion system for a hybrid switcher locomotive, which includes ultracapacitors, lead acid batteries, and a solid oxide fuel cell (SOFC) power plant, was proposed despite the finding that using only an FC system to power a locomotive is more effective [94,95]. In Europe, the refit class 321 hydrogen multiple units (HMU) train was unveiled in 2019 by French rail multinational Alstom, and UK rolling stock operating company Eversholt Rail Group manufactured 100 trains in 2021. The HMU, which has a range of 1000 km and a maximum speed of 140 km/h, was showcased as a significant development in hydrogen-powered trains [96].

In Lower Saxony, Germany, the success of HFC train trials led a public transport operator to commit to replacing all diesel trains on the line with HFC trains. Alstom delivered 14 out of the 41 trains ordered in Germany and secured orders from France for 12 dual-mode trains and from Italy for six HFC trains [97]. PKN Orlen, a Polish company, signed a contract with Alstom to supply hydrogen for their trains, with operations expected to commence within two years. Siemens Mobility has also entered the hydrogen train market, testing their train in mid-2023 and full service in Germany in January 2024 [98]. Similarly, Stadler introduced a hydrogen-powered passenger train set to operate in California in 2024 [99]. Alstom and Engie are collaborating to bring hydrogen to European rail freight, while Fortescue Future Industries successfully tested a two-stroke locomotive using blended ammonia fuel, aligning with their ambitious climate goals. Talgo in Spain has designed modular FC technology for diesel-to-hydrogen train conversions, targeting bids in Spain and abroad with their Vittal platform. These developments underscore the global push toward decarbonization and the adoption of hydrogen-powered transportation. In a separate development, Anglo American introduced a prototype HFC mining locomotive in May 2012. TransNamib recently secured funding from the German Federal Ministry of Education and Research for a hydrogen locomotive project, aiming to construct Africa's first dual-fuel hydrogen-diesel locomotives. The University of Birmingham contributed to academic research by designing and constructing the "Hydrogen Pioneer," a narrow-gauge locomotive used to assess a hydrogen hybrid railway traction vehicle's capabilities. Extensive scientific testing was conducted during this project, revealing an overall duty cycle efficiency ranging from 28% to 40%. Furthermore, a North American consortium comprising BNSF Railway Company, the US Army Engineers, and Vehicle Projects Inc. collaborated to develop a prototype hydrogen-fueled fuel cell-battery hybrid shunt locomotive tailored for urban and military base rail applications [100,101].

The configuration of FCs on trains can vary depending on the required power. For short-distance light-rail trains that have frequent starts and stops, a combination of a large battery and a small FC system is more effective. This setup benefits from a regenerative braking system that recharges the batteries during stops. On the other hand, long-distance and freight trains, which have fewer stops, are better suited for a smaller battery and a more extensive FC system [102]. Despite the fact that an HFC powertrain costs 50% more than diesel, its economic feasibility hinges on hydrogen being priced below $7 per kg. In terms of HFC technology, the lifespan of a diesel-driven train is typically around 15,000 to 20,000 hours before requiring maintenance. In contrast, an HFC stack has a lifespan of approximately 20,000 to 40,000 hours due to the absence of moving parts in an HFC engine and its simplified design [103].

5.7 Application of HFC technology in the aviation industry

HFC systems are notable for their quiet operation, lack of vibrations, and absence of NO_x emissions, making them appealing for aviation and aerospace applications. Two primary types of HFC systems garnering attention in aviation are SOFC and PEMFC. SOFC systems, functioning at high temperatures, utilize a dense ceramic layer as an electrolyte, whereas PEMFC systems, operating at lower temperatures, employ a proton-conducting membrane as an electrolyte. The integration of HFC systems can power various aircraft that rely on conventional batteries, such as flight data recorders, emergency door systems, cockpit voice recorders, floor escape path lighting, and emergency locator transmitters. The adoption of SOFC in APUs during stationary and cruising aircraft conditions could reduce fuel consumption rates by 75% and 40%, respectively. Additionally, employing SOFC in APUs could lead to an 80% reduction in ground NO_x emissions [104]. In a milestone event, Boeing demonstrated the world's first manned 20-minute flight of a glider powered by HFC with auxiliary battery support for takeoff and landing in 2008 [105]. Subsequently, the French-based Helite Company introduced the first one-seater HFC-powered aircraft capable of flying without any battery assistance one year later [106]. In 2016, Germany conducted testing on one of the earliest four-seat passenger aircraft powered by HFC. This aircraft's powertrain combined a PEMFC with a lithium battery to meet power demands during takeoff and ascent. With a maximum weight and speed of 1500 kg and 200 km/h, respectively, the aircraft had a flight range of 750 to 1500 km [107]. In 2019, ZeroAvia in California tested a small aircraft featuring an HFC powertrain, accommodating six business class seats and weighing 2 tons at takeoff. It became the largest zero-emission airplane globally without fossil fuels, capable of flying 500 miles (800 km). ZeroAvia claimed that the HFC system used in their aircraft was more commercially viable than battery-based systems and even traditional gas turbine engines, considering the overall cost [108]. Alaka'i Technologies in Massachusetts unveiled a fully autonomous HFC aircraft named Skai, capable of flying for approximately 4 hours and transporting around 1000 lb (450 kg) within a range of 400 miles (640 km) [109]. The Canadian Hydrogenics Company introduced a new lightweight aircraft powered by PEMFC, suitable for daily commuting and intercity travel within Canada [110]. In Singapore, HES Energy Systems designed a hydrogen-electric propulsion passenger aircraft for transporting four passengers over distances ranging from 310 to 3100 miles (500 to 5000 km), particularly suitable for aerial routes in rural areas between small towns, with a target to fly the prototype before 2025 [111]. In Europe, Zodiac Aerospace Company initiated a US$13 million project named HYCARUS, aiming to utilize PEMFC as a secondary power source for non-essential applications in large commercial airplanes, such as crew rest areas, lavatories, and galleys, to reduce fuel consumption and consequent GHG emissions [112]. The aerospace industry has taken note of the advantages of HFC systems, notably their production of water as a byproduct, making them appealing for space applications where the supply of food, air, and water is critical [113].

HFC technology has also been widely developed in unmanned aerial vehicles (UAVs). The primary challenges hindering the widespread use of UAVs are their weight and limited flight duration. The endurance of aircraft is a critical factor in determining how long a UAV can operate during missions. However, current UAVs often fail to meet mission duration requirements due to the low energy density of conventional batteries they use. The specific energy, or energy density, of lithium-polymer batteries commonly used in UAVs is around 200 Wh/kg, but it needs to increase to 1000 Wh/kg to significantly extend mission durations [114]. Power density (W/kg) is another crucial parameter affecting UAV performance during takeoff and climbs. Power density influences important features such as flight altitude, load capacity, and maximum speed, while energy density affects the flight time achieved by UAVs [115]. The high weight and low energy density of conventional batteries make HFC-powered UAVs more advantageous than battery-powered ones. While combustion-based propulsion systems are not feasible for UAVs due to their inefficiencies on a small scale, the modular nature of HFC systems makes them promising for small-scale applications [116]. HFC-powered UAVs can produce more power than batteries of the same weight, and they can be refueled quickly, offering longer flight times. This has led to increased interest in using HFC technology in drone propulsion systems. In 2007, AeroVironment in California introduced an efficient hybrid HFC-battery-powered lightweight aircraft equipped with solar cells in the wings. This aircraft had a wingspan of 2.6 meters, weighed 5.7 kilograms, and could fly for approximately 5 hours under remote control. Another innovative project by NASA in the United States, the Helios UAV, used a regenerative FC, PV cells, and backup batteries to achieve long flight durations and high altitudes. During the day, the FC operated as an electrolyzer to produce hydrogen using solar power, and at night, it functioned as an FC system using the stored hydrogen for flight [117]. At a commercial UAV Show in London, UK, in 2015, the Estonian-based Skycorp Company debuted a new line of intelligent HFC-powered drones incorporating advanced artificial intelligence into their operational systems. Subsequently, in 2018, the US Navy placed an order for 13 HFC propulsion systems for drone platforms from Massachusetts-based Protonex Technology [118]. In 2019, the US Army allocated a substantial $7.2 million grant specifically for HFC-powered UAVs [119]. To usher in a new era of ultralight HFC-powered drones, the Singaporean company HES Energy Systems built the Hydrogen Robotics Lab in Austin, Texas [120]. Recently, H3 Dynamics, based in Singapore, partnered with the United Arab Emirates police to deploy HFC-based drones in Dubai, enhancing security in the city known for its technological advancements [121]. Additionally, H3 Dynamics collaborated with EPH Engineering, a hydropower design firm in Brazil, to introduce a comprehensive dam inspection solution using HFC-powered Hycopter drones equipped with artificial intelligence, enabling flights lasting up to

3.5 hours to assess damages [122]. In the UK, Intelligent Energy unveiled a 2.4 kW PEMFC drone utilizing 9 liters of hydrogen stored in a 300-bar cylinder, achieving an 80-minute flight time with a 3 kg payload. Meanwhile, in South Korea, Doosan Mobility Innovation Company launched a 2.1 kW unmanned aerial hydrogen fuel cell vehicle providing over 2 hours of flight time and entered a Memorandum of Understanding with Asoa, a South Korean UAV manufacturer, to jointly deploy versatile HFC-powered drones [123]. In central Asia, an HFC-powered UAV is employed to monitor key winter wheat production areas in Uzbekistan.

Given the crucial importance of air quality in airport environments, there has been significant interest in exploring the viability of utilizing HFC systems to power various vehicles and ground support equipment (GSE) operating within airport premises. GES encompasses a range of systems like passenger stairs, baggage trolleys, cargo loaders, baggage tractors, air conditioning units, pushback tractors, and other equipment essential for passenger services, each requiring varying power capacities [5]. Since 2011, Boeing has been actively engaged in collaborative efforts with Neah Power Systems to advance the use of HFC technology in aircraft and ground support equipment, aiming for greener aviation solutions. Among the equipment used in aircraft handling, the air start unit and pushback truck are the most energy-intensive, consuming around 450 kW and 200 kW, respectively. Other equipment like air conditioning units, deck loaders, forklifts, water trucks, and baggage tractors typically require power ranging from 10 to 50 kW. In 2012, the US Department of Energy allocated $2.5 million towards developing HFCVs for transporting passenger baggage at US airports [124]. Today, HFC-powered forklifts have been implemented worldwide in airports due to their advantages, such as low maintenance, reduced noise, and zero pollutant emissions. Additionally, HFC-powered forklifts eliminate the cumbersome battery-swapping process inherent in traditional electric forklifts. Refueling an HFC system takes only a few minutes compared to the lengthy recharging time of conventional batteries. Typically, HFC-powered forklifts require refueling every eight hours, making the switch from conventional batteries to HFC systems a cost-effective solution when considering both initial capital investment and operational expenses [125]. In 2019, Linde Material Handling, a German company, unveiled the P250 tow tractor powered by HFC, boasting 25 tons of traction force tailored for airport logistics [126]. Albany International Airport in upstate New York adopted a new fleet of PEMFC GSE, including cargo tuggers, to efficiently transport packages from the airport's FedEx hub to aircraft for delivery [127]. A significant milestone occurred in 2015 when the world's first hydrogen-powered taxi fleet, operated by the Hype taxi service in collaboration with Air Liquide, was inaugurated. This initiative included establishing a refueling station at Paris Orly airport to facilitate passenger transportation to the southern part of central Paris [128]. In a move to reduce reliance on traditional battery charging methods, Hawaii's International Airport implemented hybrid HFC and battery-powered buses for seamless passenger transfers between terminals and facility locations [129]. Tokyo witnessed a similar development, with Hino Motors and Toyota Motor Corporation jointly introducing hybrid buses powered by a combination of HFC and batteries to efficiently shuttle passengers between Haneda Airport and central Tokyo [130].

Abbreviations

AFC: Alkaline fuel cell
AMFC: Alkaline membrane fuel cell
APU: Auxiliary power unit
BEV: Battery electric vehicles
CAGR: Compound annual growth rate
DOE: Department of Energy
EDLC: Electric double-layer capacitor
FC: Fuel cell
GDL: Gas diffusion layers
GHG: Greenhouse gas
GSE: Ground support equipment
HEV: Hybrid electric vehicles
HFC: Hydrogen fuel cell
HFCB: Hydrogen fuel cell bus
HFCV: Hydrogen fuel cell vehicle
HMU: Hydrogen multiple unit
ICE: Internal combustion engine
IMO: International Maritime Organization
LCA: Life cycle assessment
LDV: Light-duty vehicle
LTV: Light traction vehicles
ORR: Oxygen reduction reaction
PEMFC: Proton exchange membrane fuel cell
PV: Photovoltaic
RFC : Redox flow cell
SMR: Steam methane reforming
SCAP: Storage systems and supercapacitor
SOFC: Solid oxide fuel cell
UAV: Unmanned aerial vehicle
USCG: US Coast Guard
WTW: Well-to-wheel
ZEBA: Zero Emission Bay Area

References

[1] Pollet BG, Kocha SS, Staffell I. Current status of automotive fuel cells for sustainable transport. Curr Opin Electrochem 2019;16:90–5. https://doi.org/10.1016/j.coelec.2019.04.021.

[2] Transitions to alternative vehicles and fuels. Washington, DC: National Academies Press; 2013. https://doi.org/10.17226/18264.

[3] Larsson M, Mohseni F, Wallmark C, Grönkvist S, Alvfors P. Energy system analysis of the implications of hydrogen fuel cell vehicles in the Swedish road transport system. Int J Hydrogen Energy, vol. 40, Elsevier Ltd; 2015, p. 11722–9. https://doi.org/10.1016/j.ijhydene.2015.04.160.

[4] Von Spakovsky MR, Olsommer B. Fuel cell systems and system modeling and analysis perspectives for fuel cell development. Energy Convers Manag, vol. 43, Pergamon; 2002, p. 1249–57. https://doi.org/10.1016/S0196-8904(02)00011-0.

[5] Staffell I, Scamman D, Velazquez Abad A, Balcombe P, Dodds PE, Ekins P, et al. The role of hydrogen and fuel cells in the global energy system. Energy Environ Sci 2019;12:463–91. https://doi.org/10.1039/c8ee01157e.

[6] Roh HS, Hua TQ, Ahluwalia RK. Optimization of carbon fiber usage in Type 4 hydrogen storage tanks for fuel cell automobiles. Int J Hydrogen Energy, vol. 38, Pergamon; 2013, p. 12795–802. https://doi.org/10.1016/j.ijhydene.2013.07.016.

[7] Hosseini SE. Fundamentals of Hydrogen Production and Utilization in Fuel Cell Systems. Elsevier; 2023. https://doi.org/10.1016/C2020-0-03183-X.

[8] Elkafas AG, Rivarolo M, Gadducci E, Magistri L, Massardo AF. Fuel Cell Systems for Maritime: A Review of Research Development, Commercial Products, Applications, and Perspectives. Processes 2023, Vol 11, Page 97 2022;11:97. https://doi.org/10.3390/PR11010097.

[9] Alesker M, Page M, Shviro M, Paska Y, Gershinsky G, Dekel DR, et al. Palladium/nickel bifunctional electrocatalyst for hydrogen oxidation reaction in alkaline membrane fuel cell. J Power Sources 2016;304:332–9. https://doi.org/10.1016/J.JPOWSOUR.2015.11.026.

[10] Sharaf OZ, Orhan MF. An overview of fuel cell technology: Fundamentals and applications. Renewable and Sustainable Energy Reviews 2014;32:810–53. https://doi.org/10.1016/j.rser.2014.01.012.

[11] Choi J-H; ;, Nguyen PKT;, Kim D-J;, Yoon YS, Das G, Choi J-H, et al. Anion Exchange Membranes for Fuel Cell Application: A Review. Polymers 2022, Vol 14, Page 1197 2022;14:1197. https://doi.org/10.3390/POLYM14061197.

[12] Hussain MM, Dincer I, Li X. A preliminary life cycle assessment of PEM fuel cell powered automobiles. Appl Therm Eng 2007;27:2294–9. https://doi.org/10.1016/j.applthermaleng.2007.01.015.

[13] Gang BG, Kwon S. The proton exchange membrane fuel cell systems using methanolysis of sodium borohydride as a hydrogen source with cobalt catalysts. Int J Green Energy 2016;13:1224–31. https://doi.org/10.1080/15435075.2016.1183494.

[14] Asmatulu R, Khan WS. Nanotechnology Safety in the Energy Industry. Nanotechnology Safety, Elsevier B.V.; 2013, p. 127–39. https://doi.org/10.1016/B978-0-444-59438-9.00010-2.

[15] Rudra S, Seo HW, Sarker S, Kim DM. Supercapatteries as Hybrid Electrochemical Energy Storage Devices: Current Status and Future Prospects. Molecules 2024, Vol 29, Page 243 2024;29:243. https://doi.org/10.3390/MOLECULES29010243.

[16] Ren G, Ma G, Cong N. Review of electrical energy storage system for vehicular applications. Renewable and Sustainable Energy Reviews 2015;41:225–36. https://doi.org/10.1016/J.RSER.2014.08.003.

[17] Zhang Y, Wu L, Hu X, And HL-EM, 2008 U. Model and control for supercapacitor-based energy storage system for metro vehicles. Electrical Machines and Systems, 2008. ICEMS 2008. International Conference on. IEEE, n.d.

[18] Kaya K, Hames Y. Two new control strategies: For hydrogen fuel saving and extend the life cycle in the hydrogen fuel cell vehicles. Int J Hydrogen Energy 2019;44:18967–80. https://doi.org/10.1016/j.ijhydene.2018.12.111.

[19] Gray EMacA, Webb CJ, Andrews J, Shabani B, Tsai PJ, Chan SLI. Hydrogen storage for off-grid power supply. Int J Hydrogen Energy 2011;36:654–63. https://doi.org/10.1016/J.IJHYDENE.2010.09.051.
[20] Liu X, Reddi K, Elgowainy A, Lohse-Busch H, Wang M, Rustagi N. Comparison of well-to-wheels energy use and emissions of a hydrogen fuel cell electric vehicle relative to a conventional gasoline-powered internal combustion engine vehicle. Int J Hydrogen Energy 2020;45:972–83. https://doi.org/10.1016/j.ijhydene.2019.10.192.
[21] Qi Z. Light Traction: Fuel Cells. n.d.
[22] Home | Intelligent Energy n.d. https://www.intelligent-energy.com/ (accessed March 3, 2020).
[23] Dicks AL. PEM fuel cells. Applications. Comprehensive Renewable Energy, vol. 4, Elsevier Ltd; 2012, p. 203–45. https://doi.org/10.1016/B978-0-08-087872-0.00406-6.
[24] Hwang JJ, Wang DY, Shih NC, Lai DY, Chen CK. Development of fuel-cell-powered electric bicycle. J Power Sources 2004;133:223–8. https://doi.org/10.1016/j.jpowsour.2004.02.004.
[25] Colella WG. Market prospects, design features, and performance of a fuel cell-powered scooter. J Power Sources 2000;86:255–60. https://doi.org/10.1016/S0378-7753(99)00486-3.
[26] Plunkett JW. Automobile Industry Almanac. Houston, TX: 2007.
[27] Wipke K, Sprik S, Kurtz J, Ramsden T. Controlled Hydrogen Fleet and Infrastructure Demonstration and Validation Project: Spring 2010; Composite Data Products, Final Version March 29, 2010. 2010.
[28] Dodds PE, Ekins P. A portfolio of powertrains for the UK: An energy systems analysis. Int J Hydrogen Energy, vol. 39, Elsevier Ltd; 2014, p. 13941–53. https://doi.org/10.1016/j.ijhydene.2014.06.128.
[29] Creti A, Kotelnikova A, Meunier G, Ponssard J-P. A cost benefit analysis of fuel cell electric vehicles A cost benefit analysis of fuel cell electric vehicles Content. -; 2015.
[30] First Hyundai ix35 FCEV rolls off assembly line in Korea. Fuel Cells Bulletin 2013;2013:2. https://doi.org/10.1016/s1464-2859(13)70062-2.
[31] Brey JJ, Carazo AF, Brey R. Exploring the marketability of fuel cell electric vehicles in terms of infrastructure and hydrogen costs in Spain. Renewable and Sustainable Energy Reviews 2018;82:2893–9. https://doi.org/10.1016/j.rser.2017.10.042.
[32] Stephens I, Rossmeisl J, Science IC-, 2016 undefined. Toward sustainable fuel cells. ScienceSciencemagOrg n.d.
[33] Platinum's fuel-cell car bonanza proves elusive - Reuters n.d. https://uk.reuters.com/article/uk-platinum-fuelcells-analysis/platinums-fuel-cell-car-bonanza-proves-elusive-idUKKCN1GL1DS (accessed February 29, 2020).
[34] Rogers EM. Diffusion of innovations 2003:551. www.taylorfrancis.com/chapters/edit/10.4324/9780203887011-36/diffusion-innovations-everett-rogers-arvind-singhal-margaret-quinlan (accessed March 31, 2024).
[35] Ambrose AF, Al-Amin AQ, Rasiah R, Saidur R, Amin N. Prospects for introducing hydrogen fuel cell vehicles in Malaysia. Int J Hydrogen Energy 2017;42:9125–34. https://doi.org/10.1016/J.IJHYDENE.2016.05.122.
[36] Iida S, Sakata K. Hydrogen technologies and developments in Japan. Clean Energy 2019;3:105–13. https://doi.org/10.1093/CE/ZKZ003.
[37] Hardman S, Tal G. Who are the early adopters of fuel cell vehicles? Int J Hydrogen Energy 2018;43:17857–66. https://doi.org/10.1016/J.IJHYDENE.2018.08.006.
[38] Lipman TE, Elke M, Lidicker J. Hydrogen fuel cell electric vehicle performance and user-response assessment: Results of an extended driver study. Int J Hydrogen Energy 2018;43:12442–54. https://doi.org/10.1016/J.IJHYDENE.2018.04.172.
[39] The new Mercedes-Benz GLC F-CELL. n.d. https://www.mercedes-benz.com/en/mercedes-benz/vehicles/passenger-cars/glc/the-new-glc-f-cell/ (accessed December 2, 2018).
[40] 2018 Honda Clarity Fuel Cell – Hydrogen Powered Car | Honda n.d. https://automobiles.honda.com/clarity-fuel-cell (accessed December 2, 2018).
[41] 2018 Toyota Mirai Hydrogen Fuel Cell Vehicle | The Future of Everyday n.d. https://ssl.toyota.com/mirai/fcv.html (accessed December 2, 2018).
[42] Hyundai Nexo – Sign Up For Updates | HyundaiUSA.com n.d. https://www.hyundaiusa.com/nexo/index.aspx (accessed December 2, 2018).

[43] Tanaka S, Nagumo K, Yamamoto M, Chiba H, Yoshida K, Okano R. Fuel Cell System for Honda CLARITY FUEL CELL. ETransportation 2020;3:100046. https://doi.org/10.1016/J.ETRAN.2020.100046.

[44] Chang X, Ma T, Wu R. Impact of urban development on residents' public transportation travel energy consumption in China: An analysis of hydrogen fuel cell vehicles alternatives. Int J Hydrogen Energy 2019;44:16015–27. https://doi.org/10.1016/j.ijhydene.2018.09.099.

[45] Campíñez-Romero S, Colmenar-Santos A, Pérez-Molina C, Mur-Pérez F. A hydrogen refuelling stations infrastructure deployment for cities supported on fuel cell taxi roll-out. Energy 2018;148:1018–31. https://doi.org/10.1016/J.ENERGY.2018.02.009.

[46] Samsun RC, Rex M, Antoni L, Stolten D. Deployment of Fuel Cell Vehicles and Hydrogen Refueling Station Infrastructure: A Global Overview and Perspectives. Energies (Basel) 2022;15:4975. https://doi.org/10.3390/EN15144975/S1.

[47] Gaikwad SD, Ghosh PC. Sizing of a fuel cell electric vehicle: A pinch analysis-based approach. Int J Hydrogen Energy 2020;45:8985–93. https://doi.org/10.1016/j.ijhydene.2020.01.116.

[48] ISO 14687-2:2012(en), Hydrogen fuel — Product specification — Part 2: Proton exchange membrane (PEM) fuel cell applications for road vehicles n.d. https://www.iso.org/obp/ui/#iso:std:iso:14687:-2:ed-1:v1:en (accessed March 5, 2020).

[49] D.Sadler, A.Cargill, M.Crowther, A.Rennie, J.Watt, S.Burton, M.Haines, J.Trapps, M.Hand, R.Pomroy, K.Haggerty I. H21 Leeds City Gate: 100% hydrogen conversion feasibility study 2016. https://www.northerngasnetworks.co.uk/wp-content/uploads/2017/04/H21-Report-Interactive-PDF-July-2016.compressed.pdf (accessed March 5, 2020).

[50] Frenette G, Forthoffer D. Economic & commercial viability of hydrogen fuel cell vehicles from an automotive manufacturer perspective. Int J Hydrogen Energy 2009;34:3578–88. https://doi.org/10.1016/J.IJHYDENE.2009.02.072.

[51] Hydrogen Fueling Station Market Size, Share, Trends Report 2032 - Industry Growth Analysis n.d. https://tinyurl.com/3dcdhw58 (accessed March 31, 2024).

[52] Eudy L, Chandler K. Zero Emission Bay Area (ZEBA) Fuel Cell Bus Demonstration: Second Results Report. Golden, CO (United States): 2012. https://doi.org/10.2172/1048985.

[53] Eudy L, Post M, Jeffers M. Fuel Cell Buses in U.S. Transit Fleets: Current Status 2016. 2016.

[54] Wilson A, Marcinkoski J, Papageorgopoulos D, Kurtz J, Eudy L. DOE Hydrogen and Fuel Cell Technologies Program Record Title: On-Road Fuel Cell Stack Durability-2016. 2016.

[55] Hydrogen Fuel Cell Buses for Urban Transport in Brazil: Functional Test Results n.d. https://www.epri.com/research/products/1019311 (accessed March 31, 2024).

[56] Li X, Li J, Xu L, Yang F, Hua J, Ouyang M. Performance analysis of proton-exchange membrane fuel cell stacks used in Beijing urban-route buses trial project. Int J Hydrogen Energy 2010;35:3841–7. https://doi.org/10.1016/J.IJHYDENE.2010.01.097.

[57] Pan-Canadian Hydrogen Fuel Cell Electric Bus Demonstration And Integration Trial - Cutric n.d. https://cutric-crituc.org/marquee-projects/pan-canadian-hydrogen-fuel-cell-electric-bus-demonstration-and-integration-trial/ (accessed March 31, 2024).

[58] Powering Australia - DCCEEW n.d. https://www.dcceew.gov.au/energy/strategies-and-frameworks/powering-australia (accessed March 31, 2024).

[59] Clean Urban Transport for Europe (CUTE) - Hydrogen and Fuel Cell Buses – Policies - IEA n.d. https://www.iea.org/policies/3735-clean-urban-transport-for-europe-cute-hydrogen-and-fuel-cell-buses (accessed March 31, 2024).

[60] Alaswad A, Omran A, Sodre JR, Wilberforce T, Pignatelli G, Dassisti M, et al. Technical and Commercial Challenges of Proton-Exchange Membrane (PEM) Fuel Cells. Energies 2021, Vol 14, Page 144 2020;14:144. https://doi.org/10.3390/EN14010144.

[61] Toyota to Start Sales of Fuel Cell Buses under the Toyota Brand from Early 2017 (in Japan) - Toyota USA Newsroom n.d. https://pressroom.toyota.com/toyota-fuel-cell-buses-japan/ (accessed March 2, 2020).

[62] The Fuel Cell Industry Review 2016 FUEL CELL INDUSTRY REVIEW 2016. n.d.

[63] China's Buses Bolster Ballard's Three-Decade Quest for Hydrogen - Bloomberg n.d. https://www.bloomberg.com/news/articles/2017-03-26/china-s-buses-bolster-ballard-s-three-decade-quest-for-hydrogen (accessed March 2, 2020).

[64] Car Congress. Companies Form H2Bus Consortium to Deploy 1,000 Hydrogen Fuel Cell Buses and Infrastructure in Europe - Green Car Congress n.d. https://www.greencarcongress.com/2019/06/20190604-h2bus.html (accessed October 7, 2019).

[65] London hydrogen buses and the CHIC project 2016. https://www.all-energy.co.uk (accessed March 2, 2020).

[66] Fuel Cell Electric Buses | knowledge base n.d. https://www.fuelcellbuses.eu/ (accessed October 8, 2019).

[67] David Hart A, Howes J, Lehner -E F, Paul Dodds tech E, Hughes N, Fais B, et al. Scenarios for deployment of hydrogen in contributing to meeting carbon budgets and the 2050 target 2015. www.theccc.org.uk/wp-content/uploads/2015/11/E4tech-for-CCC-Scenarios-for-deployment-of-hydrogen-in-contributing-to-meeting-carbon-budgets.pdf (accessed March 2, 2020).

[68] Final report of the bus study: Urban buses: alternative power trains for Europe | n.d. https://www.fch.europa.eu/node/790 (accessed March 2, 2020).

[69] Heliocentris, FAUN for hybrid garbage collection vehicle. Fuel Cells Bulletin 2010;2010:3–4. https://doi.org/10.1016/s1464-2859(10)70274-1.

[70] Tesla Semi gets "peppy and quiet" hydrogen fuel cell competitor from Kenworth-Toyota n.d. https://www.teslarati.com/tesla-semi-rival-kenworth-toyota-hydrogen-fuel-cell-truck-specs/ (accessed March 3, 2020).

[71] Kenworth Trucks - The World's Best ® n.d. https://www.kenworth.com/news/news-releases/2016/august/t680-low-emissions/ (accessed March 3, 2020).

[72] Nikola Corp | Nikola One n.d. https://nikolamotor.com/one (accessed March 3, 2020).

[73] World's largest brewer orders 800 hydrogen-electric trucks n.d. https://www.edie.net/news/6/World-s-largest-brewer-orders-800-hydrogen-electric-trucks/ (accessed March 3, 2020).

[74] Third IMO GHG Study 2014 Executive Summary and Final Report. 2015.

[75] McConnell VP. Now, voyager? The increasing marine use of fuel cells. Fuel Cells Bulletin 2010;2010:12–7. https://doi.org/10.1016/S1464-2859(10)70166-8.

[76] Klebanoff LE, Pratt JW, Leffers CM, Sonerholm KT, Escher T, Burgard J, et al. Comparison of the greenhouse gas and criteria pollutant emissions from the SF-BREEZE high-speed fuel-cell ferry with a diesel ferry. Transp Res D Transp Environ 2017;54:250–68. https://doi.org/10.1016/j.trd.2017.05.012.

[77] Zero-Emission Vessels 2030. How do we get there? n.d. https://www.lr.org/en/insights/articles/zev-report-article/ (accessed February 27, 2020).

[78] Hydrogen fuel-cell ferry clears first hurdle n.d. https://www.eastbaytimes.com/2016/11/14/hydrogen-powered-ferry-national-lab-says-its-possible-though-costly/ (accessed February 29, 2020).

[79] Wang X, Zhu J, Han M. Industrial Development Status and Prospects of the Marine Fuel Cell: A Review. Journal of Marine Science and Engineering 2023, Vol 11, Page 238 2023;11:238. https://doi.org/10.3390/JMSE11020238.

[80] Hydroville | CMB.TECH n.d. https://cmb.tech/business-units/marine/hydroville (accessed April 1, 2024).

[81] VESSEL REVIEW | Hydrocat 48 – Netherlands' Windcat Workboats puts hydrogen-fuelled newbuild in operation Baird Maritime n.d. https://www.bairdmaritime.com/work-boat-world/offshore-world/offshore-operations-maintenance/crew-transfer/vessel-review-hydrocat-48-netherlands-windcat-workboats-puts-hydrogen-fuelled-newbuild-in-operation/ (accessed April 1, 2024).

[82] Video: Norway's Hydrogen-Powered Ferry Begins Service n.d. https://maritime-executive.com/article/video-norway-s-hydrogen-powered-ferry-begins-service (accessed April 1, 2024).

[83] SWITCH Maritime n.d. https://www.switchmaritime.com/ (accessed April 1, 2024).

[84] UC San Diego Receives $35 Million in State Funding for New California Coastal Research Vessel | Scripps Institution of Oceanography n.d. https://scripps.ucsd.edu/news/uc-san-diego-receives-35-million-state-funding-new-california-coastal-research-vessel (accessed April 1, 2024).

[85] Maritime Applications for Hydrogen Fuel Cells : Sandia Energy n.d. https://energy.sandia.gov/programs/sustainable-transportation/hydrogen/fuel-cells/maritime-applications/#tank-tech (accessed April 1, 2024).
[86] Yanmar Develops Maritime Hydrogen Fuel Cell System Towards a Carbon Neutral Society | Y media | YANMAR n.d. https://www.yanmar.com/global/about/ymedia/article/maritime-hydrogen-fuel-cell.html (accessed April 1, 2024).
[87] Exploring the Potential of Hydrogen for Maritime Mobility | FASTECH n.d. https://www.fastechus.com/blog/potential-of-fuel-cells-for-hydrogen-powered-ships (accessed April 1, 2024).
[88] German state thrusts hydrogen-powered hydrail into the spotlight n.d.
[89] Alstom Coradia iLint n.d. https://www.alstom.com/our-solutions/rolling-stock/coradia-ilint-worlds-1st-hydrogen-powered-train (accessed February 22, 2020).
[90] Böhm M, Fernández Del Rey A, Pagenkopf J, Varela M, Herwartz-Polster S, Nieto Calderón B. Review and comparison of worldwide hydrogen activities in the rail sector with special focus on on-board storage and refueling technologies. Int J Hydrogen Energy 2022;47:38003–17. https://doi.org/10.1016/J.IJHYDENE.2022.08.279.
[91] Gallucci M. Hydrogen trains roll into service: A new hybrid locomotive signals a growing push for zero-emission rail technologies-[News]. IEEE Spectr 2019;56:6–7. https://doi.org/10.1109/MSPEC.2019.8784110.
[92] Xu Z, Zhao N, Hillmansen S, Roberts C, Yan Y. Techno-Economic Analysis of Hydrogen Storage Technologies for Railway Engineering: A Review. Energies 2022, Vol 15, Page 6467 2022;15:6467. https://doi.org/10.3390/EN15176467.
[93] Pielecha I, Dimitrov R, Mihaylov V. Energy flow analysis based on a simulated drive of a hybrid locomotive powered by fuel cells. Rail Vehicles 2022;2022:68–76. https://doi.org/10.53502/RAIL-152703.
[94] Miller AR, Peters J, Smith BE, Velev OA. Analysis of fuel cell hybrid locomotives. J Power Sources 2006;157:855–61. https://doi.org/10.1016/j.jpowsour.2005.12.051.
[95] Guo L, Yedavalli K, Zinger D. Design and modeling of power system for a fuel cell hybrid switcher locomotive. Energy Convers Manag 2011;52:1406–13. https://doi.org/10.1016/j.enconman.2010.10.003.
[96] Hydrogen fuel cell trains to run on British railways from 2022 n.d. https://www.telegraph.co.uk/cars/news/hydrogen-fuel-cell-trains-run-british-railways-2022/ (accessed February 29, 2020).
[97] Herwartz S, Pagenkopf J, Streuling C. Sector coupling potential of wind-based hydrogen production and fuel cell train operation in regional rail transport in Berlin and Brandenburg. Int J Hydrogen Energy 2021;46:29597–615. https://doi.org/10.1016/J.IJHYDENE.2020.11.242.
[98] Saeed M, Briz F, Guerrero JM, Larrazabal I, Ortega D, Lopez V, et al. Onboard Energy Storage Systems for Railway: Present and Trends. IEEE Open Journal of Industry Applications 2023;4:238–59. https://doi.org/10.1109/OJIA.2023.3293059.
[99] Hosseini SE. Hydrogen fuel, a game changer for the world's energy scenario 2023. https://doi.org/10.1080/15435075.2023.2244050.
[100] Miller AR, Hess KS, Erickson TL, Dippo JL. Fuelcell-hybrid shunt locomotive: Largest fuelcell land vehicle. IET Seminar Digest 2010;2010. https://doi.org/10.1049/IC.2010.0046/CITE/REFWORKS.
[101] Hoffrichter A, Fisher P, Tutcher J, Hillmansen S, Roberts C. Performance evaluation of the hydrogen-powered prototype locomotive 'Hydrogen Pioneer.' J Power Sources 2014;250:120–7. https://doi.org/10.1016/J.JPOWSOUR.2013.10.134.
[102] All Aboard the U.K.'s First Hydrogen Train - IEEE Spectrum n.d. https://spectrum.ieee.org/energywise/transportation/alternative-transportation/all-aboard-uk-first-hydrogen-train (accessed February 28, 2020).
[103] Mustafa MY, Al-Mahadin A, Kanstad BE, Calay RK. Fuel cell technology application for Dubai rail systems. 2018 Advances in Science and Engineering Technology International Conferences, ASET 2018, Institute of Electrical and Electronics Engineers Inc.; 2018, p. 1–7. https://doi.org/10.1109/ICASET.2018.8376798.

[104] Baroutaji A, Wilberforce T, Ramadan M, Olabi AG. Comprehensive investigation on hydrogen and fuel cell technology in the aviation and aerospace sectors. Renewable and Sustainable Energy Reviews 2019:31–40. https://doi.org/10.1016/j.rser.2019.02.022.
[105] First e-flight-expo takes off at AERO 2009. www.e-flight-expo.com (accessed October 31, 2019).
[106] Helite unveils ultralight aircraft powered by fuel cell alone. Fuel Cells Bulletin 2009;2009:4. https://doi.org/10.1016/s1464-2859(09)70110-5.
[107] First flight of DLR's HY4 fuel cell light aircraft. Fuel Cells Bulletin 2016;2016:1. https://doi.org/10.1016/s1464-2859(16)30261-9.
[108] ZeroAvia unveils fuel cell powered prototype light aircraft, flight test. Fuel Cells Bulletin 2019;2019:6. https://doi.org/10.1016/S1464-2859(19)30368-2.
[109] Alaka'i Technologies unveils Skai hydrogen-electric VTOL aircraft. Fuel Cells Bulletin 2019;2019:6. https://doi.org/10.1016/s1464-2859(19)30232-9.
[110] Hydrogenics contract for aircraft fuel cells, Air Liquide takes stake. Fuel Cells Bulletin 2019;2019:14. https://doi.org/10.1016/s1464-2859(19)30033-1.
[111] HES unveils Element One hydrogen-electric passenger aircraft. Fuel Cells Bulletin 2018;2018:4–5. https://doi.org/10.1016/s1464-2859(18)30402-4.
[112] European HYCARUS project to demo fuel cell power in aircraft. Fuel Cells Bulletin 2013;2013:4–5. https://doi.org/10.1016/s1464-2859(13)70251-7.
[113] Sone Y, Ueno M, Naito H, Kuwajima S. One kilowatt-class fuel cell system for the aerospace applications in a micro-gravitational and closed environment. J Power Sources 2006;157:886–92. https://doi.org/10.1016/j.jpowsour.2006.03.042.
[114] office of the Secretary of Defense. Unmanned Aircraft Systems Roadmap 2005-2030 2005. https://fas.org/irp/program/collect/uav_roadmap2005.pdf (accessed November 24, 2019).
[115] Org WE, Dudek M, Tomczyk P, Wygonik P, Korkosz M, Bogusz P, et al. Electrochemical Science Hybrid Fuel Cell-Battery System as a Main Power Unit for Small Unmanned Aerial Vehicles (UAV). vol. 8. 2013.
[116] Bradley TH, Moffitt BA, Mavris D, Parekh DE. Applications - Transportation | Aviation: Fuel Cells. Encyclopedia of Electrochemical Power Sources, Elsevier; 2009, p. 186–92. https://doi.org/10.1016/B978-044452745-5.00342-7.
[117] AeroVironment's unmanned aircraft achieves record flight. Fuel Cells Bulletin 2007;2007:8. https://doi.org/10.1016/S1464-2859(07)70321-8.
[118] US Navy orders 13 Protonex fuel cells to power UAV field trials. Fuel Cells Bulletin 2018;2018:4. https://doi.org/10.1016/s1464-2859(18)30235-9.
[119] WSU, Insitu win US Army grant to develop liquid hydrogen UAV. Fuel Cells Bulletin 2019;2019:5. https://doi.org/10.1016/s1464-2859(19)30320-7
[120] In brief. Fuel Cells Bulletin 2019;2019:5–15. https://doi.org/10.1016/S1464-2859(19)30276-7.
[121] HES Hycopter drone highlighted by Dubai Police at Intersec 2019. Fuel Cells Bulletin 2019;2019:6. https://doi.org/10.1016/s1464-2859(19)30053-7.
[122] Hycopter drones from HES Energy for safety inspection of hydropower dams in Brazil. Fuel Cells Bulletin 2019;2019:1. https://doi.org/10.1016/S1464-2859(19)30355-4.
[123] Intelligent Energy showcases drone fuel cell modules in Japan. Fuel Cells Bulletin 2019;2019:4. https://doi.org/10.1016/s1464-2859(19)30319-0.
[124] DOE funds to deploy fuel cells for baggage vehicles at airports. Fuel Cells Bulletin 2012;2012:3–4. https://doi.org/10.1016/s1464-2859(12)70121-9.
[125] McConnell VP. Fuel cells in forklifts extend commercial reach. Fuel Cells Bulletin 2010;2010:12–9. https://doi.org/10.1016/S1464-2859(10)70296-0.
[126] Linde P250 fuel cell tow tractor launched to serve airport logistics. Fuel Cells Bulletin 2019;2019:4. https://doi.org/10.1016/S1464-2859(19)30362-1.
[127] Plug Power showcases fuel cell powered GSE project at local airport. Fuel Cells Bulletin 2019;2019:4. https://doi.org/10.1016/S1464-2859(19)30363-3.
[128] Air Liquide, Groupe ADP hydrogen station at Paris-Orly Airport. Fuel Cells Bulletin 2018;2018:7–8. https://doi.org/10.1016/s1464-2859(18)30017-8.
[129] US Hybrid, BYD to develop fuel cell bus for Honolulu Airport. Fuel Cells Bulletin 2018;2018:2. https://doi.org/10.1016/s1464-2859(18)30188-3.

[130] Toyota, Hino deliver fuel cell bus for Haneda airport route in Tokyo. Fuel Cells Bulletin 2011;2011:3. https://doi.org/10.1016/s1464-2859(11)70005-0.

CHAPTER 6

The role of hydrogen diplomacy in economic advancement

6.1 Introduction

Throughout history, the discovery and utilization of transformative energy sources have marked significant turning points. More than a century ago, the world experienced such a pivotal moment with the discovery of oil. This discovery not only revolutionized industries but also had a profound impact on global politics, giving rise to what is now known as oil diplomacy. Countries possessing abundant oil reserves quickly became major players in international affairs, leveraging their energy resources to wield substantial influence and shape diplomatic relations based on energy security and economic strength. Conversely, nations with advanced oil and gas technologies emerged as leaders in innovation and strategic partnerships, solidifying their roles as key drivers of the global economy. Today, we stand at the threshold of another transformative era centered around hydrogen. Much like oil transformed the world a century ago, hydrogen is poised to redefine the energy landscape, ushering in an era of sustainable development and international collaboration [1]. The parallels between the rise of oil and the emergence of hydrogen are striking. Both represent abundant and versatile energy sources capable of powering economies and fostering progress. Similar to oil, hydrogen has the potential to revolutionize industries and reshape geopolitical dynamics. Countries with access to hydrogen resources or advanced hydrogen technologies could emerge as significant players in the global energy scene, influencing international relations and strategic partnerships. Moreover, the transition to hydrogen-based energy systems has the potential to stimulate new collaborations and alliances as nations work together to develop essential infrastructure and technologies. This collaborative effort could lead to a reconfiguration of global power dynamics, with nations embracing hydrogen taking the lead in innovation and sustainable development. However, the shift towards a hydrogen-based economy presents its own set of challenges. Significant investments in infrastructure, research, and development, as well as supportive policies, are needed to overcome barriers such as high production costs and limited distribution networks. Nevertheless, the potential advantages of hydrogen, including its cleanliness, renewable nature, high

DOI: https://doi.org/10.55670/fpll.book/1

energy density, and versatility, position it as a promising alternative to fossil fuels and a key driver of the global energy transition [2]. As the world grapples with urgent issues like climate change and energy security, the ascent of hydrogen could represent a crucial moment in shaping the future of our planet (Figure 6.1).

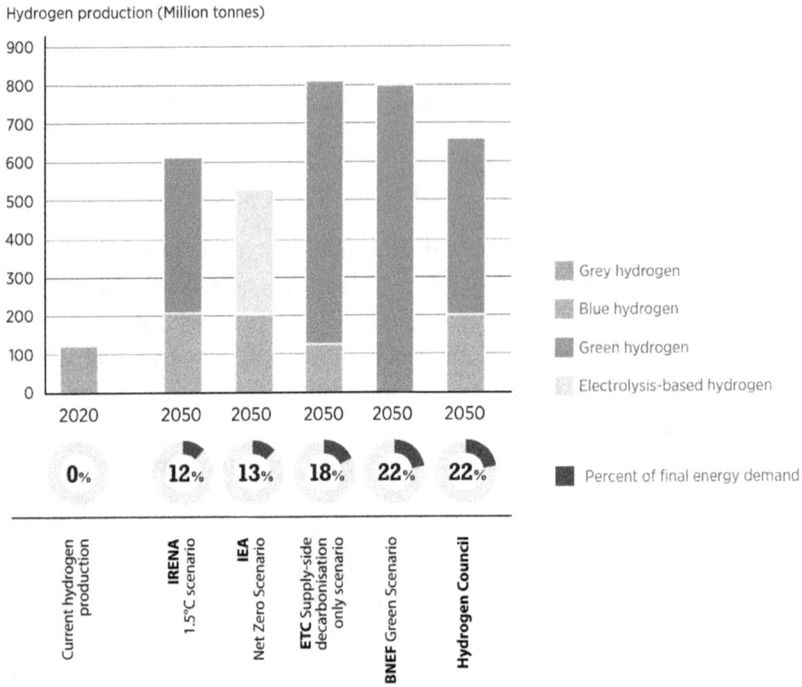

Figure 6.1 Estimates for global hydrogen demand in 2050 [3]

The advancement and implementation of hydrogen technologies have the capacity to stimulate economic expansion and job generation worldwide. Investing in hydrogen infrastructure, research, and development not only fosters innovation but also opens doors for industries and enterprises to flourish within a low-carbon economic framework. As nations shift towards hydrogen-based economies, there arises a heightened demand for skilled labor in various sectors, such as hydrogen production, storage, transportation, and utilization, leading to the creation of jobs and economic diversification. The expansion of hydrogen value chains presents complex trade-offs and dilemmas. While investing in hydrogen infrastructure is crucial for reducing overall costs, it also carries risks, especially without guaranteed supply and demand. Countries, companies, and cities that commit to the wrong pathway may face substantial losses. Conversely, those entities that achieve technological leadership stand to gain significantly. Another set of decisions revolves around the geography of hydrogen production and utilization rather than just technology. Developed nations must assess the trade-off between large-scale imports and the advantages and drawbacks of domestic hydrogen production. Conversely, countries with ample resources for inexpensive hydrogen production can choose to export hydrogen extensively or use it to attract downstream industries like iron and steel. This context gives rise to three geopolitical implications of hydrogen adoption: the potential creation of new dependencies among states if they opt for large-scale imports; a shift in energy transition interests and actor compositions if hydrogen supports fossil fuel producers and established players; and a potential increase in technological and geo-economic competition between nations [4]. In the domain of international relations, the establishment of hydrogen policies and frameworks carries significant weight. Much like how oil diplomacy shaped alliances and strategic collaborations in the past, hydrogen diplomacy is poised to redefine global cooperation and partnership efforts. Nations endowed with abundant renewable resources like solar and wind energy can tap into these sources to generate green hydrogen, thereby paving the way for energy autonomy and potential exports (Figure 6.2). Conversely, countries with advanced hydrogen technologies can capitalize on their expertise to foster alliances, share knowledge, and propel innovation globally [5]. However, the transition towards a hydrogen-centric economy presents its set of challenges. The geopolitical ramifications of this transformation could give rise to new power dynamics, potential conflicts, and apprehensions regarding green colonialism, where developing nations might be viewed primarily as providers of raw materials. To address these risks, international collaboration and governance are imperative. Setting standards, certifications, and trade agreements for hydrogen becomes crucial to ensure an equitable and sustainable global hydrogen market.

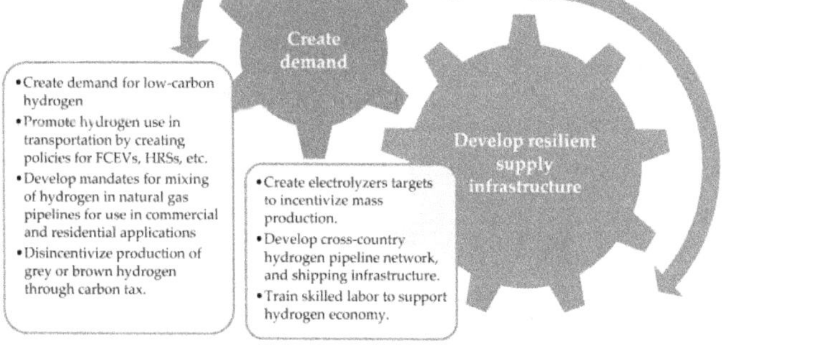

Figure 6.2 Policy frameworks for a faster transition to the hydrogen economy [6]

6.2 Geopolitical aspects of hydrogen technology

Currently, the hydrogen industry remains largely localized, with around 85% of hydrogen being produced and used on-site, primarily within refineries. To expand production, developed nations may opt to establish hydrogen facilities domestically or import hydrogen from countries with abundant renewable or fossil energy reserves. For major economies like the EU and Japan, importing green hydrogen from regions rich in cost-effective renewable resources could potentially lower the costs associated with transitioning to sustainable energy and alleviate the strain on domestic resources such as land and sea space needed for large-scale renewable deployment. However, engaging in cross-border maritime hydrogen trade could introduce new interdependencies between nations and pose additional risks related to maritime shipping. Hydrogen has the capacity to reshape the global energy trade landscape, paving the way for a new category of exporters, as illustrated in Figure 6.3. Nations such as Japan and South Korea are gearing up for substantial hydrogen imports. Conversely, countries like Australia, Chile, and New Zealand are strategically focusing on the potential for hydrogen exports. This shift could lead to the emergence of novel trade connections, and as hydrogen gradually replaces fossil fuels, it may alleviate the strain on critical maritime bottlenecks for oil, such as the Strait of Hormuz, or pivotal transit nations for natural gas like Ukraine until recently. Simultaneously, new maritime routes could gain significance in the context of global energy trade routes.

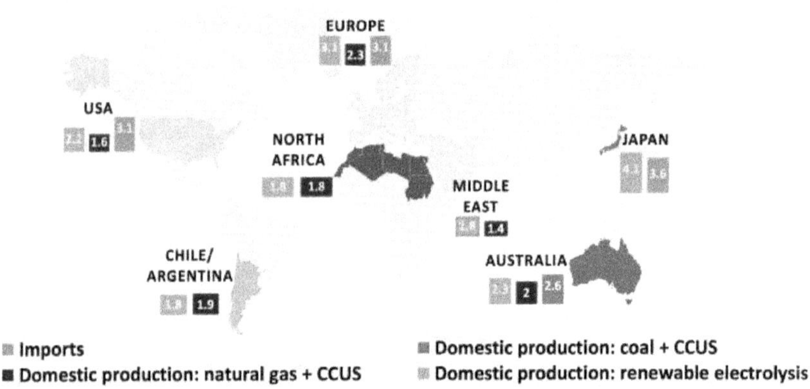

Figure 6.3 Costs of various hydrogen types by location, \$/kg of hydrogen [7]

For nations in close proximity, hydrogen can be transported via pipelines. In Northwestern Europe, for instance, there is a 900 km network of hydrogen pipelines linking Rotterdam (Netherlands), Antwerp (Belgium), and Dunkirk (France). Globally, over 4,500 km of hydrogen pipelines are already in existence [8]. Recently, German gas pipeline operators have announced plans to construct a hydrogen grid spanning approximately 5,900 km, which would be the largest worldwide. Although these regional and local networks could potentially merge into transregional networks, long-distance hydrogen pipeline transport has not yet been tested. Numerous countries are actively involved in what could be termed "hydrogen diplomacy." The Dutch government has appointed a dedicated "hydrogen envoy". Japanese diplomats and industrial stakeholders are discussing hydrogen fuel procurement with Australia, Brunei, Norway, and Saudi Arabia [9]. Germany has entered a cooperation agreement with Morocco focusing on methanol production from hydrogen [10]. South Korea is interested in collaborations with Norway, the Netherlands is eyeing Portugal as a potential hydrogen supplier [11], and industrial entities in Belgium are exploring partnerships with Oman and Chile for large-scale hydrogen imports [12].

If the current trajectory of forming bilateral partnerships persists, the hydrogen market could initially emerge from a fragmented foundation, resembling the early stages of the LNG market. Initially, LNG projects were governed by rigid, bilateral, long-term contracts with prices linked to oil, earning them the moniker floating pipelines [13]. Japan played a pivotal role in kickstarting the LNG market as the first major buyer. Its commitment to substantial hydrogen imports could position it again as a pioneer in the global gas market, this time in the realm of hydrogen. A notable difference in hydrogen trade compared to crude oil or natural gas trade is its reduced asymmetry. Hydrogen can be produced almost anywhere globally. The potential for many countries to become both producers and consumers of hydrogen, coupled with its storage capabilities, makes it challenging for exporters to exploit hydrogen trade as a strategic tool or for importers to be at the mercy of a small group of suppliers. However, unlike cross-border electricity trade, where electrons flow bidirectionally depending on supply and demand dynamics, international hydrogen trade may not be as reciprocal. Nonetheless, it will enhance the energy security of importers by serving as a backup to the electricity system.

6.3 Hydrogen policy and geo-economic competitions

Governments need to broaden their focus beyond just technologies and costs and also consider the geo-economic implications of hydrogen. The potential for centralized production and distribution of hydrogen presents opportunities to integrate segments of the fossil fuel industry into the energy transition, potentially offering a lifeline to petrostates. This potential could be strategically utilized to ensure that major oil exporters like Russia and Saudi Arabia maintain at least a minimal commitment to the goals of the Paris Agreement. In the Middle East and North Africa region, hydrogen could provide a solution to the significant challenge of diversifying economies away from dependence on oil and gas exports. These countries possess key advantages such as abundant and low-cost solar energy for green hydrogen production, options for underground carbon storage (for blue hydrogen production pathways), and a strategic geographical location to serve both

European and Asian markets [14]. Additionally, non-oil economies like Morocco could leverage their abundant renewable energy resources at low costs. However, limited freshwater availability in the region may pose a challenge, necessitating additional investments in desalination capacity, which could increase costs. Furthermore, hydrogen has the potential to engage incumbent industries in the energy transition, potentially accelerating deep and rapid decarbonization efforts. The oil and gas sectors, in particular, have shown keen interest in hydrogen due to its similarities with their core activities of producing, transporting, and distributing fuels. Additionally, existing gas infrastructure can be repurposed to some extent for hydrogen, making it a favored option among established natural gas actors, especially pipeline distribution companies.

Securing control over low-carbon energy technology value chains is crucial for enhancing economic competitiveness, bolstering national security, and achieving energy independence. Early pioneers in the hydrogen sector stand a chance to market their technologies globally. Leadership in technology can be established across various facets of the hydrogen value chain, ranging from fuel-cell membranes and precisely engineered storage tanks to pipeline materials and burners. In June 2020, Germany unveiled plans to invest 9 billion euros in expanding its hydrogen capacity, aiming to position itself as a primary supplier of this technology worldwide as part of its post-COVID-19 recovery strategy. Concurrently, German entities like RWE and Thyssenkrupp have initiated collaborations to produce green hydrogen and utilize it in steel production [15]. The pursuit of technology dominance is evident in numerous countries and industries. For instance, in the automotive sector, Japanese automakers Honda and Toyota are banking on fuel cell vehicles surpassing batteries, particularly in terms of range, while Chinese counterparts are advancing significantly in electric vehicles, and German manufacturers have long focused on enhancing the efficiency of diesel-powered combustion engines. Public funding often supports initiatives to develop hydrogen value chains, intensifying the realm of geo-economic competition. This is why the EU Commission has outlined plans to launch an EU Hydrogen Alliance, as part of its recent document on a new industrial strategy for Europe [16]. The growing cost advantage of producing electrolyzers in China, estimated to be 83% cheaper than in Western countries in 2019 and could fall 30% more by 2025, raises concerns in Europe and North America about China's potential dominance in another critical energy sector [17]. The emergence of intercontinental hydrogen value chains will heighten industrial competition among countries vying for the placement of energy-intensive industries. Nations with substantial potential to produce hydrogen from local resources, whether renewable or fossil fuels, may expand their value chains into energy-intensive sectors like chemicals and steel rather than solely exporting hydrogen to developed nations. This dynamic could introduce a new layer of geo-economic rivalry among major global powers. Furthermore, there is a risk associated with viewing developing countries solely as suppliers of raw materials, potentially leading to issues of green colonialism as the hydrogen revolution unfolds.

6.4 Building an international hydrogen economy

In the midst of the initial global energy crisis in the 1970s, the concept of a hydrogen economy was first introduced, envisioning a long-term solution that prioritizes energy independence and safety. Criticism at that time centered on issues such as fossil-fuel-based production, suboptimal fuel cell performance, and concerns regarding both cost and safety. However, recent advancements in hydrogen production, storage, and utilization technologies have sparked a renewed interest in the potential of the hydrogen economy [18]. Progress in materials, designs, and system integration has demonstrated satisfactory fuel cell performance, particularly in small-scale applications. Extensive research has been dedicated to exploring safe and cost-effective methods for hydrogen storage. These collective efforts have prompted stakeholders to launch various pilot-scale projects, opening up new avenues for academic exploration. As a result, there has been a resurgence of interest in topics related to the hydrogen economy. Tashie-Lewis et al. [19] conducted a technological review encompassing hydrogen production, distribution, storage, and power conversion within a hydrogen economy. Their conclusion emphasized that, contrary to technical challenges, the primary constraints hindering the hydrogen economy are related to technology costs. Hence, there is a pressing need to actively reduce costs through continuous investment until a specific scale is achieved.

Paired with its capability to reduce carbon emissions in the energy system, particularly when utilizing green hydrogen derived from renewable energy sources, and to ensure future security in clean energy, hydrogen is anticipated to play a central role in the ongoing transition towards achieving net-zero emissions. This potential has thrust hydrogen into prominence within industry, politics, and societies, prompting numerous countries globally to embrace national hydrogen policies and strategies that delineate pathways toward a hydrogen-based economy. Aligned with their commitment to the 2015 Paris Agreement, most industrialized nations anticipate that hydrogen will contribute significantly to attaining goals related to greenhouse gas reduction. This dedication is evident in substantial investments directed toward hydrogen technologies, positioning industrialized nations as global leaders in the realm of hydrogen research and innovation [20].

The notion of a "hydrogen economy" refers to the integration of an energy system in which hydrogen serves as a fundamental energy carrier alongside electricity. Hydrogen boasts a notably higher specific energy density per unit mass, with a lower heating value of 120 MJ/kg and a higher heating value of 144 MJ/kg, surpassing that of many traditional fuels. Its oxidation process results solely in water vapor or liquid water, thereby avoiding generating direct hazardous emissions [21]. Assuming hydrogen is produced from local, renewable energy sources, it transforms into a completely green fuel with zero equivalent CO_2 emissions. With the growing awareness of the limited nature of energy resources and the escalating concerns related to climate change and other environmental challenges, there is an intensified demand for renewable energies and enhanced energy efficiency.

The hydrogen economy experienced a resurgence in the late 1990s, driven by the potential for clean and sustainable energy. However, this renewed interest only persisted through the first half of the initial decade of the new millennium [22]. A plausible explanation for this decline is that the progress in the hydrogen chain, encompassing hydrogen production technology, fuel delivery infrastructure, and energy policy, failed to fulfill the partly unrealistic promises that were initially raised with enthusiasm. The cost of hydrogen technology remained prohibitively high, the hydrogen production chain operated with low efficiency, and there was a lack of interest in developing hydrogen-based solutions [23]. The affordability of hydrogen technology, the operational efficiency of the hydrogen production chain, and safety concerns have been historic stumbling blocks. However, ongoing research and development efforts are addressing these challenges, focusing on reducing costs and enhancing the efficiency of hydrogen technologies. Safety measures, including the development of essential elements such as pressure relief devices and leak detection, are actively being explored.

Recently, advancements in polymer-exchange membrane fuel cell (PEMFC) technology and materials that enable high-pressure hydrogen storage up to 700 bar have reignited interest in developing H_2-powered fuel cell vehicles within the automotive industry. Car manufacturers, including Toyota [24], Hyundai [25], and Honda [26] have introduced partly or fully commercialized fuel cell cars, revitalizing the relevance of the topic. The commitment to the hydrogen economy is reflected in the adoption of national hydrogen policies by various countries. Governments recognize the role of hydrogen in achieving clean energy goals and are investing in research, development, and infrastructure to facilitate the widespread adoption of hydrogen technologies. The European Union's announcement of a demand for 20 billion kg of natural hydrogen by 2030 and the significant investment by the U.S. government in hydrogen discovery underscore the global momentum towards a hydrogen-centric energy future. Hydrogen's versatility extends across various sectors, making it a promising candidate for diverse applications (Figure 6.4). From transportation, where hydrogen fuel cells power vehicles, to industrial processes and power generation, hydrogen's applications are wide-ranging. The potential for hydrogen to serve as an energy carrier, store renewable energy, and contribute to a more resilient and decentralized energy system positions it as a transformative force. Creating an international hydrogen economy is crucial for aligning with the Paris climate objectives, but it requires a coordinated effort involving multiple stakeholders. While current national hydrogen strategies provide valuable assessments from policymakers, their impact could be amplified by integrating them into regional and global strategies.

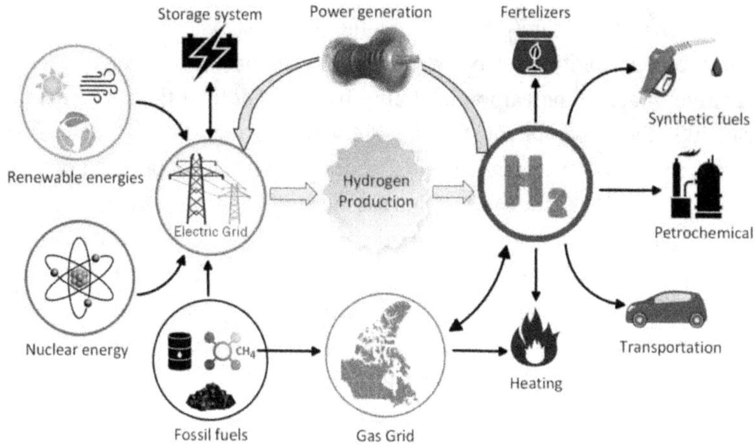

Figure 6.4 Hydrogen economy, including hydrogen production and utilization [27]

Balancing national energy security and industrial interests with the collective goal of addressing climate change and maintaining geopolitical stability is essential. National governments play a pivotal role in establishing a stable, long-term policy framework and fostering an encouraging business environment. Bringing together the various aspects of a global hydrogen economy is not a straightforward task, as there is currently no dedicated forum for such discussions. Experience from the natural gas sector shows that coordination, let alone the establishment of an international organization, can be challenging [28]. However, hydrogen holds an advantage over natural gas due to its potential link to decarbonization and its alignment with the climate agenda. Several international organizations can contribute valuable insights and expertise to discussions on hydrogen pathways and certification. For instance, the International Renewable Energy Agency specializes in renewables and green hydrogen, the International Energy Agency focuses on hydrogen from an energy security perspective, and the International Atomic Energy Agency offers insights into nuclear hydrogen production. Additionally, the International Energy Forum serves as a platform for dialogue between energy suppliers and consumers and could expand its scope to include discussions on hydrogen. Achieving collaboration among these organizations would require adept steering and leadership, which could be facilitated by groups like the G7 and G20 as potential governance bodies. Collaborative efforts are needed both among consumers to establish unified regulatory frameworks and certifications and between producers and consumers to outline the transition from fossil fuels to hydrogen pathways.

Corporations and investors are reevaluating the viability of hydrogen from a business perspective, carefully assessing the risks associated with potential setbacks against the opportunity for substantial growth that appears to be on the horizon. The imperative of achieving carbon neutrality by the middle of the century has become ingrained in their operational strategies in many instances. There is a pressing need for researchers and policymakers to delve deeper into the geopolitical dimensions of hydrogen. The emergence of cross-border maritime hydrogen trade has the capacity to reconfigure the global energy trade map, establish a new cadre of energy exporters, and restructure the geopolitical landscape, influencing relationships and alliances among nations. A cohesive effort involving national governments, international institutions, researchers, and businesses along these lines could pave the way for the hydrogen economy's advancement, potentially allowing for effective management of its geopolitical implications.

6.5 Hydrogen technology and the job market

In 2020, the global energy sector provided employment to nearly 60 million individuals across the world. By 2021, the workforce in renewable energy had grown to 12.7 million, marking an increase of 700,000 new jobs within a year despite the ongoing challenges posed by the COVID-19 pandemic. Looking ahead, as the global demand for energy is projected to double by 2050, a well-planned shift towards clean energy sources will play a crucial role in enhancing the resilience of the job market against resource scarcity and demographic shifts. The rise of hydrogen as a key 21st-century solution to meeting greenhouse gas emission targets set by the Paris Agreement underscores its potential as a significant job creator. Green hydrogen technology, in particular, is poised to generate substantial employment opportunities. It is estimated that the green hydrogen economy could satisfy around 18% of the global energy demand, leading to the creation of 30 million careers by 2050, generating a global annual revenue of $2.5 trillion. The worldwide transition to hydrogen as a sustainable energy option creates fresh job prospects within the renewable energy sector, offering various roles throughout the hydrogen value chain. The renewable energy and hydrogen sectors are anticipated to drive significant job growth, necessitating extensive infrastructure development, technological advancements, and support services. Developing new transportation and infrastructure technologies optimized for hydrogen utilization, especially in road transport, is expected to be the primary driver of job creation in a high-demand scenario. Furthermore, hydrogen diplomacy acts as a catalyst for creativity and exploration of sustainable energy technologies. As a result, novel positions are created in research institutions, universities, and tech firms dedicated to hydrogen-related progress. There is a heightened need for scientists, engineers, and technicians specializing in hydrogen fuel cells, electrolysis, and storage systems, fueling job growth in these specialized domains.

Shifting towards a hydrogen-based economy demands the implementation of workforce training and educational schemes. Both governments and industries must allocate resources to training programs, ensuring that workers are proficient in hydrogen technology, safety procedures, and infrastructure oversight. This encompasses vocational training, certification schemes, and academic curricula customized for the hydrogen industry. Consequently, hydrogen diplomacy has the potential to generate employment opportunities in education, training, and workforce development arenas. For instance, the U.S. Department of Energy has initiated programs such as the Hydrogen Education for a Decarbonized Economy (H2EDGE) program. This initiative seeks to cultivate a skilled workforce for the burgeoning hydrogen industry by offering training, educational resources, and establishing university courses and certifications [29]. Additionally, the American Institute of Chemical Engineers, in partnership with the University of Houston, delivers a "Hydrogen Economy Program" designed to provide a comprehensive understanding of the hurdles and solutions involved in building a sustainable hydrogen economy. This micro-credential program covers various aspects, including hydrogen production, transportation, safety protocols, and business prospects [30].

Countries like Germany, Japan, and the Netherlands, which are positioning themselves as frontrunners in hydrogen technology advancement, are attracting investments and partnerships from other nations. This trend is not only generating employment opportunities in technical domains such as engineering and manufacturing but also in ancillary functions like business development, finance, marketing, and legal services catering to hydrogen projects and investments [31]. The development of hydrogen and other clean energy technologies could lead to the creation of over 1.7 million fresh green jobs across Europe by 2040. Among the countries anticipated to experience significant job growth in this sector are Spain, the UK, Germany, and France [32]. Spain is poised to take the lead in green hydrogen production, expected to result in approximately 181,000 new jobs. Following closely behind, the United Kingdom is projected to create 173,000 new jobs in this sector, with Germany following at 145,000 and France at 105,000. Nevertheless, this transition will necessitate substantial reskilling and upskilling of the workforce, with approximately 60% of workers needing new skill sets. Governments and employers are urged to collaborate on training initiatives, workforce planning strategies, and inclusive programs to ensure that workers have access to emerging job opportunities in the green economy. Despite the rising involvement of women in environmentally friendly positions, their representation remains below 40% in many countries except for Spain and Italy, where it is anticipated that women will hold more than 50% of direct green jobs by 2040. The growth of additional green jobs could also contribute significantly to the gross domestic product (GDP) of the eurozone, potentially reaching 145 billion euros by 2040, translating to an average annual increase of 8.5 billion euros [33].

According to the European Commission, every billion invested in green hydrogen could result in the creation of 20,000 jobs across the entire supply chain, encompassing production, transmission, storage, and utilization. The green hydrogen economy is also forecasted to generate approximately 5.4 million upstream jobs in the EU by 2050, with a significant portion of these new jobs linked specifically to fuel cell technology, amounting to nearly three times the current employment figures in the EU chemical industry [34]. Moreover, the adoption of fuel cell electric vehicles could play a pivotal role in retaining and potentially expanding the European automotive industry, particularly within the heavy transport segment. This expectation is grounded in the fact that fuel cell vehicles involve a substantial number of components, thereby supporting production and assembly labor within Europe.

In the U.S., the extensive adoption of hydrogen energy and fuel cell technologies could lead to the generation of almost 1 million fresh employment opportunities in the United States by 2030. This study highlighted that these jobs would primarily benefit highly skilled and well-compensated technical and professional workers, playing a crucial role in fostering entrepreneurship and driving economic expansion [35]. The transition to hydrogen-based economies will result in the loss of numerous jobs within the fossil fuels sector, necessitating support for affected workers as they shift industries. Policymakers and industries must prioritize addressing the need for skills training, job creation, and climate action concurrently. This will likely involve creating new academic and corporate programs focused on education, accreditation, and vocational training. Additionally, upskilling or reskilling the current workforce will be crucial in many cases to facilitate a smooth transition. The shift to hydrogen energy presents a significant opportunity for the current global workforce and offers a chance to create a more inclusive environment for women and other underrepresented groups. Governments worldwide should collaborate closely with unions and businesses to ensure that workers have the necessary opportunities to succeed in the hydrogen economy.

Hydrogen diplomacy, characterized by international collaboration and agreements concerning hydrogen energy, significantly influences employment dynamics. This impact is wide-ranging, covering diverse facets like job generation, skill enhancement, and economic expansion. Forecasts suggest that the global demand for clean hydrogen could result in an annual creation of over 540,000 jobs by 2050 [36]. These employment opportunities will be spread across the entire spectrum of the hydrogen value chain, encompassing well-paying positions in project development, construction, engineering, operations, and various other sectors. Policy initiatives aimed at fostering investment and innovation in hydrogen, such as tax credits outlined in the Inflation Reduction Act, have the potential to propel the United States into a leading position globally. This would not only boost the country's capacity as a supplier and manufacturer of hydrogen-related technologies but also pave the way for an increased number of careers in the hydrogen sector.

6.6 Public-private partnerships

Public-private partnerships play a crucial role in driving forward the hydrogen economy, especially within the context of hydrogen diplomacy between nations. These collaborations harness public and private sectors' unique capabilities and resources to propel advancements in hydrogen production, infrastructure development, and application utilization. For example, in Saudi Arabia, the Saudi-German Hydrogen Cooperation, initiated in 2021, aims to execute bilateral hydrogen initiatives and facilitate knowledge exchange on technological advancements and regulatory frameworks, thereby establishing a comprehensive hydrogen roadmap. This partnership involves collaboration between Saudi Arabia's public sector and German expertise [37]. South Africa is actively forging public-private partnerships to jumpstart its hydrogen production efforts. For instance, a joint venture with the German Ministry of Economic Cooperation and Development has launched a €12.5 million project, aiming to craft a strategic blueprint and regulatory framework for fostering a green hydrogen economy [38]. The Hyport Duqm Project in Oman, valued at $2.5 billion, exemplifies a successful collaboration between public and private entities to develop green hydrogen and ammonia production facilities, signaling a significant investment in sustainable energy solutions [39]. The UK and South Africa have embarked on a €60 million flagship initiative under the UK-PACT international climate finance portfolio. This program aims to bolster job creation within the green hydrogen sector through collaborative efforts between public and private stakeholders [40]. These partnerships empower governments to leverage the private sector's expertise, technology, and financial backing to expedite the growth of the hydrogen economy. Simultaneously, the public sector plays a pivotal role by offering policy support, funding avenues, and coordination mechanisms to foster an enabling environment conducive to hydrogen project implementation. The synergistic blend of public and private sector strengths proves instrumental in surmounting the technical, regulatory, and economic challenges inherent in transitioning towards a hydrogen-based energy ecosystem.

Public-private partnerships facilitate the pooling of resources, expertise, and funding to achieve shared objectives effectively. In these partnerships, governments contribute public funding and resources, while private collaborators bring complementary assets to the table. The structure and goals of public-private partnerships can vary significantly, ranging from collaborative research hubs and technology consortia to initiatives focused on innovation procurement and commercialization. These partnerships serve as dynamic platforms for driving innovation, fostering economic growth, and addressing complex societal challenges through synergistic collaboration between public and private entities [41]. These collaborations offer several advantages:

- Harnessing the innovative prowess, technological advancements, and market knowledge of the private sector to tackle public sector challenges effectively.
- Providing governments with additional financial streams and financing mechanisms to bolster research, development, and infrastructure initiatives.
- Offering more adaptable and versatile policy tools compared to conventional government subsidies or tax incentives.
- Encouraging interdisciplinary cooperation and knowledge exchange among academic, industrial, and governmental researchers.
- Facilitating the enhancement of innovation capabilities and enhancing connectivity within national innovation frameworks.

Through collaboration with the private sector, governments have the opportunity to distribute the risks linked to extensive hydrogen projects and investments. This sharing of risk acts as an incentive for private sector engagement and investment in endeavors related to hydrogen. An instance is the DOE's allocation of $8 billion towards Hydrogen Hubs, alongside the tax and credit incentives outlined in the Inflation Reduction Act. These measures are anticipated to foster increased private investment within the hydrogen economy. Such public funding and backing contribute to establishing market certainty, which is essential for investors and startups to support hydrogen initiatives [42]. Through the collaboration of various stakeholders, public-private partnerships enable the sharing of insights, the alignment of common objectives, and the formulation of holistic strategies to propel the hydrogen industry forward. This cooperative method is vital in surmounting obstacles like restricted capital accessibility and insufficient awareness regarding the advantages of hydrogen technology [43].

6.7 National and international policy alignment

The legal and regulatory framework must be adjusted to endorse a hydrogen economy aligned with Paris Agreement objectives, fostering legal certainty and promoting industry growth. This involves promoting international collaboration and standardization of hydrogen certification processes and reciprocal acceptance of certificates to prevent market fragmentation. Establishing global standards, regulatory frameworks, and trade policies can effectively create a foundation for global hydrogen product exchange and trade volumes by providing clear market indications. Collaborative efforts at the national or regional levels, supported by bilateral and multilateral partnerships, can further enhance international hydrogen trade. The U.S. National Clean Hydrogen Strategy and Roadmap emphasizes the necessity of setting up regulatory structures to support large-scale clean hydrogen initiatives throughout the value chain. This involves addressing regulatory and policy deficiencies and crafting streamlined guidelines for hydrogen infrastructure permitting [44]. Governments must adopt a comprehensive policy framework with four fundamental pillars to transition hydrogen technology from a specialized resource to a widely adopted energy source:

- Each country should delineate its hydrogen ambitions, chart a path to achieve them, and serve as a reference for private sector investments.
- Policymakers should prioritize applications of hydrogen that offer the greatest value, especially in challenging industrial sectors with high emissions.
- Policies must include clear labeling of hydrogen and related products to inform consumers about emissions, thereby building trust and awareness.
- As hydrogen gains prominence, policies should integrate it into broader energy systems, with input from stakeholders such as civil society and industry.

Direct agreements with private consumers, strategies to tackle market barriers such as licensing, and investments in electrolyzer manufacturing, renewable energy, and hydrogen infrastructure are also essential. These aligned policies and incentives are critical for mitigating initial costs, attracting private investments, and stimulating demand for low-emission hydrogen, ultimately unlocking its potential in decarbonizing hard-to-abate sectors [45].

It's essential to synchronize international policies to encourage cooperation and collaboration across borders in hydrogen-related initiatives. When policies are aligned, it streamlines trade, technology transfer, and knowledge exchange among nations, fostering the evolution of a unified global hydrogen economy. Several key points emphasizing the importance of policy alignment:

- Countries have diverse regulatory frameworks for hydrogen projects and defining low-carbon hydrogen. Harmonizing these frameworks, including emissions thresholds and certification systems, is vital for smooth hydrogen trade and preventing market fragmentation [46].
- Initiatives such as the US Inflation Reduction Act and the EU's Hydrogen Bank offer financial support for hydrogen development. Coordinating these incentives globally can attract investments and scale up the hydrogen sector [31].
- Repurposing existing natural gas infrastructure for hydrogen transport and storage can cut costs significantly. Aligned policies and international collaboration are necessary to tackle technical hurdles and ensure infrastructure compatibility across borders [47].
- Collaborative research, development, and demonstrations can accelerate innovations in hydrogen production, storage, and applications. Consistent policies facilitate the sharing of knowledge and best practices.
- Establishing aligned standards, certifications, and regulations is crucial for an international hydrogen market. This ensures product interoperability and secure supply chains.

Aligning policies and regulations creates an environment conducive to the global growth of the hydrogen economy. It encourages cross-border cooperation, streamlines trade, and technology transfer, and advances the widespread adoption of hydrogen as a clean energy solution.

Efforts to align policies favoring clean energy sources like hydrogen can significantly contribute to achieving environmental objectives, including reducing greenhouse gas emissions and mitigating climate change impacts. For instance, the U.S. National Clean Hydrogen Strategy and Roadmap delineate how hydrogen integration can form a vital component of a diverse array of energy technologies, aiding the transition towards achieving net-zero emissions by 2050. On a global scale, nations are crafting legal and policy frameworks to propel the growth of the green hydrogen sector, exemplified by the EU's forthcoming regulations regarding the definition and qualification of green hydrogen. These regulatory measures aim to stimulate demand for green hydrogen, thereby fostering investment in clean energy ventures [48]. Coordinated policies that emphasize local utilization of clean hydrogen commence with smaller to medium-scale projects, and gradually expand production and applications can ensure that green hydrogen contributes effectively to sustainable development, technological progress, and job creation [49]. Engaging a wide range of stakeholders, including governments, industries, research institutions, NGOs, and communities, is essential for creating well-rounded and effective hydrogen policies that tackle environmental, equity, and sustainability issues comprehensively [50].

References

[1] Hydrogen vs Oil and Gas: Comparing Energy Sources - World-Energy n.d. https://www.world-energy.org/article/31067.html (accessed April 2, 2024).

[2] Here's How Oil And Gas Is Jumping On Hydrogen, But Is It Over-Rated n.d. https://www.forbes.com/sites/ianpalmer/2023/06/08/heres-how-oil-and-gas-companies-are-jumping-on-hydrogen-but-is-it-over-rated/?sh=2751c69b32e3 (accessed April 2, 2024).

[3] Geopolitics of the Energy Transformation: The Hydrogen Factor n.d. https://www.irena.org/Digital-Report/Geopolitics-of-the-Energy-Transformation (accessed February 2, 2024).

[4] Van de Graaf T, Overland I, Scholten D, Westphal K. The new oil? The geopolitics and international governance of hydrogen. Energy Res Soc Sci 2020;70:101667. https://doi.org/10.1016/J.ERSS.2020.101667.

[5] The Drivers of Hydrogen's Waves of Hype: Between Security and the Environment n.d. https://www.e-ir.info/2024/01/12/the-drivers-of-hydrogens-waves-of-hype-between-security-and-the-environment/ (accessed April 2, 2024).

[6] Triviño G, Agarwal R. Transition to a Hydrogen-Based Economy: Possibilities and Challenges. Sustainability 2022, Vol 14, Page 15975 2022;14:15975. https://doi.org/10.3390/SU142315975.

[7] IEA – International Energy Agency n.d. https://www.iea.org/ (accessed March 11, 2022).

[8] Hydrogen | Shell Global n.d. https://www.shell.com/what-we-do/hydrogen.html#vanity-aHR0cHM6Ly93d3cuc2hlbGwuY29tL2VuZXJneS1hbmQtaW5ub3ZhdGlvbi9uZXctZW5lcmdpZXMvaHlkcm9nZW4uaHRtbA (accessed April 2, 2024).

[9] Japan's Hydrogen Strategy and Its Economic and Geopolitical Implications | IFRI - Institut français des relations internationales n.d. https://www.ifri.org/en/publications/etudes-de-lifri/japans-hydrogen-strategy-and-its-economic-and-geopolitical-implications (accessed April 2, 2024).

[10] Germany: The German National Hydrogen Strategy and international hydr - Baker McKenzie InsightPlus n.d. https://insightplus.bakermckenzie.com/bm/projects/germany-the-german-national-hydrogen-strategy-and-international-hydrogen-partnerships (accessed April 2, 2024).

[11] South Korea joins Horizon Europe in multi-billion euro push to globalise science | Science|Business n.d. https://sciencebusiness.net/news/horizon-europe/south-korea-joins-horizon-europe-multi-billion-euro-push-globalise-science (accessed April 2, 2024).
[12] Oman, Belgium partner to develop renewable hydrogen hub n.d. https://gaspathways.com/oman-belgium-partner-to-develop-renewable-hydrogen-hub-2245 (accessed April 2, 2024).
[13] Bridge G, Bradshaw M. Making a Global Gas Market: Territoriality and Production Networks in Liquefied Natural Gas. Econ Geogr 2017;93:215–40. https://doi.org/10.1080/00130095.2017.1283212.
[14] Industry News & Market Insights | Breakbulk Events & Media | Breakbulk n.d. https://breakbulk.com/articles/north-africas-hydrogen-pathway-potential (accessed April 2, 2024).
[15] Green hydrogen for steel production: RWE and thyssenkrupp plan partnership n.d. https://www.rwe.com/en/press/rwe-generation/2020-06-10-green-hydrogen-for-steel-production/ (accessed April 2, 2024).
[16] European industrial strategy - European Commission n.d. https://commission.europa.eu/strategy-and-policy/priorities-2019-2024/europe-fit-digital-age/european-industrial-strategy_en (accessed April 2, 2024).
[17] China Leading Race to Make Technology Vital for Green Hydrogen - Bloomberg n.d. https://www.bloomberg.com/news/articles/2022-09-21/china-leading-race-to-make-technology-vital-for-green-hydrogen?embedded-checkout=true (accessed April 2, 2024).
[18] Alanne K, Cao S. Zero-energy hydrogen economy (ZEH2E) for buildings and communities including personal mobility. Renewable and Sustainable Energy Reviews 2017;71:697–711. https://doi.org/10.1016/J.RSER.2016.12.098.
[19] Tashie-Lewis BC, Nnabuife SG. Hydrogen Production, Distribution, Storage and Power Conversion in a Hydrogen Economy - A Technology Review. Chemical Engineering Journal Advances 2021;8:100172. https://doi.org/10.1016/J.CEJA.2021.100172.
[20] Maennel A, Kim HG. Comparison of Greenhouse Gas Reduction Potential through Renewable Energy Transition in South Korea and Germany. Energies 2018, Vol 11, Page 206 2018;11:206. https://doi.org/10.3390/EN11010206.
[21] Cao S, Alanne K. Technical feasibility of a hybrid on-site H2 and renewable energy system for a zero-energy building with a H2 vehicle. Appl Energy 2015;158:568–83. https://doi.org/10.1016/J.APENERGY.2015.08.009.
[22] Ball M, Weeda M. The hydrogen economy – Vision or reality? Int J Hydrogen Energy 2015;40:7903–19. https://doi.org/10.1016/J.IJHYDENE.2015.04.032.
[23] Stewart EM, Lutz AE, Schoenung S, Chiesa M, Keller JO, Fletcher J, et al. Modeling, analysis and control system development for the Italian hydrogen house. Int J Hydrogen Energy 2009;34:1638–46. https://doi.org/10.1016/J.IJHYDENE.2008.12.008.
[24] Mobility | Toyota Motor Corporation Official Global Website n.d. https://global.toyota/en/mobility/ (accessed November 17, 2023).
[25] 2023 Nexo Fuel Cell | Vehicle Overview | Hyundai USA n.d. https://www.hyundaiusa.com/us/en/vehicles/nexo (accessed November 17, 2023).
[26] Honda To Begin U.S. Production of Fuel Cell Electric Vehicles in 2024 n.d. https://hondanews.com/en-US/releases/honda-to-begin-us-production-of-fuel-cell-electric-vehicles-in-2024 (accessed November 17, 2023).
[27] Agyekum EB, Nutakor C, Agwa AM, Kamel S. A Critical Review of Renewable Hydrogen Production Methods: Factors Affecting Their Scale-Up and Its Role in Future Energy Generation. Membranes 2022, Vol 12, Page 173 2022;12:173. https://doi.org/10.3390/MEMBRANES12020173.
[28] Orttung R, Overland I. Russia and the Formation of a Gas Cartel. Problems of Post-Communism 2011;58:53–66. https://doi.org/10.2753/PPC1075-8216580305.
[29] US hydrogen workforce development programmes – Policies - IEA n.d. https://www.iea.org/policies/17838-us-hydrogen-workforce-development-programmes (accessed April 12, 2024).

[30] The Hydrogen Economy Program | AIChE n.d. https://www.aiche.org/ili/academy/courses/ble200/hydrogen-economy-program (accessed April 12, 2024).
[31] EU hydrogen diplomacy 2.0: aligning climate ambition and energy security - Elcano Royal Institute n.d. https://www.realinstitutoelcano.org/en/analyses/eu-hydrogen-diplomacy-2-0-aligning-climate-ambition-and-energy-security/ (accessed April 12, 2024).
[32] Green Hydrogen Revolution Has the Potential to Spur Significant Economic and Job Growth Across Europe n.d. https://www.manpowergroup.com/en/news-releases/news/green-hydrogen-revolution-has-the-potential-to-spur-significant-economic-and-job-growth-across-europe (accessed April 12, 2024).
[33] 1.7 million green jobs in Europe by 2040, study projects - Outsource Accelerator n.d. https://news.outsourceaccelerator.com/green-jobs-europe-2040/ (accessed April 12, 2024).
[34] Betting on green hydrogen to fulfil employment growth | CIC energiGUNE n.d. https://cicenergigune.com/en/blog/green-hydrogen-employment-growth (accessed April 11, 2024).
[35] Roger H. Bezdek. The hydrogen economy and jobs of the future. Renewable Energy and Environmental Sustainability 2019;4. https://doi.org/https://doi.org/10.1051/rees/2018005.
[36] Clean Hydrogen: Policies to Help America Lead – Third Way n.d. https://www.thirdway.org/memo/clean-hydrogen-policies-to-help-america-lead (accessed April 11, 2024).
[37] Hydrogen cooperation potential between Saudi Arabia and Germany . Federal Ministry for Economic Affairs and Climate Action (BMWK) 2022. https://www.bmwk.de/Redaktion/EN/Downloads/J/joint-study-saudi-german-energy-dialogue.pdf?__blob=publicationFile&v=1 (accessed April 12, 2024).
[38] Germany, South Africa agree to cooperate on green hydrogen projects | Reuters n.d. https://www.reuters.com/business/environment/germany-south-africa-agree-cooperate-green-hydrogen-projects-2023-06-27/ (accessed April 12, 2024).
[39] The Hurdles of Energy Transitions in Arab States - Disruptions and Dynamism in the Arab World - Carnegie Endowment for International Peace n.d. https://carnegieendowment.org/2023/05/03/hurdles-of-energy-transitions-in-arab-states-pub-89518 (accessed February 3, 2024).
[40] South Africa's Hydrogen Strategy n.d. https://www.csis.org/analysis/south-africas-hydrogen-strategy (accessed April 12, 2024).
[41] Public-Private Partnerships (PPPs): Definition, How They Work, and Examples n.d. https://www.investopedia.com/terms/p/public-private-partnerships.asp (accessed April 12, 2024).
[42] DOE's $8B Hydrogen Hub May Kickstart Venture Capital, Corporate Investments | Hart Energy n.d. https://www.hartenergy.com/exclusives/does-8b-hydrogen-hub-may-kickstart-venture-capital-corporate-investments-207196 (accessed April 12, 2024).
[43] Midwest Alliance for Clean Hydrogen: Public-private partnerships to establish regional clean hydrogen hubs | American Academy of Arts and Sciences n.d. https://www.amacad.org/climate/case-study/midwest-alliance-clean-hydrogen-partnerships (accessed April 12, 2024).
[44] U.S. National Clean Hydrogen Strategy and Roadmap 2023. www.hydrogen.energy.gov/docs/hydrogenprogramlibraries/pdfs/us-national-clean-hydrogen-strategy-roadmap.pdf (accessed January 5, 2024).
[45] Policies for green hydrogen n.d. https://www.irena.org/Energy-Transition/Policy/Policies-for-green-hydrogen (accessed April 12, 2024).
[46] Harmonising hydrogen: the need for global alignment in low carbon hydrogen regulation | Johnson Matthey n.d. https://matthey.com/science-and-innovation/expert-insights/2023/harmonising-hydrogen-the-need-for-global-alignment-in-low-carbon-hydrogen-regulation (accessed April 13, 2024).
[47] Executive summary – Global Hydrogen Review 2022 – Analysis - IEA n.d. https://www.iea.org/reports/global-hydrogen-review-2022/executive-summary (accessed April 13, 2024).

[48] Green Hydrogen | ACP n.d. https://cleanpower.org/facts/green-hydrogen/ (accessed April 13, 2024).
[49] IRENA Policy Talks 2024 – Green hydrogen for sustainable industrial development n.d. https://www.irena.org/Events/2024/Mar/IRENA-Policy-Talks-2024-Green-hydrogen-for-sustainable-industrial-development (accessed April 13, 2024).
[50] Building Stronger Community Engagement in Hydrogen Hubs - EFI Foundation n.d. https://efifoundation.org/foundation-reports/building-stronger-community-engagement-in-hydrogen-hubs/ (accessed April 13, 2024).

This page intentionally left blank

CHAPTER 7

Hydrogen business in the United States: opportunities and challenges

7.1 Introduction

Over fifty years ago, the United States' moonshot initiative successfully landed the first humans on the moon, employing hydrogen as a rocket propulsion fuel and incorporating American-manufactured fuel cells within the spacecraft. Since that historic achievement, the country has maintained its global leadership position in the field of hydrogen and fuel cells. Numerous hydrogen and fuel cell systems have been created and introduced across various industries. Funding from the U.S. Department of Energy (DOE) has yielded over 1,100 patents related to hydrogen and fuel cells, 30 commercially deployed technologies, and a portfolio of more than 65 technologies poised for commercialization in the coming years. The 2020s represent a pivotal decade for addressing climate change and preventing the irreversible consequences of the crisis, with a focus on maintaining the target of limiting the global average temperature rise to 1.5 degrees Celsius. The Biden-Harris Administration has set ambitious objectives, aiming to reduce greenhouse gas emissions by 50 to 52 percent from 2005 levels by 2030 under the Paris Agreement [1]. Additionally, the administration aims to establish a carbon pollution-free power sector by 2035 and achieve net-zero emissions by no later than 2050. A groundbreaking initiative, known as the Justice40 Initiative, has been introduced, committing to ensuring that at least 40 percent of the overall benefits from federal investments in climate and clean energy are directed to disadvantaged communities. This commitment extends to several critical hydrogen programs of the Department, including the Regional Clean Hydrogen Hubs Program and the Clean Hydrogen Manufacturing and Recycling Research, Development, and Demonstration Program. Hydrogen constitutes a crucial element within a comprehensive array of energy technologies essential for facilitating the United States' transition to a net-zero state. This transition involves leveraging regional resources and fostering fair and sustainable growth. The strategic development and utilization of hydrogen technologies will involve a consideration of various technologies spanning multiple sectors, ensuring the emergence of the most efficient, cost-effective, and sustainable

DOI: https://doi.org/10.55670/fpll.book/1

pathways through widespread market adoption. Industries that pose challenges for decarbonization through conventional methods are anticipated to emerge as primary markets for clean hydrogen. This particularly includes sectors like steel and chemicals manufacturing, heavy-duty transportation, and the production of liquid fuels. Hydrogen is not only viewed as a standalone energy solution but also as a pivotal enabling technology. It plays a crucial role in supporting renewables by facilitating long-duration energy storage. Additionally, hydrogen provides flexibility and multiple revenue streams for clean power generation, benefiting existing nuclear facilities, advanced nuclear technologies, and other innovative energy solutions.

In June 2021, the DOE initiated the Hydrogen Energy Earthshot (Hydrogen Shot) with the objective of unleashing the market potential for clean hydrogen. This ambitious initiative seeks to achieve an 80 percent reduction in the cost of clean hydrogen, bringing it down to $1 per kilogram within a decade ("1 1 1"). The Hydrogen Shot marks the inaugural effort under the DOE's Energy Earthshots, a series of initiatives designed to expedite breakthroughs in more abundant, affordable, and reliable clean energy solutions within the span of a decade. Simultaneously, these initiatives aim to generate well-compensated union jobs and foster economic growth. Expanding on this forward movement, President Biden signed the Infrastructure Investment and Jobs Act (IIJA) on November 15, 2021. Commonly referred to as the Bipartisan Infrastructure Law (BIL), this legislation represents a transformative investment in the nation's infrastructure and competitive edge, with the overarching goal of fostering a more equitable future in clean energy for the American populace. The substantial investments outlined in the BIL are poised to expedite advancements aligned with the Hydrogen Shot initiative and catalyze the emergence of fresh markets for clean hydrogen.

Besides the BIL, the Inflation Reduction Act (IRA), signed into law in August 2022, introduces a Hydrogen Production Tax Credit (PTC). This tax credit serves as an added incentive to promote the production of clean hydrogen within the United States. Furthermore, the IRA has the potential to stimulate the development of demand sectors for clean hydrogen through various supplementary programs, encompassing:

- Provision of grants and loans targeted at auto manufacturing facilities engaged in the production of clean vehicles, including Fuel Cell Electric Vehicles (FCEVs).
- Implementation of a tax credit aimed at encouraging the production of sustainable aviation fuels, a process that may involve hydrogen feedstock.
- Allocation of grants dedicated to emission reduction initiatives at ports, which could, in turn, support the deployment of fuel cells.
- Provision of grants specifically designed for clean heavy-duty vehicles, encompassing FCEVs.

These measures collectively contribute to the broader strategy of fostering clean hydrogen production and incentivizing its application across diverse sectors of the economy.

The national clean hydrogen strategy and roadmap were developed by the DOE in compliance with the directives of the BIL. In crafting this roadmap, the DOE collaborated extensively with various federal agencies and stakeholders to pinpoint essential actions that the nation should undertake. The primary goal is to facilitate the successful market integration of clean hydrogen technologies, aligning with the overarching objective of achieving a net-zero economy by the year 2050. This roadmap is an evolution of the DOE's strategic efforts spanning three decades, during which funding was strategically directed to National Laboratories, industry entities, and academic institutions. These funds supported comprehensive research, development, demonstration, and deployment (RDD&D) initiatives that played a pivotal role in bringing about the commercialization of hydrogen and fuel cell technologies. The DOE's national strategy and roadmap draw inspiration from the Hydrogen Program Plan, various analytical endeavors, and the industry-driven U.S. hydrogen roadmap published in 2020. It further builds upon existing tools and models supported by the DOE and leverages prior collaborative efforts by diverse stakeholders. The objective is to comprehensively assess emerging hydrogen markets' growth potential and ramifications. This collaborative and strategic approach underscores the commitment to fostering the widespread adoption of clean hydrogen technologies as a crucial component of the nation's energy landscape [2].

7.2 The U.S. decarbonization strategies

The current moment calls for decisive, daring, and tangible measures to fulfill the ambitious targets established by the United States in addressing the climate crisis. These objectives encompass achieving 100 percent carbon pollution-free electricity by 2035 and attaining net-zero greenhouse gas (GHG) emissions by the year 2050. The national climate strategy of the U.S. outlines a comprehensive and long-term approach, detailing pathways for the nation to meet its 2030 Nationally Determined Contribution (NDC) in alignment with global climate goals. This NDC aims for a significant 50 to 52 percent reduction in emissions compared to 2005 levels. The magnitude of the challenge, as depicted in Figure 7.1, is substantial [3]. Nevertheless, it remains within reach through a collective and concerted call to action, mobilizing all available resources. The achievement of these goals relies on a diversified portfolio of technologies and strategies that facilitate rapid scalability. This inclusive approach emphasizes the necessity of combining efforts from various sectors and leveraging a range of solutions to expedite progress toward a sustainable and low-carbon future. Attaining a net-zero economy by 2050 necessitates groundbreaking progress across all facets of the energy infrastructure and various economic sectors. Electricity provides a wide array of services across all sectors of the American economy.

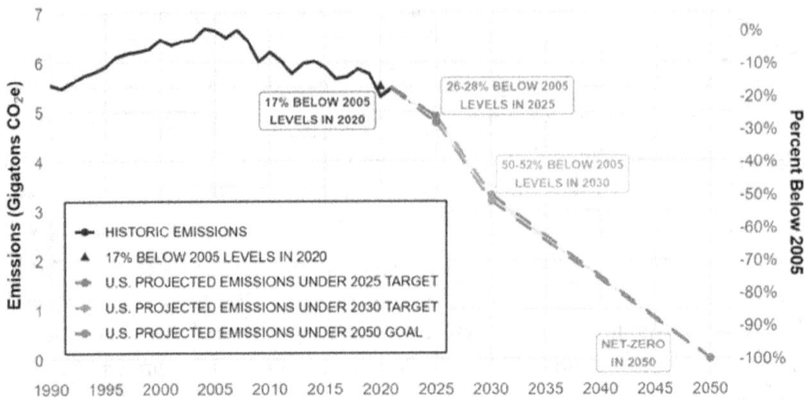

Figure 7.1 The historical trajectory of U.S. net GHG emissions 1990-2050 [3]

The shift toward a sustainable electricity system has been rapidly gaining momentum, propelled by the declining costs of solar and wind technologies, government policies at both federal and subnational levels, and increasing consumer demand. Expanding on these achievements, the United States has established the ambitious target of achieving 100% clean electricity by 2035, serving as a vital cornerstone for realizing a net-zero status by 2050 (Figure 7.2) [3]. Most of the economy can be electrified affordably and efficiently, encompassing various sectors such as cars, buildings, and industrial processes. In instances where technological challenges are posed by electrification, such as in aviation, shipping, and certain industrial processes, priority can be given to the utilization of clean fuels like carbon-free hydrogen and sustainable alternative fuels. Transitioning to cleaner energy sources becomes expedited, more cost-effective, and simpler when both existing and emerging technologies require less energy to deliver equivalent or improved services. This objective can be accomplished through a variety of established methods, including the development of new and more efficient appliances, the incorporation of efficiency measures in new and existing buildings, the adoption of sustainable manufacturing processes, and the integration of efficiency practices in diverse sectors. Hydrogen, owing to its adaptability and capacity to synergize with other sustainable technologies, emerges as a pivotal facilitator of our objective. This is particularly evident in three of the most energy-intensive and emissions-laden sectors in the United States: industry, transportation, and electricity generation. Hydrogen, functioning as a versatile carrier of energy and raw material for chemicals, presents advantages that allow the optimization of the entire array of energy resources. This encompasses renewables, nuclear power, and fossil fuels coupled with carbon capture and storage (CCS).

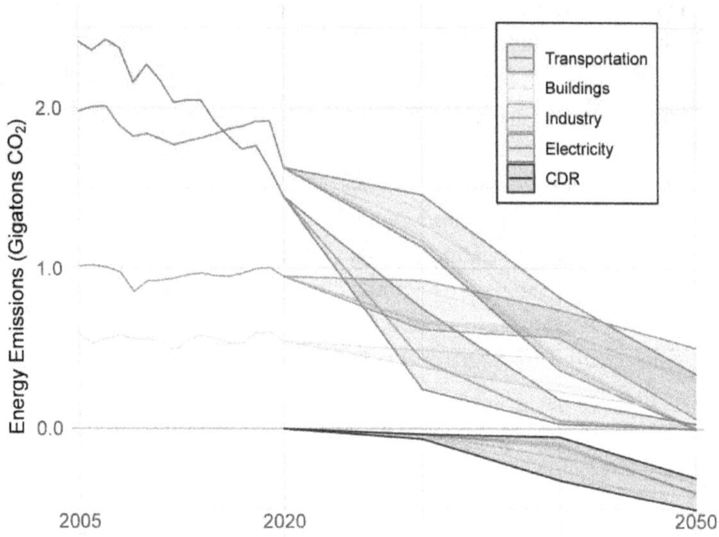

Figure 7.2 U.S. CO_2 emissions to 2050 by economic sector [3]

Hydrogen has the capacity to integrate baseload power with variable generation, providing resilience and energy storage. Subsequently, it can be deployed as a fuel or raw material for applications lacking competitive and efficient clean alternatives. While numerous opportunities exist for hydrogen utilization, the U.S. strategic approach will adopt a comprehensive perspective, considering environmental, energy justice, and equity considerations. The clean hydrogen strategy aligns with the Administration's Justice40 Initiative, committing to delivering at least 40 percent of overall benefits from federal investments in climate and clean energy to disadvantaged communities. The devised strategies and pathways will be tailored to benefit all Americans, contributing not only to emissions reduction but also positively impacting public health, fostering economic growth, generating jobs - including well-compensated union jobs- and enhancing the overall quality of life.

Renewable hydrogen production is achievable through various domestic resources, providing applicability across different sectors. The production process itself can be centralized or decentralized, connected to the grid or off-grid, thereby offering scalability, flexibility, and adaptability to regional needs. Hydrogen, in this context, introduces a multitude of options across sectors and can effectively complement existing conventional grid structures and natural gas infrastructure. This stands in contrast to the limitations of pathways solely reliant on electrons, such as those involving the electric grid and batteries. Hydrogen's capacity for storage enables its utilization in areas where electrification might pose challenges. Diverse technologies facilitate the production of clean hydrogen, ranging from electrolyzers driven by the nation's increasing share of clean energy to methane reformation with CCS, gasification, or thermal conversion of biomass and/or solid wastes with carbon capture and storage, along with various emerging technologies. Initial implementations of clean hydrogen are anticipated to leverage regional energy resources and focus on industries currently dependent on traditional natural gas-to-hydrogen technologies, excluding CCS. These sectors have the potential to achieve scale and yield immediate emissions reductions rapidly, setting the stage for the introduction of new hydrogen technologies capable of displacing fossil fuels in other sectors. The U.S. policymakers acknowledge the necessity of supplementing electrification initiatives with alternative fuels like clean hydrogen. Various studies highlight the potential of hydrogen in the U.S. energy systems, although estimates exhibit substantial divergence.

Drawing from various models and analyses specific to the United States, the DOE outlines the potential for hydrogen, as depicted in Figure 7.3 and Figure 7.4. The DOE's objective is to elevate clean hydrogen production from virtually negligible levels today to 10 million metric tons (MMT) annually by 2030, 20 MMT per year by 2040, and 50 MMT per year by 2050. While these targets are undoubtedly ambitious, they are deemed feasible, contingent on demand scenarios that presuppose the cost competitiveness of hydrogen in targeted sectors such as industrial applications, heavy-duty transportation, and extended-duration energy storage. Anticipating a fivefold surge in hydrogen production and usage by 2050, the DOE envisions a consequential reduction of approximately 10 percent in total GHG emissions in the United States relative to 2005 levels, given that all hydrogen is sourced through clean production methods. As the DOE continues to enhance and fine-tune its analyses, it remains committed to evaluating the cleanest, most sustainable, and most comprehensive pathways for hydrogen production through end-use, with a specific focus on localized and regional benefits [4].

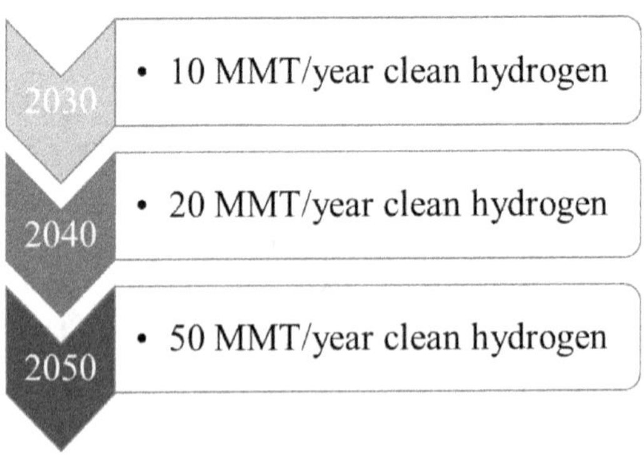

Figure 7.3 The opportunity for renewable hydrogen production in the United States

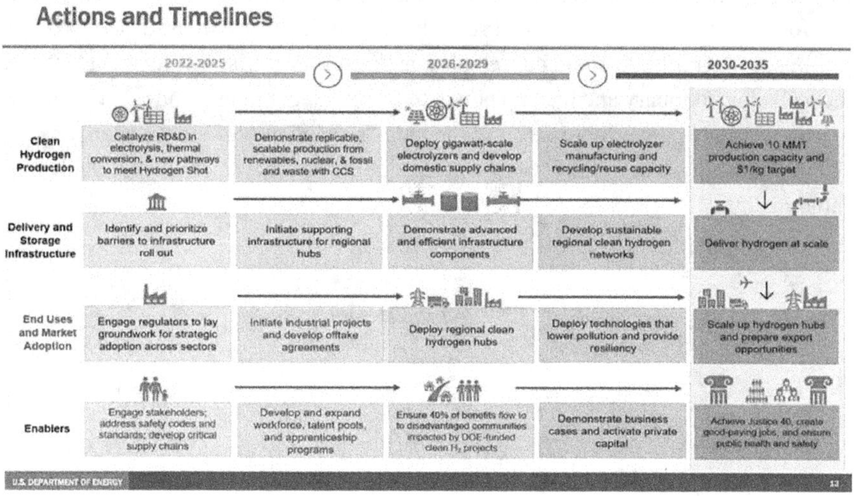

Figure 7.4 National goal of achieving net-zero emissions by mid-century using hydrogen [5]

7.3 Hydrogen production and utilization in the U.S.

The generation of clean hydrogen can be achieved through various methodologies, encompassing water-splitting facilitated by renewable or nuclear energy, deriving from fossil fuels with carbon capture and storage, and utilizing biomass or waste as feedstocks. Additionally, there are emerging pathways in preliminary developmental stages, including thermochemical, biological, and photoelectrochemical processes. The carbon emissions associated with each of these pathways are contingent on crucial factors such as carbon capture efficiency, methane leak rates, fugitive emissions, and the extent of clean electricity incorporation.

Within the United States, the industrial sector is responsible for producing approximately 10 MMT of hydrogen annually [6], as opposed to the global production of around 90 MMT per year [7]. This sizeable global production is primarily geared towards meeting the needs of petroleum refining, ammonia, and chemical industries. It's worth noting that a portion of this hydrogen is both produced and utilized within the same facility, which slightly elevates the overall hydrogen consumption. Approximately 55% of the produced hydrogen in the U.S. is utilized in the refining industry, 35% in ammonia and methanol, 2% in the steel industry, and the rest in other applications [8]. Presently, the production of hydrogen contributes to approximately 100 MMT of GHGs annually on a well-to-gate basis [9]. To support these sectors, the United States currently possesses almost 1,600 miles of specialized hydrogen pipelines and three geological caverns, including the world's largest, capable of storing 350 gigawatt-hours (GWh) of thermal energy, equivalent to providing backup for a nuclear plant for a week [10]. Beyond its traditional applications in petroleum and fertilizer production, hydrogen is gradually finding its way into various end-use sectors. This includes the deployment of over 50,000 fuel cell forklifts, nearly 50 operational retail hydrogen fueling stations, around 70 fuel cell buses, over 13,000 fuel cell vehicles, and more than 500 megawatts (MW) of fuel cells for stationary and backup power (e.g., for telecommunications), as outlined in Figure 7.5. Noteworthy projects in the industrial and energy storage domains are also establishing the United States as a significant player in global hydrogen deployment [11].

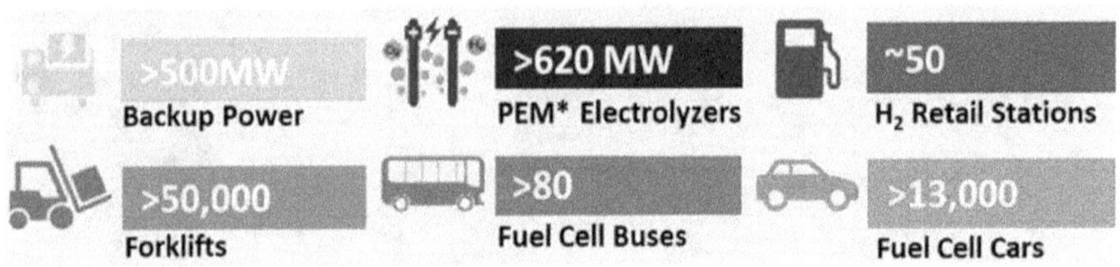

Figure 7.5 Examples of hydrogen production and utilization in the United States [11]

An example of this is the Intermountain Power Project under construction in Utah, designed to enable the generation of multiple gigawatt-hours of power using blends of natural gas and hydrogen produced through electrolysis [12]. In Louisiana, a developing clean energy complex is slated to utilize methane reforming with CCS at a remarkable 95 percent capture rate to provide clean hydrogen to local markets and for global export. This groundbreaking project is anticipated to be the world's largest carbon capture for sequestration operation, with the capacity to sequester over 5 million tons of carbon dioxide annually [13]. In October 2023, the U.S. DOE unveiled a groundbreaking initiative as part of President Biden's Investing in America agenda, dedicating $7 billion to establish seven Regional Clean Hydrogen Hubs (H2Hubs) nationwide. These H2Hubs are set to catalyze the commercial-scale deployment of low-cost, clean hydrogen, a critical component in reducing carbon emissions. The seven H2Hubs will foster the production, storage, delivery, and utilization of clean hydrogen, collectively aiming to produce three million metric tons of hydrogen each year. This production level will account for nearly a third of the 2030 U.S. production target, significantly lowering emissions from industries challenging to decarbonize and representing a significant portion of total U.S. carbon emissions (Figure 7.6). The establishment of the H2Hubs will also result in a reduction of 25 MMT of CO_2 emissions annually. This is equivalent to the annual emissions of approximately 5.5 million gasoline-powered cars. Additionally, the initiative will create and sustain tens of thousands of well-paying jobs throughout the nation, contributing to healthier communities. The $7 billion Federal investment is projected to be matched by private sector contributions, resulting in a total investment of nearly $50 billion. This substantial investment will not only strengthen local economies but also have far-reaching implications for the environment by reducing harmful emissions.

Figure 7.6 The U.S. clean hydrogen hubs announced in October 2023

7.4 Clean hydrogen market in the U.S.

As indicated in Table 7.1, the current market availability of hydrogen technologies is constrained. However, in the upcoming decade, there exists a range of potential applications for clean hydrogen, encompassing areas such as steel production, heavy-duty transportation, and the synthesis of liquid fuels for marine and aviation uses. Strategic prioritization of hydrogen deployment is essential, especially in sectors where other alternatives, such as electrification, may face challenges in terms of efficiency and cost-effectiveness. As advancements in energy technologies progress and the overall energy infrastructure undergoes decarbonization, emerging demands for hydrogen may surface. These could include applications like long-duration energy storage to facilitate a carbon-neutral electric grid or stationary heat and power generation, incorporating combined heat and power systems utilizing fuel cells and other low- or zero-emission technologies. Furthermore, the expanding availability of clean hydrogen across these sectors has the potential to drive the establishment of extensive distribution infrastructure, connecting regions with abundant and affordable supply to areas with substantial demand.

Table 7.1 Current and emerging markets for hydrogen in the U.S.

	Industrial feedstocks	Transportation	Power generation and storage	Hydrogen blending in natural gas
Existing markets	• Oil refining • Ammonia • Methanol	• Forklifts and other material-handling equipment • Buses • Light-duty vehicles	• Distributed generation: primary and backup generation • Renewable grid integration with storage and other ancillary services	• Low percentage hydrogen blending
Emerging opportunities	• Steel and cement manufacturing • Industrial heat • Bio/Synthetic fuels	• Medium and heavy-duty vehicles • Rail • Maritime • Aviation • Offroad equipment (mining, construction, agriculture)	• Hydrogen low-NO_x combustion • Long-duration energy storage • Direct/reversible fuel cells • Nuclear/hydrogen hybrids • Fossil/waste/biomass hydrogen hybrids with Carbon Capture, Utilization, And Storage (CCUS)	• High percentage hydrogen blending • Industry, building, or district heating for hard-to-electrify or limited options

To establish competitiveness in the long-term sustainable market, clean hydrogen must be attainable at a price point below a specified minimum threshold, which varies depending on the fuel and processes it aims to replace within each sector. In practical terms, especially during the transitional period, hydrogen can deliver value that extends beyond conventional cost parity perspectives, encompassing aspects such as grid services, arbitrage, and fuel flexibility. It's important to note that a cost-centric viewpoint offers a conservative estimation of market demand potential. Figure 7.7 delineates the price range at which hydrogen would rival incumbent fuels (such as diesel, natural gas, or coal) in different applications, along with the anticipated timeframe for significant-scale deployments of clean hydrogen in each sector [6]. The "willingness to pay" for each application reflects the total price at which hydrogen needs to be accessible to the end-user, incorporating production costs, distribution expenses, and additional onsite conditioning like compression, storage, and dispensing. Crucially, each sector presents distinct onsite requirements; for instance, the transportation sector may exhibit a higher willingness to pay, yet infrastructure demands such as compression and dispensing at fueling stations, along with potential liquefaction needs, can significantly contribute to the overall cost of hydrogen for the end-user.

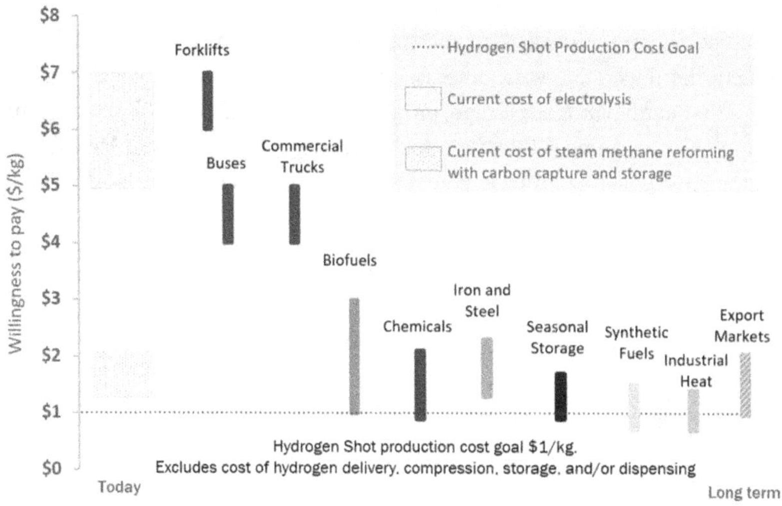

Figure 7.7 The price range at which hydrogen would rival incumbent fuels in different applications

In the United States, the specialized sector of fuel cell forklifts, catalyzed by the American Recovery and Reinvestment Act (ARRA) in 2009, set the stage for the deployment of over 50,000 fuel cell forklifts in commercial warehouses across the nation, accompanied by the establishment of more than 115 fueling stations [14]. This niche has proven to be competitive even with higher hydrogen costs, attributed to advantages such as quicker fueling times, increased operational throughput, and a smaller space requirement compared to battery forklifts. Another avenue for early market adoption lies in fuel cell trucks and buses; however, the total cost to the end-user, encompassing infrastructure, must reach approximately $5/kg for viability. Other sectors, including biofuels, chemicals, and steel, need even lower costs to ensure long-term competitiveness. The present cost of clean hydrogen production, along with the Hydrogen Shot cost target for clean hydrogen production (excluding downstream infrastructure such as delivery, storage, and dispensing), is illustrated in Figure 7.8 for contextual reference [15]. The level of hydrogen demand at the specified cost threshold within each sector will be contingent upon the evolution of competing and existing technologies and fuels. Figure 7.9 outlines potential scenarios for the anticipated demand in each sector if clean hydrogen is accessible (manufactured, transported, and dispensed) at the indicated threshold price. For example, a price point of around $5/kg for hydrogen, covering production, delivery, compression, and dispensing, would attract early adopters in the fuel cell truck market. In scenario analyses, it has been demonstrated that at approximately $4/kg,

10-14 percent of all medium and heavy-duty fuel cell trucks would generate a demand of about 5-8 MMT/year of hydrogen [16]. The lighter shaded bars in the figure depict a more optimistic demand scenario for each market illustrated. While recognizing the uncertainties in other variables such as fuel cell cost, efficiency, durability, on-board hydrogen storage, infrastructure, and the cost of incumbent fuels and technologies, ongoing refinements to analyses will be conducted. Nevertheless, these findings suggest significant potential volumes for clean hydrogen demand, contingent on the achievement of DOE targets for clean hydrogen costs.

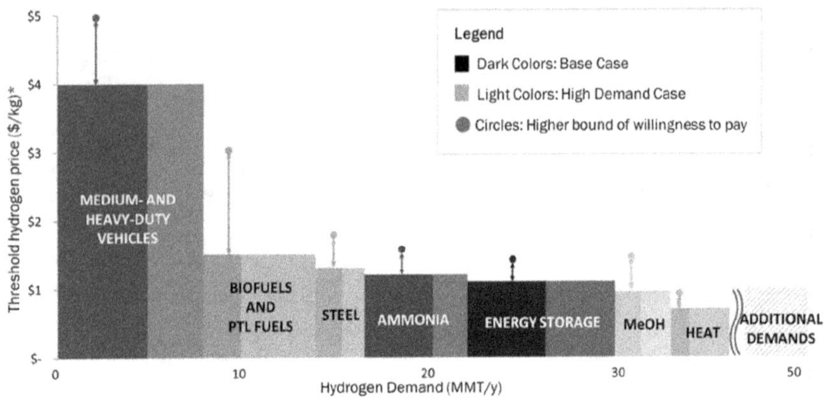

Figure 7.8 Potential scenarios for the anticipated demand in each sector if clean hydrogen is accessible

In addition to its role in fuel cells for the trucking sector, hydrogen is poised to become a crucial raw material for biofuels, encompassing sustainable aviation fuels (SAF) and power-to-liquid fuels. These applications hold the potential to decarbonize off-road vehicles and scenarios where direct electrification or fuel cells might face competitiveness challenges. Should the U.S. opt to replace all jet fuel consumption with SAF by 2050, an estimated 2-6 MMT/year of hydrogen could be necessary to manufacture 35 billion gallons of SAF from biofuels [17]. Furthermore, an additional 6 MMT/year would be requisite for generating 4 billion gallons of power-to-liquid fuels, utilizing approximately 44 MMT of carbon dioxide (roughly equivalent to the concentrated CO_2 currently available from ethanol plants in the U.S.) [18]. Hydrogen assumes a pivotal role in the decarbonization of the industrial sector, which is crucial for achieving a net-zero economy by 2050, particularly in domains like steelmaking, chemicals, and heat generation. Subject to the development of alternative solutions, hydrogen incorporation in iron refining may contribute to 10-20 percent of steelmaking by 2050, resulting in a demand of approximately 1-3 MMT/year. Moreover, ammonia plants could require an additional 4-5 MMT/year of clean hydrogen to facilitate the decarbonization of all domestic demand for traditional purposes, such as fertilizer production [19]. Given hydrogen's indispensable role as a feedstock in ammonia production, leveraging clean sources becomes imperative for effective decarbonization, making the ammonia market a prime candidate for early substantial demand for clean hydrogen.

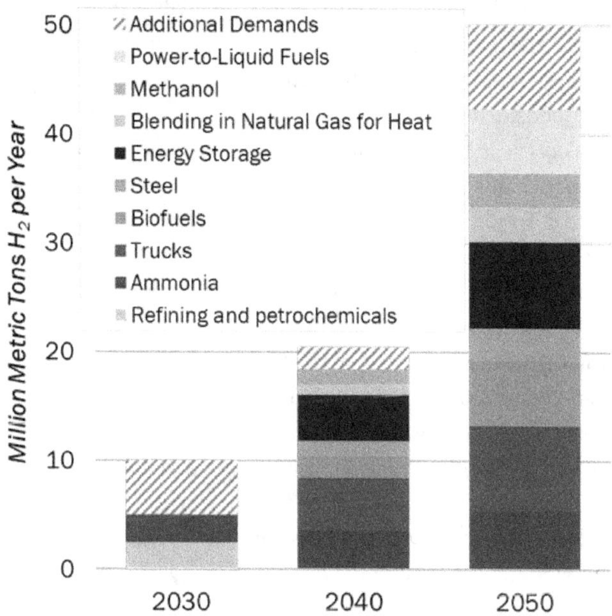

Figure 7.9 Hydrogen demand growth by sectors from 2030 to 2050 [4]

In the methanol industry, alternatives to utilizing renewable hydrogen involve the implementation of CCS technologies with traditional fossil feedstocks or the utilization of biomass feedstock. If clean hydrogen were to account for 50% of the U.S. methanol supply by 2050, meeting the demand would require approximately 1-3 million metric tons per year. Beyond its chemical attributes, hydrogen can play a role in decarbonization by substituting natural gas in sectors that necessitate high-temperature heat- a challenging application for electrification. Introducing blends of clean hydrogen and natural gas, ranging from 20% to 50% (by volume), for the production of high-temperature heat (over 550°C) in chemical and steelmaking processes could create a demand of around 1-3 million metric tons per year [19]. It's crucial to emphasize that achieving substantial emissions reduction requires high concentrations of hydrogen, considering that its energy content is only about one-third that of natural gas by volume.

Meeting the objectives of the Administration for a 100 percent clean electricity grid will generate a need for long-duration energy storage (LDES), with hydrogen playing a significant role. Assessments of the volume of LDES required in a clean grid exhibit substantial variability, contingent on factors like the extent of electrification, expansion of transmission infrastructure, and the pace of adoption of other compensatory technologies such as direct air capture. Considering diverse studies with varying assumptions concerning these factors, it is anticipated that approximately 4-8 million metric tons per year of hydrogen will be necessary by 2050 to fulfill the energy storage requirements for a completely clean grid [20].

It is crucial to highlight that these outlined hydrogen volumes for each sector are driven by cost considerations, allowing room for flexibility. Initial large-scale implementations of clean hydrogen are anticipated to concentrate on industries with established supply chains and economies of scale, such as ammonia production and the petrochemical sector. These large-scale initiatives will be complemented by smaller-scale deployments in emerging applications and expanding sectors as the infrastructure matures. Leveraging the success of early deployments and the impetus provided by the Hydrogen Shot, the United States stands poised to achieve substantial growth in clean hydrogen production, reaching 20 million metric tons per year by 2040 and 50 million metric tons per year by 2050, as depicted in Figure 7.9 [4]. This demand-driven opportunity is feasible while directing hydrogen toward decarbonizing vital sectors of the economy that pose challenges for electrification and aiding in the integration of renewables into a clean grid.

While Figure 7.9 illustrates potential scenarios for demand growth, the actual demands that materialize could diverge due to a variety of market dynamics, policy shifts (such as the production tax credit for clean hydrogen established by the Inflation Reduction Act), and advancements in technology performance and costs achievable by 2050.

In the realm of transportation, the supplementary scenarios present different assumptions concerning the cost of hydrogen fuel. For biofuels and power-to-liquid fuels, the ranges encompass strategies to optimize biofuel production from diverse feedstocks and variations in the demand for power-to-liquid fuels, considering the possibility of utilizing up to 6 MMT of hydrogen per year for power-to-liquid fuels. Regarding industrial applications, the lower end of the spectrum presupposes that ammonia is the sole market sector adopting clean hydrogen. On the higher end, it is assumed that ammonia, steelmaking, and methanol production embrace clean hydrogen to a degree consistent with the core range. Moreover, it is posited that clean hydrogen is additionally employed for petroleum refining at a rate equivalent to the current use of steam methane reforming (SMR) for this sector, approximately 6 million metric tons per year. Notably, these values do not encompass additional demand for ammonia, methanol, or other chemical hydrogen carriers for potential hydrogen export. The spectrum of hydrogen in natural gas blending is indicative of its application in decarbonizing industrial heat. The lower threshold in the sensitivity range assumes a 10 percent volume of hydrogen used in industrial sectors requiring heat exceeding 550°C, while the upper limit assumes a 50 percent volume of hydrogen used in industrial sectors requiring heat surpassing 300°C [21].

In the power sector, the dynamics influencing hydrogen utilization are intricate and interconnected. Hydrogen stands as one option for delivering adaptable, reliable, and dispatchable power, as well as serving as long-duration energy storage, including in forms such as renewable natural gas, ammonia, and other fuels. The environmental advantages of these energy carriers vary based on how they are produced, distributed, and utilized. Even if hydrogen itself is not the primary storage medium for energy, clean hydrogen is a prerequisite for renewable natural gas and other chemical storage media like ammonia. Electrolyzers offer the capacity to respond dynamically to fluctuations in renewable power, contributing not only to energy storage but also to grid services. Extensive expansions of wind, solar, and zero-emission power are imperative for establishing a clean grid. However, hydrogen and other technologies are essential for integrating renewables into a highly electrified, resilient, and equitable power system. The potential demand range for hydrogen energy storage on the grid is derived from studies modeling a clean grid with varying levels of electrification and demand-side flexibility [22]. The extent of clean hydrogen usage hinges on various barriers to market adoption, encompassing challenges related to building storage and infrastructure. By mitigating these barriers and placing a strong emphasis on addressing energy and environmental justice concerns, clean hydrogen can be deployed swiftly and safely to reduce emissions in sectors that are challenging to decarbonize.

7.5 The challenges facing the widespread adoption of hydrogen

Although there has been a substantial reduction in the costs associated with clean hydrogen technology, these technologies are still in the nascent stages of commercial deployment, presenting several significant challenges. Key obstacles include the absence of hydrogen infrastructure, limited manufacturing at scale, cost considerations, and durability, reliability, and availability challenges throughout the entire value chain [23]. Stakeholder feedback consistently highlights the cost of clean hydrogen as a primary barrier. The DOE's Hydrogen Summit in September 2021, attended by over 3,000 stakeholders from 34 countries, pinpointed various challenges in response to the question, "What is impeding widespread public acceptance and market adoption of hydrogen in the United States?" While cost emerged as the most commonly cited barrier (22% of the responses), the lack of infrastructure (19%) and the imperative for public awareness and acceptance (17%) were also recognized as substantial challenges. The need for technology advancements 11%), the lack of incentives for companies (8%), competing technologies (6%), safety concerns (6%), lack of suitable end uses (5%), and the lack of government support for research and development are the main barriers mentioned by the respondents [24].

To become a main source of energy in the U.S., a significant reduction in the levelized cost of hydrogen is imperative. As of the 2020 analysis, the expense of producing clean hydrogen through proton exchange membrane (PEM) electrolysis, particularly with renewable electricity, can exceed $5/kg [25]. The cost of electrolysis is intricately tied to the price of the electricity employed. To become competitive with other fuels, there must be an 80 percent cost reduction in low-volume PEM electrolysis-generated hydrogen [26]. While advancements in advanced and high-temperature electrolyzers are underway, challenges to their market adoption encompass cost considerations, durability, and the scale of manufacturing capacity. Furthermore, the integration and optimization of high-temperature electrolysis with thermal sources, such as nuclear power plants, are essential for enhancing efficiencies in hydrogen production. Beyond production costs, challenges in hydrogen transport, including pipelines, tube trailers, liquefaction, siting, permitting, and materials compatibility, demand attention. Operational data from California underscores that the cost of hydrogen delivered to fueling stations, encompassing compression and dispensing for vehicles, can exceed $13/kg- more than three times the cost required for competitiveness [27,28].

The efficient storage of hydrogen poses another significant challenge. Despite hydrogen offering nearly three times the energy content per unit of mass compared to gasoline, its gaseous form has a low volumetric energy density, posing challenges for compact storage in containers or tanks. There is a pressing need to reduce the weight, volume, and cost of hydrogen storage systems, with specific targets contingent on the application. Although safety has been validated through extensive testing in numerous commercial systems, ongoing efforts are essential to maintain safety standards and implement best practices.

Although it's common to store compressed hydrogen at room temperature, increasing the density of hydrogen is significantly improved by reaching lower temperatures, even at cryogenic levels. Liquid hydrogen, stored at extremely low cryogenic temperatures in well-insulated double-walled tanks, is commercially available and currently utilized for industrial-scale storage and transport. However, the incorporation of insulation, as well as addressing boil-off and venting to release accumulated pressure for safety, introduces additional costs and challenges to system performance. Comprehensive research efforts at the material, component, and system levels are indispensable to tackle these challenges. Furthermore, exploring the use of hydrogen carriers like ammonia or liquid organic hydrogen carriers (LOHCs), with a focus on understanding the carriers' cost, life cycle emissions, and toxicity, requires additional attention. Apart from the technological and cost-related challenges, there is a broader need, from an overarching energy systems perspective, to determine the optimal utilization of hydrogen in applications where more cost-effective or efficient alternatives are not available. A thorough assessment is required to understand the intricate relationship between hydrogen demands, electrification, changes in the energy grid (including the provision of clean baseload power, grid reliability, and the effectiveness of carbon capture and storage), biofuels, and sectors reliant on hydrogen as a feedstock or fuel. A detailed, region-specific approach, considering resource availability, end-uses, and supported by the funding allocated to hydrogen hubs, will guide the evolution of the hydrogen ecosystem for maximum benefit.

7.6 The U.S. strategies to develop clean hydrogen technology

While hydrogen's adaptability allows for diverse applications, the DOE concentrates on deploying clean hydrogen to decarbonize challenging-to-electrify sectors like industry and heavy-duty transportation. In these sectors, processes utilizing fossil fuels as a chemical input or generating high-temperature heat and long-duration, dispatchable power necessitate clean fuels, including hydrogen, for effective decarbonization. Notably, ammonia and methanol production contribute significantly to global GHGs in the chemicals sector, relying on natural gas as a feedstock. Clean hydrogen can achieve over 90 percent decarbonization in these processes [29,30]. Steel production, responsible for approximately 7% of global GHGs, traditionally relies on coke and natural gas [31]; transitioning to clean hydrogen as a reductant can result in emissions reductions ranging from 40-70% [32]. In current industrial practices, over half of emissions arise from directly burning fossil fuels for heat and power, with about 30% of industrial heat exceeding 300°C, likely requiring clean fuels for decarbonization. Furnaces using pure hydrogen or hydrogen-natural gas blends emerge as crucial options in such applications [33].

Since the power grid undergoes decarbonization, the significance of long-duration energy storage technologies becomes crucial for facilitating the increased use of clean electricity across various sectors. Hydrogen utilization in low-NO_x turbines or fuel cells stands out as a prominent choice to provide multi-day, dispatchable power to the grid. In scenarios marked by high electrification rates, elevated demand for clean hydrogen and other environmentally friendly fuels may arise to ensure reliable power, integrating variable renewable energy into the grid to achieve green power. In the realm of transportation, hydrogen offers compelling advantages, particularly in the trucking sector, especially for fleets featuring heavy-duty vehicles, long-distance routes exceeding 500 miles, or multi-shift operations necessitating rapid refueling. Hydrogen also serves as a vital precursor for manufacturing liquid fuels essential for large-scale energy applications, including aviation, rail, and marine sectors [34]. In the short term, clean hydrogen has the potential to replace conventional hydrogen in petroleum refining for traditional transportation. Looking ahead to the medium and long term, hydrogen can play a role in producing biofuels from biomass, enhancing fuel yield from a given feedstock and pathway, and refining fuel properties. Additionally, hydrogen can be instrumental in generating power-to-liquid fuels, displacing petroleum, especially in off-road markets.

7.6.1 Hydrogen utilization in the U.S. industry

On a global scale, the industrial sector stands as the largest consumer of energy, representing 38% of the overall energy demand. In the U.S., the overall industrial energy consumption in exajoules (EJ) will see a decline by 2050, with specific applications transitioning from fossil fuels to cleaner alternatives such as electricity, hydrogen, or biofuels (Figure 7.10) [3]. In scenarios where hydrogen production is more extensive, there is a notable increase in electricity consumption, primarily driven by the substantial electricity demand associated with the hydrogen production process. Approximately 6% of the total energy demand is allocated to hydrogen production, primarily employed in the synthesis of ammonia and other chemical compounds [35]. According to data from the International Energy Agency (IEA), the global industrial demand for hydrogen amounted to 53 MMT in 2022, constituting a significant portion of the 95 MMT utilized across all sectors [36].

Hydrogen application in steel making: Hydrogen plays a crucial role in the U.S. steel industry, a cornerstone of engineering and construction vital to societal infrastructure. With steel production relying on methods such as basic oxygen furnaces (BOFs) and electric arc furnaces (EAFs), the U.S. distinguishes itself by utilizing the EAF process in about 70% of its steelmaking, emphasizing recycling [37].

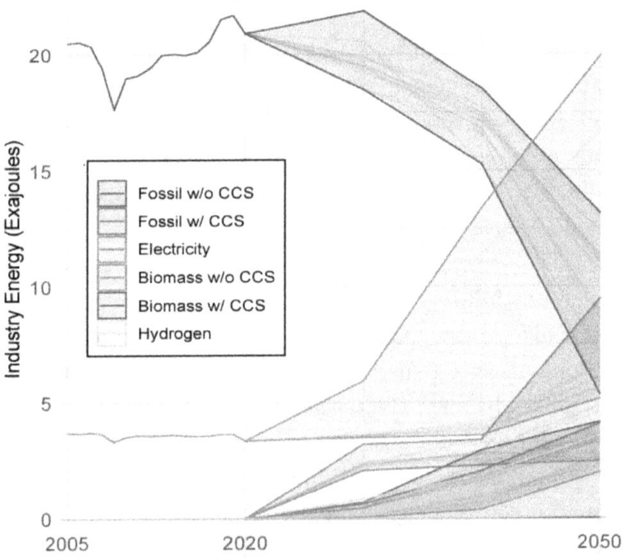

Figure 7.10 Development of hydrogen utilization in the U.S. industry by 2050 [3]

Clean hydrogen's integration as a reductant in iron ore refinement, instead of coke or natural gas, offers a remarkable opportunity to reduce life cycle emissions in primary steel production by 40-70% [38]. This shift aligns with strategies for sector decarbonization, including immediate efficiency enhancements and innovative long-term approaches. The DOE is actively advancing initiatives to integrate hydrogen into steel manufacturing, aiming to optimize processes and scale up production, thus contributing significantly to the national hydrogen strategy and fostering a cleaner, more sustainable steel industry [39,40].

Using hydrogen in chemicals: Hydrogen holds critical importance in the U.S. chemicals sector, serving as a vital feedstock in the production of ammonia and methanol. In existing plants across the United States, ammonia and methanol are conventionally produced using natural gas reforming to generate syngas, which is subsequently converted into these chemicals. The potential for decarbonization lies in replacing natural gas reforming with clean hydrogen production methods, incorporating technologies like CCS or electrolysis while mitigating methane emissions. The future integration of clean hydrogen in these processes hinges on market dynamics and the impetus to decarbonize. Currently, 88 percent of U.S. ammonia consumption is dedicated to fertilizer production, with the remaining 12% serving diverse applications like plastics, explosives, synthetic fibers, and resins [41]. Methanol, a primary building block for chemicals, relies on the overall growth of chemical production, plastics recycling rates, and the emergence of new end-uses, including its potential as a fuel or hydrogen carrier. DOE-supported initiatives encompass cost and life cycle assessments for hydrogen carriers like methanol and ammonia, with ARPA-E funding innovative approaches for ammonia production and scalable hydrogen-to-ammonia systems [42].

Using hydrogen in industrial heat: The largest energy consumer in U.S. manufacturing is process heating, predominantly reliant on fossil fuel combustion [19]. To decarbonize, options include electrification for lower heat grades (below 300°C), CCS, solar thermal, nuclear power, and using hydrogen blends or pure hydrogen for high-temperature applications in sectors like refining, chemicals, cement, and steelmaking (more than 300°C). The heat and power sector, due to cost considerations, is expected to adopt clean hydrogen at scale when widely available at a low cost or under robust decarbonization policies. Adoption necessitates advancements in low-NO_x hydrogen combustion tech and a comprehensive understanding of hydrogen's effects on infrastructure and turbine materials. The DOE's 2020-launched HyBlend initiative, uniting national labs and industry partners, addresses knowledge gaps in hydrogen-natural gas blending, encompassing materials compatibility, cost analysis, emissions, underground storage, hydrogen appliances, and low-NO_x hydrogen turbines [43]. Renewable natural gas (RNG) offers a compatible solution for decarbonizing the heat and power sector, leveraging existing infrastructure. A DOE-funded project showcased the amalgamation of an electrolyzer with a bioreactor, producing RNG from CO_2 and hydrogen [44]. This innovative bioreactor design is now transitioning to commercialization with deployments in California and the Northeast. Future RNG concepts involve catalyzing hydrogen and carbon dioxide for synthetic methane. Successful decarbonization through this method necessitates managing and mitigating fugitive methane emissions across the delivery infrastructure. Ongoing life cycle analyses within DOE's HyBlend initiative are comparing RNG to hydrogen blending. Future collaborative efforts across agencies and states aim to establish injection standards for blending hydrogen into natural gas pipelines, including defining upper blend limits. The work will also explore repurposing natural gas infrastructure for hydrogen and determining conditions necessitating new infrastructure deployment for high blend concentrations. HyBlend priorities involve mitigating risks for all communities, particularly vulnerable ones, and advocating for policies like "dig once" strategies during the installation of transmission, CCS, CO_2 pipelines, and other infrastructure.

7.6.2 Hydrogen utilization in the U.S. transportation

Hydrogen for vehicles, trucks, and buses: The transportation sector contributes about 30% of GHG emissions in the U.S., with light-duty vehicles alone responsible for 58% [45]. While battery electrification is a prominent focus for light-duty vehicles, hydrogen and fuel cells offer significant advantages for applications requiring long ranges, fast fueling, and heavy payloads. Medium and heavy-duty (MDHD) vehicles, constituting around 20% of transportation emissions, are utilized for various purposes nationwide. In the trajectory toward 2050, the total energy consumption in exajoules (EJ) for transportation experiences a decline, accompanied by a rise in the utilization of electricity and alternative fuels. This shift, encompassing biomass-derived fuels and hydrogen, is envisioned to fuel the entire U.S. transportation system predominantly (Figure 7.11) [3].

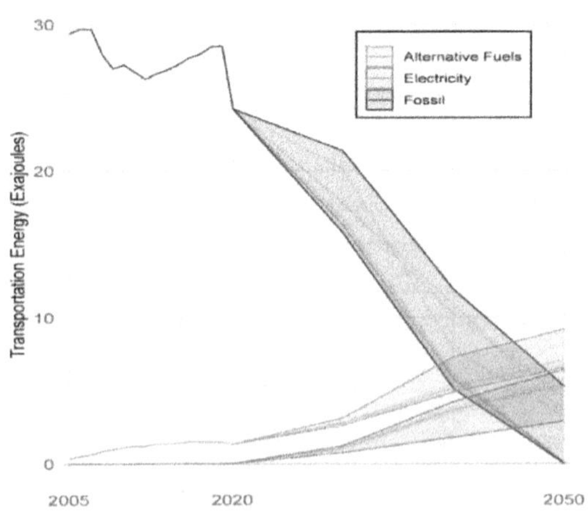

Figure 7.11 The U.S. final energy use by 2050 [3]

Collaborative efforts involving DOE, federal agencies, industry, and national laboratories through the 21st Century Truck Partnership aim to develop safe and cost-effective approaches to reduce emissions from trucks and buses [46]. The partnership focuses on both batteries and fuel cells, with fuel cells being particularly suitable for heavy-duty applications requiring fast fill times or extended driving ranges beyond 500 miles [47]. In 2020, DOE initiated the Million Mile Fuel Cell Truck Consortium (M2FCT) to address the fuel cell durability, cost, and performance essential for long-haul heavy-duty trucks [48]. Additionally, hydrogen and fuel cell truck projects were selected under the Super Truck program to showcase medium- and heavy-duty hydrogen fuel cell trucks in actual operational conditions within the next five years [49]. Supporting efforts encompass infrastructure development, fueling components, hydrogen storage technologies, and a project demonstrating 15 parcel delivery trucks in disadvantaged communities [50,51]. Transit agencies with extensive bus fleets or coach buses with substantial driving ranges can also benefit from hydrogen and fuel cells. Collaborating with the Federal Transit Administration, DOE assesses fuel cell buses and gathers real-world deployment data for future advancements. The strategy emphasizes fleets, freight, and corridors conducive to dedicated infrastructure development, reducing the risk of stranded assets and ensuring optimal utilization of emerging hydrogen fueling infrastructure [52].

Hydrogen for aviation: Preceding the COVID-19 pandemic, aviation constituted approximately 11 percent of U.S. transportation emissions, a share anticipated to expand without intensified efforts as air transport of people and goods increases [53]. The development of SAFs, encompassing biofuels and power-to-liquid fuels as substitutes for traditional jet fuel, is pivotal for decarbonizing the aviation sector [54]. In 2021, the U.S. government initiated a SAF Grand Challenge, a collaborative effort aimed at reducing costs, improving sustainability, and expanding SAF production and utilization with a targeted 50% reduction in lifecycle GHGs compared to conventional fuel [55]. The challenge outlines goals of supplying 3 billion gallons of SAFs annually by 2030 and 35 billion gallons by 2050, meeting 100 percent of aviation fuel demand by 2050. These national objectives underpin hydrogen demand in aviation, with diverse biofuel and power-to-liquid pathways under exploration. Approved aviation pathways currently utilize hydrogen as a feedstock and may co-produce renewable fuels for broader transport sector use [56]. The Net Zero Tech team, a DOE and industry collaboration through the U.S. DRIVE partnership, conducts analyses to identify fuels with significant potential. Additionally, direct hydrogen use is illustrated in specific aviation segments like short-duration flights and unmanned aerial vehicles (UAVs), with the Department of Defense (DOD) showcasing direct hydrogen fuel cells for UAVs. Industry initiatives, such as ZeroAvia and Otto's partnership to develop a 19-seat, 1,000-nautical mile-range aircraft, underscore evolving endeavors in this sector. Airbus has also unveiled design concepts integrating direct hydrogen use, including fuel cell and hydrogen combustion systems [57].

Hydrogen utilization in maritime: In addition to aviation and land vehicles, the maritime industry presents opportunities for hydrogen and its carriers, spanning inland and harbor vessels to recreational and pier-side applications. International Maritime Organization (IMO) regulations now mandate reduced sulfur content in ship fuel oil, with limits decreasing from 3.5 percent to 0.5 percent in 2020 and further to 0.1 percent in Emissions Control Areas [58]. Given these stringent regulations, hydrogen and carriers like ammonia and methanol emerge as appealing alternatives to traditional bunker fuel. Hydrogen's application in various marine vessels, drayage trucks, shore power, and cargo equipment offers the potential to cut carbon emissions and establish targeted infrastructure in key regions. Over the past decade, DOE, in collaboration with the Maritime Administration (MARAD), has been advancing hydrogen and fuel cell technologies for maritime use [59]. Collaborative efforts with state agencies and industry have led to the deployment of the first hydrogen fuel cell passenger ferry in the Western Hemisphere. Ongoing initiatives, such as an MW-scale electrolyzer project on a floating barge, demonstrate the viability of hydrogen for maritime applications, aligning with Strategy One to de-risk technologies for increased private sector investment and market adoption. Safety measures and the development of codes and standards, in coordination with organizations like IMO, MARAD, and international partners, are integral components of these activities.

Hydrogen in rail: The extensive rail network in the United States, covering over 140,000 route miles, plays a vital role in transporting goods and passengers and supporting a substantial workforce of 167,000 jobs [60]. Despite contributing only about 2% of transportation-sector emissions, the rail sector faces challenges in decarbonization due to entrenched low-cost legacy systems and inexpensive diesel fuel [61]. However, a transition to liquid fuels (including biofuels), batteries, and hydrogen offers promising avenues for achieving complete decarbonization. The economic viability of each powertrain solution will vary based on regional factors and the specific demand characteristics of each system. Globally, early demonstrations of hydrogen and fuel cell-powered trains have taken place, with hydrogen-powered trains operational in Germany since 2018 [62]. In the U.S., California's San Bernardino Transportation system is actively developing a hydrogen fuel cell passenger train set to commence service in early 2024 [63]. DOE's collaboration with the Department of Transportation's Federal Railroad Administration (FRA) includes analysis efforts to establish performance and cost targets, focusing on distinct locomotive market segments within the rail sector. Progress toward these targets will be monitored and validated in coordination with FRA [64].

7.6.3 Hydrogen for energy storage, backup power, and stationary power

As the deployment of renewable and zero-carbon generation assets increases, hydrogen becomes a versatile medium for storing energy over extended periods and providing grid services. It also presents opportunities for additional revenue by supplying hydrogen as a feedstock or fuel to various sectors. In critical facilities like hospitals and data centers requiring continuous power, hydrogen fuel cells for backup and stationary power can replace diesel generators, enhancing resilience. Remote locations with a need for consistent power, such as microgrids and telecom towers, offer promising prospects as well. Despite low utilization of backup power, transitioning from diesel to clean hydrogen marks a meaningful step towards achieving a net-zero impact. Hydrogen-powered fuel cells operate with zero emissions, are quieter, and more reliable than their diesel counterparts, bringing health and air quality benefits, particularly for disadvantaged communities in non-attainment zones. Examples of projects funded by the DOE in this sector include the world's first trigeneration system at a wastewater treatment plant, demonstrating power, heat, and hydrogen co-production through a high-temperature fuel cell. Additionally, there are projects supporting the use of hydrogen fuel cells in data centers, initiatives to reduce fuel cell costs and enhance durability, research and development on reversible fuel cells, and numerous fuel cell deployments for backup power applications [65,66].

Grid energy storage serves various purposes, including time shifting, firm capacity, avoiding the expansion of transmission lines, and providing ancillary services. Presently, pumped hydropower dominates grid energy storage, capable of discharging power for up to 12 hours [67]. The most rapidly growing form of energy storage is lithium-ion batteries, typically utilized for shorter durations, often four hours or less. As the grid transitions to 100% clean power, there is a growing need for longer-duration energy storage technologies capable of discharging power over multiple days. A potential solution involves hydrogen energy storage systems on a large scale. This approach may involve using electrolyzers to produce hydrogen during periods of excess power on the grid, storing the hydrogen in bulk (e.g., underground), and subsequently using it to generate power during times of high demand [68]. To mitigate environmental, economic, and community impacts, an optimized approach includes co-locating renewables, nuclear plants, high-temperature heat sources, and the infrastructure for storing hydrogen and carbon dioxide. This strategy is more efficient than independently building out each component of such systems.

Achieving large-scale deployments of hydrogen energy storage necessitates cost reductions in electrolyzers and fuel cells, the advancement of low-NO_x combustion technologies for hydrogen turbines, and the innovation of cost-effective bulk hydrogen storage technologies free from geographical constraints. To bolster this sector, the DOE has instituted specialized national laboratory test facilities designed to demonstrate and assess the performance of electrolyzers integrated with various power and thermal sources. These facilities serve as platforms for industry to mitigate integration risks and validate emerging technologies before actual deployment. Furthermore, the DOE is actively funding RDD&D efforts focused on low-NO_x turbines. The department has also supported numerous analysis projects and tools aimed at quantifying the economic advantages of hydrogen energy storage within specific grid conditions in collaboration with industry stakeholders [69]. In a significant development in 2022, the DOE's Loan Programs Office (LPO) announced a conditional commitment for an approximately $500 million loan guarantee for the Advanced Clean Energy Storage Project. This groundbreaking initiative represents a first-of-its-kind facility for clean hydrogen production and storage, facilitating long-term seasonal energy storage. Located in Delta, Utah, the facility will combine a 220 MW alkaline electrolyzer with salt cavern storage to enable grid-scale energy conversion and storage using hydrogen as the energy carrier. The Advanced Clean Energy Storage Project is anticipated to have a positive impact on Utah by generating up to 400 construction jobs and 25 operations jobs. Furthermore, it has the potential to stimulate long-term employment opportunities and contribute to the state's transition toward a new, clean energy economy in the future [70].

7.6.4 Development of hydrogen technology in the U.S. and environmental concerns

The carbon intensity of hydrogen production methods varies based on factors such as energy source, efficiency, and design. For fossil pathways, the carbon footprint depends on elements like carbon capture and sequestration, system energy efficiency, and fugitive emissions. In the case of electrolysis, the primary variable influencing lifecycle emissions is the carbon intensity of the electricity source, whether it comes from dedicated renewables, nuclear, or bulk grid electricity. In the U.S., the DOE is tasked with establishing an initial standard for the carbon intensity of clean hydrogen as a reference point for specific programs. This standard will be developed in collaboration with the U.S. Environmental Protection Agency (EPA) and will take into account input from industry and other stakeholders.

According to the BIL, the DOE must set a clean hydrogen production standard that supports production from designated low-carbon energy sources (e.g., fossil fuels with carbon capture, utilization, and storage (CCUS); hydrogen-carrier fuels like ethanol and methanol; renewable energy resources, including biomass; and nuclear energy). The term "clean hydrogen" is defined as hydrogen produced with a carbon intensity equal to or less than 2 kilograms of carbon dioxide equivalent per kilogram of hydrogen at the production site. The standard should also consider the "technological and economic feasibility." Furthermore, the DOE is obligated to update this standard within five years of establishing the initial benchmark.

To illustrate, emissions from Steam Methane Reforming (SMR) with CCS can vary between 3-5 kg-CO_2e/kg-H_2 in a well-to-gate scenario, depending on factors like fugitive emissions, capture rate, and the carbon intensity of the electricity grid.

Well-to-gate emissions for electrolysis are nearly zero when the electricity supply is 100 percent carbon pollution-free, aligning with the Administration's goal by 2035. However, under the current average U.S. grid mix, well-to-gate emissions for electrolysis can be twice that of SMR [71]. As the global hydrogen trade advances, there is a growing need for standardized international approaches to lifecycle analysis. This imperative was identified as a top-priority measure through a collective vote involving over 20 countries within the International Partnership for Hydrogen and Fuel Cells in the Economy (IPHE), a global government partnership initiated in 2003 to hasten advancements in hydrogen and fuel cell technologies [72]. The United States, having completed its term as Chair, currently holds the position of Vice Chair within IPHE and actively participates as a leading member in the IPHE Hydrogen Production Analysis (H2PA) task force [73]. This task force, operating under IPHE, is dedicated to devising mutually agreed-upon methods for lifecycle analysis specifically tailored to hydrogen production. Analysis guidance developed thus far has concentrated on particular hydrogen production pathways deemed relevant across more than 20 countries in the near term. Continuous efforts are underway to broaden this guidance, encompassing additional pathways and incorporating considerations for emissions associated with hydrogen distribution. While the guidance formulated by IPHE is not legally binding, it serves as an influential source for informing accounting frameworks adopted by member countries, thereby ensuring uniformity. Consequently, the United States is committed to engaging in global collaboration and coordination to expedite progress, enhance transparency, and uphold thorough analyses of emissions throughout the hydrogen value chain, taking into account potential indirect impacts from various pathways.

7.7 The U.S. strategies to accelerate hydrogen cost reduction

Addressing various challenges along the entire hydrogen value chain, the DOE emphasizes the reduction of clean hydrogen costs. Given the diverse methods of hydrogen production with varying technology readiness levels and associated environmental impacts, the DOE has focused on identifying and mitigating the primary barriers to cost reduction. A proactive approach involves prioritizing and expediting actions, fostering collaborative partnerships spanning industry, academia, and national laboratories, and maintaining a dynamic portfolio that adjusts based on performance metrics. To respond to President Biden's Climate Summit request in April 2021 and accelerate efforts to combat the climate crisis, DOE launched the Energy Earthshot initiative, which establishes ambitious goals. The Hydrogen Shot, a prominent component of this initiative, aims to significantly lower the cost of clean hydrogen while facilitating widespread deployment through initiatives such as hydrogen hubs, loan guarantees, and other strategic mechanisms. As demonstrated in Figure 7.12, the Hydrogen Shot has the potential to facilitate diverse applications and effects, capitalizing on the existing advancements across various production pathways [4]. Continuing the advancement of RDD&D efforts while concurrently minimizing costs and associated lifecycle emissions remains crucial for all hydrogen production pathways. It is anticipated that a combination of hydrogen production methods, including water electrolysis, fossil fuel-derived hydrogen with carbon capture and storage, and hydrogen production from biomass and waste feedstocks, will likely be utilized in the United States until at least 2050. Presently, thermal conversion pathways dominate the global hydrogen supply, boasting low costs but high emissions. On the other hand, electrolyzers utilizing clean energy and employing advanced methods such as photoelectrochemical and thermochemical water splitting have the potential to achieve near-zero emissions, although they currently incur higher costs at the lab scale.

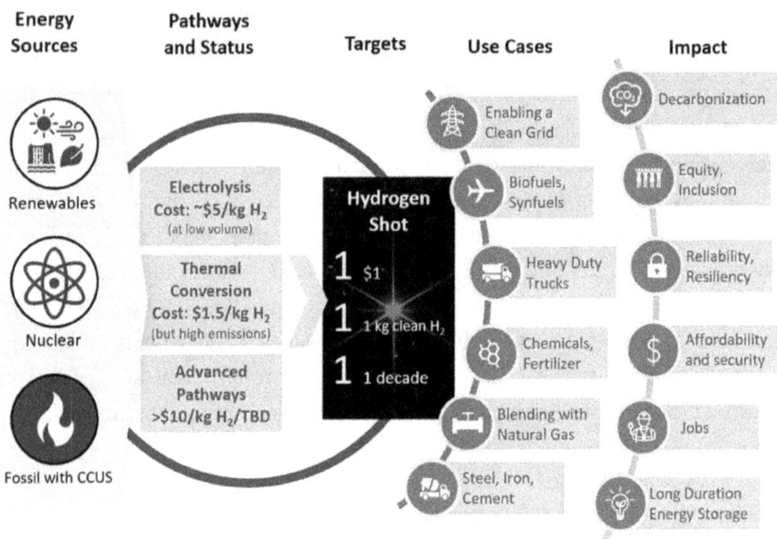

Figure 7.12 The Hydrogen Shot aims to capitalize on advancements in various pathways, facilitating a diverse range of applications and impacts [4]

7.7.1 Cost reduction in the electrolysis process

Electrolysis involves the utilization of electricity and an electrolyte or membrane to separate water into hydrogen and oxygen. Among the three predominant electrolysis technologies- alkaline, PEM, and solid oxide stem electrolyzer (SOSE)- the alkaline process stands as the most established, with a history spanning over a century. PEM electrolyzers exhibit effective operation across a range of loads with sub-second response times, making them particularly suitable for variable energy sources like solar and wind power. SOSEs, utilizing a ceramic electrolyte at elevated temperatures, are the least commercially developed among the three technologies.

Boasting higher electrical efficiency compared to PEM and alkaline systems, SOSEs prove more cost-effective in scenarios where high-temperature heat is available, such as from nuclear power plants and concentrated solar power. The current pivotal factor influencing the cost of hydrogen production from electrolysis is the expense of clean electricity, which now constitutes over half of the overall production cost. Research, development, demonstration, and deployment efforts have the potential to align with the Hydrogen Shot target by reducing the cost of renewables, enhancing electrolysis efficiency, lowering electrolyzer and balance-of-plant capital costs, and facilitating the dynamic integration of electrolyzers with the grid and renewable generators to access low-cost variable power [74,75].

Achieving a substantial reduction in the cost of clean hydrogen produced through electrolysis entails a significant decrease in capital costs, energy costs, enhancement of efficiencies, and improvement in durability and reliability to mitigate maintenance expenses. The baseline cost in 2020, set at $5/kg, represents the levelized cost of hydrogen computed using the DOE's H2A model. This calculation incorporates a conservative $1,500/kW capital cost for PEM electrolyzers (at low-volume manufacturing), a $50/MWh electricity price, and a capacity or utilization factor of 90% [76]. A comparison with today's conditions, considering a $29/MWh cost for solar and a 35 percent capacity factor based on the 2020 National Renewable Energy Laboratory (NREL) Annual Technology Baseline, results in a levelized hydrogen cost of approximately $7.50/kg, as indicated by the green arrow. This illustrates the significant impact of electricity costs on the levelized cost of hydrogen production. Access to low-cost energy with a high capacity factor, such as through integration with existing nuclear power plants, holds the potential for substantially lowering levelized costs. The presented example outlining the requirements to achieve the $2/kg target mandated by the BIL by 2026 is based on $30/MWh energy costs and $300/kW capital costs. In contrast, the Hydrogen Shot goal of $1/kg would necessitate $20/MWh energy costs and $150/kW capital costs. Across these scenarios, a 90% electrolyzer capacity factor is assumed, indicating the need for clean baseload electricity, such as nuclear power, to complement renewables. This demonstration highlights the imperative for an 80% reduction in capital costs and a 90% reduction in operating and maintenance costs. It's crucial to note that these scenarios are just potential pathways for achieving the specified cost targets. Alternative combinations of cost, efficiency, electricity prices, utilization factors, and durability, including the utilization of thermal sources for high-temperature electrolyzers, could also contribute to meeting the Hydrogen Shot goal. In 2020, the DOE initiated the establishment of Hydrogen from Next-generation Electrolyzers of Water (H2NEW), a consortium that unites national laboratories, industry partners, and academic institutions. This consortium focuses on advancing electrolyzer technologies, complementing the HydroGEN consortium, which explores diverse electrolysis technologies encompassing direct photoelectrochemical and thermochemical methods. H2NEW's mission is to expedite advancements in electrolyzer technologies with the ultimate goal of reducing costs. Achieving these cost reductions necessitates high-volume manufacturing, innovations in electrolyzer stacks and balance-of-plant components, and the integration of electrolyzers into next-generation systems [76]. No singular predominant factor drives the reduction in capital costs. Various elements, including electrolysis stacks and balance-of-plant systems, constitute the spectrum that necessitates attention. As the need for energy storage and sustainable power escalates, stakeholders must persist in investigating inventive approaches for both on-grid and off-grid integration of electrolyzers. This exploration is imperative to facilitate cost-effective access to variable clean energy.

7.7.2 Cost reduction in fossil fuel-based hydrogen production

The BIL mandates that the DOE consider and promote opportunities for hydrogen production from a diverse range of energy sources, including fossil fuels with CCS. These opportunities may arise in regions of the U.S. characterized by abundant natural gas, available reservoirs for CO_2 storage, or existing infrastructure for natural gas supply. As depicted in Figure 7.13, the current infrastructure network for natural gas and SMR plants is predominantly concentrated in the Gulf Coast region due to the abundance of natural gas and the demand for hydrogen in the petrochemical sector. Presently, hydrogen serves as a crucial feedstock in refining processes, primarily utilized for cracking heavy crude oil and desulfurizing product streams. Substituting the hydrogen used in current petroleum refineries with clean hydrogen has the potential to reduce the life cycle emissions of the refining process by approximately 12%, contingent upon the source of the hydrogen supply [4].

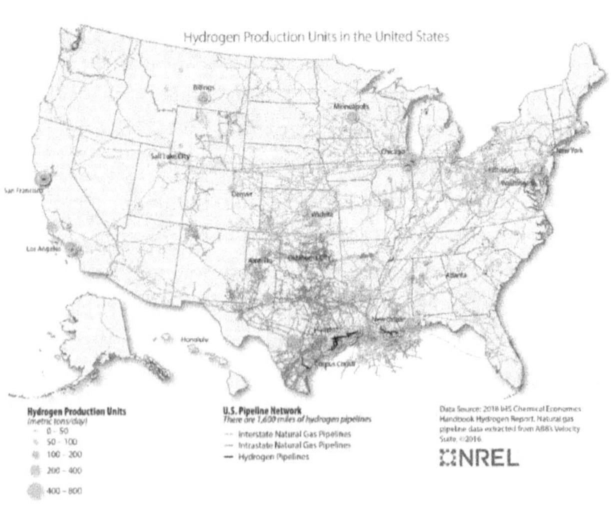

Figure 7.13 Hydrogen production units in the U.S. [4]

Efficiently capturing and storing CO_2 from SMR before its release into the atmosphere holds the potential to decrease the life cycle carbon intensity of hydrogen production by more than 50%. The extent of reduction depends on factors such as CCS rates and upstream emissions, encompassing fugitive releases during natural gas extraction, transmission, and utilization [77,78]. An alternative method for hydrogen production from natural gas, Autothermal Reforming (ATR) with carbon capture, is anticipated to be more cost-effective than traditional SMR. This approach involves integrating fuel combustion into the SMR process to enhance heat utilization. A third method, methane pyrolysis, utilizes high heat to split methane into hydrogen and solid carbon. This approach is appealing as the solid carbon generated can serve as a value-added co-product for applications like industrial rubber and tire manufacturing, as well as for specialty products such as inks, catalysts, plastics, and coatings. Recent investments by the DOE are supporting RDD&D, along with offering loans to facilitate the scaling up and deployment of pyrolysis pathways [79,80]. Presently, the cost of hydrogen production from SMR systems equipped with CCS is approximately 55% higher than that of SMR alone [78]. Achieving cost reductions in CO_2 transport and storage, variable costs, and capital costs becomes pivotal in meeting the Hydrogen Shot target. An alternative approach involves utilizing the captured carbon in various industrial processes instead of storing it underground. Emerging utilization pathways encompass the production of liquid fuels, construction materials, and chemicals.

The United States is witnessing a growing number of initiatives focused on CCSU. Notably, Air Products is currently constructing a facility in Louisiana, set to commence operations in 2026, with a daily production capacity of 750 million cubic feet of reformation-based hydrogen. This facility in Louisiana plans to leverage the state's geological features to sequester an impressive 5 million tons of CO_2 annually, marking it as the world's largest facility of its kind [81]. In Iowa, Green Plains, Inc., has revealed a carbon offtake agreement for three ethanol biorefineries, where the captured CO_2 will be transported via pipeline to underground geological structures in North Dakota for storage. Anticipated to begin operations in 2025, this project aims to sequester 10 million tons of CO_2 each year. Policies such as the 45Q tax credit for CCS play a crucial role in paving the way for the large-scale production of clean hydrogen [82].

In situations involving the utilization of fossil fuels, the DOE will give precedence to minimizing emissions throughout the entire value chain, spanning from production to end-use. Additionally, there is an emphasis on creating measurement and monitoring solutions, considering hydrogen leakage risks when making decisions regarding the expansion of hydrogen transport infrastructure, irrespective of the primary resource. Lastly, DOE is committed to prioritizing stakeholder engagement to address any potential environmental concerns and cumulative impacts on communities hosting technologies related to fossil fuel-based hydrogen and CCS.

7.7.3 Cost reduction strategies across the hydrogen value chain

Alternative methods for hydrogen production encompass biomass gasification combined with CCS, as well as SMR or ATR utilizing feedstocks like biogas derived from organic landfill materials, sewage, or agricultural residues in lieu of natural gas. Depending on the specific feedstock, these production techniques have the potential to be either low-carbon or carbon-negative. The release of CO_2 emissions when crops are utilized for hydrogen production might be considered carbon-neutral since they represent greenhouse gases captured during the growth of the feedstock. However, it is imperative to consider lifecycle emissions across the entire biomass supply chain, incorporating direct and indirect land-use changes, as well as agricultural inputs like fertilizer. When biomass pathways are coupled with CCS, their net emissions have the potential to be negative. For instance, diverting waste feedstock from landfills to produce hydrogen allows for the thermal conversion of some of the methane generated during waste processing into clean hydrogen. This signifies a reduction in atmospheric methane levels compared to regional best practices and regulations that might otherwise involve flaring.

Cost reduction efforts extend beyond hydrogen production alone, encompassing various technologies and components across the hydrogen value chain. The DOE is committed to reinforcing initiatives aimed at reducing the costs of key technologies throughout the entire value chain while also addressing vulnerabilities in the supply chain. In this regard, DOE has issued assessments of clean energy supply chains, including those related to fuel cells and electrolyzers [83]. The provisions outlined in the BIL for electrolyzer and clean hydrogen manufacturing and recycling (amounting to $1.5 billion over five years) will be utilized, alongside annual appropriations, to execute this strategic approach. The cost of delivering, storing, and dispensing hydrogen to end-users varies significantly based on the chosen mode of supply, including gaseous tube trailers, liquid tankers, pipelines for gaseous hydrogen, and chemical hydrogen carriers. R&D efforts are crucial for advancing the commercialization of chemical hydrogen carriers, focusing on enhancing their hydrogen-carrying capacity, charge-and-discharge rates, reversibility, and overall round-trip efficiency. Additionally, post-delivery, onsite conditioning of hydrogen may be necessary, and R&D efforts are essential to reduce the costs, enhance reliability, and increase the throughput of components related to compression, storage, and dispensing, particularly at hydrogen fueling stations for vehicles. Further research is also needed to decrease the cost of current onboard storage designs, such as carbon fiber overwrap, and explore innovative approaches to storage, such as insulated liquid tanks.

7.8 Regional integration and market advancements

One of the U.S. DOE strategies is centered on achieving widespread and economically feasible deployment of clean hydrogen by aligning the expansion of clean hydrogen production with a simultaneous and increasing regional demand.

Establishing large-scale clean hydrogen production facilities in close proximity to multiple end-users can catalyze the development of cost-effective hydrogen and the essential supporting infrastructure, thereby propelling the hydrogen economy forward in key market segments. The creation of regional hydrogen networks is expected to generate both immediate and sustained employment opportunities, boost tax revenues for regional economies, and contribute to emission reduction efforts. The Hydrogen Shot Request for Information (RFI), initiated in 2021, garnered over 200 responses outlining diverse resources, end-use scenarios, and potential impacts in various regions. Respondents identified specific end-use opportunities for clean hydrogen in certain regions, such as port communities or offshore wind generation. In other areas, stakeholders expressed keen interest in utilizing abundant energy resources like biomass or existing infrastructure, such as energy storage or geological caverns. Additionally, stakeholders highlighted instances where disadvantaged or tribal communities could be involved and presented potential job opportunities. The Hydrogen Shot RFI emphasized the multitude of strategic hydrogen utilization opportunities throughout the United States. In numerous instances, the existing infrastructure highlighted by respondents has the capacity to support initial regional deployment requirements. The regional clean hydrogen hub provision outlined in the BIL presents a distinctive and unparalleled opportunity for the United States to initiate a clean hydrogen economy while realizing substantial benefits at the regional and community levels. Information collected from these hubs will play a crucial role in future analyses aimed at identifying optimal approaches for market initiation. This includes strategies such as employing contracts for difference, aligning production with off-takers, generating targeted large-scale demand through anchor tenants, and utilizing existing infrastructure where applicable, including CCS and other pipeline infrastructure.

As an integral part of the strategy, the DOE persists in enhancing and revising regional analyses throughout the hydrogen value chain, encompassing factors such as the accessibility of water and other resources. Leveraging insights from national laboratory and industry analyses, DOE has computed the technical potential for hydrogen production from various domestic resources. These estimates of technical potential for renewable resources are illustrated in Figure 7.14 [84]. The most cost-effective methods for producing clean hydrogen are contingent on the availability of regional resources, and the initial developments in the market should be situated in close proximity to end-users to minimize hydrogen delivery expenses. The optimal approach for each region is determined by a combination of natural resources, existing infrastructure, and hydrogen demand opportunities, which vary across different regions. Regions like the plains, southwest, and mountains exhibit significant solar and wind resource potentials. In the midwestern, northeastern, and southeastern United States, biomass resources are abundant. The industrial heartland and the Gulf Coast, characterized by ample natural gas resources, have substantial potential for geologic CO_2 storage.

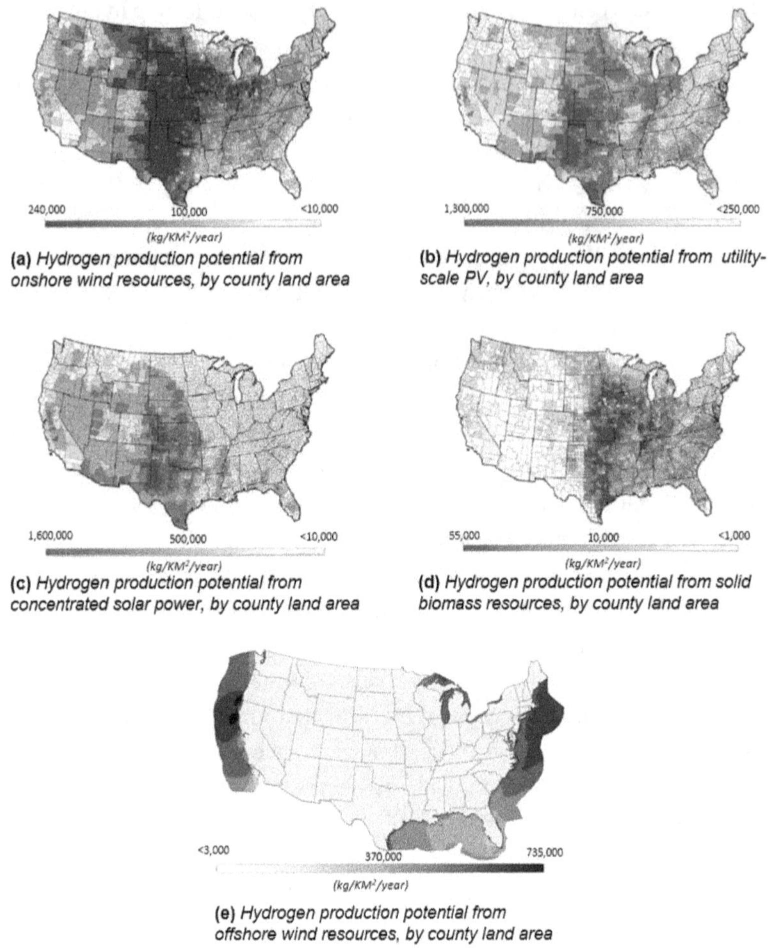

Figure 7.14 The U.S. potentials for generating clean hydrogen from onshore/offshore wind, large-scale photovoltaic solar power, concentrated solar power, and biomass resources [84]

Electrolyzers would ideally be situated in areas with abundant solar and wind resources, as well as in close proximity to nuclear power plants. With the existing nuclear fleet and emerging advanced nuclear technologies, including small modular reactors, various regions present opportunities for clean, baseload nuclear power. In regions characterized by a significant presence of renewable energy sources, electrolysis can play a crucial role in balancing variable loads on the grid. It can efficiently utilize excess capacity generated during peak production to produce hydrogen, preventing the curtailment of power. An illustrative example includes integrating electrolyzers with offshore wind systems in regions facing transmission constraints, thereby generating an additional revenue stream for renewable energy generation. As tangible hydrogen projects progress, the DOE persists in evaluating the most effective strategies and suitable locations for large-scale hydrogen storage. Hydrogen storage offers the capability to disentangle power generation from energy consumption and can achieve cost efficiencies surpassing those of alternative technologies, especially when operating at scales spanning multiple days or weeks. The storage of hydrogen can be implemented in various forms, including gaseous or liquid vessels, underground formations, or materials like hydrogen carriers, depending on the intended application. Each approach presents distinct advantages and drawbacks, prompting ongoing efforts by DOE and industry projects to diminish costs and potential emissions and enhance efficiency and storage capacity.

Commercially, tanks and liquid dewars are already in use across various industries and hydrogen fueling stations for the storage of hydrogen, spanning from hundreds of kilograms to numerous metric tons. Larger-scale vessels, while limited in deployment, have predominantly been employed to store hydrogen in liquid form, particularly for aerospace applications necessitating liquid hydrogen onboard. A notable example is the world's largest liquid hydrogen storage vessel located at Kennedy Space Center in Florida, capable of storing over 330 tons or 1.25 million gallons of liquid hydrogen [85]. On a grander scale, hydrogen storage is also facilitated by underground caverns to manage seasonal discrepancies between hydrogen supply and demand, particularly in the petrochemical sector. In the United States, three substantial geological hydrogen storage caverns exist, with the world's largest located in Beaumont, TX, storing over 7,000 tons underground [10]. Excavation of underground hydrogen storage caverns has predominantly focused on salt deposits near the hydrogen consumption point, with minimal applications in hard rock formations. Other geological formations, traditionally used for natural gas storage, hold potential for hydrogen storage, including depleted oil and gas reservoirs and aquifers. Certain areas in the United States contain industrial clusters (Figure 7.15) where multiple industries stand as potential candidates for incorporating hydrogen as either a feedstock or energy source.

The successful decarbonization of these specific industry segments hinges on the feasibility of integrating clean hydrogen, a process that needs to be evaluated on a sector-specific and region-specific basis. Nevertheless, there exists considerable potential to capitalize on networks that facilitate the establishment of hydrogen infrastructure or large-scale CCS [86]. Additionally, the development of best practices within these clusters can serve as valuable templates for application in other sectors. Efficient deployment of clean hydrogen necessitates the organization of clusters that go beyond a mere collection of disparate projects. It is imperative that these projects are strategically sized, scoped, and planned in a synchronized manner to align with each other in terms of scale, cost, and duration. Such coordination ensures the prevention of stranded assets by creating a critical mass of off-takers, capitalizing on CCS and other essential infrastructure, and maximizing the return on public investments to achieve the net-zero goal. The establishment of clean hydrogen hubs serves as a demonstration of the effectiveness of coordinating regional decarbonization efforts, bolstering the business case for these projects, and encouraging private capital investment. Moreover, these hubs offer opportunities to engage stakeholders at every stage of the process, garner public support, formulate community benefit agreements, and ensure projects progress in alignment with environmental, health, and equity objectives. Industries that already have a significant hydrogen demand, such as ammonia production, are likely to be early adopters of clean hydrogen due to their existing supply chains and economies of scale.

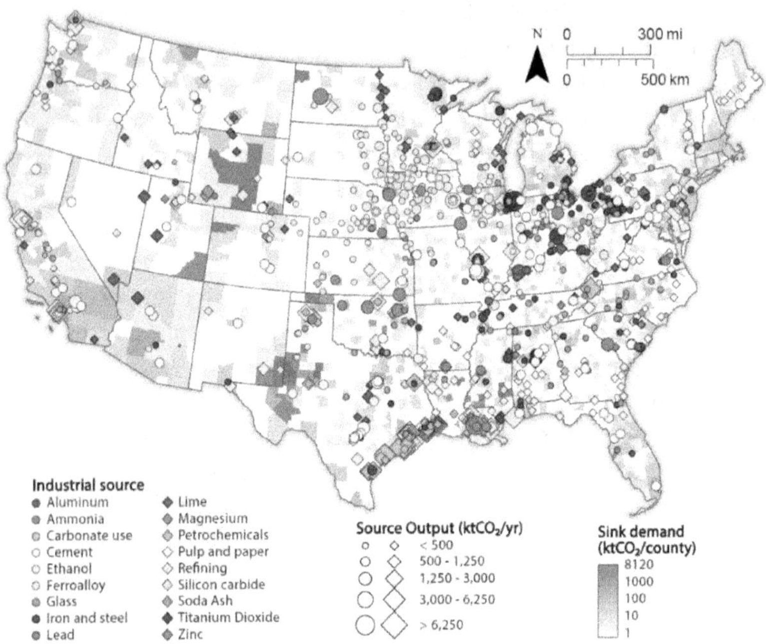

Figure 7.15 The U.S. potential regions for decarbonization and development of hydrogen technology [87]

7.9 The U.S. hydrogen roadmap

A comprehensive roadmap for planned actions has been delineated, covering the near-term until 2025, mid-term until 2029, and longer-term initiatives extending to 2035. Federal agencies are committed to partnering with states, Tribal governments, communities, and other stakeholders to identify regulatory gaps and devise effective strategies to address these gaps. In line with this commitment, the DOE has announced specific actions aimed at supporting the near-, mid-, and long-term goals of achieving clean, affordable, and sustainable hydrogen production [87]:

Near-term goals (2022-2025):

- Evaluate various pathways comprehensively, considering lifecycle analysis, sustainability, cost implications, regional aspects, and equity considerations. This assessment will aid in prioritizing strategies, identifying gaps, and establishing interim goals.
- Institute a Clean Hydrogen Production Standard to provide a benchmark and guidelines for the production of clean hydrogen, aligning with environmental and economic objectives.
- Demonstrate diverse clean hydrogen production technologies, encompassing pyrolysis, waste utilization, renewables, and nuclear-based methods. These demonstrations aim to showcase the viability and versatility of different pathways.
- Decrease the cost of electrolyzers on a large scale by investing in RDD&D efforts focusing on manufacturing, stack improvement, and Balance of Plant (BOP) component enhancements.
- Lower the cost of thermal conversion technologies through RDD&D efforts concentrating on modular designs and intensified processes, contributing to more cost-effective and efficient systems.
- Create low-cost, durable membranes and separation materials to enhance the performance and longevity of components in clean hydrogen production technologies.
- Identify opportunities for standardization of components, reduce reliance on critical materials, and promote the development of a resilient supply chain to ensure stability and scalability.
- Design and implement accelerated stress testing techniques to assess and enhance the durability of clean hydrogen production technologies, ensuring longevity and reliability.
- Publish case studies providing insights into different pathways, emissions, and costs. Additionally, update the Greenhouse Gases, Regulated Emissions, and Energy Use in Transportation (GREET) model for increased user-friendliness, transparency, and inclusion of additional pathways in support of 45V.
- Develop a robust data collection and monitoring framework for future deployments, enabling comprehensive tracking and analysis of clean hydrogen production initiatives.
- Identify necessary worker competencies and establish consensus-based, industry-accepted training credentials, particularly in disadvantaged communities. This includes training programs covering safety, codes, and standards for the clean hydrogen workforce.

- Promote higher education and apprenticeship programs, with a focus on attracting individuals to join the hydrogen workforce. This effort includes raising awareness about career opportunities and benefits within the hydrogen industry.
- Enhance and revise sophisticated analytical models and tools for evaluating delivery and storage pathways, identifying gaps, and prioritizing strategies.
- Innovate technologies for precise monitoring and mitigation of hydrogen leaks and boil-off during storage and transportation.
- Evaluate the compatibility of pipeline and component materials with hydrogen, including hydrogen blends with natural gas.
- Progress novel approaches for cost-effective, high-efficiency hydrogen liquefaction and strategies to mitigate boil-off.
- Engage in the discovery and development of materials for hydrogen carriers suitable for bulk storage and distribution.
- Identify geologic formations suitable for bulk hydrogen storage, along with their associated development and operational requirements.
- Develop and optimize infrastructure designs for hydrogen applications in key sectors, such as industry and energy storage.
- Innovate technologies for high-throughput dispensing of hydrogen, particularly for heavy-duty vehicles.
- Create and standardize fueling protocols for heavy-duty and off-road vehicles where hydrogen is the optimal solution.
- Accelerate RDD&D efforts to reduce the cost of high-pressure and liquid hydrogen storage tanks, including those made of carbon fiber composite.
- Establish a data monitoring and collection framework to assess upstream and on-site emissions.
- Lay the regulatory groundwork for large-scale clean hydrogen deployments, spanning production, processing, delivery, storage, and end-use.
- Collaborate across industries (e.g., nuclear, renewables, fossil, CCS, energy storage) to identify regulatory and policy gaps, implementing strategies such as "Dig Once" approaches for co-locating conduits.
- Develop streamlined guidance for hydrogen pipeline and large-scale project permitting with stakeholder engagement, addressing environmental, energy, and equity priorities.
- Devise market structures and offtake agreements to expedite progress in the clean hydrogen sector.
- Initiate the transition to clean hydrogen for challenging-to-decarbonize industrial applications, specifying potential scale-up locations (e.g., ammonia, refineries, steel).
- Progress efficient end-use technologies (fuel cells/other power conversion with low/zero emissions) and select for scaling up.

- Execute robust modeling and enhance data collection to quantify climate impacts of hydrogen leakage.
- Develop best practices and guidance for assessing life cycle emissions in real-world clean hydrogen deployments, informing "guarantees of origin" and certification schemes.
- Establish safety, risk, and reliability data monitoring and collection frameworks.
- Foster long-term, credit-worthy offtake, including from emerging sectors on the verge of adopting clean hydrogen.
- Develop and implement frameworks for extensive and inclusive community engagement, incorporating environmental and energy justice, disadvantaged communities, Tribes, Tribal organizations, labor unions, industry, academia, national laboratories, and Federal, state, and local governments to ensure broad participation. Hold listening sessions to gather stakeholder feedback.
- Incorporate Community Benefit Plans into funding opportunities, requiring applicants to describe and commit to community and labor engagement, investing in creating good jobs, furthering diversity, equity, inclusion, and accessibility, and meeting Justice 40 goals.
- Identify metrics for diversity, equity, inclusion, accessibility, and other key priorities for teams, organizations, and geographical/community locations for Federally funded demonstrations.
- Launch tools and platforms (e.g., H2Matchmaker) to facilitate partnerships, inclusion, and market success.
- Develop retraining programs for workers (e.g., from fossil industries) to enable both near- and long-term well-paying jobs.
- Create recruitment and career programs for students from underrepresented communities, fostering diversity, equity, inclusion, and accessibility.
- Develop and implement sustainability frameworks and NEPA best practices.
- Develop educational resources to support hub community outreach and engagement strategies.
- Improve data collection on regional priorities (e.g., criteria pollution) and identify applications to inform clean hydrogen deployments.

Mid-term goals (2026-2029):

- Implement large-scale deployment of clean hydrogen derived from renewables, nuclear, and fossil with carbon capture and storage.
- Facilitate the production of clean hydrogen through electrolysis at a cost target of $2 per kilogram.
- Establish domestic electrolyzer manufacturing capacity at a multi-gigawatt scale.
- Showcase catalysts and components, minimizing the use of critical materials while achieving competitive performance and durability.

- Optimize the integration between electrolyzers and clean energy sources to enhance efficiency, reduce costs, and improve resilience.
- Progress promising lab-scale concepts for hydrogen production, including thermochemical, photoelectrochemical, and biological approaches.
- Gather data from real-world demonstrations to inform RDD&D and continually enhance performance and durability.
- Continuously refine and update pathway assessments to target the most sustainable, equitable, resilient, and cost-effective approaches.
- Utilize rigorous analyses, lessons learned, best practices, and broad stakeholder feedback to identify pathways for scale-up with maximum benefits.
- Review and refine work competencies and industry-accepted training standards to align with industry needs.
- Validate and refine analyses, models, and tools for prioritizing delivery and storage pathways across various applications.
- Demonstrate the efficient and reliable operation of hydrogen pipeline compressors.
- Quantify loss rates from gaseous and liquid hydrogen infrastructure to determine mitigation requirements in large-scale deployments.
- Develop designs for commercial-scale, high-efficiency systems for hydrogen liquefaction.
- Advance promising concepts for hydrogen carriers and design reliable, low-cost regenerator systems.
- Commence regional demonstrations for bulk hydrogen storage, including underground approaches, ensuring local and regional benefits.
- Demonstrate novel, efficient, and cost-effective approaches to bulk hydrogen delivery.
- Deploy scalable hydrogen fueling stations to support early fleet markets, such as heavy-duty trucks and buses.
- Establish monitoring systems and data collection for potential hydrogen and other emissions/releases.
- Design sustainable and equitable regional clean hydrogen networks in key locations to maximize benefits, ensuring energy and environmental justice and equity. Facilitate international alignment of codes and standards pertaining to hydrogen technologies.
- Tackle regulatory obstacles to enhance the accessibility of electrolyzers to renewable and nuclear energy sources.
- Disseminate safety best practices and insights gained from initial deployments through publicly accessible platforms.
- Establish a minimum of two Regional Clean Hydrogen Hubs, showcasing hydrogen applications in challenging-to-decarbonize sectors like industry and heavy-duty transport.

- Develop nationwide guidelines for hydrogen blending limits.
- Provide clean hydrogen to generate a minimum of 3 billion gallons of sustainable aviation fuels from biomass and waste by 2030.
- Enhance the efficiency and cost-effectiveness of recovering and recycling raw materials from electrolyzers, fuel cells, and other components along the hydrogen value chain to ensure reduced dependence on foreign imports.
- Aggregate and analyze safety, risk, and reliability data to derive early insights that can shape future deployments.
- Continuously refine and improve community engagement and inclusion strategies, incorporating lessons learned.
- Cultivate public-private partnerships to foster inclusion and expedite progress.
- Formulate and implement community benefit agreements with disadvantaged communities in Hub regions.
- Initiate deployments of hydrogen technologies that mitigate criteria pollution in nonattainment areas, offering resilience, employment opportunities, and other key benefits for local and disadvantaged communities.
- Conduct impact assessments of hydrogen technologies on regional water supply and other local resources.
- Identify and apply lessons learned for environmental and risk assessments, fostering global and regional collaborations.
- Collaborate with unions to develop and expand registered apprenticeship programs for hydrogen technologies.
- Establish education and engagement pathways for first responders and code officials.
- Leverage platforms like H2Tools to share best practices and lessons learned in the hydrogen domain.

Long-term goals (2030-2035):

- Achieve a minimum production of 10 million metric tons (MMT) per year of clean hydrogen by 2030.
- Enable the production of clean hydrogen at a cost of $1 per kilogram from various resources.
- Showcase electrolysis stacks that minimize the utilization of critical materials while achieving targeted performance and durability.
- Exhibit novel and commercially viable methods for hydrogen production, leveraging diverse feedstocks, such as wastewater or high-temperature heat, at a large scale.
- Establish resilient and sustainable domestic supply chains for all employed production pathways, promoting independence from imports.
- Continue gathering data from real-world deployments to inform RDD&D efforts, identifying remaining gaps and refining strategies.

- Apply best practices, lessons learned, and rigorous analyses, including global collaboration and sustainability frameworks, to advance the most sustainable, equitable, resilient, and cost-effective approaches, maximizing benefits.
- Sustain training programs at universities, community colleges, and unions to support a robust hydrogen workforce.
- Design networks of hydrogen infrastructure optimized for regional supply and demand, collaborating with local communities and stakeholders to maximize benefits and address energy, environmental, and equity goals.
- Demonstrate advanced liquefaction with double the efficiency of current concepts.
- Develop a long-term storage plan/strategic hydrogen reserve to ensure the resilience of the hydrogen supply.
- Deploy Regional Clean Hydrogen Hubs with advanced, low-cost clean hydrogen storage and infrastructure.
- Collect data, including emissions data, from demonstrations of bulk hydrogen distribution (e.g., through pipelines or carriers) in real-world environments to inform RDD&D, reducing costs and improving reliability.
- Continue collecting data to inform the scaling up of optimal delivery and storage pathways through RDD&D.
- Share any safety or other best practices related to hydrogen infrastructure across diverse stakeholders to enable continuous improvement.
- Leverage global collaborations on hydrogen infrastructure to inform long-term investment plans and explore opportunities for hydrogen exports.
- Establish market structures and regulatory guidance conducive to the export of clean hydrogen.
- Apply insights from large-scale deployments to identify priority sectors for future growth, emphasizing holistic approaches that align with the most efficient, affordable, and climate-aligned goals, prioritizing public health, safety, and environmental considerations.
- Demonstrate and quantify the advantages of hydrogen in enhancing the resilience of future clean energy systems and addressing disaster mitigation, including aspects like microgrids, cybersecurity, and serving remote communities.
- Demonstrate ultra-low-NO_x turbine operation and low-precious-group-metal (PGM) fuel cell operation using 100% hydrogen for power generation by 2030.
- Launch at least one Regional Clean Hydrogen Hub demonstrating the use of hydrogen in energy storage for a clean grid, quantifying opportunities for hydrogen to contribute to achieving a carbon pollution-free grid by 2035, considering regional factors.
- Continue collecting and analyzing safety, risk, and reliability data, developing insights for continuous improvement.
- Quantify benefits derived from deployments and identify additional policy or program priorities to expedite progress in targeted, no-regrets areas.

- Establish manufacturing facilities for clean hydrogen technologies in disadvantaged communities to generate local and regional benefits.
- Evaluate the techno-socioeconomic impact of Regional Clean Hydrogen Hubs.
- Develop and refine market structures to distribute costs and benefits associated with new technologies equitably.
- Incorporate adaptation, cybersecurity, resilience, and other mitigation approaches into strategic plans for scaling up.
- Update and refine sustainability frameworks and best practices to inform future hydrogen deployments.
- Leverage global collaborations and initiatives to optimize success across the RDD&D pipeline, ensuring an equitable transition to clean energy.

7.10 Development of hydrogen utilization in the U.S.

The initial deployment of clean hydrogen applications in the U.S. has been initiated by targeting established markets that currently have limited alternatives for achieving decarbonization through the use of clean hydrogen. This primarily involves pre-existing facilities in refining and ammonia production. The strategic approach involves setting up industrial clusters that integrate large-scale production with end-use applications, facilitating cost reduction, and establishing infrastructure that can be later utilized for other markets in subsequent phases. Specific hydrogen applications include (Figure 7.16):

- Transit buses, particularly in regions requiring extended travel distances and high operational uptime, present an attractive use case. Transit agencies with extensive bus fleets, where individual battery electric vehicle charging may pose challenges, could benefit.
- Forklifts and material handling equipment in warehouses, ports, and industrial sites, characterized by high utilization, predictable refueling locations, and a need for rapid refueling. The U.S. DOE has already catalyzed this niche by enabling thousands of systems and fostering a burgeoning infrastructure.
- Heavy machinery in mining, construction, and agriculture, characterized by substantial power needs, quick refueling requirements, and potential operation far from power grids, could gain from fuel cell propulsion.
- Long-haul heavy-duty trucks, accounting for a significant portion of transportation-sector greenhouse gas emissions in the United States, have high utilization, energy requirements, and demand fast refueling [88].
- Refineries, currently the largest hydrogen market, lack alternatives for heavy crude oil cracking and desulfurization. The transition to clean hydrogen usage will generate immediate demand, leading to prompt emissions reduction.
- Ammonia production, which currently relies on carbon-intensive hydrogen, can seamlessly shift to clean hydrogen without requiring retrofitting. As the second-

largest captive market for hydrogen after refining, ammonia can provide a stable demand for clean hydrogen.

By supporting demonstrations and infrastructure for these diverse markets, the widespread adoption of hydrogen in specific regions will be facilitated, offering tangible benefits to disadvantaged communities and workers who would otherwise be exposed to fossil fuel emissions.

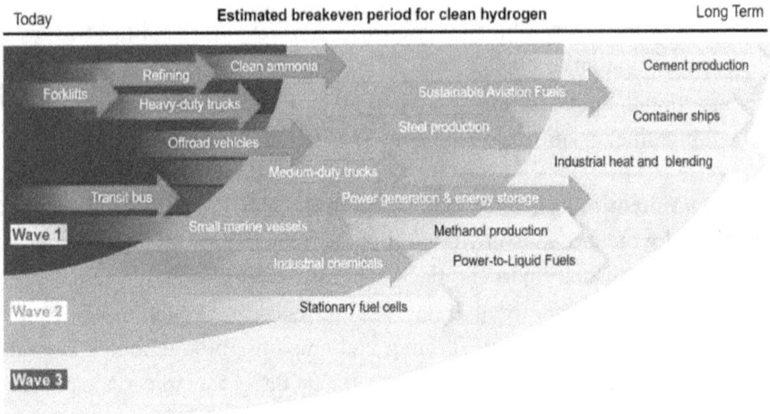

Figure 7.16 The U.S. hydrogen roadmap (2022-2035) [89]

Applications of hydrogen in the U.S. in the second phase encompass use cases where clean hydrogen presents a progressively growing economic value proposition underpinned by industry commitments and advancing policy initiatives. This stage incorporates a broader spectrum of transportation applications and extends to encompass more extensive use of industrial fuel and feedstock. Additional examples beyond those in the initial phase include:

- Regional ferries, propelled by fuel cells, stand to be cost-competitive with internal combustion engines for transporting people or goods over short distances, particularly as hydrogen and fuel cell costs decrease.
- Decarbonization in steel production can be achieved with clean hydrogen when applied to iron ore-based steel production requiring carbon-free reductants and high temperatures, where electrolytic production may not yet be feasible.
- Energy storage and power generation can transition to gas turbines fueled by mixtures of hydrogen and natural gas or pure hydrogen as technologies develop to produce low nitrogen oxides. Fuel cells can serve as power conversion technology. Clean hydrogen plays a crucial role in seasonal storage, essential for fully decarbonizing the grid and eliminating fossil-based generation.

- Medium-duty trucks utilizing hydrogen fuel cells are anticipated to become more widely available at scale, especially as heavy-duty transport paves the way for the expansion of hydrogen distribution and refueling infrastructure.
- Certain industrial chemical production processes, such as those in the plastics industry, necessitate high-temperature heat challenging to achieve with electricity or rely on hydrogen feedstock from fossil sources. These sectors can achieve decarbonization through the use of clean hydrogen for heat generation and as a feedstock.
- Aviation can shift to sustainable fuels produced using green hydrogen, aligning with the DOE goal of 3 billion gallons of sustainable aviation fuel. The large-scale production of clean hydrogen will also lay the groundwork for power-to-liquid production in the long term. Industry feedback suggests that certain market segments could directly utilize hydrogen, although cryogenic storage may be necessary due to energy density requirements.

Applications in the third phase will attain competitiveness as clean hydrogen production undergoes substantial scaling, costs decrease, and infrastructure becomes more widely available. Examples include:

- Methanol produced with clean hydrogen can be directly utilized as a fuel or fuel supplement in applications such as container ships, rail transport, or other maritime uses. Methanol serves as an energy carrier.
- Backup power and stationary power derived from fuel cells can replace diesel generators, providing resilience to critical 24/7 facilities like hospitals and data centers. This transition offers advantages to disadvantaged communities and contributes to improved air quality. Backup power, distinct from energy storage, focuses on providing resilience for a specific customer or microgrid, while energy storage supports the broader macro grid.
- In cement production, clean hydrogen can be employed to reduce direct CO_2 emissions in cases where electrification is not viable due to high heat requirements.
- Container ships, responsible for approximately 90% of global trade by volume, contribute around 3% of global carbon emissions and a significant portion of sulfur dioxide emissions. Potential alternatives in the third wave encompass clean ammonia, clean methanol, and liquefied clean hydrogen.
- Blending with the available natural gas networks can facilitate the targeted decarbonization of high-temperature heating systems, particularly in the industrial sector where specific processes, such as those in the chemical industry, demand elevated temperatures. While this application can commence even during the first wave, substantial cost reductions are essential for economic viability.

The progression of clean hydrogen deployment phases heavily relies on technological development, research advancements, and the establishment of supportive policy frameworks. However, prioritizing sectors that are commercially viable, lack viable decarbonization alternatives, and exhibit industry momentum will enhance the sustainability of public investments.

7.11 International collaboration

The United States remains committed to collaborating with other nations to establish an economical, environmentally friendly, and sustainable global hydrogen economy, aligning with the collective climate objectives of the DOE. During the launch of the Hydrogen Breakthrough Agenda at COP26 in November 2023 in Dubai, various government representatives engaged in discussions about a potential framework for global hydrogen coordination. This proposed coordination framework aims to bring together diverse organizations and initiatives to prevent redundancy, optimize resource utilization, and expedite the effective expansion of clean hydrogen technologies. The DOE hydrogen goals actively collaborate with the European Union and other international partners to enhance coordination efforts. The program will also maintain pivotal roles in numerous multi-lateral and bilateral hydrogen partnerships. As specific activities and mutually agreed-upon priorities are established, the DOE's Hydrogen Program will continue to assume a leadership role, fostering collaboration, information sharing, and expediting actions toward tangible and successful outcomes. The U.S. government has consistently demonstrated strong leadership by actively engaging and supporting its counterparts in numerous countries. Serving as co-leads for hydrogen initiatives under both the Clean Energy Ministerial and Mission Innovation, along with their roles as former chair and current vice chair of the IPHE and substantial contributions to the IEA's hydrogen and fuel cell programs, the DOE has been acknowledged for its pivotal role in driving progress and achieving tangible results. Notable initiatives include the launch of the H_2 Twin Cities program, promoting collaborations among cities globally in deploying hydrogen technologies with a focus on equity and justice. The U.S. DOE has also co-led efforts to facilitate international trade, develop a standardized methodology for assessing the carbon footprint of hydrogen, harmonize codes and standards, and establish an early career network entirely managed by students and early career professionals from over 34 countries. As the global momentum for clean hydrogen continues to grow, the DOE remains committed to advancing these initiatives and introducing additional concrete actions, including [90]:

- Cooperation in technological advancements and synchronization in the standardization of regulations, codes, and standards.
- Encourage global collaboration through shared research and development efforts, specifically enhancing hydrogen safety and optimizing infrastructure supply chains.
- Examination and assessment of hydrogen's capabilities across various sectors, addressing its potential to mitigate CO_2 emissions and other pollutants.
- Communication, education, and outreach initiatives.

These pillars serve as guiding principles to foster global cooperation and advance the responsible development and utilization of hydrogen technologies.

7.11.1 The U.S. hydrogen diplomacy

Hydrogen diplomacy in the United States is poised to play a pivotal role in shaping the nation's role in the global energy landscape, with a focus on hydrogen production, delivery, and utilization, including fuel cell systems. As the U.S. endeavors to position itself as a leader in the burgeoning hydrogen economy, effective diplomatic strategies are essential to navigate international collaborations, address challenges, and capitalize on the vast potential of hydrogen as a clean energy carrier. At the heart of U.S. hydrogen diplomacy is a recognition of the transformative power of hydrogen in achieving decarbonization goals and fostering sustainable energy solutions. This diplomatic endeavor involves engaging with nations worldwide to promote collaboration, share technological advancements, and collectively address global challenges associated with the production, delivery, and utilization of hydrogen. Within the realm of hydrogen production, the U.S. must engage in diplomatic initiatives to foster international partnerships that align with the nation's decarbonization strategies. Whether it be the production of green hydrogen through renewable energy sources or blue hydrogen with carbon capture and storage, diplomatic efforts should aim to establish bilateral and multilateral agreements. These agreements can encompass technology transfer, research collaborations, and the creation of a global framework for sustainable hydrogen production. By actively participating in international discussions, the U.S. can contribute to the development of standardized practices, ensuring the global consistency of hydrogen production methods. The emergence of a clean hydrogen market in the U.S. opens avenues for economic growth and diplomatic influence. As the U.S. positions itself as a major player in the global hydrogen trade, effective hydrogen diplomacy becomes imperative. Diplomatic channels should be utilized to negotiate trade agreements, facilitate international investments, and establish transparent mechanisms for hydrogen transactions. The U.S. can leverage its economic influence to promote fair trade practices, contributing to the growth of a global hydrogen market that is both sustainable and economically beneficial.

As it navigates the complex landscape of hydrogen diplomacy, the United States can strategically engage with various regions to leverage its unique strengths and contribute to a global hydrogen ecosystem. Distinguishing between blue and green hydrogen and focusing on technological collaboration with Japan and Korea provides a comprehensive approach to harnessing the potential of hydrogen in a sustainable manner. The Middle East, endowed with abundant natural gas resources, can play a significant role in the production of blue hydrogen through methane reforming coupled with CCS. The U.S. can forge strategic partnerships with Middle Eastern countries to collaborate on large-scale blue hydrogen projects. Diplomatic efforts should center on negotiating agreements for the export of blue hydrogen, creating a symbiotic relationship that leverages the Middle East's resource wealth and the U.S.'s technological expertise in CCS. This collaboration not only ensures a stable and cost-effective supply of hydrogen but also fosters economic ties and promotes shared goals of decarbonization. By addressing concerns related to carbon capture and storage, the U.S. can build trust and form a resilient partnership that benefits both regions.

Europe, with its ambitious renewable energy targets, presents an opportune alliance for the U.S. in the realm of green hydrogen. Collaborating on the production of green hydrogen through electrolysis powered by renewable energy sources can be a cornerstone of this partnership. The U.S. can engage in diplomatic discussions to facilitate technology transfer, joint research, and investment in renewable energy projects. Additionally, exploring collaboration with North African countries blessed with abundant solar resources can enhance the green hydrogen supply chain. Diplomacy should focus on developing intercontinental agreements for green hydrogen production, transportation, and trade. By aligning strategies and sharing expertise, the U.S., Europe, and North Africa can collectively contribute to the growth of a sustainable global hydrogen market, fostering economic development and environmental stewardship. In addition, Australia, rich in renewable resources, emerges as a crucial partner for the U.S. in establishing a green hydrogen production hub. Diplomacy should prioritize the development of agreements to facilitate the export of Australian green hydrogen to the U.S. This collaboration can involve joint investments in large-scale renewable energy projects coupled with advanced electrolysis technologies. By leveraging Australia's vast renewable potential, the U.S. can ensure a stable supply of green hydrogen, fostering energy security and reducing dependence on fossil fuels. Additionally, diplomatic efforts can focus on aligning regulatory frameworks, establishing common standards, and promoting research collaboration to drive down the costs of green hydrogen production. A strategic alliance with Australia positions the U.S. as a key player in the rapidly expanding global green hydrogen market.

Japan and Korea, renowned for their technological prowess, are pivotal allies for the U.S. in advancing hydrogen technologies. Diplomacy should concentrate on fostering collaborations in research and development, knowledge-sharing, and the deployment of innovative hydrogen technologies. Joint initiatives can include the development of hydrogen fuel cells for various applications, advancements in storage and transportation technologies, and the establishment of international standards. The U.S. can benefit from technological exchanges, incorporating Japanese and Korean expertise to enhance the efficiency and reliability of hydrogen technologies. Additionally, diplomatic efforts should seek to harmonize regulations, facilitating the seamless integration of hydrogen technologies across borders. By positioning itself at the forefront of hydrogen innovation, the U.S. can solidify its role as a global leader in the hydrogen technology sector.

Abbreviations

ARRA: American Recovery and Reinvestment Act
ATR: Autothermal Reforming
BIL: Bipartisan Infrastructure Law
BOP: Balance of Plant
CCS: Carbon capture and storage
CCUS: Carbon capture, utilization, and storage
DOD: Department of Defense
DOE: Department of Energy
EPA: Environmental Protection Agency
FCEV: Fuel Cell Electric Vehicles
FRA: Federal Railroad Administration
GHG: Greenhouse gas
GREET: Greenhouse Gases, Regulated Emissions, and Energy Use in Transportation
H2Hubs: Regional Clean Hydrogen Hubs
H2NEW: Hydrogen from Next-generation Electrolyzers of Water
H2PA: Hydrogen Production Analysis
IEA: International Energy Agency
IIJA: Infrastructure Investment and Jobs Act
IMO: International Maritime Organization
IPHE: International Partnership for Hydrogen and Fuel Cells in the Economy
IRA: Inflation Reduction Act
LDES: Long-duration energy storage
LOHC: Liquid organic hydrogen carriers
LPO: Loan Programs Office
M2FCT: Million Mile Fuel Cell Truck Consortium
MARAD: Maritime Administration
MDHD: Medium and heavy-duty
MMT: Million metric tons
NDC: Nationally Determined Contribution
NREL: National Renewable Energy Laboratory
PEM: Proton exchange membrane
PGM: Precious-group-metal
PTC: Production Tax Credit
RDD&D: Research, development, demonstration, and deployment
RFI: Request for Information
RNG: Renewable natural gas
SAF: Sustainable aviation fuels
SMR: Steam methane reforming
SOSE: Solid oxide stem electrolyzer
UAV: Unmanned aerial vehicle

References

[1] Hosseini SE. The US hydrogen fuel industry today and future. Future Energy 2022;1:1. https://doi.org/10.55670/fpll.fuen.1.1.7.

[2] US Hydrogen Road Map — Fuel Cell & Hydrogen Energy Association n.d. https://www.fchea.org/us-hydrogen-study (accessed November 24, 2023).

[3] The Long-Term Strategy of the United States, Pathways to Net-Zero Greenhouse Gas Emissions by 2050 2021:1–61. www.whitehouse.gov/wp-content/uploads/2021/10/US-Long-Term-Strategy.pdf (accessed November 24, 2023).

[4] Department of Energy US. DOE National Clean Hydrogen Strategy and Roadmap 2022:1–120. www.hydrogen.energy.gov/docs/hydrogenprogramlibraries/pdfs/clean-hydrogen-strategy-roadmap.pdf?Status=Master (accessed November 24, 2023).

[5] Department of Energy n.d. https://energy.gov/ (accessed December 15, 2017).

[6] Connelly E, Elgowainy A, Ruth Peer Reviewed By M, Vickers J, Randolph K, Levi P, et al. DOE Hydrogen and Fuel Cells Program Record 19002: Current Hydrogen Market Size: Domestic and Global 1900. www.hydrogen.energy.gov/docs/hydrogenprogramlibraries/pdfs/19002-hydrogen-market-domestic-global.pdf?Status=Master (accessed November 25, 2023).

[7] International Energy Agency I. Global Hydrogen Review 2021 n.d. https://iea.blob.core.windows.net/assets/5bd46d7b-906a-4429-abda-e9c507a62341/GlobalHydrogenReview2021.pdf (accessed November 25, 2023).

[8] Hydrogen Industry Market Research | S&P Global n.d. https://www.spglobal.com/commodityinsights/en/ci/products/hydrogen-chemical-economics-handbook.html (accessed November 25, 2023).

[9] Comparison of commercial, state-of-art, fossil-based hydrogen production technologies. National Energy Technology Laboratory 2022. www.netl.doe.gov/projects/files/ComparisonofCommercialStateofArtFossilBasedHydrogenProductionTechnologies_041222.pdf (accessed November 25, 2023).

[10] Michael J. G. Air Liquide 2-10-22 SENR Cmte Hrg n.d. www.energy.senate.gov/services/files/C00CE119-046B-4E3C-8C7C-B534B4A1674B (accessed November 25, 2023).

[11] Satyapal S, Department of Energy U. 2022 AMR Plenary Session Introduction-Energy, Market, and Policy Context 2022. www.hydrogen.energy.gov/docs/hydrogenprogramlibraries/pdfs/review22/plenary4_satyapal_2022_o.pdf (accessed November 25, 2023).

[12] DOE Announces First Loan Guarantee for a Clean Energy Project in Nearly a Decade | Department of Energy n.d. https://www.energy.gov/articles/doe-announces-first-loan-guarantee-clean-energy-project-nearly-decade (accessed November 25, 2023).

[13] Landmark U.S. $4.5 Billion Louisiana Clean Energy Complex n.d. https://www.airproducts.com/energy-transition/louisiana-clean-energy-complex (accessed November 25, 2023).

[14] Devlin P, Moreland G, Joseck F, Satyapal S. DOE Hydrogen and Fuel Cells Program Record Title: Industry Deployed Fuel Cell Powered Lift Trucks Fuel Cell Powered Lift Truck Deployment Data/Assumptions/Calculations 2018. https://www.energy.gov/sites/prod/files/2014/03/f10/fuel_cell_mhe_cost.pdf.

[15] Satyapal S, Howell D, Marcinkoski J, Vijayagopal R, Adams J, James B, et al. Title: Hydrogen Class 8 Long Haul Truck Targets Originator: Jason Marcinkoski Reviewed by: Fuel Cell Technologies and Vehicle Technologies Offices; Members of the 21 st Century Truck Partnership; DOE Advanced Truck Technologies Subsection of the Electrified Powertrain Roadmap 1 DOE Technical Targets for Advanced Truck Technologies 1900. http://cta.ornl.gov/vtmarketreport/index.shtml,.

[16] TEMPO: Transportation Energy & Mobility Pathway Options Model | Transportation and Mobility Research | NREL n.d. https://www.nrel.gov/transportation/tempo-model.html (accessed November 25, 2023).

[17] FACT SHEET: Biden Administration Advances the Future of Sustainable Fuels in American Aviation | The White House n.d. https://www.whitehouse.gov/briefing-room/statements-

releases/2021/09/09/fact-sheet-biden-administration-advances-the-future-of-sustainable-fuels-in-american-aviation/ (accessed November 25, 2023).

[18] Argonne GREET Publication : Assessment of Potential Future Demands for Hydrogen in the United States n.d. https://greet.anl.gov/publication-us_future_h2 (accessed November 25, 2023).

[19] Dollinger C, States Department of Energy U. Industrial Decarbonization Roadmap 2050. www.energy.gov/sites/default/files/2022-09/Industrial%20Decarbonization%20Roadmap.pdf (accessed November 25, 2023).

[20] Solar Futures Study | Department of Energy n.d. https://www.energy.gov/eere/solar/solar-futures-study (accessed November 26, 2023).

[21] McMillan CA, Ruth M. Using facility-level emissions data to estimate the technical potential of alternative thermal sources to meet industrial heat demand. Appl Energy 2019;239:1077–90. https://doi.org/10.1016/J.APENERGY.2019.01.077.

[22] Net-Zero America Project n.d. https://netzeroamerica.princeton.edu/?explorer=year&state=national&table=2020&limit=200 (accessed November 26, 2023).

[23] Satyapal S. Testimony of Dr. Sunita Satyapal Director for a Hearing on Hydrogen, U.S. Senate Energy and Natural Resources Committee n.d. www.energy.senate.gov/services/files/FE1C53B0-3925-46E3-B1D3-B8E2C0DD92B6 (accessed November 26, 2023).

[24] Satyapal S. High-level Recap and Menti Questions Results 2021. www.energy.gov/sites/default/files/2021-09/h2-shot-summit-closing-plenary-recap.pdf (accessed November 26, 2023).

[25] Satyapal S, Litynski J, Horton L. Hydrogen Shot Overview. 2021.

[26] Hydrogen Shot | Department of Energy n.d. https://www.energy.gov/eere/fuelcells/hydrogen-shot (accessed November 26, 2023).

[27] Cost to refill | Hydrogen Fuel Cell Partnership n.d. https://h2fcp.org/content/cost-refill (accessed November 26, 2023).

[28] Satyapal S, Department of Energy Hydrogen U, Cell Technologies Office F. DOE Hydrogen Program Overview: 2021 AMR Plenary Session. 2021.

[29] Zang G, Sun P, Elgowainy A, Wang M. Technoeconomic and Life Cycle Analysis of Synthetic Methanol Production from Hydrogen and Industrial Byproduct CO2. Environ Sci Technol 2021;55:5248–57. https://doi.org/10.1021/ACS.EST.0C08237/SUPPL_FILE/ES0C08237_SI_001.PDF.

[30] Liu X, Elgowainy A, Wang M. Life cycle energy use and greenhouse gas emissions of ammonia production from renewable resources and industrial by-products. Green Chemistry 2020;22:5751–61. https://doi.org/10.1039/D0GC02301A.

[31] Iron and Steel Technology Roadmap – Analysis - IEA n.d. https://www.iea.org/reports/iron-and-steel-technology-roadmap (accessed November 27, 2023).

[32] Elgowainy A, Sun P, Zang G, Bobba P, Wang M. Technoeconomic and Life Cycle Analysis of Synthetic Fuels and Steelmaking 2021. www.hydrogen.energy.gov/docs/hydrogenprogramlibraries/pdfs/review21/sa174_elgowainy_2021_o-pdf.pdf?Status=Master (accessed November 27, 2023).

[33] Epa U, Change Division C. Inventory of U.S. Greenhouse Gas Emissions and Sinks: 1990 2020 – Main Text 1990. https://www.epa.gov/ghgemissions/draft-inventory-us-greenhouse-gas-emissions-.

[34] Hosseini SE. Hydrogen and Fuel Cells in Transport Road, Rail, Air, and Sea. Comprehensive Renewable Energy, Second Edition: Volume 1-9 2022;1–4:317–42. https://doi.org/10.1016/B978-0-12-819727-1.00005-4.

[35] International Energy Agency I. Global Hydrogen Review 2021 n.d. https://iea.blob.core.windows.net/assets/5bd46d7b-906a-4429-abda-e9c507a62341/GlobalHydrogenReview2021.pdf (accessed November 27, 2023).

[36] Global Hydrogen Review 2023 – Analysis - IEA n.d. https://www.iea.org/reports/global-hydrogen-review-2023 (accessed November 27, 2023).

[37] Steel Production - American Iron and Steel Institute n.d. https://www.steel.org/steel-technology/steel-production/ (accessed December 1, 2023).

[38] Elgowainy A, Sun P, Zang G, Bobba P, Wang M. Technoeconomic and Life Cycle Analysis of Synthetic Fuels and Steelmaking 2021. www.hydrogen.energy.gov/docs/hydrogenprogramlibraries/pdfs/review21/sa174_elgowainy_2021_o-pdf.pdf?Status=Master (accessed December 1, 2023).

[39] H2@Scale | Department of Energy n.d. https://www.energy.gov/eere/fuelcells/h2scale (accessed December 1, 2023).

[40] FACT SHEET: The United States and European Union To Negotiate World's First Carbon-Based Sectoral Arrangement on Steel and Aluminum Trade | The White House n.d. https://www.whitehouse.gov/briefing-room/statements-releases/2021/10/31/fact-sheet-the-united-states-and-european-union-to-negotiate-worlds-first-carbon-based-sectoral-arrangement-on-steel-and-aluminum-trade/ (accessed December 1, 2023).

[41] Nitrogen Statistics and Information | U.S. Geological Survey n.d. https://www.usgs.gov/centers/national-minerals-information-center/nitrogen-statistics-and-information (accessed December 1, 2023).

[42] REFUEL | arpa-e.energy.gov n.d. https://arpa-e.energy.gov/technologies/programs/refuel (accessed December 1, 2023).

[43] HyBlend: Opportunities for Hydrogen Blending in Natural Gas Pipelines | Department of Energy n.d. https://www.energy.gov/eere/fuelcells/hyblend-opportunities-hydrogen-blending-natural-gas-pipelines (accessed December 1, 2023).

[44] NREL Marks Partner Forum With Dedication of Bioreactor | News | NREL n.d. https://www.nrel.gov/news/program/2019/nrel-marks-partner-forum-with-dedication-of-bioreactor.html (accessed December 1, 2023).

[45] Epa U, of Transportation O, Quality A, Division S. Fast Facts: U.S. Transportation Sector Greenhouse Gas Emissions, 1990-2019 (EPA-420-F-21-076, December 2021) 1990. https://nepis.epa.gov/Exe/ZyPDF.cgi?Dockey=P1013NR3.pdf (accessed December 1, 2023).

[46] 21st Century Truck Partnership | Department of Energy n.d. https://www.energy.gov/eere/vehicles/21st-century-truck-partnership (accessed December 1, 2023).

[47] Hydrogen in Transportation | US EPA n.d. https://www.epa.gov/greenvehicles/hydrogen-transportation (accessed December 1, 2023).

[48] Home | M2FCT n.d. https://millionmilefuelcelltruck.org/ (accessed December 1, 2023).

[49] DOE Announces Nearly $200 Million to Reduce Emissions From Cars and Trucks | Department of Energy n.d. https://www.energy.gov/articles/doe-announces-nearly-200-million-reduce-emissions-cars-and-trucks (accessed December 1, 2023).

[50] Adams J, Department of Energy U. Technology Acceleration Overview 2021. www.hydrogen.energy.gov/docs/hydrogenprogramlibraries/pdfs/plenary10_adams_2021_o.pdf?Status=Master (accessed December 1, 2023).

[51] Hanlin J. Fuel Cell Hybrid Electric Delivery Van Project 2021. www.hydrogen.energy.gov/docs/hydrogenprogramlibraries/pdfs/review21/ta016_hanlin_2021_o-pdf.pdf (accessed December 1, 2023).

[52] Fuel Cell Electric Bus Evaluations | Hydrogen and Fuel Cells | NREL n.d. https://www.nrel.gov/hydrogen/fuel-cell-bus-evaluation.html (accessed December 1, 2023).

[53] FACT SHEET: Biden Administration Advances the Future of Sustainable Fuels in American Aviation | The White House n.d. https://www.whitehouse.gov/briefing-room/statements-releases/2021/09/09/fact-sheet-biden-administration-advances-the-future-of-sustainable-fuels-in-american-aviation/ (accessed December 1, 2023).

[54] Aviation Climate Action Plan | Federal Aviation Administration n.d. https://www.faa.gov/sustainability/aviation-climate-action-plan (accessed December 1, 2023).

[55] Sustainable Aviation Fuel Grand Challenge | Department of Energy n.d. https://www.energy.gov/eere/bioenergy/sustainable-aviation-fuel-grand-challenge (accessed December 1, 2023).

[56] D7566 Standard Specification for Aviation Turbine Fuel Containing Synthesized Hydrocarbons n.d. https://www.astm.org/d7566-21.html (accessed December 1, 2023).

[57] Airbus Hydrogen Plane With Turboprop Design Gaining Favor For Zero Emission Jet - Bloomberg n.d. https://www.bloomberg.com/news/articles/2021-02-11/airbus-turboprop-

design-gaining-favor-as-first-hydrogen-plane?leadSource=uverify%20wall (accessed December 1, 2023).

[58] Merk O. Shipping Emissions in Ports n.d. www.itf-oecd.org/sites/default/files/docs/dp201420.pdf (accessed December 1, 2023).

[59] Maritime Environmental and Technical Assistance (META) Program | MARAD n.d. https://www.maritime.dot.gov/innovation/meta/maritime-environmental-and-technical-assistance-meta-program#Fuel%20Cells (accessed December 1, 2023).

[60] Freight Rail Overview | FRA n.d. https://railroads.dot.gov/rail-network-development/freight-rail-overview (accessed January 3, 2024).

[61] Fast Facts on Transportation Greenhouse Gas Emissions | US EPA n.d. https://www.epa.gov/greenvehicles/fast-facts-transportation-greenhouse-gas-emissions (accessed January 3, 2024).

[62] Alstom's Coradia iLint hydrogen train runs for the first time in France | Alstom n.d. https://www.alstom.com/press-releases-news/2021/9/alstoms-coradia-ilint-hydrogen-train-runs-first-time-france (accessed January 3, 2024).

[63] Green-Tech for the US: Stadler Signs First Ever Contract for Hydrogen-Powered Train - SBCTA n.d. https://www.gosbcta.com/green-tech-for-the-us-stadler-signs-first-ever-contract-for-hydrogen-powered-train/ (accessed January 3, 2024).

[64] Ahluwalia RK, Peng J-K, Cetinbas F, Papadias DD, Wang X, Kopasz J, et al. Rail, Aviation, and Maritime Metrics 2021. www.hydrogen.energy.gov/docs/hydrogenprogramlibraries/pdfs/review21/ta034_ahluwalia_2021_o-pdf.pdf (accessed January 3, 2024).

[65] Tri-Generation Success Story: World's First Tri-Gen Energy Station - Fountain Valley n.d. www.energy.gov/sites/default/files/2016/12/f34/fcto_fountain_valley_success_story.pdf (accessed January 3, 2024).

[66] Early Markets: Fuel Cells for Backup Power Overview n.d. www.energy.gov/sites/prod/files/2014/10/f19/ftco_early_mkts_fc_backup_power_fact_sheet.pdf (accessed January 3, 2024).

[67] Frazier AW, Cole W, Denholm P, Machen S, Gates N, Blair N. Storage Futures Study: Economic Potential of Diurnal Storage in the U.S. Power Sector 2020. https://www.nrel.gov/docs/fy21osti/77449.pdf.

[68] Denholm P, Cole W, Frazier AW, Podkaminer K, Blair N. The Four Phases of Storage Deployment: A Framework for the Expanding Role of Storage in the U.S. Power System n.d. www.nrel.gov/docs/fy21osti/77480.pdf (accessed January 3, 2024).

[69] Advanced Turbine Systems | Department of Energy n.d. https://www.energy.gov/fecm/advanced-turbine-systems (accessed January 3, 2024).

[70] Satyapal S, Schrecengost R, Marcinkoski J, Vetrano J, Shrader T. DOE Hydrogen Program Panel Discussion 2022. www.hydrogen.energy.gov/docs/hydrogenprogramlibraries/pdfs/review22/plenary5_program_panel_2022_o.pdf?Status=Master (accessed January 3, 2024).

[71] Argonne GREET R&D Model n.d. https://greet.anl.gov/ (accessed January 4, 2024).

[72] Home International Partnership for Hydrogen&Fuel Cells in the Economy n.d. https://www.iphe.net/ (accessed January 4, 2024).

[73] IPHE WP Methodology Doc Oct 2021 | iphe n.d. https://www.iphe.net/iphe-working-paper-methodology-doc-oct-2021 (accessed January 4, 2024).

[74] Mayyas A, Ruth M, Pivovar B, Bender G, Wipke K. Manufacturing Cost Analysis for Proton Exchange Membrane Water Electrolyzers 2019. www.nrel.gov/docs/fy19osti/72740.pdf (accessed January 4, 2024).

[75] Peterson D, Vickers J, Desantis D, Ayers K, Hamdan M, Harrison A A Approved K, et al. DOE Hydrogen and Fuel Cells Program Record Title: Hydrogen Production Cost From PEM Electrolysis-2019] Originators 2020. www.hydrogen.energy.gov/docs/hydrogenprogramlibraries/pdfs/19009_h2_production_cost_pem_electrolysis_2019.pdf?Status=Master (accessed January 4, 2024).

[76] Vickers J, Peterson D, Randolph K, Irwin L, Desantis D, Hamdan M, et al. DOE Hydrogen and Fuel Cells Program Record Title: Cost of Electrolytic Hydrogen Production with Existing

Technology 2020. www.hydrogen.energy.gov/docs/hydrogenprogramlibraries/pdfs/20004-cost-electrolytic-hydrogen-production.pdf?Status=Master (accessed January 4, 2024).
[77] Argonne GREET Publication : Assessment of Potential Future Demands for Hydrogen in the United States n.d. https://greet.anl.gov/publication-us_future_h2 (accessed January 4, 2024).
[78] Energy Analysis | netl.doe.gov n.d. https://www.netl.doe.gov/energy-analysis/details?id=ed4825aa-8f04-4df7-abef-60e564f636c9 (accessed January 4, 2024).
[79] Open For Business: LPO Issues New Conditional Commitment for Loan Guarantee | Department of Energy n.d. https://www.energy.gov/lpo/articles/open-business-lpo-issues-new-conditional-commitment-loan-guarantee (accessed January 4, 2024).
[80] U.S. Department of Energy Announces $28 Million to Develop Clean Hydrogen | Department of Energy n.d. https://www.energy.gov/fecm/articles/us-department-energy-announces-28-million-develop-clean-hydrogen (accessed January 4, 2024).
[81] Landmark U.S. $4.5 Billion Louisiana Clean Energy Complex n.d. https://www.airproducts.com/energy-transition/louisiana-clean-energy-complex (accessed January 4, 2024).
[82] The Section 45Q Tax Credit for Carbon Sequestration 2023. https://sgp.fas.org/crs/misc/IF11455.pdf (accessed January 4, 2024).
[83] Securing America's Clean Energy Supply Chain | Department of Energy n.d. https://www.energy.gov/policy/securing-americas-clean-energy-supply-chain (accessed January 4, 2024).
[84] Connelly E, Penev M, Milbrandt A, Roberts B, Gilroy N, Melaina M. Resource Assessment for Hydrogen Production 2040. www.nrel.gov/docs/fy20osti/77198.pdf (accessed January 4, 2024).
[85] Fesmire JE, Swanger A. Overview of the New LH 2 Sphere at NASA Kennedy Space Center 2021. www.energy.gov/sites/default/files/2021-10/new-lh2-sphere.pdf (accessed January 5, 2024).
[86] Psarras PC, Comello S, Bains P, Charoensawadpong P, Reichelstein S, Wilcox J. Carbon Capture and Utilization in the Industrial Sector. Environ Sci Technol 2017;51:11440–9. https://doi.org/10.1021/ACS.EST.7B01723/SUPPL_FILE/ES7B01723_SI_006.XLSX.
[87] U.S. National Clean Hydrogen Strategy and Roadmap 2023. www.hydrogen.energy.gov/docs/hydrogenprogramlibraries/pdfs/us-national-clean-hydrogen-strategy-roadmap.pdf (accessed January 5, 2024).
[88] Ledna C, Muratori M, Yip A, Jadun P, Hoehne C. Decarbonizing Medium- & Heavy-Duty On-Road Vehicles: Zero-Emission Vehicles Cost Analysis 2022. www.nrel.gov/docs/fy22osti/82081.pdf (accessed January 5, 2024).
[89] Department of Energy n.d. https://energy.gov/ (accessed April 14, 2024).
[90] Fifth Hydrogen Energy Ministerial Meeting Held n.d. https://www.meti.go.jp/english/press/2022/1007_001.html (accessed January 5, 2024).

CHAPTER 8

The European Union's hydrogen strategy: building a sustainable future

8.1 Introduction

Since the ratification of the groundbreaking Paris Agreement in 2015, the worldwide dedication to decarbonization has notably intensified. Prominent economies such as The United States, China, Canada, the European Union (EU), Japan, and the United Kingdom have enacted legislation mandating the attainment of net-zero emissions goals within the next four decades. These policy commitments have directed attention towards the profound decarbonization of all energy sectors, encompassing those with challenging-to-mitigate emissions, reigniting enthusiasm for clean hydrogen. The EU is grappling with the challenge of achieving its carbon emissions reduction targets, with energy production and consumption contributing to 75% of the region's greenhouse gas emissions. Furthermore, the EU heavily relies on imports for 58% of its energy supply, predominantly in the form of oil and gas. To address these issues, the European Commission has identified hydrogen as a pivotal element in its clean energy policy. In July 2020, the Commission proposed a strategy aimed at expediting the development of green hydrogen sourced from renewables, with the goal of integrating it into the region's energy supply by 2050 [1]. Hydrogen constitutes approximately 2% of the EU's energy mix, and 95% of it is categorized as brown hydrogen. This type of hydrogen is derived from natural gas or oil, contributing to an annual emission of 70-100 million tons of CO_2. In contrast, green hydrogen, produced through electrolysis using renewable electricity, holds the potential to represent up to 20% of the EU's energy supply by 2050. Research indicates that this could fulfill 20-50% of the transport sector's demand and 5-20% of industrial demand.

The EU's strategy on hydrogen, coupled with the REPowerEU plan (the foundation of the REPowerEU plan rests on three fundamental pillars: conserving energy, generating clean energy, and broadening the European Union's energy sources), presents a comprehensive framework aimed at promoting the integration of renewable and low-carbon hydrogen. This initiative seeks to facilitate the decarbonization of the EU in a cost-effective manner while concurrently

DOI: https://doi.org/10.55670/fpll.book/1

diminishing reliance on imported fossil fuels. The EU's primary objective revolves around advancing renewable hydrogen, with a target to domestically produce 10 million tons and import an additional 10 million tons by the year 2030 [2]. In establishing this objective, the EU acknowledged the importance of importing green hydrogen sourced from global renewable energy reservoirs to supplement its domestic production and release renewable power capacities for direct electrification. Subsequently, through its Renewable Energy Directive mandating that 42.5% of industrial hydrogen usage should be green by 2030, the EU indicated a substantial increase in the demand for environmentally friendly products in the coming five years. Regarded as the most ambitious green hydrogen strategy globally, the EU's hydrogen strategy aims to achieve 40 GW of green hydrogen production by 2030 (Figure 8.1).

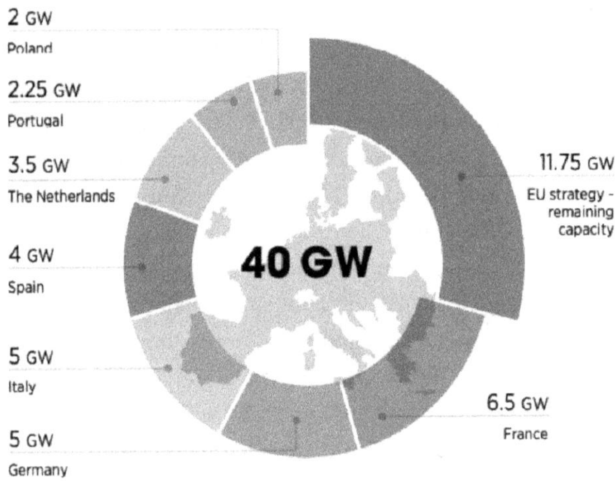

Figure 8.1 Electrolyzer capacity goals in EU hydrogen strategies, 2030 [1]

Presently, the EU finds itself at a pivotal juncture. Undoubtedly, it occupies a prominent position in the international hydrogen competition, with approximately 80% of recently declared large-scale hydrogen projects situated within Europe. Additionally, EU member states spearheaded the surge in the worldwide deployment of electrolyzers in 2020 [3]. However, concurrently, there is an imperative for the EU to promptly formulate and execute a unified, long-term strategy to establish competitive and secure hydrogen markets, ensuring the continuation of its leadership role. While the EU's strategy forms a robust groundwork for developing a hydrogen economy, it explicitly outlines targets only until 2030, allowing for varied interpretations regarding the potential structure of hydrogen markets post-2030. The EU has yet to address a pivotal concern: identifying competitive and dependable sources for large-scale renewable hydrogen. Although certain resource-abundant EU member states, such as Spain, may become regional exporters, none are poised to emerge as global export leaders. However, potential regional or long-distance partners, such as Morocco and the United States, could play a substantial role in EU hydrogen markets by providing competitively priced supplies of renewable hydrogen [4]. On the other hand, the Russian conflict with Ukraine has accelerated Europe's plans to bolster energy security [5]. Consequently, the EU has intensified its endeavors to decarbonize the energy sector through the revised "Fit for 55" initiative, now named "RePowerEU," envisioning 1200 GW of Renewable Energy Sources (RES) capacity by 2030. Hydrogen assumes a crucial role, with the program aiming to deploy 6 GW of electrolysis capacity for renewable hydrogen generation by 2024 and 40 GW by 2030 [6]. While the EU's hydrogen strategy initially emphasizes natural gas-based hydrogen using steam methane reforming (SMR) and carbon capture and storage (CCS) technology in the short and medium term, it prioritizes non-fossil, renewable hydrogen in the long term [7]. However, this approach prompts pertinent questions about the specific implementation and development of a hydrogen industry in Europe and the feasibility of repurposing existing transport routes from natural gas for hydrogen trade.

8.2 European Union policies to develop the hydrogen economy

Renewable hydrogen is identified as a crucial element for replacing natural gas, coal, and oil in challenging-to-decarbonize sectors like industries and transportation. In this context, the EU Commission has outlined several actions [8]:

- Urges the European Parliament and the Council to align the sub-targets for renewable fuels of non-biological origin in the Renewable Energy Directive for industry and transport with the REPowerEU goals (75% for industry and 5% for transport) and swiftly conclude the revision of the hydrogen and gas market package.
- Plans to augment investments in Horizon Europe for the Hydrogen Joint Undertaking by EUR 200 million to double the number of Hydrogen Valleys.
- Releases two Delegated Acts on the definition and production of renewable hydrogen for public feedback.
- Encourages the industry to expedite efforts in establishing hydrogen standards, particularly for hydrogen production, infrastructure, and end-use appliances.
- Commits to regularly reporting, in collaboration with Member States, starting in 2025, on hydrogen uptake and the utilization of renewable hydrogen in challenging-to-decarbonize applications in industry and transport.

Swift action is imperative to establish a hydrogen infrastructure capable of producing, importing, and transporting 20 million tons of hydrogen by 2030. While cross-border hydrogen infrastructure is still in its early stages, the foundation for planning and development has been laid through the incorporation of hydrogen infrastructure into the revised trans-European networks for energy. The total estimated investment requirements for key hydrogen infrastructure categories range from EUR 28 to 38 billion for EU-internal pipelines and 6 to 11 billion for storage. To facilitate the importation of up to 10 million tons of renewable hydrogen, the Commission will actively support the creation of three major hydrogen import corridors spanning the Mediterranean, the North Sea area, and, when conditions permit, Ukraine. Green Hydrogen Partnerships will play a role in streamlining the importation of green hydrogen while concurrently aiding decarbonization efforts in the partner countries. Various forms of fossil-free hydrogen, particularly nuclear-based, will also contribute to substituting natural gas.

In the transportation sector, a combination of electrification and the utilization of fossil-free hydrogen will be employed to substitute traditional fossil fuels. To bolster the adoption of hydrogen and electrification in industrial sectors, the Commission is taking several measures:

- Implementing carbon contracts for difference and establishing dedicated REPowerEU windows within the Innovation Fund. This initiative aims to facilitate a complete shift in existing hydrogen production within industrial processes from natural gas to renewables and the transition to hydrogen-based production methods in emerging industrial sectors like steel production.
- Providing guidance to Member States on renewable energy and power purchase agreements (PPAs).
- Collaborating with the European Investment Bank (EIB) to create a technical advisory facility under the InvestEU Advisory Hub, which will offer support for PPA-financed renewable energy projects. A specific REPowerEU window will be established to back (1) innovative applications of electrification and hydrogen in industry, (2) cutting-edge clean tech manufacturing (such as electrolyzers, fuel cells, innovative renewable equipment, energy storage, or heat pumps for industrial purposes), and (3) mid-sized pilot projects for validating, testing, and optimizing highly innovative solutions.

Concurrently with legislative advancements, the European industrial sector has established a robust portfolio of hydrogen initiatives. The European Clean Hydrogen Alliance has identified more than 840 hydrogen pipeline projects spanning the entire value chain covering hydrogen production and its application in sectors like chemicals, refining, steel, as well as transportation, specifically in heavy-duty road and maritime transport throughout Europe (Figure 8.2) [9]. Sixteen EU Member States have embraced national hydrogen strategies, collectively aiming for 40 GW of electrolyzer capacity by 2030, equating to 5.6 million tons of renewable hydrogen. Addressing infrastructure requirements, the initial Union list of projects of common interest and mutual interest, encompassing hydrogen and electrolyzers, is in progress through the implementation of the trans-European energy networks policy [10]. Global public funding for hydrogen projects has reached approximately USD 308 billion, with the United States taking the lead due to tax credits introduced through the Inflation Reduction Act.

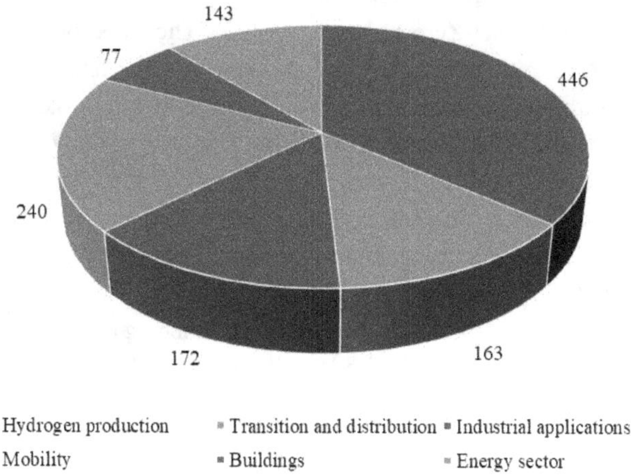

Figure 8.2 Hydrogen pipeline projects in Europe from various value chains

Although the overall European funding is comparable to that of the United States, the distinction lies in the nature of the funding mechanisms. In Europe, grants constitute the predominant form of financial support; however, project delays often occur due to the time-consuming approval process for these grants. Europe is home to more than 30% of globally proposed hydrogen investments. Europe holds a prominent position in the production of electrolyzers, and a collaborative statement aimed at expanding electrolyzer manufacturing capacity underscores the preparedness of EU industries to contribute [11]. Producers of electrolyzers aim to achieve an annual capacity of 25 GW by the year 2025. A significant milestone in this endeavor is the recent inauguration of the largest electrolyzers gigafactory in Europe, established through a collaborative venture between Siemens and Air Liquide, located in Berlin [12]. However, electrolysis-based hydrogen production is currently limited in scale and lacks competitiveness compared to traditional hydrogen production methods. Leveraging these competitive strengths and the early establishment of regulatory structures, numerous developers of hydrogen projects are contemplating investments in Europe.

8.2.1 European Hydrogen Bank

In her State of the European Union address on September 14, 2022, President von der Leyen announced the establishment of the European Hydrogen Bank [13]. The primary goal of the bank is to bridge the investment gap and establish connections between the future supply of renewable hydrogen and the targeted demand of 20 million tons of renewable hydrogen. The European Hydrogen Bank (Figure 8.3) is designed to facilitate both the production of renewable hydrogen within the EU and its imports, aligning with the RepowerEU objectives and the transition to climate neutrality [14]. The Hydrogen Bank aligns with the goals outlined in the Green Deal Industrial Plan and the Net-Zero Industry Act [15]. The expansion of electrolyzer manufacturing for renewable hydrogen production is expected to enhance the competitiveness and resilience of European industries, including steel and fertilizer production, as well as the shipping industry. The scaling up of the European hydrogen market will position European companies to take a leading role in the emerging global hydrogen market, creating new growth opportunities and high-quality jobs. This communication on the Hydrogen Bank accompanies a legislative proposal for a Net-Zero Industrial Act. Implemented by the European Commission, the European Hydrogen Bank will feature two new financing mechanisms to support renewable hydrogen production both within the EU and internationally. Additionally, it will enhance transparency regarding hydrogen demand, supply, flows, and prices while playing a coordination role and facilitating integration with existing financial instruments to support hydrogen projects. For the European Hydrogen Bank to achieve success, it is imperative to promptly finalize the regulatory framework governing the production and utilization of hydrogen. The Commission urges the European Parliament and the Council to expedite the implementation of delegated acts elucidating the qualification criteria for hydrogen and hydrogen-based fuels as renewable.

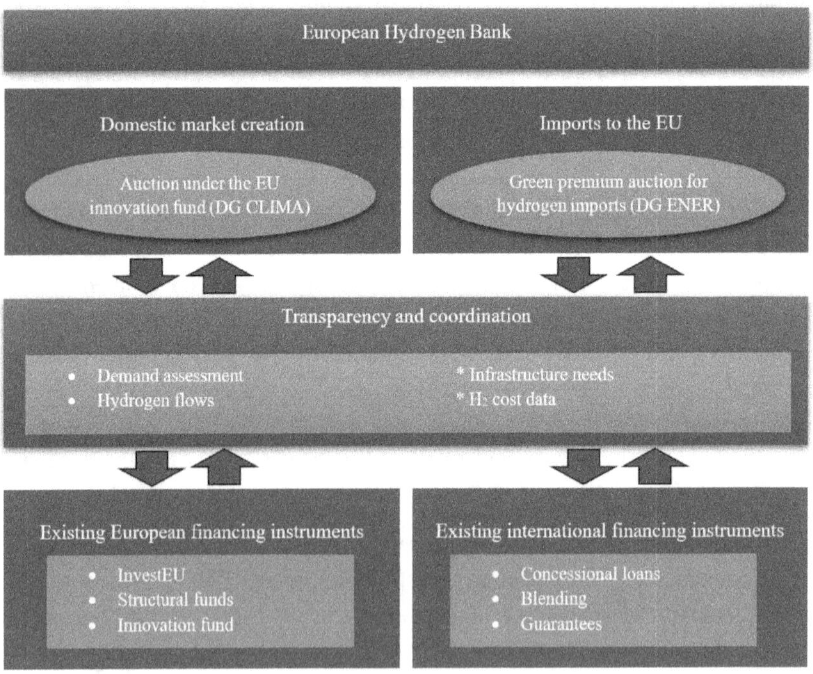

Figure 8.3 European Hydrogen Bank activities

Simultaneously, the Commission calls for the swift conclusion of negotiations and adoption of the proposed Renewable Energy Directive, along with future regulations for efficient hydrogen markets, including the definition of low-carbon hydrogen [14]. While the Commission is actively working towards accelerating the expansion of renewable hydrogen, the RePowerEU initiative recognizes the role of other forms of fossil-free hydrogen, particularly nuclear-based, in replacing natural gas. Adherence to the Water Framework Directive is crucial to address the additional freshwater requirements in locations where additional renewable and low-carbon hydrogen production capacities are being deployed. The European hydrogen market encounters four key investment challenges: increasing the manufacturing capacities for electrolyzers, expanding new hydrogen production capabilities, creating opportunities in new sectors for renewable and low-carbon hydrogen, and establishing dedicated hydrogen infrastructure. This necessitates a growing skilled workforce, emphasizing the need for substantial investments in re-skilling and upskilling [16].

The European Hydrogen Bank (Figure 8.3) is built on four core pillars to be executed by the European Commission. It will comprise two novel financing mechanisms to support renewable hydrogen production in the EU and globally. Additionally, the Bank will enhance demand visibility by establishing connections with off-takers, concurrent Member State initiatives, and existing data centers. Lastly, the Bank will play a pivotal role in coordination, facilitating integration with current financial instruments to bolster hydrogen projects. Its significance lies in unlocking private sector investments, fostering early market development, and contributing to price discovery by fostering competition for financing. This process aims to enhance investor confidence and promote knowledge acquisition in project finance within the private financing sector. Over time, the bank aspires to diminish the cost gap to a level where private off-takers find it feasible and are willing to cover the expense. The European Hydrogen Bank will operate in accordance with international commitments, whether at a bilateral or multilateral level, both domestically and internationally.

8.3 Investment in hydrogen infrastructure

As of now, approximately 8 million tons of hydrogen are consumed in the EU, with the majority being produced from natural gas (without CCS). The current production of electricity-based hydrogen within the EU is less than 0.3 million tons, and the installed electrolyzer output capacity is around 160 MW. Achieving the production target of 10 million tons of renewable hydrogen would demand an electrolyzer output capacity of 80-100 GW, equivalent to approximately 150-210 GW of additional renewable capacity generating low-cost electricity to ensure the competitiveness of renewable hydrogen compared to its fossil counterparts. The anticipated total investment required to produce, transport, and consume 10 million tons of renewable hydrogen is expected to range between EUR 335-471 billion.

Among these, EUR 200-300 billion is earmarked for additional renewable electricity production. The projected investments for key hydrogen infrastructure categories by 2030 include EUR 50-75 billion for electrolyzers, EUR 28-38 billion for EU-internal pipelines, and EUR 6-11 billion for storage. The upscaling of electrolyzer manufacturing capacities is estimated to necessitate investments of up to EUR 1.2 billion. To enable the import of 10 million tons of renewable hydrogen, including derivatives, international value chains will require an additional EUR 500 billion in investments [14]. Notably, a significant portion of the investment in the hydrogen sector is expected to be sourced from private capital. The climate taxonomy delegated act already directs private funding toward activities related to the manufacturing of equipment for hydrogen production and use, as well as the production and storage of hydrogen. Simultaneously, all pertinent EU funds are actively deployed to facilitate an expedited expansion of the hydrogen market in Europe. Horizon 2020 has played a role in supporting the upscaling of electrolyzers by backing the development of the initial 100 MW scale electrolyzers. This complements the research and demonstration projects funded under the Clean Hydrogen Joint Undertaking. To bring hydrogen technology to market, the EU Emissions Trading System (ETS) Innovation Fund has awarded eight large-scale hydrogen production or consumption projects. The entire project pipeline outlined in the 2021 large-scale call for the Innovation Fund would result in the construction of approximately 2.6 GW of electrolyzer capacity. Moreover, the 2022 Large Scale call for proposals, concluding on March 16, allocates a EUR 1 billion call to hydrogen and industry electrification, along with EUR 700 million to clean tech manufacturing, encompassing electrolyzers. State-aid financing has been greenlit for two rounds of Important Projects of Common European Interest (IPCEIs) aimed at supporting the production and utilization of renewable and electricity-based hydrogen, with a total funding of EUR 10.6 billion, expected to unlock an additional EUR 15.8 billion in private investments. Additionally, various Member States are establishing competitive bidding schemes to facilitate the development of renewable hydrogen projects at the least cost. Over EUR 10 billion has been allocated under the Recovery and Resilience Plans, with EUR 4.9 billion designated for IPCEIs. Concerning infrastructure, the Connecting Europe Facility for Energy retains a budget of EUR 3.3 billion to support selected energy infrastructure categories, including hydrogen, up to 2027.

Cohesion policy funding, notably through the European Regional Development Fund (ERDF) and the Just Transition Fund (JTF), along with the InvestEU Fund, will play a substantial role in supporting Member States and regions throughout the entire hydrogen supply chain. This spans from innovation to piloting and initial deployment. InvestEU allows for the financing of such investments, potentially under the EIB's Thematic Green Transition Financial Product or through other Implementing Partners such as the European Bank of Reconstruction and Development or the Nordic Investment Bank. InvestEU offers a versatile framework for effective top-ups or blending with other EU instruments, including the Innovation Fund. Furthermore, the Joint Undertaking on Clean Hydrogen is investing in enhancing efficiency and reducing the cost of producing, storing, and distributing renewable hydrogen. The aim is to increase its competitiveness with hydrogen produced from fossil fuels and expedite its substitution for renewable hydrogen. The EIB, serving as the EU's climate bank, has committed to surpassing 50% of its overall lending by 2025 for climate sustainability and environmental issues, utilizing its own resources. Over the past decade, the EIB has supplied over EUR 1 billion in financing directly associated with hydrogen projects. This commitment was recently complemented by the EIB's REPowerEU package of EUR 30 billion, intending to mobilize up to EUR 115 billion by 2027 to drive investments leading to the decarbonization of the EU industry. Nevertheless, a green premium still exists, signifying higher costs for entities opting for hydrogen over fossil fuels. Targeted allocation of public funds to cover this green premium can catalyze private sector investments by mitigating the risks associated with renewable hydrogen production. Estimates indicate that hydrogen imports by ship, where hydrogen must be imported as ammonia, methanol, liquid organic hydrogen carriers, or e-fuels, would necessitate a market premium in the range of EUR 3-5/kg (inclusive of transport, storage, and distribution). As renewable hydrogen is not yet available globally, the green premium becomes essential to secure production capacity. An estimated budget of EUR 1 billion is projected to facilitate the creation of 0.04-0.06 million tons of renewable hydrogen production capacity per year. Post-2025, the market premium is anticipated to decrease due to declining production costs and heightened demand for green products produced using renewable hydrogen. Additionally, the EU Emissions Trading Scheme and the proposed European Taxation Directive will enhance the attractiveness of renewable and low-carbon hydrogen compared to its fossil fuel counterpart. Anticipating a decline in production costs and a surge in demand for renewable hydrogen, approximations suggest a total green premium of around EUR 90-115 billion for both domestic production and the import of a total of 20 million tons of renewable hydrogen. The sooner the production of renewable hydrogen scales up, the more likely this green premium will diminish. In the short term, additional resources at the EU level will be imperative to fortify efforts in deploying hydrogen and supporting the expansion of manufacturing for other critical net-zero technologies. The Innovation Fund's legal foundation, particularly the new provisions for EU-level competitive bidding auctions, presents a promising and cost-efficient avenue for bolstering the manufacturing and deployment of renewable hydrogen and other pivotal net-zero technologies in Europe. This contributes to reinforcing Europe's sovereignty in key technologies for climate action and energy security. The Net-Zero Industry Act lays the groundwork for organizing strategic projects, offering substantial scope to amplify the financial leverage of the Innovation Fund. This consideration aligns with the higher ETS revenues and the agreement stipulating that Member States invest all their ETS revenues in climate and energy objectives.

The same as the legislative initiatives in the United States supporting the expansion of hydrogen infrastructure, the EU is taking steps to allocate public funding to facilitate the growth of the sector within the bloc. One notable example is the dedication of $3.18 billion from the Innovation Fund to launch the European Hydrogen Bank, aiming to stimulate investment and boost hydrogen demand. Notably, this EU subsidy surpasses the proposed incentives in the United States, providing production credits of up to $4.78 per kilogram of hydrogen over a 10-year period. This financial support complements existing schemes that offer assistance to the sector, even though they are not specifically tailored to hydrogen [17]. Similar to the hub concept in the United States, the EU has established the Clean Hydrogen Partnership, formerly named the Fuel Cells and Hydrogen Joint Undertaking [18]. This initiative aims to develop as many as nine hydrogen valleys, offering subsidies of up to €105.4 million for their establishment. These valleys are designed to be cross-border, fostering collaboration among diverse regional governments to promote the development of an all-encompassing and sustainable hydrogen ecosystem. The EU expects that the private sector will significantly boost the funding for these valleys, reaching a multiple of five, resulting in over $636.7 billion directed towards the expansion of green hydrogen [19]. Besides EU strategies for hydrogen technology deployment, each EU member country has developed a unique hydrogen strategy tailored to its specific energy transition goals and industrial landscape. These hydrogen policies reflect a concerted effort to align national priorities with EU-wide goals for sustainable hydrogen development and carbon neutrality. In this section, we briefly explore the hydrogen policies of four key European countries: Germany, France, Spain, and Italy. Delving into the unique strategies adopted by each nation provides insights into their approaches toward hydrogen development, shedding light on the diverse initiatives undertaken to advance the shared goals of sustainable energy and carbon neutrality within the broader European context.

8.3.1 Germany's hydrogen leap

In Germany, domestic hydrogen consumption currently stands at approximately 55 TWh. The primary demand for hydrogen is associated with material production processes in the industrial sector, with a balanced distribution between the basic chemicals (ammonia, methanol production, etc.) and the petrochemicals sector (conventional fuel production). The majority of hydrogen used in these processes is classified as 'grey' hydrogen. Electrolysis processes contribute to meeting about 7% of the demand (3.85 TWh). Since certain hydrogen used in the petrochemicals industry is available as a by-product of other processes (e.g., catalytic reforming) and doesn't need to be produced independently, the current hydrogen consumption level of around 55 TWh cannot be entirely substituted with 'green' hydrogen. The future trajectory of the hydrogen market in Germany, as well as globally, hinges significantly on the ambition demonstrated by climate action policies and corresponding strategies. Aligned with the Paris Agreement and the Federal Government's commitment to achieving greenhouse gas neutrality by 2050, the hydrogen market is anticipated to witness the following trends [20]: By 2030, with increased market penetration, hydrogen demand will experience an initial upswing, particularly in the industrial sector (chemicals, petrochemicals, steel), and to a lesser extent, in the transport sector. Conservative estimates suggest a 10 TWh rise in industry demand. Additionally, a surge in demand is foreseen from fuel-cell-driven electric vehicles, with a potential future uptake in other sectors like heating. To achieve the 2050 carbon neutrality goal, Germany will need to rely significantly on hydrogen technologies. Various studies have analyzed the potential effects on the entire energy system in the case of a 95% reduction of greenhouse gases against the 1990 baseline. These studies forecast consumption of electricity-based energy sources in 2050 ranging from 110 TWh to roughly 380 TWh. This demand will arise in the industrial, transport, and transformation sectors. The development of overall demand and consumption levels in these sectors will be decisively impacted by the future policy framework, particularly in terms of climate action ambitions and the strategies being pursued [21,22]. Germany's National Hydrogen Strategy, released in June 2020, focuses on achieving climate goals while also recognizing the economic opportunities of a growing hydrogen market. The strategy aims to make Germany a leader and exporter of green hydrogen technologies, with a long-term goal of achieving carbon neutrality by 2050 (Figure 8.4). The supply of hydrogen is crucial for Germany's goal of achieving climate neutrality, and the country is taking steps to ensure the increased use of hydrogen in various sectors, including industry, heavy commercial vehicles, and air and sea transport (Figure 8.5) [23]. The development of a hydrogen infrastructure, including a hydrogen network comprising more than 1,800 kilometers of new and converted pipelines, is also part of the strategy. In the transport sector, the demand for hydrogen is primarily driven by heavy trucks with fuel cells, which are expected to consume about 63% (73 TWh) of the 117 TWh of hydrogen in 2045. This high demand is due to the dominance of these trucks in freight transport on roads. Additionally, around 32 TWh of hydrogen are projected to be used in cars and light-duty vehicles in 2045, with fuel cell cars accounting for 30% of the passenger transport demand, while battery electric vehicles dominate individual mobility. To supply this hydrogen demand, Germany will need to focus on domestic hydrogen production via electrolysis and the import of green hydrogen. By 2030, it is estimated that about 29 TWh of hydrogen will be produced domestically via electrolysis, requiring an electrolyzer capacity of 15 GW. In addition to domestic production, around 45 TWh of green hydrogen are expected to be imported into Germany. In 2045, approximately 47% of the hydrogen used in Germany is imported from other European nations or Northern Africa through pipelines or ships (Figure 8.6).

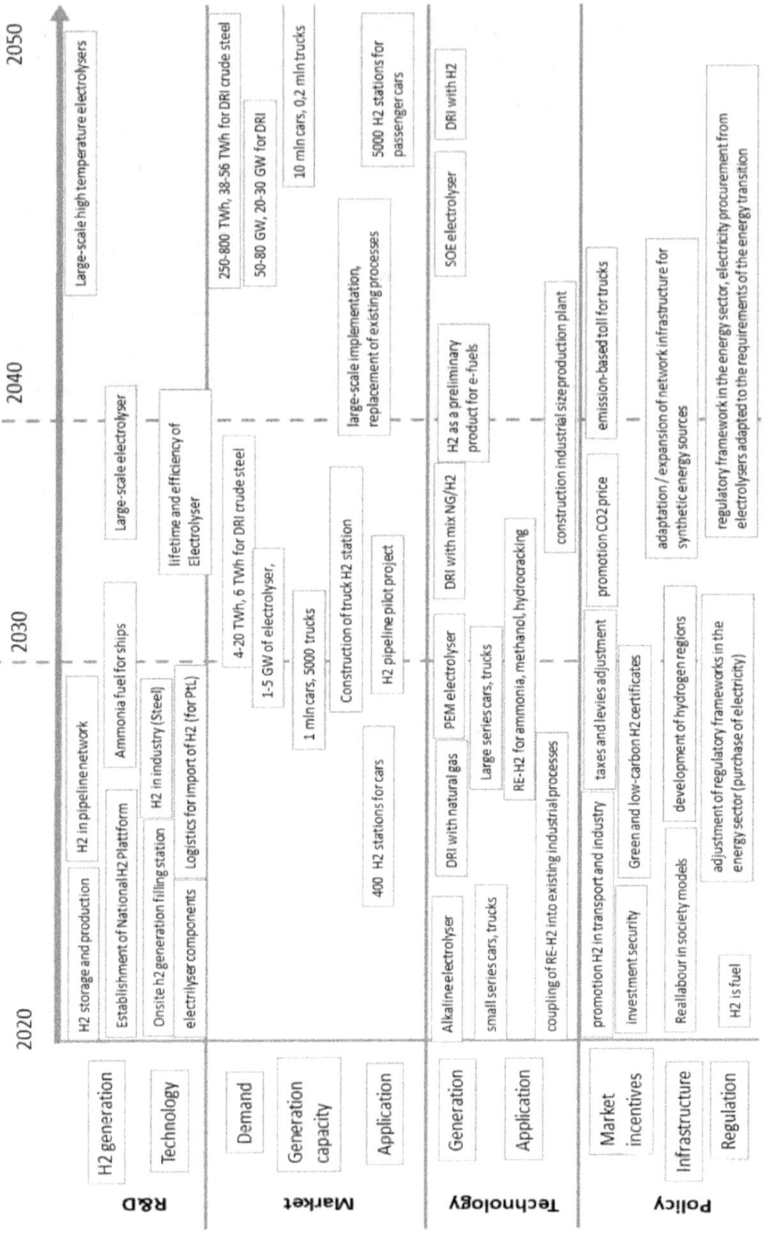

Figure 8.4 Germany hydrogen roadmap 2050 [23]

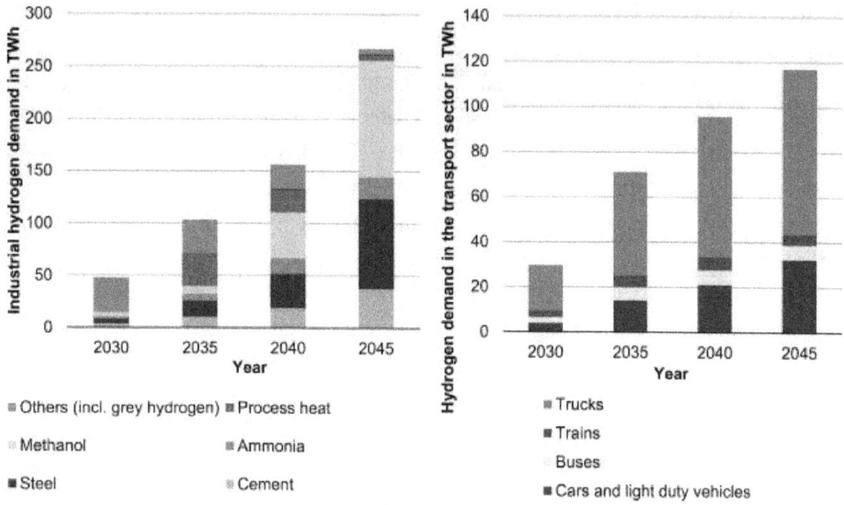

Figure 8.5 Evolution of hydrogen consumption in the industrial sector (left) and the transport sector (right) based on usage [23]

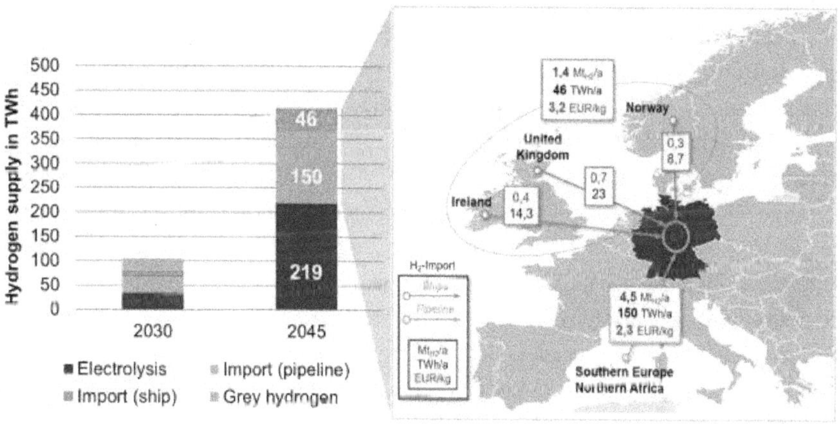

Figure 8.6 Evolution of the German hydrogen provision through in-country electrolysis and imports via pipelines and ships to Germany (left), and the routes and associated costs of hydrogen imports in the year 2045 [23]

The majority, amounting to 150 TWh, is imported via pipelines from Southern Europe, primarily Spain, or Northern Africa, predominantly Algeria, utilizing repurposed natural gas pipelines. The average import cost through pipelines from these regions is projected to be 2.3 €/kgH_2 in 2045. The import potential via pipelines from Southern Europe and Northern Africa is capped at 150 TWh due to existing pipeline network limitations, with no expansion considered in this analysis. Additionally, 46 TWh of hydrogen is imported in a liquid state via ships from Northern Europe, including the United Kingdom, Ireland, and Norway, with an average cost of 3.2 €/kgH_2. Domestically, about 219 TWh of hydrogen is produced through electrolysis, requiring 71 GW_{H2} of electrolyzer capacity in 2045. This results in an electricity demand of about 311 TWh for domestic hydrogen production, constituting over 25% of the total electricity demand. To meet this demand, a substantial expansion of wind turbines and photovoltaic plants is crucial, with installed capacities increasing significantly from 62 GW and 51 GW in 2020 to approximately 285 GW and 449 GW, respectively, in 2045. Achieving these cost-optimal renewable energy capacities necessitates a substantial acceleration of the current expansion rates. The expansion rate for wind turbines must increase from an average of about 3.3 GW per year between 2012 and 2021 to around 8.9 GW per year, while the expansion rate for photovoltaic plants must rise from about 2.5 GW per year to approximately 15.8 GW per year [24]. The German Federal Government acknowledges its global responsibility for reducing greenhouse gas emissions and recognizes the pivotal role of hydrogen in contributing to worldwide climate change mitigation efforts.

A crucial initial step towards accelerating hydrogen technology adoption is the establishment of a robust and sustainable domestic market for hydrogen production and utilization within Germany. This domestic focus aims to serve as a model, inspiring other nations to embrace hydrogen technology. To incentivize the rapid deployment of hydrogen technology, especially electrolyzers, the German government plans to design compatible incentives aligned with the broader energy transition goals. Anticipating a demand of 90 to 110 terawatt-hours (TWh) of hydrogen by 2030, Germany aims to establish up to 5 gigawatts (GW) of generation capacity, generating 14 TWh of green hydrogen while requiring 20 TWh of renewables-based electricity. Stringent monitoring mechanisms embedded in the National Hydrogen Strategy will track the development of green hydrogen demand, ensuring that increased electricity demand from electrolyzers does not lead to a rise in carbon emissions. An additional 5 GW of capacity is targeted, preferably by 2035 but no later than 2040. Nevertheless, relying solely on domestically produced green hydrogen will not meet the entirety of the growing demand. Consequently, a substantial portion of the required hydrogen will need to be sourced through imports. Numerous locations within the European Union boast significant capacities for producing green hydrogen due to the abundance of renewable-based electricity. The German Federal Government is committed to harnessing this potential and expanding generation capacities in collaboration with European Member States, particularly those sharing borders with the North and Baltic Seas, as well as southern European nations. A key focus will be the utilization of offshore wind energy, with efforts directed at establishing a reliable regulatory framework for offshore wind energy in collaboration with neighboring states. The government also aims to systematically develop production sites in partner countries, potentially through development cooperation initiatives [25]. To provide assurance for suppliers, consumers, and investors in Germany and abroad, the government plans to initiate an investment and innovation campaign, collaborating closely with partner countries.

The Hydrogen Strategy will be instrumental in promoting the establishment of production capacity, fostering new supply chains, and sharing technology and solutions with partner countries, ultimately contributing to job creation and long-term economic growth in both Germany and its partners. The acceleration of hydrogen technology rollout will be guided by the Action Plan for the National Hydrogen Strategy [26], alongside current budget and financial estimates, with a commitment to adapt the strategy based on evaluations if hydrogen demand surpasses initial expectations. The role of hydrogen technology and its associated alternative energy sources is fundamental to the success of the energy transition. Certain sectors, including air and maritime transport, as well as industries with inherent process-related emissions, face significant challenges in electrification, especially in the long term. Hence, there is a crucial need to substitute fossil inputs and fossil fuels with renewable-based alternatives. Hydrogen serves as a crucial foundational material for Germany's industrial sector, finding applications in industries such as chemicals and steel production. Approximately 55 terawatt-hours (TWh) of hydrogen, predominantly derived from fossil energy sources, are annually utilized for industrial purposes in Germany. There is an imperative shift towards green hydrogen-based production for these applications. Moreover, hydrogen and hydrogen-based Power-to-X (PtX) products must play a key role in decarbonizing emission-intensive industrial processes, opening up new avenues for hydrogen and PtX applications. For instance, achieving greenhouse gas-neutral German steel production by 2050 would necessitate over 80 TWh of hydrogen, while around 22 TWh of green hydrogen would be required for the transition of German refinery and ammonia production to hydrogen. With existing and anticipated demand, Germany's industrial sector is poised to significantly accelerate the market adoption of hydrogen, positioning itself as a global leader in hydrogen technology [27]. Germany is poised to assume a pivotal role in the global competition for the advancement and export of hydrogen technologies. The diverse network of German stakeholders in the hydrogen technology sector, coupled with their robust international connections, not only stands as a crucial element for the successful domestic market expansion of hydrogen technologies but also enhances the prospects of German enterprises in this forward-looking market. The production of components for hydrogen generation, utilization, and supply will foster regional value creation, fortifying companies engaged in these domains. To ensure widespread regional benefits from the emerging potential of the hydrogen economy, particular emphasis is placed on the implementation of the hydrogen strategy, especially in funding measures. Facilitating the market expansion of hydrogen technologies not only contributes significantly to addressing the economic fallout of the coronavirus pandemic but also lays a foundational pillar for the sustainable trajectory of the German industry.

8.3.2 France's hydrogen horizon

France has prioritized hydrogen through its national strategy, directing attention to industry and heavy-duty transport and allocating substantial public funding of EUR 7 billion. The government's emphasis lies in fostering the growth of domestic industry and enhancing French proficiency in this emerging technology. In 2020, France unveiled an ambitious national strategy to develop a decarbonized hydrogen sector, with a commitment to invest 9 billion euros in public funding to expand the sector on an industrial scale throughout the country. The goal was to achieve 6.5 GW of installed electrolyzer capacity by 2030, leading to an annual reduction of 6 million tons of CO_2 emissions [28]. The estimated job creation ranges from 50,000 to 150,000, suggesting that the hydrogen sector has the potential to support a substantial workforce in the country. In 2021, France Hydrogène published a study that outlined two scenarios within this strategy: the "Ambition 2030" scenario, targeting an annual production of 680,000 tons of low-carbon or renewable hydrogen, and the "Ambition+ 2030" scenario, which aims for a significant increase to 1,090,000 tons of low-carbon or renewable hydrogen to meet the regulatory requirements of the 'Fit for 55' package. The strategy also identifies seven large clusters that will serve as cornerstones for the widespread implementation of hydrogen projects, enabling cost reductions through the combined production and various end uses of hydrogen. The strategy has received support from over 100 signatories, including major companies such as EDF and Engie, as well as regional governments [29]. However, there have been discussions and differing views with France's European neighbors regarding the use of "pink hydrogen", which is produced by nuclear reactors. The French strategy focuses on domestic hydrogen production to decarbonize the industry and promote hydrogen mobility, with an emphasis on applications in refineries, the chemical industry, steel production, and the mobility sector. The political priority is to develop a domestic industry sized to meet national demand, which is seen as a more secure sourcing strategy than relying on imports. France aims to lead in green hydrogen production and has set ambitious targets to achieve this, aligning with its long-term climate and energy goals.

France aims to have 20-40% of its total hydrogen consumption sourced from low-carbon and renewable sources by 2030. This objective is part of the French government's broader strategy to develop a domestic hydrogen industry and decarbonize various sectors, including industry and mobility. The plan includes significant investments and targets for the production and consumption of low-carbon and renewable hydrogen. The goal is to reduce carbon emissions and promote the use of green hydrogen in line with France's long-term climate and energy objectives. The strategy involves a combination of legislative measures, financial incentives, and the development of hydrogen infrastructure to support the transition towards a more sustainable hydrogen economy. France Hydrogène, a key player in the French hydrogen sector, has conducted an extensive data collection effort across the 12 regions of mainland France, encompassing over 250 projects and local distribution networks. This exercise involved semi-structured interviews with France Hydrogène's regional delegations and Conseils régionaux (local authorities at the regional level) to gather, analyze, and verify the data. The study incorporated a large volume of data from projects at various stages of completion, with some confidential projects not being included in the data. Additionally, France Hydrogène utilized open-source materials for different data categories, such as reports, studies, regional and national monitoring units, and regional land-use plans. The combined projects listed in the study are projected to consume approximately 1,070,000 tons of low-carbon and renewable hydrogen annually between the present and 2030 in France. This quantity exceeds the target set in the National Hydrogen Strategy and aligns closely with the goals of the Ambition+ 2030 scenario. The primary areas where hydrogen is expected to be utilized are concentrated around industrial zones [30].

France is actively scaling up its hydrogen developments, particularly in the industrial sector and at ports, as well as in the gas grids. One notable project, the HyDeal Ambition, involves European gas TSOs (Transmission System Operators) preparing to facilitate hydrogen trade through the creation of dedicated hydrogen pipelines. These pipelines will be laid alongside existing gas pipelines across Western Europe and North Africa, enabling the transportation of hydrogen to support the region's energy transition. This initiative reflects France's commitment to advancing its hydrogen infrastructure and promoting the use of hydrogen as a clean energy source in collaboration with its European partners [31].

In terms of rail transport, the integration of hydrogen-powered trains into the regional, non-electrified rail network aligns with a broader strategy for local development. Fourteen hydrogen-powered Régiolis trains have been commissioned by four regional governments and are set to be operational by 2025, replacing existing diesel-powered trains. These new trains ensure zero emissions, maintaining passenger comfort and performance levels comparable to their predecessors. Featuring a 1,000 km range, with 600 km from an onboard hydrogen tank containing 184 kgH_2 and a 300-kW fuel cell system. In various sea and river transportation applications, hydrogen and fuel cells prove versatile, demonstrated by approximately thirty projects across France. These projects include powering passenger shuttle boats, small harbor vessels, inland waterway boats, and quayside refueling devices. Developing solutions encompass power generation from 200 kW to 2.5 MW, with onboard reservoirs holding substantial quantities of compressed or liquid hydrogen. Despite ongoing technical and regulatory challenges, decarbonizing the aviation industry is crucial for its future sustainability. The French aeronautics industry is exploring multiple approaches to achieve this, including hydrogen utilization for propulsion through turbojet or electric motor (powered by a fuel cell), operating Auxiliary Power Units, providing emergency backup power, or substituting kerosene with e-kerosene derived from low-carbon or renewable hydrogen [32,33]. Hydrogen also plays a crucial role in energy-related applications, serving as a practical solution to address the flexibility needs of the electrical grid. Additionally, it finds applications in off-grid scenarios, including stand-alone power systems, remote areas, electric generators, and emergency backup systems. With the evolving energy landscape and the growing significance of renewable energy in the French electricity generation mix, there is a potential surge in hydrogen demand in the short term. This demand arises from hydrogen's capability to provide flexibility to the grid, aligning with the concurrent expansion of hydrogen transportation and storage infrastructure.

8.3.3 Italy's hydrogen revolution

Italy emerges as an attractive market for green hydrogen development, leveraging its extensive renewable energy infrastructure and nationwide gas transport network. The strategic positioning of Italy in the Mediterranean makes it an ideal hub for hydrogen trade between potential major exporters in Africa and the Middle East and consumers in northern Europe. Projections indicate that by 2050, hydrogen could constitute nearly a quarter of Italy's overall energy demand. Despite this potential, regulatory challenges pose obstacles to the full realization of green hydrogen. Specific legislative frameworks governing the authorization process for electrolysis-based hydrogen production and incentives for Power-to-Gas (P2G) plants are lacking, hindering economic viability. Overcoming these hurdles is crucial for Italy to assert itself as a leader in the hydrogen sector, supporting the country's transition to an integrated, flexible, and energy-efficient national system. Notably, Italy has taken steps toward energy transition, establishing the Ministry for Ecological Transition in 2021 and preparing a "Hydrogen, Mobility, Energy-Efficiency Decree" [34]. The Ministry of Economic Development (MISE) envisions the need for up to €10 billion in investments from 2020 to 2030 to establish a low-carbon hydrogen economy in Italy, aligning with national hydrogen penetration targets. This investment plan encompasses various areas, including €5-7 billion for hydrogen production, €2-3 billion for hydrogen distribution and consumption infrastructure (e.g., hydrogen-powered trains, trucks, and refueling stations), €1 billion for Research and Development, and funding for infrastructure integration [35]. Notably, these figures exclude approximately €2 billion earmarked for developing the hydrogen supply chain under the Next Generation EU initiative, aimed at mitigating the economic impact of the COVID-19 pandemic.

To align with the European Green Deal's 2050 coal phase-out targets, Italy has embraced an ambitious Energy and Climate National Integrated Plan (PNIEC). This plan aims to reduce emissions by 55% by 2040, enhance renewable energy production, and actively promote the production and utilization of hydrogen generated from renewable electricity through research and development activities. Recognizing the limitations of short-term electricity storage options like pumping and electrochemical accumulation, the PNIEC emphasizes the need for long-term storage technologies, including P2G. These technologies will enable the storage of surplus energy from non-dispatchable renewable sources in clean energy carriers, fostering efficiency and potentially merging the gas and electricity sectors into a unified energy sector. In November 2020, the MISE released guidelines for the National Hydrogen Strategy, identifying sectors where green hydrogen is expected to be competitive in the short term. The MISE is actively engaged in an IPCEI focused on the strategic hydrogen value chain, reflecting Italy's collaboration with other EU Member States and the European Commission in industrial policy initiatives. The MISE has established a hydrogen panel comprising over 70 national stakeholders involved in the development, implementation, and utilization of green hydrogen across various sectors, emphasizing a comprehensive approach to hydrogen initiatives. To stimulate the growth of the hydrogen market, the Italian Government envisions installing around 5 GW of electrolysis capacity by 2030. However, the existing regulatory framework for hydrogen production in Italy exclusively addresses production using fossil fuels, leading to protracted and cumbersome administrative procedures. Consequently, industry players are awaiting specific regulations tailored to green hydrogen production through electrolysis (Figure 8.7) [36].

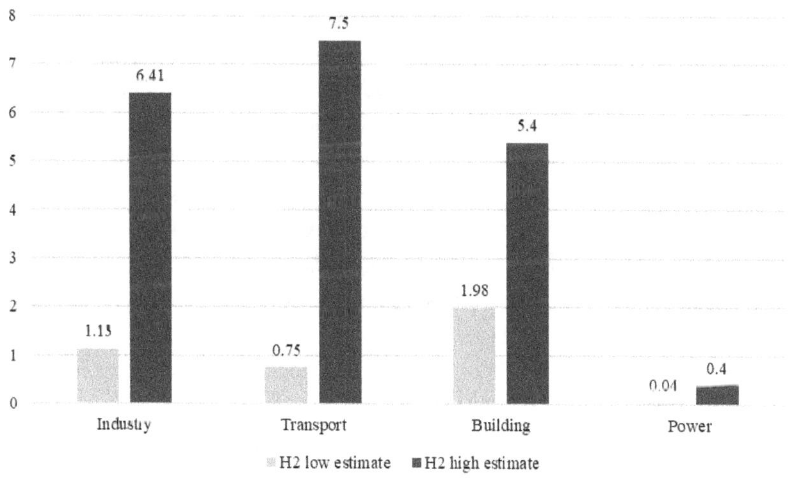

Figure 8.7 Italy clean H_2 demand estimates by 2030 (Italy), TWhH

Such regulations could incorporate a support mechanism to offset the initial high costs of the process, especially over the next decade, and streamline the current authorization procedures. A desirable enhancement would involve implementing distinct authorization regulations based on plant size and differentiating between hydrogen production processes with greenhouse gas emissions and those with minimal or no emissions. Careful consideration of the production plant's location is also imperative to evaluate the most suitable environmental verification procedures. Particularly during the initial testing phase, many plants might be situated in industrial areas, facilitating proximity to end-use points and enabling the implementation of experimental production, transportation, and consumption systems centered around hydrogen, often referred to as "Hydrogen Valleys". Therefore, it is crucial for green hydrogen to be endorsed as a preferred energy carrier, especially in the long term, by relevant authorities and administrations at all levels. This endorsement is essential to prevent uncertainties and obstacles that could impede its development.

The MISE Guidelines establish a minimum objective for the expansion of conventional gas transportation networks to accommodate hydrogen, estimating that by 2030, an average of up to 2% of distributed natural gas could be substituted with hydrogen. In April 2019, Snam, a pioneer in Europe, conducted a successful experiment by injecting a 5% hydrogen and natural gas mix into its transmission network in Contursi Terme, Salerno. This trial, known as the hydrogen-gas mix ("H2NG"), supplied industrial companies, including a pasta factory and a mineral water bottling company, with the blended gas for approximately a month. In December 2019, the Contursi experiment was duplicated, with the hydrogen percentage increasing to 10%. Multiple studies, including those by Snam and other industry players, affirm that blending hydrogen with natural gas up to at least 10% is a feasible and secure solution [34]. The PNIEC allows for pilot projects aimed at assessing the functionality, cost-effectiveness, and replicability of various technological solutions for green hydrogen use. Consequently, in February 2020, the Italian Independent Regulatory Authority for Energy, Networks, and Environment (ARERA) initiated a public consultation process regarding pilot projects to optimize the management and innovative applications of existing natural gas transmission and distribution networks. ARERA focused on potential pilot projects related to (a) the integration of renewable gases into networks and (b) applications of P2G and Power-to-Hydrogen (P2H) technology [37]. P2H technology is critical for utilizing hydrogen produced through electrolysis in various ways:

- Employing it as a storage carrier to generate electricity with reversible fuel cell systems (Power-to-Power or "P2P").
- Transporting it through the natural gas grid (blending with natural gas).
- Directing it into dedicated infrastructure for use as a fuel, such as in transportation.

ARERA anticipates that P2G plants will be economically viable by 2030, primarily due to the need for substantial amounts of low-cost electricity to enhance competitiveness or incentivize this technology. From a regulatory perspective, a precise definition of P2G plants, typically categorized as production facilities, is essential. During the early stages of an incompletely developed market, transport or distribution network operators will play a pivotal role in planning and executing fully integrated P2G initiatives within the transmission network. This scenario will only materialize when the market is mature and fully competitive, necessitating the adoption of tariff regulation and specific regulatory requirements, including financial separation from other regulated activities. The gas transport system is poised to be a crucial component of the national energy system. Achieving this goal should be feasible without the need for additional significant infrastructure development beyond those already planned or under construction. Instead, technological innovations to the existing gas transport system are expected to be the driving force.

In the transportation sector, hydrogen's initial application is envisioned for public transit, particularly for long-distance travel, commercial fleets, and segments of the rail network lacking electrification. Regional or local transport companies will initially oversee hydrogen refueling stations based on a thorough assessment of fleet demand and distinct requirements for both light and heavy vehicles. The development of hydrogen fuel cell trains is being considered for non-electrified commercial rail routes. Approximately 4,763 km of Italy's primary rail network still relies on diesel technology, and certain hydrogen train applications are already demonstrating cost competitiveness [38]. Over the long term, hydrogen may emerge as a viable option for decarbonizing the aviation and maritime sectors by producing liquid synthetic kerosene or other synthetic fuels. Concerning the maritime sector, new technologies incorporating hydrogen as fuel in gas turbines are entering the market, enabling the exploration of combined cycle propulsion systems with high efficiency and low emissions. The success of hydrogen in marine transport hinges on the availability of an extensive refueling network. Vessels in the maritime sector, except for local ferries and smaller units, necessitate long travel ranges with limited refueling opportunities. Despite being liquefied, hydrogen requires significantly higher volumes than fossil fuels. Nevertheless, hydrogen offers a potential compromise, maintaining comparable range capabilities to fossil fuels while ensuring a total reduction in emissions at the expense of reducing available storage capacity. The future introduction of hydrogen refueling stations at major national and international ports could potentially reduce on-board fuel storage volumes. Given the sector's complexity, diverse ship types, and operations, there is currently no established and unequivocal strategy. However, hydrogen and hydrogen-rich fuels are expected to play a crucial role in this context.

8.3.4 Spain's hydrogen surge

Spain's expansive and sparsely populated territory, which receives over 2,500 hours of sunshine annually, provides an ideal environment for the production of green hydrogen, thanks to its abundant wind and solar energy resources. The Spanish government has formulated the "Hydrogen Roadmap: A Commitment to Renewable Hydrogen" plan to contribute to achieving climate neutrality and a 100% renewable electricity system. The roadmap sets objectives for 2030 and a vision for 2050, aiming to ensure that renewable hydrogen plays a significant role in the country's climate neutrality by 2050. The European Strategy outlines three phases of renewable hydrogen development. The first phase (2020-2024) aims to have at least 6 GW of electrolyzers installed in the EU, along with up to one million tons of renewable hydrogen produced. Spain plans to have at least 4 GW of electrolyzers installed by 2030, as part of the measures to prepare the regulatory framework and goals for the next decade. The country also aims to reach 25% of industrial hydrogen consumption from renewable sources by 2030. Additionally, Spain is planning to introduce renewable hydrogen-powered transportation vehicles, including a fleet of at least 150 buses, 5,000 light and heavy vehicles, and two commercial train lines. The roadmap also envisions the implementation of a network with a minimum of 100 renewable hydrogen stations and green hydrogen-powered handling machinery at the main five ports and airports. Spanish companies are already investing in renewable hydrogen projects, with Enagás, a Spanish energy company and transmission system operator, planning the largest hydrogen plant in Spain. This project involves a 32 MW electrolyzer powered by a 150 MW photovoltaic plant and is expected to require an investment of USD 309 million. To achieve the ambitious objectives of the Spanish Hydrogen Roadmap, an estimated investment of over USD 10 billion will be required during the 2020-2030 period [39].

While certain sectors are transitioning towards electricity, others pose more challenges. Hydrogen proves advantageous in providing high temperatures essential for heavy industry compared to electricity. In sectors like automobiles and heating, electricity remains effective. Electrification is crucial for certain sectors, but it's equally essential to reduce fossil fuel usage in areas where electricity isn't as straightforward. Globally, around 80% of energy consumption is not electrified, emphasizing the need for an immediate replacement of fossil fuels. HyDeal, a green hydrogen company led by Lepercq, aims to replace high-emission processes in steel manufacturing and ammonia production by building solar farms and electrolyzers in northern Spain. This initiative aligns with the broader goal of reducing reliance on coking coal, blast furnaces, and natural gas. As per the Hydrogen Roadmap, the current hydrogen consumption in Spain stands at approximately 500,000 tons per year, primarily consisting of grey hydrogen. This grey hydrogen serves as a raw material, predominantly in refineries (constituting around 70%) and chemical product manufacturing (50%), with the remaining consumption distributed across sectors like metallurgy. Despite this current scenario, the Hydrogen Roadmap envisions Spain harnessing its potential to become a technological frontrunner in the production and utilization of green hydrogen. The favorable climate and expansive areas available for renewable energy projects, whether solar or wind, are key advantages. The Sapin hydrogen roadmap illustrated in Table 8.1 sets ambitious targets aligned with the three phases outlined in the EU Hydrogen Strategy. The regulatory measures outlined in the Hydrogen Roadmap include a primary package aimed at overcoming existing regulatory barriers associated with considering hydrogen production as an industrial activity, specifically categorized under the chemical industry for inorganic gas production [40].

Table 8.1 The Spain hydrogen roadmap vs the EU hydrogen strategies from 2020 to 2050

	EU hydrogen strategy	Spain hydrogen roadmap
First phase (2020-2024)	Deployment of 6 GW of electrolyzers and the generation of 1 million tons of renewable hydrogen within the European Union.	Installation of 300-600 MW of electrolyser plants.
Second phase (2025-2030)	Deployment of 40 GW of electrolyzers and the production of up to 10 million tons of renewable hydrogen within the European Union.	• Implementation of no less than 4 gigawatts (GW) of electrolyzer facilities. • Industrial Sector: Ensure a minimum of 25% contribution from renewable hydrogen to the total hydrogen consumption across all industries by 2030. • Transportation: • Deployment of 150-200 fuel cell buses by 2030. • Introduction of 5,000-7,500 light and heavy-duty fuel cell vehicles for freight transport by 2030. • Establishment of 100-150 publicly accessible hydrogen stations by 2030. • Power Sector/Energy Storage: Operationalization of commercial hydrogen projects by 2030, serving the purposes of electricity storage and/or utilizing surplus renewable energy.
Third phase (2030-2050)	The advancement and widespread implementation of technologies for generating renewable hydrogen at a large scale.	• Economy centered around the production and utilization of renewable hydrogen. • Competitiveness of hydrogen production through renewable energy in comparison to alternative production technologies. • Achieving societal carbonization by the year 2050. • Enhanced controllability of renewable energy sources. • Provision of a high-quality, sustainable, and economically viable energy supply.

This classification limits the implementation of hydrogen production infrastructure to industrial-zoned land, subjecting it to rigorous environmental impact assessments, irrespective of the use of renewable technologies. To address these challenges, the roadmap proposes modifications to the classification of on-site renewable hydrogen production at refueling stations, streamlined procedures for small-scale green hydrogen production facilities, and the development of regulatory measures facilitating dedicated electricity lines and hydroducts for renewable hydrogen transport. Furthermore, a Guarantee of Origin System is suggested to trace hydrogen production methods collaboratively with other European countries. Additionally, addressing high production costs, the roadmap advocates considering the positive environmental impacts of green taxation, including indirect taxation, as part of broader tax reform initiatives. Notably, the Automotive Industry Value Chain Tax Plan is exploring reforms in key vehicle taxes to introduce greater environmental considerations.

8.4 European Union international collaboration for hydrogen economy deployment

The EU is firmly dedicated to international collaboration to hasten the green transition, including the advancement of the hydrogen market both within the EU and globally. Supporting the establishment of renewable hydrogen economies in EU Partner countries is seen as a means to expedite their journey toward climate neutrality, fostering broader social and economic development. Investments in renewable energy and hydrogen have the potential to enhance the appeal of investing in countries facing high investment costs, a low share of renewables, and underdeveloped energy infrastructure while also creating opportunities for export diversification. Simultaneously, it is crucial to address the sustainability aspects of renewable hydrogen production to prevent exacerbating water stress and potential adverse effects on electricity and water accessibility. The recent agreement among co-legislators on a Carbon Border Adjustment Mechanism (CBAM) further bolsters domestic decarbonization efforts in partner countries, with hydrogen imports into the EU subject to carbon content-based adjustments from 2026 [41]. The REPowerEU Communication suggests importing an additional 10 million tons of renewable hydrogen by 2030, supplementing the 10 million tons of renewable hydrogen produced within the EU, positioning the EU as the world's largest aspiring market for renewable hydrogen. The majority of export projects aim to transport hydrogen and its derivatives by ship, necessitating the expansion of shipping and port infrastructure in Europe, while international hydrogen trade through pipelines is anticipated to develop towards the end of this decade. Preliminary market analysis and input from stakeholders indicate the necessity for a dedicated mechanism to facilitate the import of hydrogen from third countries. In the absence of an established market, EU Member States can play a role in supporting European off-takers, facilitating essential off-take agreements with producers in third countries. It is crucial for this mechanism to align with energy diplomacy and trade policies, ensuring that the EU avoids creating new strategic dependencies. Incorporating the sustainable development goals into the framework of EU support for imported renewable hydrogen would grant entry to the European market and foster the sustainability of local production and consumption of renewable electricity and hydrogen. This approach aims to address social, economic, and environmental needs, benefiting EU partner countries and their populations. Several Member States have formulated strategies to back the import of hydrogen from third countries. Consequently, the Commission is exploring the design of the international component of the European Hydrogen Bank to encourage a cohesive EU strategy for renewable hydrogen imports. The overarching goal is to bridge the cost disparity between renewable hydrogen produced in third countries and transported to the EU, and the fossil fuels it could replace within the EU.

The European Hydrogen Bank has enhanced visibility into hydrogen flows, transactions, and pricing through its initiatives. Specifically, the Commission aims to coordinate information to bolster confidence in the evolving hydrogen market, utilizing data from European and international off-take agreements to offer transparent price details and establish price benchmarks. While there is a growing wealth of information on renewable hydrogen demand and supply globally and within the EU, the Bank will provide additional value by consolidating and complementing such information. This will involve leveraging existing sources and structures, including data from the Commission, such as Eurostat's hydrogen data collection efforts [42], initiatives like the European Clean Hydrogen Alliance [43], Renewable and Low-Carbon Fuels Alliance [44], and Zero Emission Aviation Alliance [45], along with insights from existing project financing mechanisms and industry monitoring. The Bank will also draw on resources like the fuel cells and hydrogen observatory under the Clean Hydrogen Joint Undertaking [46], which currently offers a public portal for European Hydrogen data. Moreover, the Bank will collaborate with organizations like the IEA and IRENA for additional information. With these resources, the Bank will be well-equipped to assess EU demand for both domestic and imported renewable hydrogen, potentially organizing a "call for interest" as a voluntary and non-binding expression of interest from EU off-takers. The swift establishment of dedicated hydrogen infrastructure within and connected to the EU, linking supply and demand, is pivotal for achieving the EU decarbonization goals. The revised TEN-E* regulation serves as a unique tool for planning European energy infrastructure [47]. The European Hydrogen Bank will contribute to infrastructure planning by providing information on hydrogen flows obtained through its activities, including details on delivery points. Additionally, the Bank will facilitate the coordination of the Memoranda of Understanding (MoUs) at both the Member State and company levels with third countries and foreign hydrogen producers. It will also lend support to the incorporation of dedicated provisions related to hydrogen within the energy and raw materials trade chapters in free trade agreements or sustainable investment facilitation agreements (SIFA) [48]. Green Hydrogen Partnerships and MoUs involving the European Commission, Member States, and European companies, including ports, can offer insights into potential hydrogen flows from third countries. Coordinated by the European Hydrogen Bank, this information will provide clarity on where and when hydrogen infrastructure will be required. The European Commission, representing the EU, has entered into MoUs and/or Partnerships with Egypt, Japan, Kazakhstan, Morocco, Namibia, and Ukraine. Provisions focusing on hydrogen cooperation have been included in the updated EU-Chile Association Agreement, and discussions are ongoing for free trade agreements with India and Australia. These collaborative efforts aim to support the green transition in partner countries, fostering the development of the renewable energy sector and industrial supply chains while reinforcing regulatory frameworks, technology deployment, and sustainable renewable hydrogen production. The MoUs acknowledge natural resource constraints and demonstrate mutual political commitment to environmental, social, and economic sustainability standards. Ongoing diplomatic outreach with the Kingdom of Saudi Arabia aims to establish a partnership on hydrogen [49].

The updated EU-Chile Association Agreement stands out as the first agreement specifically addressing hydrogen, recognizing Chile's potential for hydrogen exports to the EU [50]. The EU takes a leading role in international hydrogen initiatives, participating in forums like the Clean Energy Ministerial Hydrogen Initiative, Mission Innovation on clean hydrogen, and the International Partnership for Hydrogen in the Economy. Publicly available data indicates that, from 2021 to 2022, EU Member States and companies have entered into hydrogen cooperation MoUs with at least 30 countries globally. The European Hydrogen Bank could play a role in improving transparency and coordination of renewable hydrogen transactions and negotiations between the EU and third countries, streamlining efforts on both sides. To ensure a diverse range of renewable hydrogen sources, the Commission is exploring an instrument inspired by the transparency provisions of the EU Energy Platform under Council Regulation (EU) 2002/2576 [51]. This instrument could contribute positively to the emerging renewable hydrogen market by providing intelligence and potentially allowing the Commission to issue recommendations on hydrogen investment coordination.

8.5 Green hydrogen imports by the European Union

To achieve its decarbonization objectives, the EU should reinforce regulatory signals indicating its intention to capitalize on opportunities for importing renewable hydrogen, continuing to incorporate imports into its long-term energy transition strategy. The EU has made significant strides in establishing supportive frameworks for global hydrogen suppliers, establishing targets for deploying green fuel in industrial and transport sectors, and implementing programs to offset green premiums for end-users. However, additional funding is needed for these incentive schemes to expedite the transition to alternative fuels. While the EU has committed to procuring and distributing renewable hydrogen imports for European end-users, there is a need to intensify investments in infrastructure at ports and industrial sites, concurrently providing clarity on rules governing infrastructure access. The European Commission (EC) has outlined specific criteria for hydrogen products to qualify as "renewable" within the bloc, yet it must promptly endorse clear certification schemes and align verification conditions with carbon market requirements to alleviate uncertainty and bureaucratic challenges for potential suppliers. The EU, particularly in Northwest Europe, will require imports of hydrogen to fulfill the demand for green fuel in 2030 and beyond. Several key trading partners are well-positioned to provide competitive early supplies. Despite the EU's progress in shaping its green hydrogen policy, with initiatives like REPowerEU and the mobilization of funding by Member States for early procurement, the reliance on imports remains crucial for strategic reasons. While Europe has the motivation and financial resources to deploy new renewables within its borders, various challenges, such as land use restrictions and lengthy grid-connection queues, can impede development. The integration of new renewables is essential for enabling direct electrification of the EU's power grid, buildings, and road transport. To mitigate domestic demand for renewables from heavy industry and transport, the EU can leverage the importation of abundant and cost-effective renewable energy produced abroad, distributing green hydrogen to industrial end-users clustered around ports. Importation ensures European manufacturers access a secure and reliable green energy supply, facilitating fuel switching without disruptions and without relying on constrained renewable power sources within the region. An evaluation of projections for green hydrogen supply in Europe for 2030 indicates that many Member States constrained by renewables will not be on track to meet local demand through domestic production alone. Figure 8.8 illustrates that when assessing the projected hydrogen demand of EU Member States against their respective domestic project pipelines, nearly all countries will fall short of meeting demand solely through local production in 2030 [52].

Even with the potential development of intra European trade, significant supply gaps are anticipated. This scenario will exert pressure on heavy industry and transport players to achieve decarbonization targets, necessitating the sourcing of products from external markets. The EU will have a variety of international options for suppliers, as numerous stakeholders are developing export-focused projects in regions such as Australia, Southeast Asia, Africa, the Middle East, and North and South America. Several countries have formally committed to supplying green hydrogen to specific EU Member States through initial exports. The EU has the opportunity to enhance its energy import resilience by engaging with multiple potential suppliers to secure renewable fuel. These potential supplier countries share a common feature: the availability of cost-effective renewable power that can significantly reduce hydrogen production costs within their markets, making them competitive with domestic production costs in Europe. Since hydrogen production costs are primarily influenced by levelized electricity costs (LCOE), countries like Chile and Australia, with optimal solar and wind conditions, can offer the EU a supply at a very low base cost. Similarly, parts of the United States with favorable conditions, supported by government incentives for renewable hydrogen production, can provide Europe with green hydrogen (converted to ammonia) at a globally competitive base cost. However, the production costs of green hydrogen abroad do not directly determine the end-user cost of hydrogen in Europe. Additional expenses, including storage, ammonia (NH_3) conversion, transportation, offloading, and re-conversion, will contribute to a "landed cost" for renewable hydrogen upon arrival in EU ports. Achieving economies of scale across ammonia transport, synthesis, and cracking processes can help reduce these costs.

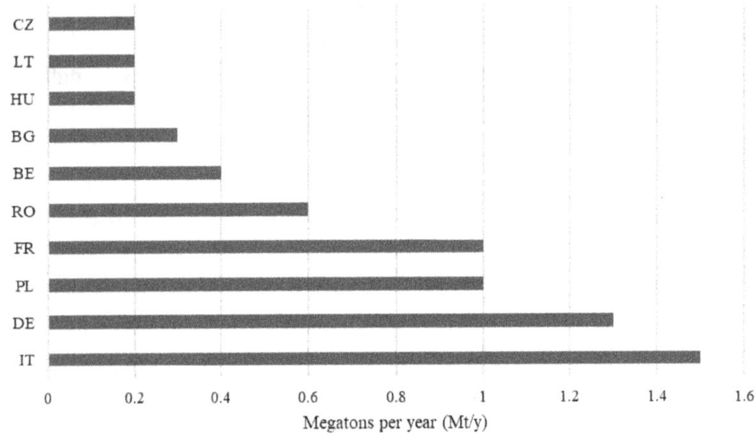

Figure 8.8 Discrepancy between anticipated demand for green hydrogen in EU Member States and the projected supply (derived from announced projects) by 2030

The readiness of candidate suppliers to progress, such as the ongoing construction of export-oriented projects by companies in Saudi Arabia and partnerships with European off-takers and existing ammonia export infrastructure owners, varies. In contrast, Algeria, despite expressing aspirations to supply hydrogen to Europe, has not advanced concrete hydrogen production projects. The combination of export market readiness and landed product costs will determine which suppliers can offer the most secure and cost-effective supply to European end-users by 2030. Ultimately, the EU will need to secure renewable hydrogen imports from various partners to ensure a reliable supply of green energy for heavy industry and transportation. The most ambitious potential suppliers, such as the United States and Australia, are actively preparing to export approximately 2 million tons per year to Europe by 2030, underlining the critical role of supplies from various markets in meeting the anticipated demand. By importing renewable hydrogen from diverse and well-prepared suppliers, the EU can obtain a consistent source of green fuel without jeopardizing its energy security while benefiting from economically advantageous international imports. As green hydrogen markets continue to develop in Africa and South America, Europe can further enhance the diversity of its fuel imports by collaborating with new suppliers in closer geographic proximity. In the short term, the United States is strategically positioned to provide Europe with cost-effective green hydrogen by 2030. The US Gulf Coast, in particular, is actively investing in hydrogen storage and conversion infrastructure to facilitate trade, showcasing a high level of export readiness. Recent tax credits established under the Inflation Reduction Act (IRA), specifically for clean hydrogen and clean electricity production, will offer substantial incentives to US producers. The combined impact of these credits is expected to significantly reduce the levelized cost of green hydrogen over the lifespan of a new plant, reaching the Final Investment Decision (FID) in 2030, estimated at approximately \$2.83 per kilogram. Consequently, US exports arriving at selected European ports by 2030 will be more cost-effective than local production, even compared to projects receiving EU subsidies from initiatives like the European Hydrogen Bank (EHB) [52]. Figure 8.9 demonstrates that by 2030, the imported green hydrogen from the US Gulf Coast into underserved European ports will be more cost-effective than locally available alternatives. The availability of a reliable and cost-competitive supply of green hydrogen from the United States and other supplier countries will act as a catalyst for the swift transition to hydrogen-based fuels within heavy industries concentrated around ports.

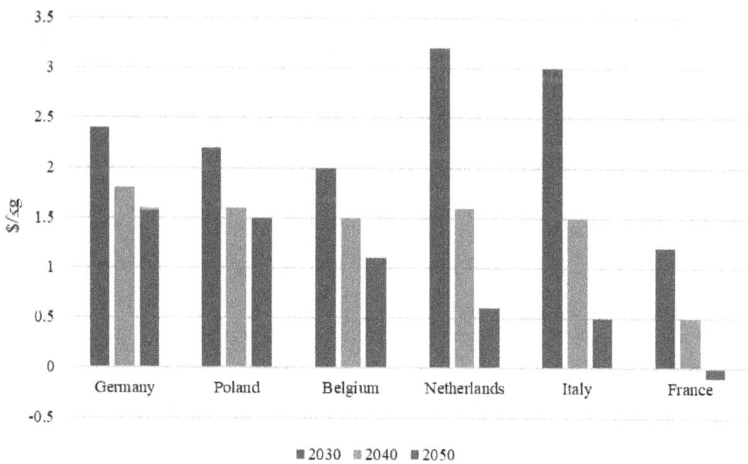

Figure 8.9 Price differential between domestically generated and imported green hydrogen from the US Gulf Coast to specific EU ports, 2030-40

This increased demand will necessitate substantial investments in new infrastructure for hydrogen storage and distribution, contributing to cost reductions for local European hydrogen producers. These producers can expand their operations while industries and transportation sectors take advantage of the initial lower-cost imports.

As domestic hydrogen production scales up, the prices of renewable power in Europe will stabilize, and the costs of electrolyzers will decrease. At this point, strategically deploying imported hydrogen across the bloc as an affordable complement to domestic products tailored to regional needs becomes feasible. Certain European countries, such as Germany, the Netherlands, and Belgium, will continue to have competitive imports of green hydrogen, playing a crucial role in addressing supply gaps. Importing hydrogen will remain a reliable source of green fuel in industrial hubs around Northern European ports, where land constraints create ongoing competition for renewable power. This region will benefit from developing infrastructure for importing green hydrogen, enabling consistent utilization of international products to enhance long-term energy security for its industries, which are anticipated to drive some of the highest levels of demand for green hydrogen in Europe. In contrast, Southern Europe will experience significant benefits from the scaling of local hydrogen markets and technology cost reduction over time. Leveraging low-cost solar and wind power, Spain and Portugal are projected to produce domestic green hydrogen at lower costs by the mid-2030s, surpassing even subsidized imports by 2040. These locally produced hydrogen products can be deployed to meet local demand or transported through pipelines to nearby EU markets where supply gaps for green hydrogen are expected to persist. European policymakers can address these market variations among member states by strategically investing in regions poised to derive the most benefit from green hydrogen imports, both in 2030 and in the long run. Focusing public support on specific ports can facilitate a synergistic relationship between markets better served by imports from third countries and those more suited to local production or intra-European hydrogen trade. Ports are pivotal drivers of the EU's hydrogen market evolution, acting as gateways to industrial hubs fueling the demand for green fuel in their decarbonization efforts. Port infrastructure, including ammonia import capacity, storage facilities, ammonia crackers, and hydrogen pipelines, will play a crucial role in facilitating early imports. Table 8.2 provides an assessment and ranking of European ports as potential hubs for the green hydrogen trade. Data on current and planned ammonia import capacity, also depicted in Table 8.2, highlights that to address expected supply gaps in most markets effectively, countries will need to invest significantly in ammonia import capacity and critical port infrastructure components like crackers, storage facilities, and hydrogen pipelines. All EU Member States with import aspirations currently lack the necessary infrastructure to secure ample international hydrogen to meet their projected supply gaps. In order to achieve the ambitious hydrogen deployment goals, set by the EU, it is essential to incorporate imports into a holistic energy strategy. This approach not only fosters energy security but also provides domestic industries with the necessary time to mature. A crucial aspect is ensuring these imports are entirely decarbonized, aligning with Europe's objectives for meeting decarbonization targets.

Table 8.2 Country readiness to facilitate hydrogen trade in 2030 [52]

Country	Ideal port for imports	Cost differential ($/kg)	Country supply gap (Mt)[1]	Planned countrywide Ammonia import capacity (kt/y-H_2 equivalent)	Insights
Germany	Hamburg; Brunsbüttel	2.36	1.31	7	Germany stands out as a suitable hydrogen importer, given the substantial cost difference compared to domestic supply, a significant gap in hydrogen demand, and the presence of current and proposed ammonia import infrastructure.
Netherlands	Rotterdam; Terneuzen	3.15	-0.34	29	Despite ambitious targets for domestic production, the Netherlands is strategically positioned to capitalize on affordable hydrogen imports, particularly due to its extensive planned import infrastructure.
Italy	Ravenna; Venice	3.00	1.50	29	As Italy scales up its domestic market for hydrogen, there will be a significant supply gap. To take advantage of more cost-effective imports, Italy would need to enhance its import capacity to fully leverage short-term opportunities.

Poland	Police; Świnoujście	2.23	1.03	53	Poland is facing a situation similar to Italy, with a lower supply gap and cost advantage. This makes the business case for imports relatively less attractive in the short term.
France	Le Havre; Dunkirk; Rouen	1.29	0.98	75	In the short term, France is anticipated to encounter a substantial hydrogen supply gap as domestic projects scale up. However, the country can utilize its robust ammonia import infrastructure to secure a cost-effective supply from abroad.
Belgium	Antwerp; Ghent	1.92	0.38	109	Locally generated hydrogen in Belgium will continue to be more expensive than imported hydrogen, and funding has been allocated to its ports for expanding infrastructure investments.
Spain	Avilés; Gijón; Bilbao	1.75	-0.46	609	Spain, possessing an optimal renewables profile, is poised to produce affordable hydrogen domestically by 2030. Although the country has moderate ammonia import capacity, it can still capitalize on its advantageous renewable energy conditions for cost-effective hydrogen production.

Portugal	Lisbon;Sines	0.97	-0.23	862	Portugal, endowed with an optimal renewables profile, is set to generate cost-effective hydrogen domestically by 2030. However, the country has limited ammonia import capacity.

[1]Supply gap: Weighing forecast hydrogen demand against project pipelines (negative supply gap value indicates oversupply)

8.6 The European Union's hydrogen plans for 2050

Industry experts project that green hydrogen could fulfill approximately 24% of the global energy demand by the year 2050, translating to annual sales reaching around €630 billion [53]. The proportion of hydrogen in Europe's energy composition is anticipated to rise from its current level of under 2% to approximately 13-14% by the year 2050. The strategy delineates a roadmap titled 'Towards a Hydrogen Ecosystem in Europe: A Roadmap to 2050', structured around three distinct phases [54]:

- In the initial phase spanning 2020-2024, a primary goal is the establishment of a minimum of 6 GW of renewable hydrogen electrolyzers within the EU, facilitating the production of up to one million tons of renewable hydrogen. This phase underscores the imperative to scale up the manufacturing of electrolyzers, including units with capacities of up to 100 MW.
- The subsequent phase, spanning 2025-2030, aims for hydrogen to become an integral component of a unified EU energy system, with the installation of at least 40 GW of renewable hydrogen electrolyzers within the EU by 2030, resulting in the production of up to 10 million tons of renewable hydrogen. During this phase, hydrogen gradually achieves cost competitiveness, necessitating demand activation policies to expedite its adoption in novel applications such as steel manufacturing. Additionally, local hydrogen clusters or hydrogen valleys are envisioned to evolve.
- The concluding phase, covering 2030-2050, envisions the maturity and widespread deployment of renewable hydrogen technologies, extending their reach to encompass all challenging-to-decarbonize sectors where viable alternatives are either not feasible or entail higher costs.

The comprehensive investment requirements for the period spanning 2020 to 2030 encompass various domains:

- Investments in electrolyzers – Estimated between €24-42 billion.

- Expansion of solar and wind production/connection capacity by 80-120 GW- Projected at €220-340 billion.
- Investments in retrofitting approximately half of the existing plants with carbon capture and storage – Approximating €11 billion.
- Investments in hydrogen transport, distribution, storage, and hydrogen refueling stations- Roughly €65 billion.

Investments in production capacities between 2020 and 2050:

- Estimated within the range of €180-470 billion in the EU.
- The strategy underscores that substantial investments will be imperative for adapting end-use sectors to hydrogen consumption and hydrogen-based fuels. For instance, it highlights that converting a typical EU steel installation reaching the end of its life to hydrogen would necessitate an investment of €160-200 million.
- Aviation fuel suppliers are mandated to guarantee that all fuel provided to aircraft operators at EU airports includes a specified minimum proportion of Sustainable Aviation Fuel (SAF) starting in 2025. Additionally, beginning in 2030, a minimum portion of synthetic fuels must be integrated, with both SAF and synthetic fuel percentages progressively escalating until 2050. The obligation entails incorporating 2% SAF by 2025, escalating to 6% by 2030, and reaching 70% by 2050. As of 2030, 1.2% of fuels must also consist of synthetic fuels, increasing to 35% by 2050.

Abbreviations

ARERA: Italian Independent Regulatory Authority for Energy, Networks, and Environment
BCAM: Carbon Border Adjustment Mechanism
CCS: carbon capture and storage
EC: European Commission
EIB: European Investment Bank
EHB: European Hydrogen Bank
ERDF: European Regional Development Fund
ETS: Emissions Trading System
EU: European Union
FID: Final Investment Decision
IPCEI: Important Projects of Common European Interest
IRA: Inflation Reduction Act
JTF: Just Transition Fund
MISE: Ministry of Economic Development
MoUs: Memoranda of Understanding
LCOE: Levelized electricity costs
P2G: Power-to-Gas
P2H: Power-to-Hydrogen
P2P: Power-to-Power
PNIEC: Energy and Climate National Integrated Plan
PPA: Power purchase agreements
PtX: Power-to-X

RES: Renewable Energy Sources
SAF: Sustainable Aviation Fuel
SIFA: Sustainable investment facilitation agreements
SMR: steam methane reforming

Reference

[1] Sadik-Zada ER. Political economy of green hydrogen rollout: A global perspective. Sustainability (Switzerland) 2021;13. https://doi.org/10.3390/su132313464.
[2] Hydrogen n.d. https://energy.ec.europa.eu/topics/energy-systems-integration/hydrogen_en (accessed January 7, 2024).
[3] World Energy Investment 2020 – Analysis - IEA n.d. https://www.iea.org/reports/world-energy-investment-2020 (accessed January 7, 2024).
[4] The Geopolitics of Renewable Hydrogen in Low-Carbon Energy Markets. Geopolitics, History, and International Relations 2020;12:9–44. https://www.ceeol.com/search/article-detail?id=877414 (accessed January 7, 2024).
[5] Hosseini SE. Transition away from fossil fuels toward renewables: lessons from Russia-Ukraine crisis. Future Energy 2022;1:2–5. https://doi.org/10.55670/fpll.fuen.1.1.8.
[6] Wolf A, Zander N. Green Hydrogen in Europe: Do Strategies Meet Expectations? Intereconomics 2021;56:316–23. https://doi.org/10.1007/S10272-021-1008-3/METRICS.
[7] EUR-Lex - 52020DC0301 - EN - EUR-Lex n.d. https://eur-lex.europa.eu/legal-content/EN/TXT/?qid=1594897267722&uri=CELEX:52020DC0301 (accessed January 7, 2024).
[8] EUR-Lex - 52022DC0230 - EN - EUR-Lex n.d. https://eur-lex.europa.eu/legal-content/EN/TXT/?uri=COM%3A2022%3A230%3AFIN&qid=1653033742483 (accessed January 15, 2024).
[9] Project pipeline - European Commission n.d. https://single-market-economy.ec.europa.eu/industry/strategy/industrial-alliances/european-clean-hydrogen-alliance/project-pipeline_en (accessed January 15, 2024).
[10] Consultation on the list of candidate Projects of Common Interest in all infrastructure categories n.d. https://energy.ec.europa.eu/consultations/consultation-list-candidate-projects-common-interest-all-infrastructure-categories_en (accessed January 15, 2024).
[11] European Clean Hydrogen Alliance | European Electrolyser Summit 2022. https://ec.europa.eu/docsroom/documents/50357 (accessed January 15, 2024).
[12] European Clean Hydrogen Alliance | 7th Hydrogen Forum Nov 2023- final report 2023. https://ec.europa.eu/docsroom/documents/57115?locale=en (accessed January 15, 2024).
[13] State of the Union Address by President von der Leyen n.d. https://ec.europa.eu/commission/presscorner/detail/en/speech_22_5493 (accessed January 16, 2024).
[14] EUR-Lex - 52021PC0804 - EN - EUR-Lex n.d. https://eur-lex.europa.eu/legal-content/EN/TXT/?uri=COM%3A2021%3A804%3AFIN&qid=1640001545187 (accessed January 16, 2024).
[15] A Green Deal Industrial Plan for the Net-Zero Age n.d. https://commission.europa.eu/system/files/2023-02/COM_2023_62_2_EN_ACT_A%20Green%20Deal%20Industrial%20Plan%20for%20the%20Net-Zero%20Age.pdf (accessed January 16, 2024).
[16] Home - Green Skills for Hydrogen n.d. https://greenskillsforhydrogen.eu/ (accessed January 16, 2024).
[17] EU raises European Hydrogen Bank bid ceiling to €4.5/kg — but most green H2 projects will have to settle for less | Hydrogen news and intelligence n.d. https://www.hydrogeninsight.com/policy/eu-raises-european-hydrogen-bank-bid-ceiling-to-4-5-kg-but-most-green-h2-projects-will-have-to-settle-for-less/2-1-1509268 (accessed January 18, 2024).

[18] Clean Hydrogen Joint Undertaking | European Union n.d. https://european-union.europa.eu/institutions-law-budget/institutions-and-bodies/search-all-eu-institutions-and-bodies/clean-hydrogen-joint-undertaking_en (accessed January 18, 2024).
[19] Nine hydrogen valleys projects to be funded by the Clean Hydrogen Partnership - H2 News n.d. https://www.hydrogenfuelnews.com/hydrogen-valleys-europe/8557163/?expand_article=1 (accessed January 18, 2024).
[20] For Germany, climate technologies are the best strategy for rapid energy independence | Breakthrough Energy n.d. https://breakthroughenergy.org/news/for-germany-climate-technologies-are-the-best-strategy-for-rapid-energy-independence/ (accessed January 17, 2024).
[21] The Role of Hydrogen in Germany's Energy Transition | n.d. https://www.energypartnership.cn/home/the-role-of-hydrogen-in-germanys-energy-transition/ (accessed January 17, 2024).
[22] Germany's Hydrogen Industrial Strategy n.d. https://www.csis.org/analysis/germanys-hydrogen-industrial-strategy (accessed January 17, 2024).
[23] Neue Ziele auf alten Wegen? Strategien für eine treibhausgasneutrale Energieversorgung bis zum Jahr 2045 - JuSER n.d. https://juser.fz-juelich.de/record/908382 (accessed January 17, 2024).
[24] Schöb T, Kullmann F, Linßen J, Stolten D. The role of hydrogen for a greenhouse gas-neutral Germany by 2045. Int J Hydrogen Energy 2023;48:39124–37. https://doi.org/10.1016/J.IJHYDENE.2023.05.007.
[25] Germany, UK seek to expand cooperation on offshore wind-based hydrogen | Clean Energy Wire n.d. https://www.cleanenergywire.org/news/germany-uk-seek-expand-cooperation-offshore-wind-based-hydrogen (accessed January 17, 2024).
[26] Jülich P. The National Hydrogen Council The Hydrogen Coordination Office n.d. www.wasserstoffrat.de/fileadmin/wasserstoffrat/media/Dokumente/EN/2021-07-02_NWR-Hydrogen_Action_Plan.pdf (accessed January 17, 2024).
[27] The National Hydrogen Strategy 2020. www.bmwk.de/Redaktion/EN/Publikationen/Energie/the-national-hydrogen-strategy.pdf?__blob=publicationFile&v=6 (accessed January 17, 2024).
[28] France's Hydrogen Strategy: Focusing on Domestic Hydrogen Production to Decarbonise Industry and Mobility | IDDRI n.d. https://www.iddri.org/en/publications-and-events/report/frances-hydrogen-strategy-focusing-domestic-hydrogen-production (accessed January 17, 2024).
[29] Industrial groups push for support in France's National Hydrogen strategy | Reuters n.d. https://www.reuters.com/business/energy/industrial-groups-push-support-frances-national-hydrogen-strategy-2023-05-31/ (accessed January 17, 2024).
[30] France | Green Hydrogen Organisation n.d. https://gh2.org/countries/france (accessed January 17, 2024).
[31] The French Hydrogen Strategy - WFW n.d. https://www.wfw.com/articles/the-french-hydrogen-strategy/ (accessed January 18, 2024).
[32] Alstom's hydrogen train tested on the French rail network | La French Fab, the Fabulous French Industry! n.d. https://lafrench-fab.com/news/alstoms-hydrogen-train-tested-on-the-french-rail-network/ (accessed January 18, 2024).
[33] France in the vanguard for green hydrogen | La French Fab, the Fabulous French Industry! n.d. https://lafrench-fab.com/news/france-in-the-vanguard-for-green-hydrogen/ (accessed January 18, 2024).
[34] Hydrogen law and regulation in Italy | CMS Expert Guides n.d. https://cms.law/en/int/expert-guides/cms-expert-guide-to-hydrogen/italy (accessed January 18, 2024).
[35] Italy – HyResource n.d. https://research.csiro.au/hyresource/policy/international/italy/ (accessed January 19, 2024).
[36] Giuli M. Italy in the International Hydrogen Economy 2022. https://www.iai.it/en/pubblicazioni/italy-international-hydrogen-economy (accessed April 14, 2024).
[37] Arera EN: [Translate to English:] Homepage n.d. https://www.arera.it/en (accessed January 19, 2024).

[38] FNM and Alstom present Italy's first hydrogen-powered train | Alstom n.d. https://www.alstom.com/press-releases-news/2023/10/fnm-and-alstom-present-italys-first-hydrogen-powered-train (accessed January 19, 2024).

[39] Spain Renewable Hydrogen Roadmap n.d. https://www.trade.gov/market-intelligence/spain-renewable-hydrogen-roadmap (accessed January 18, 2024).

[40] The Spanish Hydrogen Strategy - WFW n.d. https://www.wfw.com/articles/the-spanish-hydrogen-strategy/ (accessed January 18, 2024).

[41] Carbon Border Adjustment Mechanism - European Commission n.d. https://taxation-customs.ec.europa.eu/carbon-border-adjustment-mechanism_en (accessed January 16, 2024).

[42] New energy statistics 2022 to support the Green Deal - Products Eurostat News - Eurostat n.d. https://ec.europa.eu/eurostat/web/products-eurostat-news/-/cn-20220128-1 (accessed January 17, 2024).

[43] European Clean Hydrogen Alliance - European Commission n.d. https://single-market-economy.ec.europa.eu/industry/strategy/industrial-alliances/european-clean-hydrogen-alliance_en (accessed January 17, 2024).

[44] Renewable and Low-Carbon Fuels Value Chain Industrial Alliance - European Commission n.d. https://transport.ec.europa.eu/transport-themes/clean-transport/alternative-fuels-sustainable-mobility-europe/renewable-and-low-carbon-fuels-value-chain-industrial-alliance_en (accessed January 17, 2024).

[45] Alliance for Zero Emission Aviation n.d. https://defence-industry-space.ec.europa.eu/eu-aeronautics-industry/alliance-zero-emission-aviation_en (accessed January 17, 2024).

[46] Clean Hydrogen Joint Undertaking | European Union n.d. https://european-union.europa.eu/institutions-law-budget/institutions-and-bodies/search-all-eu-institutions-and-bodies/clean-hydrogen-joint-undertaking_en (accessed January 17, 2024).

[47] TEN-E (Trans-European Networks for Energy): CCUS – Policies - IEA n.d. https://www.iea.org/policies/12670-ten-e-trans-european-networks-for-energy-ccus (accessed January 17, 2024).

[48] EU/Angola-first-ever Sustainable Investment Facilitation n.d. https://ec.europa.eu/commission/presscorner/detail/en/ip_23_5774 (accessed January 17, 2024).

[49] COP27: Saudi Arabia targets Europe, Asia-Pacific in global hydrogen push | S&P Global Commodity Insights n.d. https://www.spglobal.com/commodityinsights/en/market-insights/latest-news/electric-power/111522-cop27-saudi-arabia-targets-europe-asia-pacific-in-global-hydrogen-push (accessed January 17, 2024).

[50] EU-Chile agreement sent to Council n.d. https://ec.europa.eu/commission/presscorner/detail/en/ip_23_3682 (accessed January 17, 2024).

[51] Regulation - 2022/2576 - EN - EUR-Lex n.d. https://eur-lex.europa.eu/eli/reg/2022/2576 (accessed January 17, 2024).

[52] The Value of Green Hydrogen Trade for Europe - RMI n.d. https://rmi.org/insight/the-value-of-green-hydrogen-trade-for-europe/ (accessed January 7, 2024).

[53] Why Hydrogen Is Essential in the Global Journey Towards Net Zero | Environment | Sustainability | Feature Story | Panasonic Newsroom Global n.d. https://news.panasonic.com/global/stories/995 (accessed January 19, 2024).

[54] EUR-Lex - 52020DC0301 - EN - EUR-Lex n.d. https://eur-lex.europa.eu/legal-content/EN/TXT/?uri=CELEX:52020DC0301 (accessed January 19, 2024).

CHAPTER 9

Hydrogen technology in Asia and the Pacific: A vision for a cleaner future

9.1 Introduction

Asian nations have traditionally relied on coal and natural gas for their electricity needs, but a notable shift is occurring. Asia contributes to around 60% of global CO_2 emissions and is confronted with a pressing requirement to shift its energy composition by integrating renewable energy and green hydrogen [1]. Many Asian countries have committed to ambitious net-zero and decarbonization objectives. While there has been significant growth in conventional renewable energy projects like onshore wind, offshore wind, solar, and hybrid wind-and-solar in recent years, these alone may not be adequate to achieve the region's net-zero and decarbonization targets. Faced with these challenges, Asian countries are exploring alternative approaches to reduce carbon emissions and decrease dependence on fossil fuels. Similar to other parts of the world, Asia is well-positioned for the advancement of green hydrogen- produced through water electrolysis powered entirely by renewable energy- and green (or low-carbon) ammonia- produced through water electrolysis and separation of nitrogen from the air- to play significant roles. Robust government policies and support, coupled with private sector commitment and investment, propel the emerging green hydrogen sector forward in the Asia Pacific region, encompassing nearly 50 countries at different stages of their hydrogen journey (Table 9.1) [2].

As the prices of renewable electricity continue to decrease, green hydrogen is anticipated to reach cost competitiveness with gray hydrogen and other conventional energy sources. Certain countries in Asia and the Pacific possess substantial renewable energy capacities, such as Australia and India, and exhibit considerable potential for green hydrogen production. By 2030, the cost of green hydrogen in India is projected to decrease to approximately $2 per kilogram, achieving parity with gray hydrogen and other fossil fuels in various applications [3]. Blue hydrogen, while offering some advantages, is not inherently carbon-free and is not expected to undergo the cost reductions predicted for green hydrogen. Therefore, considering the regional environmental and economic conditions, Asia should prioritize investments in green hydrogen production over gray and blue hydrogen.

Hydrogen Diplomacy
DOI: https://doi.org/10.55670/fpll.book/1

Table 9.1 Hydrogen technology strategy in various countries in Asia and the Pacific

Countries	National hydrogen strategies	Highlights
Australia	Available	**Vision 2030:** • 1.3 billion AUD public investment • Cost of clean hydrogen below $2/kg
China	In preparation	• 100 Mt of green hydrogen by 2060 • Deliver 2% of final energy consumption by hydrogen
India	Available	**Vision 2040:** • 5 Mt of green hydrogen by 2030 • Cost of hydrogen below $2/kg
Japan	Available	**Vision 2050:** • 20 Mt hydrogen per year • Power generation using hydrogen at ¥ 17/kWh • 80% GHG emission reduction
South Korea	Available	**Vision 2040:** • 5.26 Mt hydrogen per year • Power 10% of the country with hydrogen by 2030

There is a complementary relationship between the production of renewable energy and green hydrogen, where hydrogen acts as a means of storing energy and can help balance variations in renewable energy generation. For example, it is anticipated that by 2050, the cost of green hydrogen produced from offshore wind resources in the southwest of England may be just half of the current cost of natural gas in the area. If combined with offshore wind energy generation, there's a possibility that green hydrogen could meet all of the region's energy requirements. Considering the unique topographical characteristics of countries in the Asia and Pacific region, nations with significant potential for offshore wind energy production, such as Australia, may experience similar developments [4]. Shifting from traditional fossil fuel-based grey hydrogen to low-carbon hydrogen is crucial for achieving the global climate targets outlined in the Paris Agreement. This transition can be achieved through the adoption of green hydrogen, produced using renewable electricity, and blue hydrogen, which involves natural gas combined with carbon capture, utilization, and storage (CCUS). Despite the growing interest from policymakers and industries worldwide, the market for low-carbon hydrogen is still in its early stages. Currently, only 0.3 Mt out of the total 51 Mt in the industry sector is supplied with low-carbon hydrogen, while the majority comes from grey hydrogen. The production of green and blue hydrogen offers broader benefits for the entire energy system [5]. Green hydrogen aids in integrating intermittent renewable energy sources like solar and wind energy in the electricity sector, with electrolyzers serving as assets for providing flexibility. Moreover, the excess heat generated during green hydrogen production can be utilized in district heating systems, diversifying heat sources. Blue hydrogen production aligns with existing industries, including oil and gas, facilitating a just energy transition by allowing the transfer of infrastructures, skills, and jobs. Additionally, the CO_2 captured in the process can find various applications, serving as a carbon point source for the production of electrofuel via Power-to-X technology, supporting enhanced oil recovery in field production, or finding use in the food industry [6]. The Asia and the Pacific (APAC) region presently represents half of the global industrial hydrogen demand, with the People's Republic of China (PRC) alone accounting for 33% (17 MtH_2) for the production of ammonia and methanol. Japan has been at the forefront of hydrogen technology development and has assumed a leadership role in the advancement of the hydrogen industry. Japan took the lead in 2017 by announcing its national hydrogen strategy encompassing various sectors, making it the world's first country to establish a comprehensive national hydrogen framework. Subsequently, other APAC countries, including the PRC, the Republic of Korea, New Zealand, and Australia, have also formulated their respective strategies to support the hydrogen industry. Some of these APAC countries' national policies also address the industrial applications of hydrogen. Figure 9.1 demonstrates hydrogen demand in various APAC countries compared to the whole world.

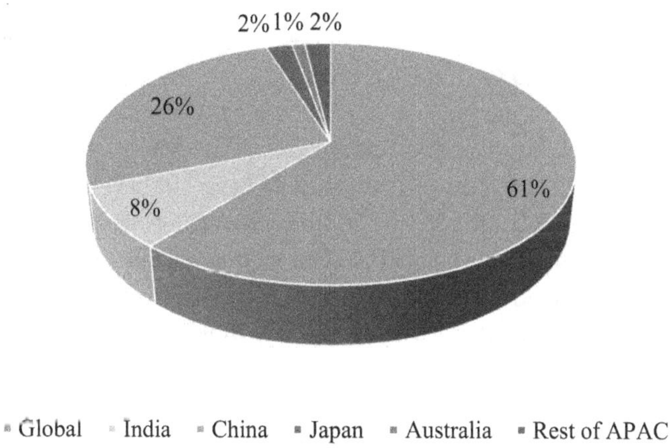

Figure 9.1 Hydrogen demand in various APAC countries compared to the whole world

The APAC region is emerging as one of the rapidly expanding markets for low-carbon hydrogen, with numerous countries actively investing in the advancement and utilization of this environmentally friendly energy carrier. The increasing demand for hydrogen in the APAC region is propelled by a blend of factors, including the imperative to curtail greenhouse gas (GHG) emissions, enhance energy security, and foster economic growth. The APAC region holds a pivotal position in the global hydrogen market. Among the APAC countries, the PRC leads as the largest consumer, boasting an annual consumption of 23.9 Mt, followed by India, the Russian Federation, Japan, Indonesia, and the Republic of Korea. In the

PRC, a significant portion of the produced hydrogen serves as feedstock in oil refineries and chemical production, primarily derived from fossil fuels, with approximately 60% sourced from coal and 25% from natural gas. In India, nearly 99% of grey hydrogen finds application in petroleum refining and ammonia production for fertilizers [7]. Notably, two major refiners, Reliance Industries Limited and Indian Oil Corporation Limited, contribute to over 70% of the national hydrogen production in India [8]. The APAC region stands out as one of the most rapidly expanding markets for low-carbon hydrogen. Notably, this region hosts some of the world's largest hydrogen markets, including Japan, the Republic of Korea, the People's Republic of China (PRC), and Australia, along with burgeoning markets in Southeast Asia and India. The early considerations of export/import capabilities have already left a notable imprint on the formulation of national hydrogen policies in countries marked by significant industrialization and energy needs, like Japan and the Republic of Korea, as well as those with robust export-oriented foundations, such as Australia. This intricate interplay between export and import strategies is reshaping the global energy landscape, forging new trade relationships and routes among nations.

Chile presently boasts the world's most cost-effective production of green hydrogen, ranging from €3.5 to €3.75 per kilogram. Over the next thirty years, a substantial increase in the number of countries capable of producing renewable-derived hydrogen for just one euro per kilogram is anticipated. In the APAC region, Australia, the PRC, and India are projected to have the lowest production costs, ranging between 1 and 1.25 euros per kilogram by 2050. Comparatively, Japan and the Republic of Korea are expected to have relatively higher costs, ranging from 2.50 to 2.75 euros per kilogram [9]. While the PRC is poised for rapid development in renewable energy-based green hydrogen due to its extensive renewable power capacity, the country may prioritize self-consumption, potentially neither exporting nor importing hydrogen given its substantial domestic demand. As the demand for clean hydrogen rises, certain APAC countries are anticipated to emerge as significant hydrogen suppliers to other nations globally. Australia, with abundant natural gas, coal reserves, and renewable energy potential, presents promising prospects for both green and blue hydrogen. India, aiming to establish itself as a major player in the global hydrogen market, plans to leverage its experience in renewable energy, particularly solar power, for low-carbon hydrogen production. The country's extensive, cost-effective workforce and proficiency in renewable energy technology development position it as an attractive source of green hydrogen for international partners. India has already forged partnerships for importing green hydrogen from countries like Australia and Japan. On the other hand, Japan and the Republic of Korea, being major hydrogen importers in the APAC region, lack the natural resources for low-carbon hydrogen production and rely on imports to meet their escalating demand for clean energy. Both nations are actively exploring import avenues, particularly from countries like Australia and Malaysia, endowed with abundant natural resources suitable for low-carbon blue and green hydrogen production. As the APAC region intensifies its investments in the hydrogen economy, the growth of hydrogen trading is expected to gain momentum. The increased demand for low-carbon hydrogen, recognized as a clean and sustainable energy source, is set to propel the expansion of the hydrogen market. Considering the prevailing hydrogen consumption patterns, renewable energy generation capacities, geographical positioning, and industrial frameworks of each nation, it is evident that the PRC, India, Japan, the Republic of Korea, and Australia are poised to assume pivotal roles in the hydrogen value chain within the APAC region. Consequently, the subsequent sections will delve into an examination of various key countries in the APAC region, focusing on their low-carbon hydrogen production and consumption. This exploration will encompass insights into national hydrogen strategies and ongoing project developments.

9.2 China, the largest hydrogen producer in the world

The PRC holds the position of the world's largest hydrogen producer, producing approximately 33 MtH_2 per year, with 12 MtH_2 annually meeting industrial quality standards, as indicated by the National Development and Reform Commission (NDRC) in 2022. Nearly 80% of China's annual hydrogen production originates from environmentally impactful coal gasification and steam methane reforming (Figure 9.2) [10]. The remaining 20% is derived from industrial by-products such as hydrogen generated during chlorine and caustic soda production, coke oven gas, and propane dehydrogenation [11].

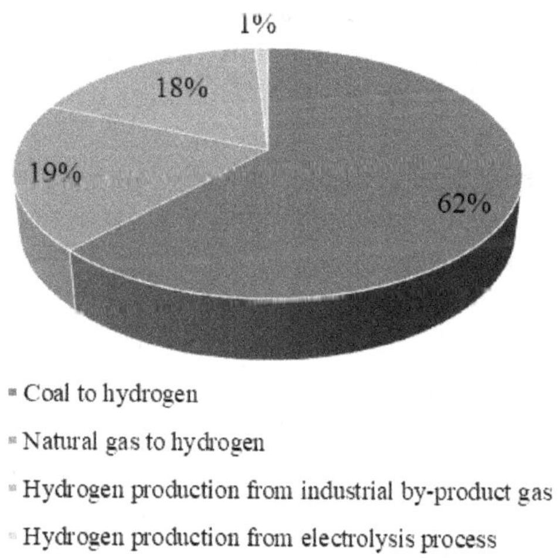

Figure 9.2 Percentage of various hydrogen production processes in China [10]

Presently, the majority of hydrogen is generated from fossil fuels, serving as feedstocks in refineries or chemical facilities. Following a gradual initiation, China has emerged as a frontrunner in deploying electrolyzers. In 2020, China's contribution to global electrolyzer capacity for dedicated hydrogen production was below 10%, primarily seen in small-scale demonstration initiatives. However, by 2022, China's installed capacity surpassed 200 MW, constituting 30% of the global capacity, inclusive of the world's largest electrolysis project, boasting a capacity of 150 MW. Anticipated to achieve 1.2 GW in installed electrolyzer capacity by the close of 2023, accounting for half of the global capacity, China is set to maintain its leading role, having initiated another groundbreaking electrolysis project this year with a capacity of 260 MW. Furthermore, China is positioned to solidify its dominant position in electrolyzer deployment, with over 40% of globally approved electrolysis projects being attributed to the country.

In March 2022, the Chinese government unveiled the inaugural long-term development plan for hydrogen in the PRC, outlining strategies for the period 2021-2035 [12]. The plan outlines robust strategies and objectives for the hydrogen market in China, with focusing on [13]:

- Ensuring the sustainable development of the hydrogen market.
- Formulating strategies for the industrial by-product H_2 and renewable H_2.
- Facilitating the orderly development of the market for Fuel Cell Vehicles (FCVs).
- Exploring diverse, industry-decarbonized applications for hydrogen.

The Chinese plan outlines a phased strategy for developing a domestic hydrogen industry and advancing technologies and manufacturing capabilities, with a key emphasis on the country's commitments to carbon peaking and neutrality. This long-term plan extends and builds upon hydrogen-related visions and strategies outlined in recent documents, including the 14th Five-Year Plan (2021–2025), where hydrogen is recognized as a "frontier" sector and one of the six industries targeted for focused advancement. China is increasingly fostering the production and adoption of lower-emission hydrogen to meet energy demands while concurrently decarbonizing its economy. China's pursuit of clean hydrogen marks a significant stride toward achieving its climate objectives. As the current largest global producer of hydrogen, generating approximately 33 million tons (Mt), the majority of which is produced from fossil fuels in refineries or chemical facilities, China is now steering toward cleaner alternatives. In China, the growth in hydrogen usage is relatively moderate, approximately 0.5%. Nevertheless, China retains its status as the largest individual consumer of hydrogen by a significant margin, representing nearly 30% of worldwide hydrogen consumption, which is more than twice the consumption of the second-largest consumer, the United States. China dominates the global industrial consumption of hydrogen, accounting for 35% of the total usage, with the Middle East following at 14%, North America at 10%, and India at 9% [14].

The China Hydrogen Alliance envisions China's hydrogen demand reaching 35 Mt by 2030 (constituting at least 5 percent of the Chinese energy supply) and escalating to 60 Mt by 2050 (representing 10 percent) [15]. The plan adopts a phased approach to foster the growth of the domestic hydrogen industry, advance technologies, and enhance production capabilities. China aims to produce 100,000 to 200,000 tons of green hydrogen annually and put around 50,000 hydrogen-fueled vehicles on the roads by 2025 [16]. Significantly, the 5+1 city alliances and regions, backed by national funding-namely Beijing, Shanghai, Guangdong, Hebei, and Henan for the FCV Demo program, and Shandong for the "Hydrogen Community" program- aim to introduce 48,000 hydrogen fuel cell vehicles by the close of 2025. In an optimistic scenario, China could potentially embrace 80,000 to 100,000 FCVs by 2025 if all local governments successfully attain their FCV targets. However, with the unveiling of the national plan, it is evident that certain local targets may fall short or not materialize as anticipated [13]. The newly introduced Chinese policy is set to operate for an initial four-year period, focusing its subsidies on specific cities and city clusters. Each city cluster has the potential to receive subsidies of up to 1.7 billion Yuan (equivalent to US$250 million), contingent upon their success in achieving various targets related to technology application and the promotion of fuel cell vehicles [17].

Currently, China stands as the world's leading market for fuel cell trucks and buses and the third-largest market for fuel cell electric vehicles. China's emphasis on heavy-duty vehicles and its significant contribution to the deployment of fuel cell trucks results in a scenario where, despite having only 20% of all Fuel Cell Electric Vehicles (FCEVs), the country consumes over 50% of the hydrogen used in road transport [18]. In 2022, China saw the addition of more than 200 fuel cell cars, a notable development considering the country's previous focus on deploying FCEVs primarily in heavier vehicle segments. As of June 2023, China has become the leading market for fuel cell light commercial vehicles, boasting over 800 units in deployment.

The number of fuel cell trucks witnessed rapid growth, surging by over 60% in 2022 globally and surpassing 7,100 by the year's end. In the first half of 2023, this stock further increased, reaching over 8,000. The predominant market for these trucks is China, which now commands over 95% of the global fuel cell truck market. This significant growth in China is attributed largely to a more than fivefold increase in heavy-duty fuel cell trucks, driven by favorable policies and supporting infrastructure. The global inventory of fuel cell buses experienced a comparable growth rate to that of light-duty vehicles, marking a roughly 40% increase in 2022 compared to the preceding year. As of June 2023, the global count stands at approximately 7,000 fuel cell buses, with China housing about 85% of this total. China contributed around 1,300 new fuel cell buses in 2022. Following China, Europe boasts the second-largest stock, trailed by Korea and the United States.

As of June 2023, there are approximately 1,100 operational hydrogen refueling stations (HRS) worldwide, and plans are in place for the construction of several hundred additional stations. Out of the current stations, China hosts over 300, Europe has about 250 HRS, and both Korea and Japan each have around 180 (Figure 9.3). In response to the increasing demand for mobile fuel cells, Hyundai inaugurated its initial hydrogen fuel cell manufacturing facility beyond Korea in 2023, located in Guangzhou, China, and has an annual production capacity of 6,500 hydrogen fuel cell systems [19].

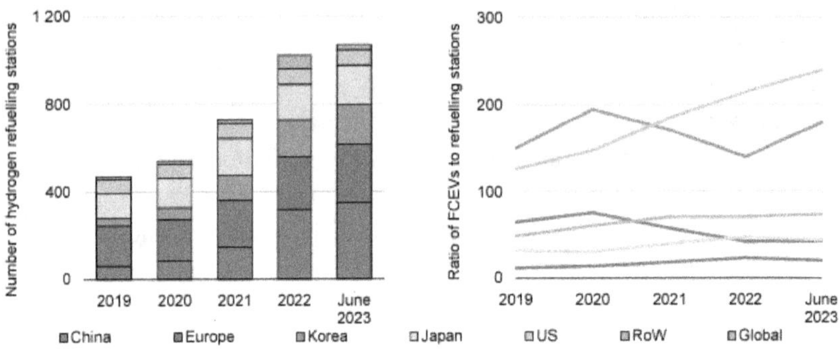

Figure 9.3 Distribution of hydrogen refueling stations across regions and the ratio of fuel cell electric vehicles to refueling stations, spanning the period from 2019 to June 2023 (RoW: rest of the world) [14]

The year 2021 was notable, marked by the initiation of a significant project in China boasting an electrolysis capacity of 150 MW. In 2023, China is poised to maintain its leadership in deploying the largest electrolyzers, as a 260 MW facility by Sinopec commenced operations in July 2023, with additional projects totaling over 550 MW currently in the construction phase [20]. While China currently excels in alkaline water electrolysis (AWE) technology, which is cost-effective but less compatible with intermittent renewable energy sources than proton exchange membrane (PEM) electrolyzers, the plan raises questions about China's ability to acquire expertise in advanced technologies and surpass other economies in hydrogen-related innovation race [21]. Manufacturing capacity is another key focus outlined in the Chinese long-term plan.

In 2020, China represented 8 percent of the global electrolyzer stock and 35 percent of the global manufacturing capacity for electrolyzer equipment and components [22]. A comparison of four various methods of water electrolysis technology is given in Table 9.2 [23]. As of the end of 2022, alkaline electrolyzers constituted 60% of the global installed capacity, with PEM electrolyzers making up approximately 30%. Future trends suggest a shift in market share favoring PEM electrolyzers, although specific technologies for many upcoming projects remain undisclosed.

Table 9.2 A comparison of four various methods of water electrolysis technology in China

Electrolytic diaphragm	30% KOH asbestos film	Proton Exchange Membrane	Anion exchange membrane	Solid state oxide
Hydrogen production purity	≥99.8%	≥99.99%	≥99.99%	–
Operation difficulties	Need to control differential pressure, need to dealkalize	Fast start/stop, water vapor only	Fast start/stop, water vapor only	No change in start/stop, water vapor only
Maintainability	Strong alkali corrosive	Non-corrosive media	Non-corrosive media	–
Environmental requirements	Asbestos air pollution	Non-polluting	Non-polluting	–
Technology Maturity	Industrial Scale-up	Rapid Commercialization	Initial demonstration run	Laboratory phase
Hydrogen production efficiency (kWh/kg of hydrogen)	50–70%	50–83%	–	45–55%
System Cost (USD/kW)	500–1000	700–1400	–	2000

Solid oxide electrolyzers (SOEC) presently contribute less than 1% to the overall installed capacity. Geographically, China and Europe each held a one-third share of the installed capacity by the end of 2022, while the United States and Canada collectively represented 10%. Since the commencement of 2023, China has made substantial strides in water electrolysis-based hydrogen production, with the anticipated installed electrolyzer capacity reaching 1.1 GW by the year's end, constituting 50% of the global share. In June 2023, the initial demonstration of direct electrolysis of seawater took place on an offshore platform in Xinghua Bay, China [24]. A notable development is the commissioning of one of the world's largest electrolyzer plants, the Chifeng Ammonia Demonstration Project (completed by the end of 2023), boasting an annual hydrogen production of 24,200 t H_2/year (approximately 140 MW of electrolyzer capacity). By 2024, eight electrolytic hydrogen projects currently under construction with electrolyzer capacities exceeding 125 MW are slated to commence operations in China. Among these, the Sinopec-planned electrolytic hydrogen project at its Ordos chemical facility could rank among the world's largest, targeting 30 kt of annual hydrogen production (equivalent to about 350 MW of electrolysis, assuming a 50% capacity factor).

Project announcements for electrolyzer developments in China align with broader global trends, anticipating significant growth in the coming years. In comparison to 2023, the expected installed electrolyzer capacity in China is projected to triple to 3.3 GW by 2024 and nearly reach 5.4 GW by 2025 (Figure 9.4).

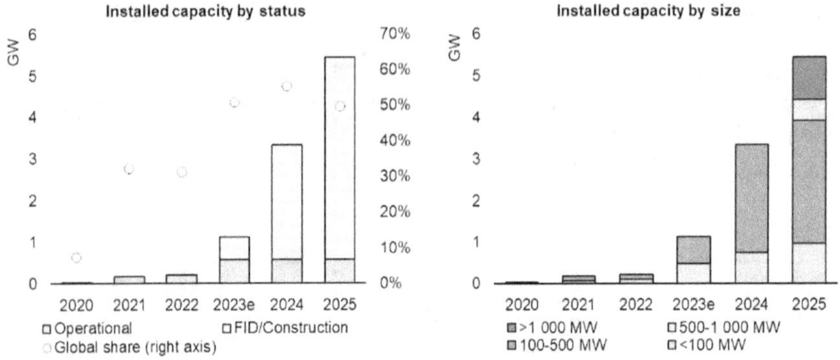

Figure 9.4 Electrolyzer capacity in China [14]

The average size of hydrogen projects is experiencing rapid growth, particularly in China, where an increasing number of projects fall within the 100-500 MW capacity range. Approximately 60% of the projects announced in 2023 in China fall into this category. A notable example is the Kuqa project by Sinopec, boasting a 260 MW capacity and having commenced operations in June 2023. Furthermore, plans are in place to install larger-scale facilities ranging from 500-1,000 MW starting in 2025. By that year, gigawatt-scale projects are expected to constitute almost 20% of the capacity that has reached at least the Final Investment Decision (FID) stage. In addition to advancements in electrolyzer deployment, Chinese companies have developed more extensive and efficient electrolyzers. In December 2022, Peric Hydrogen Energy Technologies introduced the largest single-unit hydrogen alkaline electrolyzer, featuring a hydrogen capacity of 2,000 Nm^3/h or about 9 MW. Furthermore, Longi Hydrogen unveiled a new alkaline model in February 2023 with a direct current electricity consumption of 4 kWh/Nm^3 (equivalent to 44.5 kWh/kg H_2), demonstrating greater efficiency compared to most electrolyzers available in the market [25]. The question arises as to whether China will emerge as a dominant global manufacturer of electrolyzer equipment and components, especially for more advanced technologies. Additionally, the potential emergence of a geo-economic issue between China and the West in the global supply chain for hydrogen technologies, akin to other clean energy technologies, remains uncertain. Despite the absence of a national strategy, several state-owned enterprises (SOEs) in China have been investing in hydrogen projects, and various subnational governments have identified hydrogen as a key economic priority or formulated hydrogen development plans. The long-term commercial viability of these SOE projects and the alignment of subnational plans with national policymaking for shaping the country's hydrogen economy remain unclear [26,27]. Employing low-carbon methods for hydrogen production and integrating clean hydrogen across diverse industrial sectors holds the potential to alleviate China's carbon emissions. An illustrative example is the Chinese government's projection that achieving its 2025 goal for renewable-based hydrogen could lead to an annual reduction in the country's carbon emissions, ranging from one to two million tons [28].

9.2.1 Challenges in green hydrogen production in China

More than 70% of the expenses associated with green hydrogen production are attributed to electricity generation. When the cost of renewable energy-derived electricity drops to 0.15 yuan (RMB) per kilowatt-hour (kWh), the price of green hydrogen could decrease to 16 RMB per kilogram, aligning with the cost of blue hydrogen. With electricity priced at 0.4 RMB per kWh, the production cost of hydrogen from AWE is around 30 RMB per kg, while the hydrogen production cost from PEM is approximately 40 RMB per kg. Table 9.3 summarizes the cost structure of AWE and PEM.

Table 9.3 Cost structure of green hydrogen production by AWE vs. PEM

Hydrogen production method	Electricity cost	Equipment depreciation cost	Operation and maintenance cost	Construction depreciation cost	Other costs
AWE	85.7%	6.6%	5.2%	0.4%	2.1%
PEM	53%	27.9%	17.7%	1.3%	0.1%

There are two primary drivers for reducing green hydrogen costs. Firstly, lowering the expense of renewable electricity is pivotal in reducing the overall cost of green hydrogen. The second step involves eliminating technical barriers to green hydrogen development [29]. Breakthroughs in crucial electrolyzer technologies, such as catalysts and PEMs, will not only decrease electrolyzer prices through independent development but will also reduce dependence on imports. Improving the efficiency of electricity utilization holds significance. Energy losses constitute a considerable portion of the green hydrogen cost, with energy losses exceeding 30% during electrolysis and reaching up to 50% in hydrogen fuel cell applications [23]. Reducing the cost of green hydrogen hinges on autonomously developing fundamental materials like PEM, Perfluorosulfonic Acid Resins (PFSR), and catalysts, along with essential components such as the gas diffusion layer and bipolar plate. The manufacturing of PEM and PFSR poses technical challenges [30]. In China, the timeframe for large-scale PEM production research and development is relatively short, leading to the importation of PEM electrolyzers, PEMs, and

PFSRs. Most PEM electrolyzers utilize DuPont Corporation's perfluorinated sulfonic acid proton exchange membrane. Sulfonated polymer membranes form the basis for perfluorosulfonic acid PEMs [31]. Their performance is influenced by four factors: good proton conductivity, electro-permeability of water molecules in the membrane, gas permeability in the membrane, and electrochemical stability [32]. To minimize proton polarization during the transfer process, sustain electrochemical performance, and enhance mass transfer capacity, the membrane must be ultra-thin (up to 10 μm thick) and robust to prevent fatigue damage, chemical degradation, and even membrane failure under high load conditions. Membrane preparation techniques need to balance proton conductivity, hydrogen-oxygen cross-permeation, and mechanical stability under high-pressure differentials while reducing membrane thickness [33]. China still faces technical barriers in synthesizing and sulfonating perfluorinated substances, as well as challenges with polymer denaturation, degradation, and proton polarization during the membrane formation process. Consequently, there are gaps in key technical indicators such as membrane thickness, energy consumption per unit, and the stability and reliability of domestically manufactured PEMs. The catalysts essential for hydrogen production through PEM electrolysis significantly contribute to the high cost of green hydrogen [34]. The electrolyzer's anode operates in a highly acidic environment, requiring a catalyst with corrosion resistance and overpotential reduction [35].

In China, PEM electrolyzers commonly employ precious metals like platinum and iridium as anode oxygen precipitation catalysts, which poses challenges in ensuring the effectiveness of domestically synthesized catalysts. The high cost of these precious metal catalysts, constituting 10–40% of the total hydrogen production cost from electrolytic water, poses a significant constraint on the future growth of the green hydrogen industry, as the import of expensive platinum and iridium reserves is required [36]. Addressing issues such as reducing precious metal concentrations, enhancing catalytic functionality, maintaining stable performance, and extending catalyst lifespan has become a focal point in scientific research within China's green hydrogen electrocatalyst field. The ongoing expansion of the green hydrogen industry is anticipated to drive demand for PEM, PFSR, and related chemical materials and equipment, subsequently bolstering technological development and localization capabilities. For instance, Shandong Dongyue Future Hydrogen Energy Materials Co. has achieved mass production of PFSR and PEM, with its PEM demonstrating performance comparable to imported membranes [37]. In 2021, the China National Power Investment Group Hydrogen Energy Technology Development Co. inaugurated a hydrogen fuel cell PEM production line in Wuhan.

In China, the green hydrogen infrastructure lacks a unified system, operating within the framework of a hazardous chemical management system despite hydrogen energy being categorized as renewable energy. This adherence to hazardous chemical regulations complicates the coordination of planning and construction for essential support infrastructures like hydrogen refueling stations and transmission pipelines. Consequently, the efficient transmission and utilization of hydrogen energy, especially for large-scale cross-regional deployment, are severely hampered. Green hydrogen-supporting facilities are typically situated in remote areas or industrial parks for hazardous chemicals due to these regulatory constraints, leading to increased investment costs for storing and transporting green hydrogen.

Additionally, administrative permission procedures for the green hydrogen industry in China are intricate, involving challenges from infrastructure construction to project operation. Lengthy approval processes and complex procedures, exacerbated by the absence of specific policy regulations and technical standards, contribute to delays. The green hydrogen sector requires four key administrative licenses- fire protection approval, special equipment permits, hydrogen sales and operation permits, and a hydrogen chemical park production permit- issued by different authorities, such as the Housing and Construction, Market Supervision, Emergency Management, and Energy departments. These procedures involve environmental evaluations, land planning and use, safety assessments, and other demonstration aspects [10].

Uneven deployment is observed in China's green hydrogen industry chain, necessitating a well-organized industrial chain encompassing renewable energy generation, hydrogen production, storage, and consumption. Geographical mismatches between the production and consumption ends of the green hydrogen industry chain pose challenges to effective industrial system integration [38]. For instance, manufacturing plants are often located in regions abundant in renewable energy, like central and western China, while the main consumption demands lie in the southeast coastal region. Current green hydrogen storage and transportation methods primarily rely on high-pressure gaseous approaches, like tube trailers, with limited storage capacity and high risk. Mainstream transportation methods, such as tube trailers, face efficiency challenges and long-distance restrictions, limiting the expansion of the industrial chain. The operational or under-construction renewable energy hydrogen refueling stations are concentrated in specific regions like the Yangtze River Delta, Pearl River Delta, and the Beijing-Tianjin-Hebei region, indicating that the support infrastructure is still in its early stages.

Lastly, comprehensive industrial convergence applications for green hydrogen are in the nascent stages in China. The majority of hydrogen usage is concentrated in petrochemical industries, particularly synthetic methanol, synthetic ammonia, and petroleum refining. In contrast, the proportion of hydrogen used in transportation, construction, or daily consumption is relatively low, with less than 0.1% utilized in fuel cells.

9.2.2 China's goals for green hydrogen development

China aims to establish clear regulations specifying the proportion of green hydrogen utilization across major industries while progressively reducing dependence on fossil energy. Simultaneously, there is a need to iteratively enhance laws and regulations governing the green hydrogen sector, including safety protocols for hydrogen production and technical standards for fuel cell vehicles. China must devise a comprehensive plan for the green hydrogen energy industry, systematically advancing the construction of a green hydrogen production, supply, storage, and marketing system. This foundational work can be coupled with coordinated industry and regional development planning. Within the framework of the dual carbon goals, China envisions the evolution of its green hydrogen industry through three key stages:

- Start-up phase (Present–2030): During this phase, reliance on grey hydrogen and industrial by-product hydrogen will form the basis of support for innovative hydrogen technologies. The initiation of the green hydrogen energy industry will gradually take place.
- Accelerated development phase (2030–2050): As green hydrogen becomes more affordable and sees improvements in trading markets and infrastructure, this phase will witness breakthroughs in key technologies related to hydrogen production, storage, and transportation. Application scenarios for green hydrogen will expand into sectors such as steelmaking, clean transportation, and energy storage. Diversification of end consumption methods is expected. The proportion of fossil energy hydrogen production will progressively decrease to zero, with new technologies like renewable green hydrogen taking the lead. This phase will witness the establishment of an integrated system for green hydrogen energy supply, heat supply, and power supply.
- Marketization phase (2050–2060): In this mature phase, green hydrogen technology will see extensive application across various challenging decarbonization fields. The green energy strategy will become a pivotal component of the national low-carbon energy strategy. Grey hydrogen will be phased out from the hydrogen end-consumption market, and hydrogen supply at refueling stations, especially in megacities like Beijing and Shanghai, will transition to predominantly blue hydrogen and green hydrogen, reaching 100% in these key refueling locations [39].

To meet the dual carbon objectives, it is imperative to explore the developmental trajectory of China's green hydrogen industry. This involves advancing green hydrogen technologies, refining industry planning, fostering the entire industrial chain, and enhancing applications. The initial focus should be on enhancing catalyst efficiency, optimizing platinum utilization, minimizing platinum usage, seeking alternatives to platinum catalysts, and diminishing dependence on imports. Additionally, there is a target to reduce platinum catalyst consumption to below 0.25 g/kW by 2030, with a future aim of achieving less than 0.1 g/kW. Technological breakthroughs in materials like PFSR and PEMs are crucial, with increasing demand anticipated as the cost of locally substituted PEMs decreases and economies of scale take effect. Within the incremental market scenario, the cost of PEMs is expected to gradually decrease to $214/ton by 2050.

China has outlined intentions to increase its solar and wind generation capacity, with a target of doubling the capacity from around 600 gigawatts (GW) in 2020 to 1,200 GW by 2030 [40]. The China Hydrogen Alliance has additionally forecasted that the production of hydrogen based on renewable sources could potentially reach 100 million tons by 2060, constituting 20 percent of China's final energy consumption. There is a possibility of substantial growth in renewable-based hydrogen production in China between 2025 and 2060.

Regarding the industrial sector, the PRC hydrogen plan aims to augment the application of hydrogen to replace fossil fuels and facilitate the transition of industries like synthetic ammonia, synthetic methanol, refining, coal-to-gas, and others from high-carbon to low-carbon processes. Additionally, the plan emphasizes research and development in the Direct Reduced Iron (DRI) pathway within the iron and steel industry.

The green hydrogen sector faces challenges in developing storage and transportation technologies that are simultaneously safe, efficient, and cost-effective. The extensive existing natural gas pipeline network in the PRC also serves as a crucial means of hydrogen transportation. The policy further encourages pilot demonstrations involving the blending of hydrogen into the existing gas pipeline networks. Global hydrogen pipelines, with a total length of only 5,000 km, significantly lag behind the extensive 3 million km of natural gas pipelines, prompting efforts to explore the safe and efficient repurposing of natural gas pipelines for hydrogen transport, particularly following a German court decision allowing this [41]. In China, experimental initiatives have been implemented to deliver hydrogen-blended natural gas through pipelines [42]. Despite the comprehensive natural gas transmission network, the pure hydrogen pipeline network spans only about one hundred kilometers. Noteworthy projects, such as the construction of regional hydrogen transmission pipeline networks in Weifang City, Shandong Province, are underway. According to the China Natural Gas Development Report 2022, the total length of natural gas pipelines has surpassed 116,000 km, demonstrating a well-connected major pipeline network. However, the integration of advanced technology and the widespread adoption of natural gas blending with hydrogen are still at early stages. Addressing the need for large-scale application requires improvements in technical systems, operational standards, and other essential aspects. Policymakers are encouraged to eliminate barriers to research and development in hydrogen transmission technology. In the realm of hydrogen energy applications, a critical aspect is hydrogen refueling stations, where imported compressors are predominantly used. There is a call to expedite technological advancements in the domain of liquid hydrogen refueling stations.

China's inaugural cross-regional hydrogen pipeline is incorporated into its 2023 Implementation Plan for the Construction of a National Oil and Gas Network. This pipeline is designed to convey hydrogen from Ulanqab in North China's Inner Mongolia Autonomous Region to Beijing, spanning a distance of over 400 km and boasting a capacity ranging from 100,000 to 500,000 tons/year [43]. The state-owned China Petroleum Pipeline Engineering Corporation proposed in 2023 that the nation could potentially require 6,000 km of hydrogen pipelines by the year 2050 [44].

The Chinese government's recent plan underscores a growing emphasis on research and innovation in hydrogen-related technologies. In August 2023, China's National Standards Commission issued comprehensive standard guidelines encompassing safety and operational aspects related to hydrogen production, storage, transport, and end-use. This emphasis reflects China's ambition to address existing deficiencies in expertise, experience, and infrastructure for producing low-carbon hydrogen. Nevertheless, there is a lack of a distinct development policy and industrial roadmaps for the green hydrogen sector aligned with China's national and local dual carbon goals. Specifically, there is a deficiency in specific measures outlining the phased development of the green hydrogen industry, especially given the current prevalence of grey hydrogen. The allocation and utilization of green hydrogen across major industries at various stages of the dual carbon goals lack a clear definition. Moreover, there is an absence of a unified national roadmap for the development of the green hydrogen industry. Local government policies pertaining to green hydrogen, often driven by local economic interests, fall short of fully recognizing the pivotal role of green hydrogen in reshaping the energy landscape, fostering economic growth, and influencing lifestyles [10]. Neglecting the varying levels of local economic and industrial development, the influx of the green hydrogen industry may lead to redundant infrastructure and unhealthy competition, hindering the establishment of a robust green hydrogen industry chain. The development trajectory of the green hydrogen industry should be tailored to local conditions due to distinct industrial bases. However, many local authorities are currently fixated on hydrogen fuel cell vehicles.

9.2.3 Improvement of the clean hydrogen industry chain in China

As outlined in the Medium- and Long-Term Plan for the Development of the Hydrogen Energy Industry (2021–2035), China envisions a pivotal role for hydrogen energy in its future national energy system, with a focus on realizing the benefits of green hydrogen in the broader energy structure transformation. The effective promotion of hydrogen station construction, integration of the hydrogen industry, hydrogen production, establishment of hydrogenation stations, and infrastructure development are deemed essential. Local governments, for instance, can enhance administrative approval management by implementing streamlined procedures such as Green Channels or One-Stop processes to boost efficiency and allocate space for hydrogen refueling facilities. Shanghai's corresponding plan (2022–2035) encourages the provision of space for hydrogen refueling facilities within newly constructed gas refueling and charging stations. Public facilities, including standalone hydrogen refueling stations, fall within the scope of consideration. In cases where existing refueling, gas filling, and charging stations meet specified standards and safety criteria, there is no need for separate approval procedures to obtain land and construction licenses for hydrogenation stations. The green hydrogen industry chain encompasses the entire process from preparation, storage, and transportation to refueling and utilization, allowing green hydrogen facilities to be integrated into existing land [45]. Comprehensive support policies and planning for the entire industrial chain of the green hydrogen sector should be devised, treating hydrogen energy as a distinct energy category. Improvements are sought in regulations, testing mechanisms, and overall support. The integration of green hydrogen with renewable energy sources, substantial carbon reduction in industries, and diverse applications in transportation, aerospace, energy, and other industrial sectors should be actualized [46].

The optimization of the approval process for the entire green hydrogen industry chain, along with the construction of supporting facilities, is essential. Collaborative efforts among China's Housing and Urban-Rural Development, Emergency Management, Market Supervision, and other relevant authorities are needed to establish administrative regulations and a list of Administrative Approval Powers aiming to enhance administrative approval efficiency. Legal conditions and procedures for fire protection approval, filling, hydrogen station operation, and hydrogen production should be clearly outlined. Improved collaboration among different administrative departments is crucial. Subsequently, the issuance of pertinent policies can augment financial support for green hydrogen businesses [47]. Enterprises engaged in the production of green hydrogen or its use for local hydrogen filling stations could receive subsidies based on the price difference between the average price of industrial byproduct hydrogen and the annual hydrogen supply. This standard subsidy should be periodically adjusted in line with national policies and the prevailing local development context. To alleviate the costs associated with renewable energy generation and power grid connection, support policies for green hydrogen enterprises should be introduced. Additionally, local industry associations play a vital role in facilitating negotiations on electricity prices between the power sector and entities within the green hydrogen industry. Collaboration between the government, banks, insurance companies, and research organizations could result in the implementation of preferential loan policies for enterprises or institutions involved in the green hydrogen chain, as well as those engaged in innovative technological research, development, and equipment [48].

Another step to improve China's clean hydrogen industry chain is the enhancement of the financial support mechanism. This stage involves the implementation of tax incentives for enterprises or research institutions involved in the research and development of green hydrogen technology. Specifically, these incentives include targeted grants for green hydrogen equipment and projects. As outlined in the Catalogue of Preferential Income Tax for Public Infrastructure Projects released by the State Taxation Administration (STA), the green hydrogen industry qualifies for a three-fold exemption and a three-hundred-percentage-point income tax preference policy. The Income Tax Preference Policy entails a waiver of corporate income tax for enterprises during the initial three years of operation, followed by a 50% reduction in tax from the fourth to the sixth year. The primary objective of this policy is to stimulate business investments in the essential infrastructure and environmental projects crucial for the advancement of the green hydrogen industry. Consequently, the implementation of a preferential policy for green hydrogen can significantly support local entrepreneurs in fostering industrial development [49,50].

In addition to the tax incentives, a crucial measure to be implemented involves enhancing the safety supervision of the entire green hydrogen industry chain. Given hydrogen's low density and small molecule size, which can lead to potential leaks and explosions, prioritizing safety is paramount for the future progress of the green hydrogen sector [51]. This necessitates bolstering the safety supervision framework throughout the entire industrial chain by rigorously adhering to relevant regulations and standards. Notably, many policies governing the measures for the supervision and management of safety in the hydrogen industry have been formulated by local authorities rather than the central government. Rigorous monitoring should encompass green hydrogen equipment design, facility construction, transportation, and application. Green hydrogen equipment should be equipped with comprehensive safety features, including automated safety control mechanisms, immediate alarm systems, activation of exhaust devices, and intelligent techniques for an instantaneous shutdown of the production process [52].

9.3 Japan, a hydrogen technology leader

With the Paris Agreement, the international community committed to decreasing GHGs to limit the global mean temperature increase to well below 2 °C. In alignment with this, Japan submitted its Nationally Determined Contributions (NDCs), targeting a reduction in emissions by 15%–17% by 2030. Japan's NDCs were recently revised, setting a more ambitious goal of an 80% reduction in GHGs by 2050, with a commitment to achieve carbon neutrality in the latter half of the century. Notably, Japan's Government has further elevated this commitment by pledging to establish a net-zero emission society by 2050 [53]. In its pursuit of decarbonizing the energy system, Japan has implemented various policy measures aimed at restructuring the feed-in-tariff system, boosting green energy production, and enhancing overall energy security [54]. Alongside increased support for renewable energy sources, Japan envisions the establishment of a “Hydrogen Society” by the mid-century, involving the widespread adoption of FCEVs, the substitution of hydrogen-based power generation for fossil power generation, and a shift to hydrogen and synthetic gases in the industrial sector. To meet the future demand for hydrogen, the country plans to support the domestic production of green hydrogen through electrolysis and aims to participate in international hydrogen markets. Japan aims to grow its hydrogen market from the current two million tons per year to three million tons per year by 2030 and a substantial 20 million tons per year by 2050. The strategy is to leverage scale to significantly reduce the cost of hydrogen to approximately one-third of the current level by 2030 [55]. Regarding global hydrogen markets, Japan aspires to import approximately 5-10 million tons of hydrogen by 2050, with a significant portion sourced from Australia [56]. However, the substantial reliance on imported energy carriers poses a challenge to Japan's national goal of enhancing energy security. The Japanese energy system is currently heavily dependent on imported fossil fuels, resulting in an 87% dependency on fossil fuels in primary energy consumption. Given its limited domestic fossil resources, Japan has a low self-sufficiency rate of 9.6% compared to other OECD countries. Conversely, Japan boasts the second-highest global installed capacity of solar photovoltaic (PV) plants and ranks third in generation capacity [57].

Japan, characterized by its high level of industrialization and scarcity of hydrocarbon resources, recognizes the multifaceted benefits of utilizing hydrogen. These advantages encompass energy security, industrial competitiveness, and the reduction of carbon emissions. In 2017, Japan introduced the pioneering Basic Hydrogen Strategy, marking the world's inaugural national hydrogen plan. This initiative spurred 26 countries and economies, including Japan, to develop their hydrogen strategies by 2022. Subsequently, Japan hosted the Hydrogen Energy Ministerial Meeting (HEM) in the following year, aiming to instigate top-down hydrogen policies and has since been at the forefront of shaping a global hydrogen community. Under this strategic framework, Japan achieved significant milestones, such as commercializing the world's first FCVs, expanding the use of fuel cells in households, and securing a leading position with a notable number of related patents globally. By leveraging advancements in research and development across various hydrogen-related sectors like transportation, power generation, and industrial heating, Japan has demonstrated a succession of diverse hydrogen-related technologies, maintaining its technological leadership on the global stage [58].

A significant hydrogen research complex is situated in Fukushima Prefecture known as the Fukushima Hydrogen Energy Research Field (FH2R), inaugurated in March 2020. Spanning 0.18 km^2, FH2R harnesses solar power generation with a capacity of 20 megawatts on-site and grid electricity to electrolyze water in a 10-megawatt-class hydrogen production unit- the largest globally. FH2R serves as a testing ground for various experiments to identify optimal operation control technologies, integrating power grid demand response with hydrogen supply and demand response. As Japan explores diverse technological options and hydrogen sources, Japanese companies actively engage in hydrogen supply chain projects with nations worldwide, including Australia, Brunei, Indonesia, Malaysia, Singapore, Saudi Arabia, Russia, the United Arab Emirates, and the United States. Notable initiatives include:

- Saudi Arabia: In September 2020, Japan received the world's first shipment of ammonia derived from fossil fuels from Saudi Arabia [59]. Ammonia has the potential to serve as a fuel for gas turbines. Successful experimentation was conducted in Japan in 2022, demonstrating the direct use of 100% ammonia in a 2 MW gas turbine. Ongoing initiatives aim to advance the development of a 40 MW turbine designed for the exclusive utilization of pure ammonia. The combustion of ammonia can potentially result in the emission of nitrous oxide (N_2O), a potent GHG. However, in the 2 MW demonstration project conducted in Japan, it was feasible to achieve a 99% reduction in overall GHG emissions (combining CO_2 and N_2O) when comparing the ammonia-fired gas turbine to a gas turbine using natural gas.
- Brunei: A demonstration involving the shipping of natural gas-based hydrogen from Brunei to Japan in the form of methylcyclohexane (MCH) took place [60]. Japan received the inaugural shipment from Brunei in May 2020, utilizing the supply for domestic power generation [61,62].
- Australia: The first shipment of liquefied hydrogen derived from lignite from Australia to Japan took place in 2021. The $500 million, decade-long project, which involved the extraction of hydrogen from Victorian coal, marked several technological breakthroughs, including the first shipment of liquefied hydrogen to Japan [63,64]. The hydrogen was produced from Australian brown coal and was transported to Japan, where it arrived as a large quantity of liquefied hydrogen, signifying a significant milestone in the efforts to develop hydrogen as a future alternative carbon-free fuel [65]. The project has been described as a major milestone in the Hydrogen Energy Supply Chain, with the first international shipment of liquified hydrogen being a key achievement. A Japanese-Australian venture involved in the production of hydrogen from brown coal has entered its commercial demonstration phase, indicating progress in this area [66].

9.3.1 Japan's short and long-term strategies to develop hydrogen technology

Previously, Japan's hydrogen strategy was formulated to establish a domestic hydrogen market ahead of the global curve by advancing hydrogen technology. However, considering Japan's energy demand, the growth potential of the domestic hydrogen market seems limited. In contrast, the global hydrogen market is anticipated to generate annual revenues of $2.5 trillion and create 30 million jobs by 2050. Recognizing the anticipated rapid expansion of the global market, the strategy needs to be reoriented toward addressing overseas markets. This revised approach should encompass both domestic and international market developments, presenting a roadmap for achieving carbon neutrality by 2050, aligning with the shared vision of the public and private sectors. The strategy acknowledges the challenges faced by the country, outlines policies to address them, and conveys the nation's ambition to expedite the transition toward a hydrogen-based society. Additionally, alongside the overarching hydrogen policy, the Basic Hydrogen Strategy will introduce two key components. The first is the Hydrogen Industry Strategy, outlining policies to enhance the competitiveness of Japanese industries in hydrogen markets. The second is the Hydrogen Safety Strategy, focusing on ensuring the safe use of hydrogen. The strategy will undergo revisions as needed but is typically reviewed every five years [67]. In June 2023, the Japanese government approved an amendment to its Basic Hydrogen Strategy, emphasizing the expanded utilization of hydrogen as a fuel and outlining intentions to allocate over $100 billion for hydrogen investments in the upcoming 15 years. The roadmap introduced fresh objectives for the specifications of fundamental technologies and cost breakdowns, outlining the necessary measures to attain these goals. For instance, mobility goals encompassed the deployment of 200,000 FCVs by 2025 and 800,000 by 2030, along with the establishment of 320 fueling stations by 2025 and 900 by 2030 [68].

Japan confronts inherent challenges in terms of energy security due to its status as an island nation with limited natural resources and an absence of international pipelines or power connections. The potential for expanding renewable energy capacity in the country is restricted. The major contributors to Japan's overall energy consumption are fossil fuels- oil (37%), coal (27%), and gas (20%) [69]. With negligible domestic production, Japan heavily relies on imported crude oil and liquefied gas, a significant portion of which is procured from the Middle East. The Middle East accounted for more than 94% of Japan's oil imports, with major suppliers being Saudi Arabia and the United Arab Emirates. Additionally, 10% of the country's gas imports, primarily from Qatar, originate from the same region [70]. Russia's incursion into Ukraine has heightened uncertainty in the energy market, especially in Europe, prompting an increased focus on hydrogen production and utilization as an alternative to natural gas [71]. For Japan, which faces challenges in terms of readily available energy resources, hydrogen presents an opportunity to enhance energy security for various reasons. Firstly, hydrogen can be produced from renewable energy sources, allowing for in-country production and storage. Additionally, Japan can secure hydrogen from diverse sources in Asia and the Indo-Pacific region, contributing to a more varied supply chain. This diversification further bolsters energy security. In the context of energy security, it is essential to implement policy measures to ensure the reliable procurement and recycling of rare metals and rare-earth minerals crucial for water electrolysis equipment and fuel cells. Simultaneously, there is a need to develop innovative technologies aimed at reducing the consumption of rare metals. Currently, fossil fuels like coal and petroleum maintain economic advantages over hydrogen. However, this continued reliance on fossil fuels is contentious in the context of global efforts toward decarbonization due to sustainability concerns. According to the Energy Technology Perspectives 2020 by the IEA, hydrogen produced through electrolysis utilizing renewable electricity is anticipated to become cost-competitive with existing fossil fuels as renewable energy costs decrease. Additionally, the price of hydrogen derived from renewable energy has shown relatively minor fluctuations amid resource price variations following Russia's invasion of Ukraine. As carbon pricing initiatives gain momentum in Japan and elsewhere, converting the environmental value of carbon into specific product prices, hydrogen, as a non-fossil fuel, is poised to become a commercially sustainable and appealing energy source, providing economic stability [72]. To ensure a stable, large-scale, and cost-effective supply of hydrogen and ammonia over the long term, it is crucial to stimulate the demand for these substances. Meeting this demand requires not only establishing a foundation for domestic hydrogen production but also increasing the utilization of hydrogen sourced internationally. Different countries and economies possess varying conditions, such as renewable energy production potential and market size. Presently, Japan aims to utilize up to 3 million tons per year by 2030 and approximately 20 million tons per year by 2050, fostering the shift towards a hydrogen and ammonia-based society. An additional target is set for the utilization of 12 million tons of hydrogen (including ammonia) annually by 2040 (Table 9.4) [73].

Table 9.4 Japan's hydrogen strategy

Targets	Current	Strategy (revised in June 2023)
Volume (use)	2 Mt/pa	3 Mt/pa by 2030 12 Mt/pa by 2040 20 Mt/pa by 2050
Volume (domestic production)	2 Mt/pa	No numeric target
Volume (imports)	-	No numeric target, but assumes large volume of supply to be via import
Low-carbon intensity	-	Hydrogen: 3.4kg-CO_2e/kg-H_2 (well-to-gate) Ammonia: 0.84kg-CO_2e/kg-NH_3 (gate-to-gate)
Price efficiency	100 yen/Nm^3	30 yen/Nm^3 by 2030 20 yen/Nm^3 by 2050
Refueling stations	180	1,000 by 2030
FCV uptake	7,700+	Equivalent 800,000 passenger vehicles by 2030 (no separate targets for e.g. buses)
Co-firing %	Hydrogen: 30% Ammonia: 20%	Hydrogen: 30%+ (demonstration) by 2030 Ammonia: 20%+ (commercial) by 2030 (dependent on size) Longer term aspirations of 50%.
Installed home fuel cells	-	3 million by 2030
CCS storage capacity		Between 6-12 million tons annually by 2030

Ensuring the availability of affordable hydrogen and ammonia is a prerequisite for enhancing their utilization. Japan's targets for hydrogen supply costs are 30 yen/Nm^3 (approximately 334 yen/kg) by 2030 and 20 yen/Nm^3 (approximately 222 yen/kg, aiming to make hydrogen power generation costs lower than gas-fired power generation costs) by 2050. The ammonia supply cost target for 2030 has been set at the upper 10-yen level/Nm^3 (equivalent in calorific value to hydrogen). In recent years, significant fluctuations in fossil fuel prices have occurred. Consequently, the hydrogen supply cost target aligns with the recent fossil fuel price. The Japanese government has established low-carbon targets that are no less ambitious than those of other countries, following a specified calculation methodology. These targets drive the widespread adoption of hydrogen in alignment with the set objectives. More specifically, a stringent carbon intensity target has been defined, aiming for 3.4 kg-CO_2 or less of carbon emissions per 1 kg of hydrogen production, spanning from the well to the production gate. This target is considered technically feasible, given the current level of technology. Hydrogen meeting this carbon intensity target will be recognized as low-carbon hydrogen. Similarly, ammonia derived from hydrogen with emissions of 0.84 kg-CO_2e/kg-NH_3 or less per 1 kg gate-to-gate (including hydrogen production) will be designated as low-carbon ammonia [74].

In terms of enhancing energy security, establishing domestic hydrogen production and supply systems is crucial. Despite some opinions suggesting that domestically produced hydrogen in Japan may be more expensive than imported hydrogen due to the higher costs associated with renewable energy production and carbon capture and storage (CCS) in Japan, there is potential for cost reduction. This can be achieved by leveraging controllable renewable energy output, particularly when surplus electricity prices decrease. Hydrogen is anticipated to play a role in balancing energy and facilitating the integration of renewable energy into the market. To fully exploit opportunities for domestic hydrogen production, it is imperative to narrow the price gap between hydrogen fuel and conventional fuels. In support of this goal and to bolster energy security, selective and substantial support should be directed toward domestic businesses with the potential for significant future price reductions and competitiveness.

In the realm of hydrogen production, the demand for water electrolyzers is projected to grow continuously, given their industrial advantage in producing hydrogen from renewable energy. By 2030, the world is expected to have 134 GW of installed electrolyzer capacity. To enhance Japan's presence in the global energy supply market, it is essential for Japanese companies to enter this upstream segment of the hydrogen supply chain, both domestically and internationally. Consequently, a new water electrolyzer capacity target of approximately 15 GW is set for Japanese-related companies (including parts and materials manufacturers) in combined domestic and international markets by 2030. These targets will be subject to timely and appropriate reviews, considering technological developments and the evolving landscape of global demand.

9.3.2 Development of hydrogen fuel cell technology in Japan

Figure 9.5 demonstrates the number of HFCVs on the road in Japan and selected countries by 2030 [75]. Despite Japan having a smaller share of global hydrogen transportation compared to other regions, the country is projected to have the highest number of FCVs by 2050, approximately 37.3 million passenger vehicles. This is attributed to the relatively lower distance traveled by each vehicle, which is around one-third of their counterparts in the United States. Japan, with approximately 78.1 million passenger vehicles currently in use, is expected to play a significant role in hydrogen adoption by 2050. In Western Europe, an estimated 30 million FCVs are anticipated, while North America is expected to deploy 20.4 million FCVs by 2050 within other OECD nations. Among non-OECD nations, Southeast Asia is projected to deploy around 20.2 million FCVs, and China is expected to deploy approximately 12.2 million FCVs by 2050 [76].

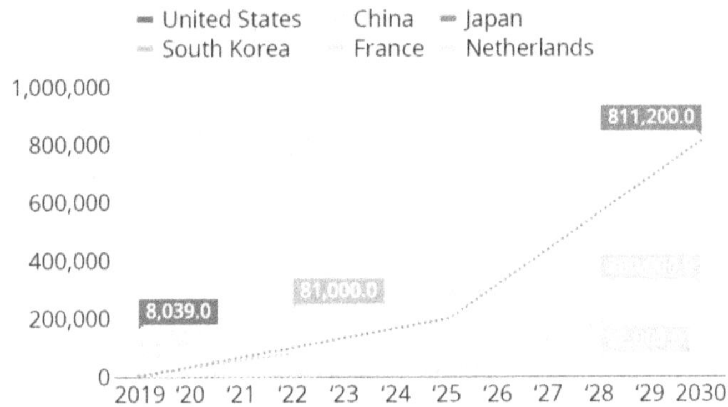

Figure 9.5 The number of HFCVs on the road in Japan and selected countries by 2030 [75]

The continuous advancements in vehicle technology have already led to the presence of FCVs in the market, and their efficiency and affordability are rapidly improving. Currently, there are around 11,200 FCVs in operation globally, and projections indicate a substantial increase to approximately 2.5 million vehicles within the next decade. By 2050, the global deployment of FCVs is expected to exceed 120 million, necessitating sustained and increased deployment rates post-2030, reflecting improvements in affordability and competitiveness with traditional vehicle types [75]. For the substantial deployment of FCVs in the future transport sector, additional investments in hydrogen infrastructure are essential, particularly in storage, distribution, and refueling infrastructure. Hydrogen-importing nations, such as Japan, will need to establish and maintain a maritime import-export network. Despite the majority of hydrogen being produced from fossil fuel sources in the maximum penetration scenario, Japan, which is likely to import all its future hydrogen, may offset its CO_2 emissions and achieve a low-carbon transport sector while producer nations reap economic benefits at the expense of the environment. The distribution of hydrogen production and consumption is a notable aspect. In 2050, although non-OECD nations generate approximately one-third of the global hydrogen supply, their consumption is only one-tenth, with 69% of the surplus being exported to OECD nations, contributing significantly to the estimated deployment of FCVs. In addition to the prominent role of hydrogen in passenger cars, it also moderately contributes to the mass transportation sector by facilitating the deployment of FCV buses by 2050. The efficiencies and costs of FCV passenger cars and buses are derived from publicly available fuel cell data from Toyota Motor Corporation, as utilized in the Mirai FCV. Considering Japanese stakeholders' preferences for future car purchases, there is a growing inclination toward considering FCV purchases, reaching approximately one-third of stakeholders by 2030 compared to less than one-fifth in 2019. If this increasing acceptance of FCVs persists and promised improvements in infrastructure and performance materialize, projecting a fleet comprising around 50% FCVs by 2050 in Japan is an ambitious yet reasonable estimate, especially considering similar research outcomes in other regions. Hybrid vehicle preferences remain robust throughout the tested time periods in Japan, supporting the assumption that gasoline-hybrid vehicles will continue to play a significant role, displacing gasoline vehicles. While the model outcomes represent an optimized energy system in terms of total system cost and achieving CO_2 reduction targets outlined in the Paris Agreement, there are also complementary outcomes related to stakeholder preferences. The stakeholders are optimistic about the environmental and energy efficiency aspects of FCVs, aligning with comparative studies highlighting the advantages of FCVs in terms of CO_2 emissions per kilometer and overall energy efficiency. Due to these advantages, it is predicted that by 2050, FCV affordability will surpass that of gasoline vehicles, potentially alleviating concerns expressed by stakeholders regarding FCV costs. In the minimum hydrogen penetration scenario, despite an overall increase in hydrogen allocated to transport, the global energy system shifts toward a local production and consumption approach, primarily due to non-preferential carbon reduction targets for non-OECD nations. Under this arrangement, even Japan produces all its hydrogen from High-Temperature Gas-Cooled Reactor (HTGR) nuclear facilities; however, a much smaller quantity is used in Japan (17.1 MTOE versus 106 MTOE for the maximum scenario). Non-preferential carbon reduction targets result in no imports, and hydrogen is only allocated to passenger transportation for FC buses at the same level as the maximum scenario.

9.4 South Korea's hydrogen deployment policies

With the rising global demand for renewable energy sources to combat climate change, South Korea is striving to emerge as a leading force in establishing a hydrogen-based economy [77]. The Korean government has underscored its dedication to achieving net-zero carbon emissions by 2050 through initiatives such as the "Hydrogen Economy Roadmap" (January 2019), the establishment of the "Hydrogen Economy Promotion and Safety Management Act" (Hydrogen Act) (February 2020, enforced on February 5, 2021), and the implementation of the "Green New Deal" (July 2020) [78]. These endeavors have accelerated the development and adoption of hydrogen vehicles, the creation of an ecosystem for hydrogen production and distribution, and advancements in fuel cell technology. Private investment plans in the hydrogen economy, coupled with government support, were recently endorsed during the third meeting of the Korean Hydrogen Economy Council. Despite Korea's investments in photovoltaics and wind energy, its onshore wind and solar potentials are limited. However, the Korean hydrogen industry boasts a well-established presence with decades of experience in industrial energy applications using hydrogen as a feedstock. The Korean Green New Deal is driven by the goals of sustainability and regional equality. Scholars highlight that major global Green New Deal initiatives serve as climate-focused economic stimulus policies fostering investments, yielding economic output, and generating jobs in the short term. Moreover, these initiatives serve as a foundation for long-term innovation and economic development aligned with environmental constraints, narrowing the gap in regional inequalities within a country or between countries. In line with these principles, the Korean Green New Deal, announced on July 14, 2020, along with the "Hydrogen Economy Roadmap" (January 2019), articulates the dual objectives of significantly advancing the hydrogen economy and mitigating resulting regional inequalities within the country [79]. South Korea heavily relies on imports to satisfy nearly 98% of its fossil fuel consumption due to limited domestic resources. The absence of international oil and natural gas pipelines compels South Korea to depend on tanker shipments of liquefied natural gas (LNG) and crude oil to meet its energy demands. In July 2020, South Korea introduced its Green New Deal as part of a broader economic initiative aimed at reducing GHGs, expanding renewable energy capacity, enhancing energy efficiency in electricity infrastructure, transitioning to a low-carbon and decentralized energy supply, and investing in innovations within the green industry sector [80]. Despite being the world's seventh-largest energy consumer in 2021, South Korea's economic growth, primarily driven by exports of automobiles, ships, semiconductors, and petrochemicals to Asian trading partners, was robust, with a 4.1% increase in real GDP [81]. Renewables, accounting for only 3% of South Korea's primary energy consumption, demonstrated a consistent growth trend since 2015 when they comprised less than 1% of total energy consumption. However, the Ministry for Trade, Industry and Energy (MOTIE) adjusted its 10th Basic Plan for Long-Term

Electricity Supply and Demand at the end of 2022, lowering the previous target of renewables contributing 30% to electricity generation to 22% by 2030. The plan also aims for a significant increase in nuclear power's share, reaching 32% in 2030 and 35% in 2036, marking a departure from the previous administration's plan to phase out nuclear power [82]. To accelerate its hydrogen economy, MOTIE has outlined a roadmap focusing on increasing hydrogen demand in transportation, electric power, and industrial sectors. The government envisions achieving goals such as having 30,000 commercial hydrogen vehicles, establishing 70 hydrogen refueling stations by 2030, and having hydrogen account for 7% of power generation by 2036 [83]. In September 2022, South Korean company Posco unveiled its intentions to introduce hydrogen co-firing power generation technology, marking the country's inaugural hydrogen hybrid power generation initiative with a target of producing over 950 GWh per year. This company aims to establish an infrastructure for green hydrogen supply with an annual capacity of 1.26 million tons by 2035 [84]. Additionally, South Korea aims to establish substantial domestic and international production hubs. The country recently initiated its first large-scale hydrogen demonstration project in Jeju City, with a goal of generating 1,176 tons per year of hydrogen at a 60% utilization rate [85]. Plans are underway to develop distribution infrastructure to boost the utilization of clean hydrogen. This involves the creation of a hydrogen power generation bidding market in the coming year, the enactment of legislation for hydrogen businesses, and the implementation of a clean hydrogen certification system in 2024. South Korea is also focused on securing essential technologies throughout the entire hydrogen value chain, encompassing advancements in seven key areas: water electrolysis, liquid hydrogen carriers, trailers, charging stations, fuel cells in mobility, fuel cells in power generation, and hydrogen turbines. At present, the majority of hydrogen in South Korea is produced in facilities specializing in petrochemical processes, chlor-alkali operations, and steam-reforming plants (Figure 9.6) [86]. Regulations impeding business activities will be eliminated, and domestic production will be commercialized for export to global markets. The country aspires to nurture 600 specialized hydrogen companies by 2030, a significant increase from the current 52 firms [87]. Furthermore, South Korea has ambitious plans to construct the world's largest liquid hydrogen plant and expand the number of liquefied refueling stations.

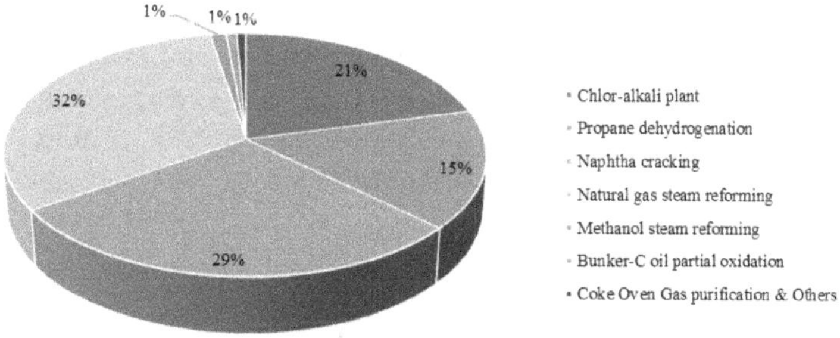

Figure 9.6 South Korea's hydrogen production pathway

It will establish a 4 million tons per year ammonia receiving terminal and a 100,000 tons per year liquid hydrogen receiving terminal, along with installing a dedicated hydrogen pipeline network. Following South Korea's Hydrogen Economy Roadmap, the government established the Hydrogen Economy Council, led by the Prime Minister and composed of eight ministries, including Trade, Industry, and Energy (MOTIE); Economy and Finance (MOEF); Interior and Safety (MOIS); Science and ICT (MSIT); Environment (MOE); Land, Infrastructure, and Transport (MOLIT); Oceans and Fisheries (MOF); and SMEs and Startups (MSS). Since 2021, the government has also formed five committees of hydrogen experts to promote research and development (R&D) collaboration among academic institutes, companies, and organizations. These committees cover production, storage/transportation, utilization for transportation, utilization for energy, and safety/infrastructure, with approximately 70 experts leading each committee. They segment target domains for detailed analyses and propose research fields for future investment. As of the first half of 2023, South Korea continues to hold the position of the world's largest market for fuel cell cars, boasting a fleet exceeding 32,000 FCVs. Hydrogen vehicle prices are anticipated to compete with those of internal combustion engine vehicles, with a durability of 500,000 km, priced at 40 million won (USD 41,000) by 2025. For personal transportation, the roadmap envisions the introduction of 80,000 hydrogen taxis and 40,000 hydrogen buses by 2040. It also includes the deployment of 30,000 hydrogen trucks over the same period for freight transport. Similar to the adoption of electric vehicles, the limited number of refueling stations hinders the widespread adoption of hydrogen-powered vehicles. From 2019 onward, the ratio of FCEVs per HRS in South Korea has consistently ranged between 140 and 200. In comparison, other significant markets such as China, Japan, and Europe have fewer than 50 FCEVs per hydrogen refueling station. Korea's roadmap calls for an increase in the number of refueling stations from 310 in 2022 to 1200 in 2040, respectively. According to the government's plan, various subsidies will be offered for different types of installations. In urban settings, the Ministry of Environment will grant subsidies equivalent to 50% of the installation cost per station, up to 1.5 billion won. Additionally, the Ministry of Land, Infrastructure, and Transport will provide 750 million won per station for installations along highways. South Korea also intends to relax regulations concerning the location and approval process for hydrogen refueling station installations. Furthermore, the establishment of a regulatory sandbox will be utilized to facilitate installations in the short term. In terms of power generation and supply, the roadmap aims to elevate production from 307.6 MW to 15 GW by 2040 using fuel cells. According to the roadmap, 8 GW of the generated power would be allocated for domestic consumption, while 7 GW would be earmarked for exports. To promote power generation through fuel cells, the government plans to establish an exclusive LNG price for fuel cells and maintain renewable energy certificates (RECs) for a defined period to offer investors a level of certainty. Additionally, it aims to enhance the value of RECs for green hydrogen projects and is contemplating the implementation of a long-term, fixed-price system for power generated from fuel cells. The roadmap also outlines a target of supplying 2.1 GW of power from fuel cells to households and other buildings by 2040. To achieve this, a combination of incentives, such as a special electricity tariff and mandatory measures, like replacing cooling systems at public institutions with fuel cell systems, will be employed. If the demand for hydrogen aligns with South Korea's hydrogen roadmap, the hydrogen demand is anticipated to rise from 130,000 tons in 2019 to 5.36 million tons per year by 2040. Presently, South Korea meets its hydrogen demand through byproducts in the petrochemical process. As demand increases in the initial stages, the country aims to produce hydrogen using LNG. However, for the longer term, there are considerations for generating green hydrogen. This involves utilizing surplus renewable energy and hydrogen produced in tandem with the establishment of new large-scale renewable energy projects. The roadmap also envisions South Korea commencing the import of green hydrogen by 2030. To facilitate the domestic transportation of imported hydrogen, South Korea is exploring the installation of pipelines at the importation point. These potential pipeline initiatives, along with others, would supplement South Korea's existing 200-km hydrogen pipeline. The Korea Gas Corporation (KOGAS) has developed its own hydrogen plan as well. The state-owned KOGAS plans to invest 4.7 trillion won ($4.06 billion) by 2030 in research and development and the construction of 25 hydrogen production facilities, along with the addition of 700 km of new pipeline [88]. Additionally, the domestic price of a kilogram of hydrogen is expected to decrease from 6500-7500 won (USD 5.42-6.25) to 3000 won (USD 2.5) by 2040 [77]. South Korea's shift toward a hydrogen economy has the potential to establish a new domestic market and support the growth of the nation's advanced industries. According to research conducted by McKinsey & Company, the substantial scale of the hydrogen industry is expected to generate significant employment opportunities. To meet the demand solely within the domestic market, South Korea's hydrogen industry could potentially employ around 600,000 people by the year 2050, with a predominant focus on the automotive sector. The research also indicates that domestic revenues across the entire energy value chain, spanning hydrogen generation, distribution, and end-user applications like hydrogen vehicles, could reach KRW 70 trillion (USD 60.47 billion) annually by 2050 (Figure 9.7) [89].

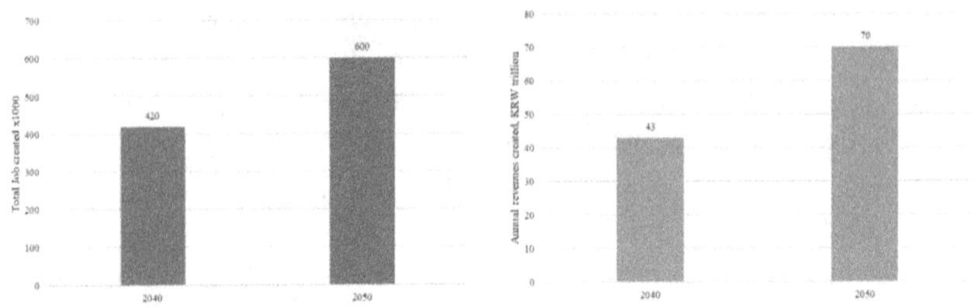

(a) (b)

Figure 9.7 South Korea's hydrogen economy (a) new jobs (b) annual revenues

9.4.1 Hydrogen technology and the role of the private sectors

The Hyundai Motors Group and its affiliated conglomerates play a leading role in advancing hydrogen as a viable fuel source. Hyundai took a significant step in 2013 with the launch of the Tucson iX35, the first mass-produced FCEV, and has since prioritized the development of this technology as a key element of its strategic vision. In 2018, Hyundai unveiled its "FCEV Vision 2030" outlining ambitious plans for investment. The Hyundai Motor Group and its suppliers intend to commit 7.6 trillion won ($6.7 billion) to expand facilities and research and development. The overarching goal is to produce 700,000 fuel cell systems annually by 2030, with 500,000 allocated for FCEVs and the remainder designated for applications such as rolling stocks, drones, vessels, and forklifts. These investments are anticipated to generate 51,000 jobs [90]. Having trailed behind Toyota in recent years, Hyundai has emerged as the world's largest seller of hydrogen vehicles. The Nexo SUV, currently available in markets like South Korea, the USA, and the European Union, has contributed to global sales of 4,987 Nexo FCEVs, with 4,194 units sold in South Korea [91]. As part of its 2030 plan, Hyundai aims to escalate sales to 40,000 FCEVs in 2022, 130,000 in 2025, and an ambitious target of 500,000 by 2030. Hyundai envisions that at an annual sales volume of 100,000 units, FCEVs will achieve price competitiveness with electric vehicles. Moreover, the company anticipates that strengthened environmental regulations will level the playing field among fuel cell, electric, and internal combustion engine vehicles in terms of total cost. The broader adoption of hydrogen as a fuel source in society hinges on three key factors: enhancing safety measures, advancing related technologies to drive cost reduction, and heightening public awareness of hydrogen's merits as an alternative energy carrier. To further the proliferation of this technology, Hyundai is exploring the sale of hydrogen fuel systems to competitors, aiming to move beyond reliance on the sales of its own vehicles [92]. While there is potential for economic synergies between passenger and commercial vehicle production, Hyundai's plans for commercial vehicles are less detailed. In 2019, the company established a joint venture, Hyundai Hydrogen Mobility, with H_2 Energy in Switzerland, with the goal of providing 1600 heavy-duty trucks for leasing by 2025. Although the initial focus is on Switzerland, Hyundai envisions expanding to other parts of Europe over time [93]. In 2020, Hyundai entered into a memorandum of understanding with the U.S. Department of Energy to collaboratively assess the viability of hydrogen vehicle technology [94]. Similar to its efforts in South Korea, Hyundai's success in the USA will likely depend on government support, either from the U.S. federal government or state governments, particularly in the construction of refueling stations.

Hyundai is not the sole investor in hydrogen fuel cell technology; Doosan also spun out its fuel cell division into a separate entity for accelerated growth and is poised to complete construction on South Korea's first liquid hydrogen plant in 2021 [95]. Doosan's Mobility Division has also developed a hydrogen fuel cell drone tailored for deliveries in areas with inadequate infrastructure. Hanwha Energy achieved a significant milestone by completing the world's first by-product hydrogen power plant in July 2020 at the Daesan Industrial Complex. Utilizing fuel cells manufactured by Doosan, this plant represents a notable advancement [96]. Posco, a dominant steel producer in South Korea, is actively involved in both fuel cell production and the manufacturing of components for fuel cells. The South Korean government has played a significant role in fostering hydrogen infrastructure by establishing a special-purpose corporation called HyNet, comprising 13 companies. This consortium invested 135 billion won ($118.7 million) in constructing 100 refueling stations in 2022 and will continue to manage these stations through 2028.

9.5 India's hydrogen strategies

India is ranked as the world's third-largest energy consumer, following China and the United States. Additionally, with a population of 1.4 billion people, India stands as the second-most populous country globally. The country's increasing energy demands are driven by both population growth and the ongoing process of modernization. In his Independence Day speech in August 2021, the Indian Prime Minister introduced the National Hydrogen Mission, signaling the pivotal role of green hydrogen in India's efforts to address climate change, with aspirations to position the country as a global hub for green hydrogen production and exports [97]. In January 2023, the Ministry of New and Renewable Energy unveiled the National Green Hydrogen Mission document [98], outlining the following key points:

- The mission aims to develop capabilities to produce a minimum of 5 million tons of green hydrogen by 2030, with the potential to scale up to 10 million tons annually, contingent on growth in export markets [99].
- Establishing a robust domestic electrolyzer manufacturing ecosystem is identified as a crucial objective.
- The mission adopts a phased development approach:

- **Phase 1 (2022-23 to 2025-26):** Initial emphasis on green hydrogen development in sectors already utilizing hydrogen, such as refineries, fertilizer manufacturing, and city gas networks. This phase also includes advancing research and development, regulatory/standards development, and pilot projects in areas like steel production, long-haul heavy-duty mobility, and shipping.
- **Phase 2 (2026-27 to 2029-2030):** Building on Phase 1, this phase involves extending initiatives to new sectors of the economy, such as railways and aviation.

- Mandates for publicly owned companies:
  - The Shipping Corporation of India is mandated to retrofit at least two ships to operate on green hydrogen or a derivative fuel by 2027.
  - Government-run oil and gas companies (public sector undertakings or PSUs) are required to charter at least one ship, each powered by green hydrogen or a derivative fuel by 2027.
- To support refueling, the mission aims to establish green ammonia bunkers and refueling facilities at a minimum of one port by 2025, with the goal of extending this infrastructure to all major ports by 2035 [100].

The Union Cabinet granted approval for the National Green Hydrogen Mission with a budget of Rs. 19,744 crore. The Strategic Interventions for Green Hydrogen Transition (SIGHT) Programme, featuring an allocation of Rs. 17,490 crore, stands as a crucial financial support initiative under this Mission. The program outlines two distinct financial incentive mechanisms aimed at promoting the domestic manufacturing of electrolyzers and fostering the production of green hydrogen [101]. In June 2023, the guidelines for implementing the 'SIGHT Programme-Component I: Incentive Scheme for Electrolyzer Manufacturing' were unveiled, covering the fiscal years 2025-26 to 2029-30, with a designated budget of Rs. 4,440 crore [102]. In January 2024, the capacity allocation for selecting electrolyzer manufacturers to establish manufacturing capacities in India was disclosed under the SIGHT scheme (Tranche-1). Simultaneously, in June 2023, the guidelines for implementing the "SIGHT Programme- Component II: Incentive Scheme for Green Hydrogen Production (under Mode 1)" were introduced for the fiscal years 2025-26 to 2029-30, with an allocated budget of Rs. 13,050 crore for all modes under this component. In June 2023, the World Bank sanctioned US$1.5 billion in financing to expedite India's development of low-carbon energy, supporting the implementation of the National Green Hydrogen Mission through the First Low-Carbon Energy Programmatic Development Policy Operation. In August 2023, aligning with the National Green Hydrogen Mission, the Ministry of New and Renewable Energy issued an Office Memorandum on the Green Hydrogen Standard for India. Subsequently, in October 2023, the Ministry of New and Renewable Energy unveiled the final R&D Roadmap for the Green Hydrogen Ecosystem in India [103].

The current cost of hydrogen produced through electrolysis ranges from approximately $7/kg to $4.10/kg, depending on various technology options and associated soft costs. This cost is relatively high, making it challenging to compete with the existing prices of grey or brown hydrogen. However, India boasts some of the most competitive levelized costs of electricity (LCOE) for solar and wind globally despite being a net importer of natural gas. Considering the anticipated decline in electrolyzer costs and LCOE, it is more advantageous to prioritize the expansion of green hydrogen production in India over the production of grey or blue hydrogen [104].

Anticipating a decrease in costs for both electrolyzers and renewables, our analysis suggests that in the optimal scenario, the price of green hydrogen could reduce to around $1.60/kg by 2030 and $0.70/kg by 2050. Irrespective of the scenario, the key takeaway is evident: Green hydrogen has the potential to compete with grey hydrogen by 2030, if not sooner. Other factors, such as the imposition of a potential carbon price on fossil fuels, could further contribute to the cost competitiveness of green hydrogen.

In India, the current utilization of hydrogen is almost evenly distributed between refining and ammonia production, with a minor portion allocated to methanol production. A small amount of hydrogen, around 0.3 million tons, is presently being used in steel production. India had a potential for emerging demand in heavy-duty, long-haul freight transportation and, to a limited extent, in the power sector and niche applications such as industrial forklifts, cell phone towers, and city gas distribution, as well as speculative and early-stage applications in aviation, shipping, and cooking. The estimation of hydrogen demand is based on a scenario assuming high adoption rates of technology and policies facilitating the transition to green hydrogen. This scenario considers factors such as increased use of green hydrogen in various sectors, greater penetration of fuel cell trucks, a rapid decline in electrolyzer and renewable costs, and financing options for this transition. The demand for green hydrogen is evaluated within the overall hydrogen demand, considering the cost parity of end-use products based on green hydrogen compared to those based on grey/brown hydrogen. The advent of cost-competitive green hydrogen introduces the potential for market expansion, particularly in sectors already reliant on grey hydrogen. The extent of green hydrogen adoption will hinge on production costs relative to alternative hydrogen sources, the proportion of hydrogen costs in the final product cost, and any external efforts to stimulate demand in the short term. Based purely on cost competitiveness, green hydrogen is anticipated to become the predominant player in the hydrogen market over the long term. Even by 2030, green hydrogen is poised to play a substantial role in both existing brownfield consumption and new greenfield investments. The projection suggests that nearly 94% of hydrogen demand in 2050 can be satisfied by green hydrogen, a significant increase from the 16% expected in 2030. The cumulative value of the green hydrogen market in India is estimated to reach $8 billion by 2030 and $340 billion by 2050.

This shift carries significant implications for the greenhouse gas emissions of challenging-to-decarbonize sectors. Over the period from 2020 to 2050, India has the potential to reduce CO_2 emissions by 3.6 gigatons compared to a scenario with limited hydrogen adoption. While industrial feedstock presents a more accessible market, the primary source of long-term decarbonization potential lies in steel, followed by heavy-duty trucking due to their substantial demand scale. From an energy security standpoint, the production of domestically sourced green hydrogen could result in net energy import savings ranging from \$246 billion to \$358 billion cumulatively between 2020 and 2050, with \$3 billion to \$5 billion savings projected between 2020 and 2030 alone. This reduction is attributed to decreases in both natural gas imports as grey hydrogen is substituted with green hydrogen and oil imports as long-haul freight transitions to hydrogen fuel cell trucks.

India's domestic market for electrolyzers has the potential to reach \$31 billion by 2050, indicating a demand of 226 GW. Anticipated demand by 2030 is estimated to be around 20 GW. In 2050, green hydrogen produced in India is expected to be highly competitive with hydrogen from nations like Australia and the United States, which are already exploring the possibility of ocean shipping for hydrogen transport. Even by 2030, Indian green hydrogen could demonstrate competitiveness in certain regions. The outlook for pipeline trade to major ports and energy trading hubs in the region, such as Singapore, is plausible if end-use sectors like shipping and the airline industry, in addition to refining, increase their utilization of hydrogen.

To achieve India's ambitious net-zero targets by 2070, substantial efforts are required domestically from both the public and private sectors. Public sector enterprises like Indian Oil Corporation Limited (IOCL), traditionally focused on coal power plants and oil refining, are diversifying their energy portfolios. Major companies outlined investment plans for green hydrogen production and electrolyzer manufacturing even before the formalization of the policy framework in India. Adani Group, in partnership with France-based TotalEnergies, committed \$50 billion over the next decade to green hydrogen and associated ecosystems, aiming to develop a hydrogen capacity of 1 MMT per annum through 2030 [105]. Reliance Industries Limited, with a net-zero goal by 2035, disclosed a \$10 billion investment for clean energy infrastructure (hydrogen, solar cells, fuel cells, and batteries) in collaboration with Denmark's Stiesdal A/S for electrolyzer manufacturing [106]. Additionally, Indian firms like PTC India, L&T, Greenko, and Renew Power have entered memoranda of understanding or joint ventures with international private companies for projects related to green hydrogen production and export, large-scale manufacturing of alkaline water electrolyzers, and the establishment of a center of excellence for facilitation of green hydrogen. Many of these Indian companies have formed partnerships in countries such as Singapore, Egypt, Norway, and Belgium. Greenko's commitment to produce and supply green hydrogen and ammonia is noteworthy, with targets of around 3 MMT of green ammonia to cater to both domestic and export demand [107]. Avaada Group secured approximately $1 billion from Brookfield Renewable to fund its green ammonia and hydrogen ventures in India [108]. Several pilot projects initiated by public sector enterprises focus on green hydrogen production and utilization. NTPC showcased India's first green hydrogen blending project with natural gas in the piped natural gas network for residential consumption and plans to establish green hydrogen fueling stations for cleaner mobility in high-altitude areas like Ladakh. IOCL, through global tenders, is in the process of setting up its inaugural refinery project. Indian Railways aims for carbon neutrality by 2030, with the anticipation of the first hydrogen-fueled train rolling out in 2024. India is actively engaging in international collaboration on clean hydrogen through government-to-government efforts, notably participating in discussions within various international groups such as the Quad's clean hydrogen strategic initiative. Bilateral engagements with the United States and Germany have been established, involving the formation of hydrogen task forces. The focus of cooperation encompasses the creation of public-private partnerships in the clean hydrogen sector, fostering advancements in technology, conducting collaborative studies and pilot programs, promoting innovation and research and development, and aligning regulations, codes, and standards related to clean hydrogen. The European Investment Bank has outlined plans to contribute one billion euros in support of India's green hydrogen ecosystem. In addition to cooperation within the European Union, Japan, Singapore, and South Korea stand out as primary export destinations for green hydrogen from India. Exploratory talks are underway for bilateral agreements with the EU, targeting 10 MMT per annum of green hydrogen, and with Singapore, aiming for 5 MMT per annum of green ammonia [109].

9.6 Australia's hydrogen technology ambitions

Australia's inaugural National Hydrogen Strategy, ratified by all Australian governments and disclosed in 2019, delineated 57 actions and principles that outlined the preliminary measures for Australia's venture into establishing a large-scale domestic hydrogen industry. Embracing a flexible approach to hydrogen industry development, the Energy and Climate Change Ministerial Council (ECMC) endorsed a review of the National Hydrogen Strategy, referred to as the Strategy Review, on February 24, 2023. The primary objective of the review is to ensure that the national strategy positions Australia to emerge as a global hydrogen leader by 2030, encompassing both export activities and the decarbonization of Australian industries. Despite being the third nation to unveil a hydrogen strategy in 2019, over 30 nations have since introduced hydrogen strategies, reflecting the growing international enthusiasm for the hydrogen opportunity [110]. Australia's hydrogen strategy is centered around the pivotal role hydrogen technology must assume to fulfill the nation's pledge of achieving net-zero emissions by 2050 and reducing greenhouse gas emissions by 43% below 2005 levels by 2030. Additionally, various state and territory jurisdictions have expressed their own ambitions toward achieving a Net Zero emissions target. Figure 9.8 depicts Australia's energy mix target for 2070 [111] and Figure 9.9 illustrates the forecast cost ranges of hydrogen based on various production methods by 2050 [112].

Australia currently boasts an extensive investment pipeline featuring 35 projects dedicated to the production of green ammonia and other hydrogen chemical derivatives, with a combined value exceeding $150 billion. The year 2022 has witnessed a doubling of proposed projects in this sector within Australia, and a significant milestone was achieved with the financial investment decision on the inaugural 10 MW electrolyzer project in the country. Energy security concerns in Europe have led to competitive pricing of green ammonia at various points in 2022. The dual factors of fuel security apprehensions and aspirations for decarbonization have spurred substantial interest from Japan and South Korea in importing hydrogen chemical derivatives, such as methanol and ammonia, from Australia. The decarbonization of existing ammonia production presents a relatively straightforward opportunity, given that the industry already utilizes 425,000 tons of hydrogen annually, produced from unabated fossil fuels.

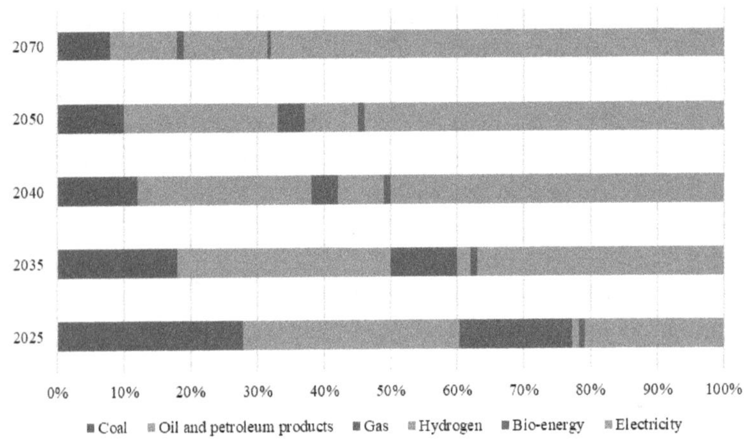

Figure 9.8 Australia's energy mix 2025- 2070

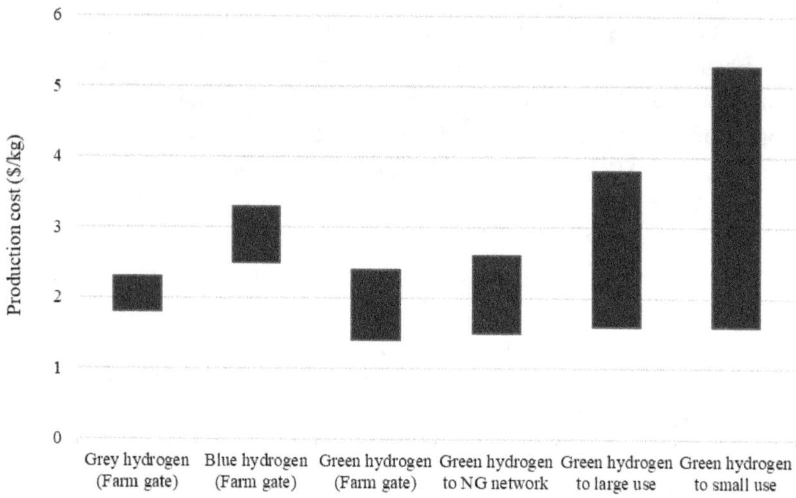

Figure 9.12 Hydrogen cost range in Australia by 2050

The current CO_2 emissions from domestic production, catering to mining explosives and the agricultural sector, amount to approximately 4 million tons per year. Beyond the direct reduction in carbon emissions, a widespread transition in this industry would rapidly enhance the economics of Australia's hydrogen production projects, with over 35 projects specifically geared toward ammonia production for both domestic and international markets. This significant shift could also mitigate investment risks in the domestic electrolyzer manufacturing sector, as it would ultimately necessitate around 7,000 megawatts of electrolysis- a stark contrast to the cumulative global installed capacity of electrolyzers, which stood at 513 megawatts at the close of 2021 [113].

Almost half of Australia's end-use energy is attributed to its industrial sector, with around half of this energy being utilized for process heat generation. This is predominantly achieved through the direct combustion of fossil fuels [114]. Hydrogen presents a viable alternative to natural gas in various industrial processes, particularly those demanding elevated temperatures. This substitution is applicable in sectors such as industrial heating for iron pellet production, alumina refining, and food processing [115–117]. By transitioning to renewable hydrogen for calcination in alumina refining alone, annual savings of 3.5 $MtCO_2e$ could be realized. Collectively, these changes could contribute to reducing Australia's emissions from industrial processes and product use by 23.7%, equivalent to removing 1.6 million cars from the nation's roads annually [111].

Hydrogen can actively contribute to the decarbonization and restructuring of national electricity grids by providing a schedulable load, effectively storing excess energy during periods of surplus generation, and releasing it to meet peak electricity demand when needed. This strategy aligns well with the characteristics of intermittent renewable electricity generation. Australia stands out as a leading proponent of using hydrogen for grid firming, exemplified by the early initiatives such as the 200 MW hydrogen-fired power station in Whyalla by the South Australian Government, the ongoing construction of the Tallawarra B gas generator, and the Hunter Power Project [118].

Australian companies are actively engaged in the production and testing of the initial fleet of hydrogen buses, while the exploration of hydrogen-powered trucks and mining vehicles is underway to facilitate the reduction of greenhouse gas emissions from the heavy road transportation sector, responsible for nearly 40% of emissions in Australia's transport sector. Additionally, Australia is at the forefront of global initiatives in designing hydrogen-powered ferries [119]. Currently, Australia's contribution to fuel oil for international maritime shipping is relatively small compared to its share of global seaborne cargo (by weight). This is attributed to the common practice of bunkering at ports with more cost-effective fuel oil sources. The maritime industry is making significant strides in decarbonizing its operations. In a future where ships utilize hydrogen or equivalent derivatives like ammonia or methanol, bunkering practices may shift to occur in Australia, capitalizing on the country's cost-effective hydrogen production. An example of this trend is the 2022 Collaboration Agreement between Yara Clean Ammonia and Pilbara Ports Authority, aiming to promote the adoption of ammonia as a marine fuel in Western Australia's Pilbara region [120].

Numerous mining and remote communities in Australia currently depend on diesel or natural gas generators for their electricity needs. Studies highlight that hydrogen has achieved cost parity with battery-diesel hybrid systems in remote power services [112]. Substituting diesel with hydrogen presents a tangible short-term opportunity, delivering health and air quality benefits, along with potential synergies with associated heavy vehicle fleets. Australian startups are actively innovating in this domain, creating new prospects for domestic manufacturing. Beyond catering to the energy needs of rural and remote areas in Australia, such technologies could support the transition from diesel generators to clean hydrogen generators in the Pacific Islands and certain ASEAN nations, incorporating clean hydrogen into renewable microgrids. Both internationally and domestically, hydrogen is already in use for micro-grid projects serving civilian and defense applications.

Hydrogen aviation is gaining momentum in Australia as a promising solution for decarbonizing the aviation industry. Several initiatives and collaborations are underway to develop hydrogen-powered aircraft and related infrastructure. For instance, the Hydrogen Flight Alliance (HFA), a nine-partner consortium in the state of Queensland, is targeting 2026 for the nation's first hydrogen-electric commercial flight, a 70-minute journey from Brisbane to Gladstone. The alliance aims to develop the green hydrogen flight ecosystem and ensure Australia's leading role in the aviation industry's transition towards net-zero emissions by 2050 [121,122]. Additionally, companies like Stralis Aircraft are working on the design and manufacture of emission-free hydrogen electric aircraft, with plans to modify a Beechcraft 1900D for hydrogen power by 2026 [123]. These efforts reflect a growing focus on hydrogen aviation as a key enabler for achieving net-zero emissions in the aerospace industry. The critical factor for Australia lies in attaining extremely low-cost renewable energy generation on a large scale to power both the direct air capture devices and the electrolyzers.

9.6.1 Australia's hydrogen diplomacy

Australia, with its abundant renewable resources, is well-positioned to play a crucial role in the global hydrogen landscape. Strategic hydrogen diplomacy can strengthen Australia's partnerships with key regions such as Europe, the U.S., and others, fostering international collaboration in both hydrogen production and technology development (Figure 9.10) [124]. Australia and Europe can forge a formidable partnership centered around green hydrogen production. European countries, driven by ambitious renewable energy targets, can benefit from Australia's vast solar and wind resources. For instance, the Australia-Germany Hydrogen Accord aims to produce the cheapest green hydrogen in the world, with a focus on technological innovation, research, development, and deploy [125,126]. Diplomatic efforts should focus on establishing long-term agreements for the export of Australian green hydrogen to Europe. Collaborative initiatives, including joint investments in renewable energy projects and the development of large-scale electrolysis facilities, can ensure a stable and cost-effective supply of green hydrogen. Bilateral discussions on regulatory harmonization and the creation of common standards will be crucial for seamless integration into the European market. EU policymakers are looking to quadruple existing targets for green hydrogen, with half of the 20 million tons targeted for 2030 expected to be imported, and Australia is seen as a prime market for this import [127]. Such collaboration not only aligns with Europe's decarbonization goals but also enhances Australia's standing as a major contributor to the global green hydrogen market.

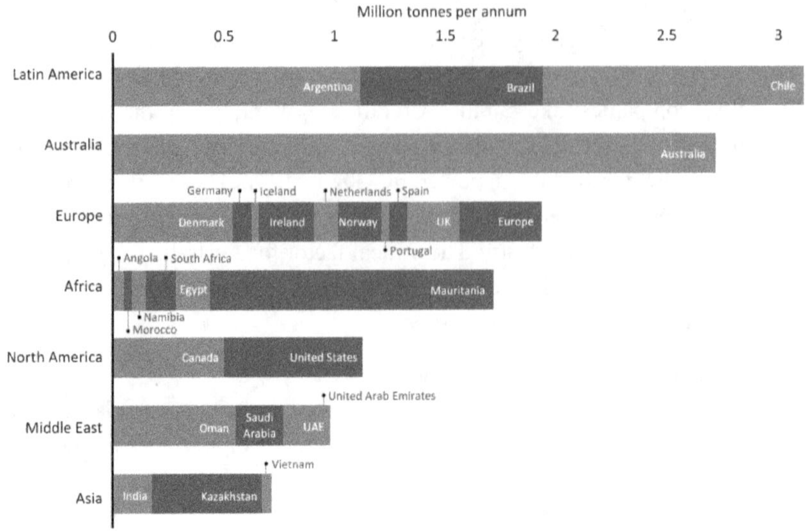

Figure 9.10 Australian hydrogen projects targeting export markets [124]

Collaboration between Australia and the U.S. in the hydrogen sector holds significant promise, particularly in the context of technology development and trade. One significant example is the formal partnership to accelerate the development and deployment of zero-emissions technology, as well as cooperation on critical minerals and clean energy supply chains [128]. Additionally, an Australia-based company has outlined plans to expand its research into hydrogen fuel in New Mexico, demonstrating practical cooperation between the two countries [129]. Australia's potential to become a green hydrogen production hub aligns with the U.S.'s increasing focus on clean energy and hydrogen utilization. Diplomatic efforts should concentrate on facilitating technology transfer, research collaboration, and the establishment of joint ventures between Australian and U.S. companies. The U.S. can benefit from Australia's expertise in large-scale renewable projects and electrolysis, while Australia can leverage U.S. advancements in hydrogen storage, transportation, and utilization technologies. Trade agreements can ensure a reciprocal flow of expertise, resources, and hydrogen products. As both nations work towards energy security and environmental sustainability, a strategic alliance can be mutually beneficial.

Australia's geographic proximity to the Asia-Pacific region positions it as a key player in fostering hydrogen cooperation with neighboring countries. Diplomatic initiatives should focus on creating a regional framework for hydrogen trade, technology exchange, and collaborative research. Australia's green hydrogen exports can meet the growing demand in countries like Japan and South Korea, which are actively pursuing hydrogen as part of their energy transition strategies. By engaging in multilateral agreements, Australia can contribute to the development of a robust regional hydrogen market. For instance, Australia and Singapore have agreed to pursue a Memorandum of Understanding on low-emissions technologies, including hydrogen cooperation [130]. In addition, Australia and Japan released a joint statement just two months after the launch of the National Hydrogen Strategy, highlighting their cooperation on hydrogen and fuel cells. One of the key projects underpinning the Australia-Japan relationship is the Hydrogen Energy Supply Chain project, which aims to produce hydrogen from brown coal in Australia and export it to Japan [131,132]. These initiatives demonstrate Australia's proactive approach to international engagement in the hydrogen sector, aiming to secure future export markets, address global climate objectives, and enhance economic cooperation and trade within the Asia-Pacific region. Joint ventures, knowledge-sharing initiatives, and capacity-building programs can further strengthen ties within the Asia-Pacific community, facilitating a collective effort toward sustainable and clean energy development.

Australia's hydrogen diplomacy can extend to the Middle East, where there is significant potential for collaboration in blue hydrogen production. Diplomatic efforts should focus on creating partnerships for the export of Australian technology and expertise related to carbon CCS. The Middle East, with its vast natural gas resources, can benefit from Australia's knowledge in developing large-scale CCS projects. Simultaneously, Australia can gain access to Middle Eastern markets for the export of blue hydrogen. Establishing a collaborative framework will involve negotiations on technology transfer, investment agreements, and shared research initiatives. Such collaboration can position Australia as a global leader in blue hydrogen technology and contribute to the establishment of a sustainable hydrogen value chain [133]. The Middle East offers opportunities for Australian investors interested in leveraging the region's potential for green hydrogen production. For example, Fortescue Future Industries has signed a partnership agreement with Jordan to explore the production of green hydrogen, indicating the growing interest and opportunities for collaboration in the region [134].

Australia can explore hydrogen cooperation with African nations, leveraging its experience in renewable energy projects and green hydrogen production. The continent of Africa is uniquely positioned to become a major producer of green hydrogen, given its vast potential for renewable energy, particularly solar and wind power [135]. Diplomatic efforts should aim at creating partnerships for joint ventures, capacity-building programs, and the establishment of pilot projects. Australia's expertise in large-scale electrolysis, coupled with Africa's potential for renewable energy projects, can contribute to the development of a vibrant hydrogen market. Trade agreements and technology transfer initiatives will be crucial components of this collaboration. By engaging with African nations, Australia can

contribute to the expansion of green hydrogen production globally, fostering sustainable development and creating economic opportunities on the continent.
The East Africa Hydrogen Corridor is an example of the potential for green hydrogen in Africa, with several countries working to interconnect their evolving hydrogen ecosystems [136]. International collaboration is seen as key to accelerating clean hydrogen, with the Australian government already allocating significant funds to support clean hydrogen initiatives [137]. By engaging with African nations, Australia can contribute to the expansion of green hydrogen production globally, fostering sustainable development and creating economic opportunities on the continent. This collaboration has the potential to benefit both Australia and African nations, contributing to the global effort towards a more sustainable and low-emission energy future.

Australia's leadership in hydrogen diplomacy should extend to multilateral platforms and global initiatives. Active participation in forums such as the Hydrogen Initiative of the Clean Energy Ministerial, the International Partnership for Hydrogen and Fuel Cells in the Economy, and the Mission Innovation Hydrogen Challenge can strengthen Australia's diplomatic standing and facilitate international collaboration [138–140]. By contributing to the development of international standards, advocating for supportive policies, and fostering research partnerships, Australia can play a central role in shaping the global hydrogen agenda.

9.7 Hydrogen, next black gold in the Middle East

The transition to renewable energy is perceived to heighten security risks and geopolitical tensions, particularly in relation to critical materials and cybersecurity, and the nations previously reliant on hydrocarbon exports are likely to face significant challenges and emerge as the primary losers from this energy transition [141]. Hence, Middle Eastern countries find their future significantly affected by ongoing energy transitions. These nations possess some of the world's largest reserves and have economies that heavily rely on income generated from hydrocarbon rents, whether through export revenue in exporting economies or foreign remittances from nationals employed in exporting states. Consequently, a decline in global demand for hydrocarbons has consistently posed a substantial existential threat to the region. Having experienced a prolonged period of prosperity driven by fossil fuels, these countries are now exploring the potential of hydrogen to transition their economies and address the climate crisis comprehensively.

Clean hydrogen, whether produced through blue or green methods, presents significant possibilities for export and economic diversification for Middle Eastern countries. These nations, leveraging their established expertise in the hydrocarbon sector, possess a distinct advantage in blue hydrogen, benefiting from well-established trade routes and markets. In the context of blue hydrogen production, the Middle East boasts the lowest cost globally, amounting to USD 0.43 per kilogram [142]. The Middle East, with a potential comparative advantage in renewable energy, may also find opportunities in green hydrogen. While the global role of clean hydrogen in the energy landscape remains uncertain, its potential market, including derivatives like ammonia, is estimated at $400–$700 billion, with the Gulf Cooperation Council (GCC) states potentially capturing $70–$200 billion of that market [143]. By 2050, the region's hydrogen consumption is projected to range from 25 to 50 megatons, with substantial demand expected in Saudi Arabia and Oman, along with robust markets in Qatar, Bahrain, and the United Arab Emirates (UAE) [144]. The hydrogen market is poised to cause one of the most significant disruptions in the energy sector, and the Middle East, known for being among the world's most cost-effective producers of oil, natural gas, and renewables, is strategically positioned to engage in competitive production of both blue and green hydrogen. The Middle East had been slower in embracing hydrogen strategies in comparison to Europe, advanced Asian economies, Australia, and the United States. With well-equipped port facilities and essential infrastructure for hydrogen exportation, Saudi Arabia, the UAE, and Oman are prominent contenders in hydrogen production and exportation (Figure 9.11). Oman, in particular, is anticipated to emerge as the leading hydrogen exporter in the Middle East in the current decade [145]. Recognizing the potential, Middle Eastern countries are increasingly embracing the idea of establishing themselves as major players in the emerging hydrogen industry.

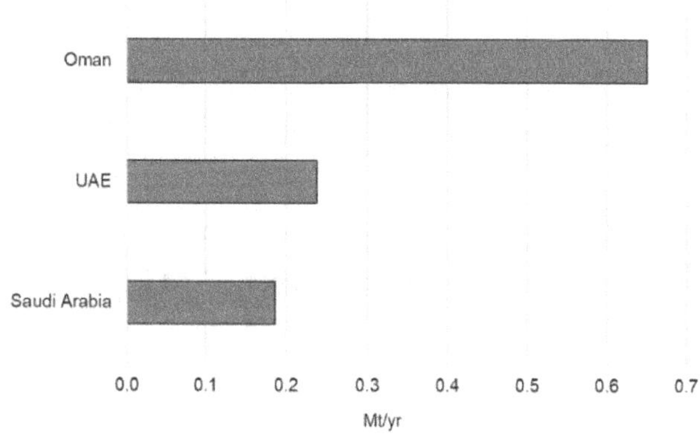

Figure 9.11 Hydrogen export in the selected Middle Eastern countries by 2030

9.7.1 Oman

In October 2022, Oman's government introduced its green hydrogen strategy, outlining a production target of 1 million to 1.25 million tons of green hydrogen by 2030. This target is expected to increase to 3.25 million to 3.75 million tons annually by 2040 and further escalate to 7.5 million to 8.5 million tons per year by 2050. The government has projected that a total investment of USD 140 billion until 2050 would be necessary to achieve these goals, in addition to the USD 190 billion cumulative investments required for the National Strategy for an Orderly Transition to Net Zero. Additionally, Oman established an independent entity, Hydrom, to guide the accelerated development of the hydrogen sector [146]. Oman has committed to achieving Net Zero Emissions by 2050 and has taken significant steps to lead this transition, including the establishment of the Oman Sustainability Centre. This center will oversee and implement plans and programs aimed at carbon neutrality. The energy minister emphasized the crucial role of green hydrogen in Oman's pursuit of decarbonization, economic development, and energy security. Oman aspires to be one of the world's largest green hydrogen producers and exporters, with a target of one million tons annually by 2030. Throughout 2022, the Ministry of Energy and Minerals formulated a green hydrogen strategy and outlined the sector's structure. To spearhead the country's green hydrogen strategy, the government has founded Hydrom Oman, a fully owned autonomous subsidiary of Energy Development Oman SAOC (EDO), regulated by the Ministry of Energy and Minerals. Hydrom's responsibilities include delineating government-owned land areas, structuring large-scale green hydrogen projects, allocating land to developers, and facilitating the development of common infrastructure and associated industries and hubs in collaboration with the Public Authority for Special Economic Zones and Free Zones (OPAZ). According to the IEA's analysis of the existing project pipeline, Oman has the potential to emerge as the sixth-largest global exporter of hydrogen by 2030 and the leading exporter in the Middle East region. Oman's share would constitute 61% of total hydrogen exports from the Middle East, equivalent to 1.1 million tons of hydrogen per year, with the United Arab Emirates contributing 20% and Saudi Arabia 16%. To provide context for Oman's current hydrogen goals, it's helpful to compare them to the existing level of LNG trade. In 2021, Oman was the third-largest LNG exporter in the Middle East-North Africa region, shipping almost 14 billion cubic meters (560 petajoules [PJ]) of LNG. According to the IEA's analysis of the present project pipeline, the projected 78 PJ (0.65 Mt) hydrogen exports in 2030 would amount to only one-seventh of the current LNG exports. However, achieving the renewable hydrogen production targets for 2040 would already represent 80% of the current LNG export volumes in terms of energy equivalence. Meeting the 2050 target of 1,020 PJ (8.5 Mt) would significantly surpass current LNG volumes. If Oman successfully achieves its 2050 hydrogen production target, the country could secure a 1.8% share of the global hydrogen market. This percentage is greater than Oman's current contribution to the global supply of oil and natural gas, which stands at 0.8% and 0.3%, respectively.

By leveraging Oman's abundant high-quality renewable resources alongside the global cost reductions in electrolyzers, PV, and wind technologies, the cost of renewable hydrogen is anticipated to witness a substantial decline by 2030. Assuming the continued alignment of hydrogen deployment with global climate objectives, the levelized cost of renewable hydrogen derived from solar PV and wind by 2030 could potentially reach as low as USD 1.6/kg H_2 in Oman, USD 1.4/kg H_2 in Chile, USD 1.7/kg H_2 in the United States, and USD 1.9/kg H_2 in Australia. In contrast, countries and regions like Central Europe, Japan, and South Korea may experience relatively higher levelized costs ranging between USD 2.6/kg H_2 and USD 3.8/kg H_2 (Figure 9.12) [147]. These elevated production costs and more restricted production potentials due to higher population densities are likely to position these countries or regions as global importers of low-emission hydrogen and its derivatives. To fulfill Oman's renewable hydrogen production targets, achieving the 2030 goal of 1 million tons of H_2 production would necessitate 10 million tons of purified water. The 2040 objective of 3.25 to 3.75 million tons of H_2 would require 33 to 38 million tons of purified water, and the 2050 target of 7.5 to 8.5 million tons of H_2 would demand 75 to 85 million tons of purified water. The international trade of ammonia and the infrastructure essential for its storage and transport are firmly established and technologically mature. Countries endowed with abundant natural gas resources export ammonia globally, primarily for use as a feedstock in fertilizer production. Ammonia is traded in its anhydrous form, existing as a gas under normal conditions, and transported in fully refrigerated liquefied tankers at -33°C.

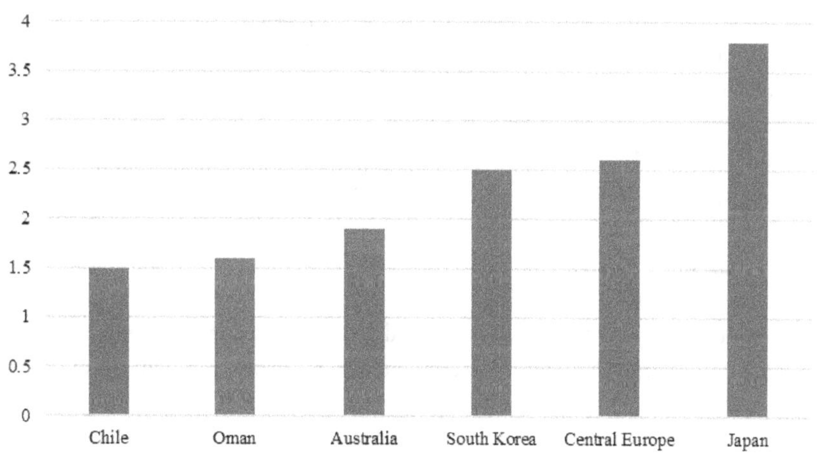

Figure 9.12 Green hydrogen production costs in selected potential export countries and import markets in 2030 ($)

Despite its importance, global trade is relatively small compared to production, as other ammonia derivatives, particularly urea, offer a more convenient means of transporting nitrogen for fertilizers in solid bulk cargo form. Oman currently produces 2.2 Mt of ammonia, primarily for export purposes. In 2020, approximately one-tenth of the ammonia produced was directly exported, while the majority (1.9

Mt) underwent further processing to produce and export urea. Oman has set a target to produce 1 Mt to 1.25 Mt of hydrogen from renewable electricity by 2030, catering to both domestic demand and export. Since renewable hydrogen exports are expected to be directed mainly to Europe and some countries in the Asia Pacific region, shipping will be the primary mode of transport. Given the current maturity of technologies and supply chain status, exports are anticipated to take the form of ammonia. As of 2022, Oman had only one terminal in the port of Sur for handling ammonia. However, a new export terminal located in the Salalah Free Zone in Dhofar Governorate became operational in September 2022. If Oman were to export all its renewable hydrogen in the 2030 target as ammonia, it could amount to up to 6 Mt, more than 30 times the country's current ammonia exports. This would necessitate substantial growth in new port infrastructure, including storage tanks and dedicated deepwater jetties. When planning the location of export terminals, the technical safeguards and safety distance requirements for ammonia handling should be considered, allowing for adequate space for new terminals and their potential expansion beyond 2030. Based on existing announcements regarding the location of renewable hydrogen projects for export, Oman would need ammonia export terminals at the Port of Duqm and in Salalah by 2030. Salalah could potentially export up to 0.7 Mt per year of ammonia, utilizing the recently commissioned ammonia export facility, with the potential for expansion if needed. The port of Duqm could export about 2.8 Mt per year of renewable ammonia, requiring an installed storage capacity of at least 160 kt and the construction of several large storage tanks. In total, exporting 6 Mt to 7 Mt of renewable ammonia would necessitate at least 400 kt of ammonia storage across different ports.

New hydrogen projects should take into account the proximity to a deepwater port capable of hosting an ammonia export terminal or the construction of hydrogen or ammonia pipelines leading to such a port. Additionally, given that the completion of ammonia storage tanks typically takes more than three years, planning and construction of export facilities would need to commence in the coming years to meet the 2030 targets. Oman's expected ammonia exports by 2030, based on its targets, would be roughly 30% of today's global ammonia trade, requiring 12 to 14 very large ammonia tankers dedicated year-round for transportation [148]. As few countries can manufacture these tankers, securing shipbuilding capacity well in advance is crucial, considering the competing demand for other types of liquefied gas tankers and vessels in general.

Oman's ambition to become a major exporter of green hydrogen in the form of ammonia is driving significant developments in its export infrastructure. The country currently exports around 200,000 tons of ammonia per year, but this capacity will need to increase significantly to realize its hydrogen targets. According to the IEA, Oman could annually export over 670,000 tons of renewable hydrogen by 2030, mostly in the form of ammonia, which would require 20-30 times more export capacity than it has today [149]. Out of Oman's five industrial ports (Duqm,

Muscat, Salalah, Sohar, and Sur), the ports of Salalah, Sohar, and Sur have existing ammonia infrastructure and host ammonia terminals. Since hydrogen exports are expected to primarily target Europe and some countries in the Asia Pacific region, shipping will be the primary mode of transport. Given the current state of technologies and the supply chain, exports are anticipated to be in the form of ammonia. While Oman currently exports approximately 0.2 million tons per year of ammonia, this capacity must expand to realize its hydrogen targets in the form of ammonia [150]. Depending on the proportion of domestic hydrogen use, Oman may require up to 20-30 times more ammonia export capacity than today by 2030. Achieving this would necessitate significant expansion in new export infrastructure, particularly in ammonia storage tanks and dedicated deepwater jetties. Low-emission methanol could be another potential carrier for hydrogen, but it requires a sustainable source of CO_2 in addition to hydrogen. The projected levelized cost of renewable ammonia in Oman for 2030 stands at USD 400/ton. Taking into consideration shipping distances ranging from 10,000 km to 20,000 km, the comprehensive supply cost of ammonia, encompassing production, conversion, and transport, is elevated from USD 440/t NH_3 to USD 520/t NH_3 for Australia, Chile, and Oman. This positions Oman as a competitive provider of renewable ammonia, with supply costs on par with the higher range of ammonia market prices observed between 2010 and 2020 and significantly below the peak levels surpassing USD 1000/ton experienced globally in 2022 due to surges in natural gas prices [151].

Increasing domestic renewable supply can position Oman as an appealing destination for new energy-intensive industries looking to decarbonize their supply chains. Anticipated growth in demand for low-emission products aligns with the goals of governments and companies striving to achieve net-zero emissions. To maintain competitiveness while embracing decarbonization, these industries will seek access to both cost-effective and low-emission energy sources, potentially considering relocation to regions where such resources are readily available. Oman, with its potential for abundant, low-cost renewable electricity and hydrogen, coupled with extensive available land, stands out as one of the prime candidates to attract industry relocations amid the ongoing global energy transition.

9.7.2 United Arab Emirates

With the objective of achieving climate neutrality, the National Hydrogen Strategy of the UAE incorporates initiatives to boost low-carbon local industries by fostering the growth of the hydrogen economy (Figure 9.13) [142]. In September 2022, the UAE Ministry of Energy and Infrastructure (MoEI) entered into a partnership with Australia's GHD, collaborating with the German-based Fraunhofer-Gesellschaft to craft a National Hydrogen Strategy. This strategic initiative aims to position the UAE as one of the world's leading hydrogen producers by 2031, aiming to produce 1.4 million tons of low-carbon hydrogen annually through various production processes [152].

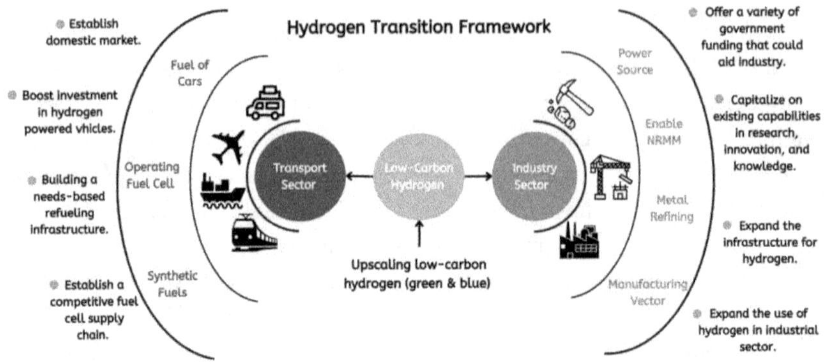

Figure 9.13 The UAE's hydrogen transition framework

The strategy envisions sectoral demand in the UAE reaching 2.1 million tons per year by 2031, along with an additional export potential of 0.6 million tons per year, under favorable market conditions. Concurrently, the strategy seeks to stimulate and democratize investments in hydrogen research and development. Looking ahead to a defossilized future, the Hydrogen Strategy authors from Fraunhofer CINES and GHD project that the UAE's low-carbon hydrogen production capacity could escalate to 7.5 million tons per year by 2040 and nearly 15 million tons per year by 2050 [153].

These plans include the production of both blue and green hydrogen, with the Dubai Electricity and Water Authority (DEWA) initiating the region's first industrial-scale green hydrogen project in collaboration with Siemens Energy [154]. The UAE's National Hydrogen Strategy also outlines the establishment of "hydrogen oases" or clusters, with the aim of reducing emissions in hard-to-abate industries. Additionally, the country has launched a "Hydrogen Alliance" to produce blue and green hydrogen using natural gas and renewable energy sources, with the intention of marketing it both domestically and overseas [155]. Most recently, in 2024, the Abu Dhabi government announced a new industrial complex to produce and store green hydrogen, further demonstrating the country's commitment to hydrogen production [156]. These initiatives reflect the UAE's strategic move towards sustainable energy production and its efforts to play a significant role in the global energy transition. The strategy outlines specific measures, including the establishment of "hydrogen oases"- hydrogen production hubs that will catalyze the domestic hydrogen market [157]. Additionally, a supportive policy framework will be implemented to foster the growth of the hydrogen industry and facilitate collaboration with neighboring countries in hydrogen production, transport, and utilization, paving the way for a global hydrogen market. A significant proportion of the UAE's energy is derived from natural gas and oil, utilized for electricity generation and various industrial processes demanding substantial heat. In the realm of power generation, the predominant share is held by gas, constituting 96.7%, while low-carbon sources contribute a comparatively modest 2.7% [142]. The researchers anticipate substantial domestic and export demand for hydrogen, driven by the UAE's 2050 net-zero target and global demand for eco-friendly hydrogen, sustainable aviation fuel, and environmentally sustainable chemical products.

To optimize the role of hydrogen in achieving net-zero emissions, the UAE is aiming for hydrogen consumption to reach 212 million metric tons by 2030. The country has set an ambitious target of capturing 25% of the global low-carbon hydrogen market by the same year. Currently, the Abu Dhabi National Oil Company (ADNOC) produces over 300 kt of hydrogen annually, with plans to expand production to 500 kt annually, particularly for decarbonizing high-polluting heavy sectors. The increased production of low-carbon hydrogen, for both domestic use and exportable goods like ammonia, is expected to require relatively modest additional investment. The UAE has aligned its policies with international organizations to accelerate hydrogen growth within the country. Collaborative projects, such as the waste-to-hydrogen initiative between British waste-to-energy company Chinook Sciences and Emirati waste management company Bee'ah, exemplify this commitment. The UAE envisions becoming a major exporter of green hydrogen to the South Asia market, facilitated by a potential pipeline connection. Consequently, large-scale solar applications are deemed a promising investment for the UAE to emerge as a leading green hydrogen exporter in the future [158]. Based on electrolyzer cost assumptions, solar predictions, and learning rates, UAE industries' green hydrogen production cost is projected to be between USD 0.95/kg and USD 1.35/kg between 2032 and 2038 [159].

In the realm of blue hydrogen, the country benefits from reliable and cost effective hydrocarbons, extensive facilities for ammonia and hydrogen production, and well-characterized underground formations suitable for carbon dioxide storage. Favorable conditions for green hydrogen are underpinned by outstanding solar generation capabilities and a low-levelized cost of solar power generation. The existing infrastructure, including versatile port facilities, adaptable LNG export and import terminals, gas pipelines, salt domes for hydrogen storage, a central geographical location connecting critical markets, and a consistently business-friendly and innovative approach, all contribute to the success of hydrogen projects. Furthermore, the UAE boasts an advanced network of petrochemical and refining facilities, including the Ruwais ammonia plant, and envisions potential expansions into methanol and synthetic fuel manufacturing [160]. The country is recognized for its reliable and sustainable investment environment, with a robust history of successful public-private collaborations involving both domestic and foreign companies across the energy and industrial sectors.

9.7.3 Saudi Arabia

Saudi Arabia's keen interest in hydrogen stems primarily from its strategic goal of ensuring economic security. The utilization of hydrogen aligns with the objectives outlined in Saudi Vision 2030, the country's ambitious plan to diversify exports, enhance local content within existing sectors' supply chains, and foster the development of new industrial segments [161]. The Kingdom aspires to position itself as the leading global hydrogen supplier, a strategic move that could reduce its dependence on oil as the primary income source in international markets. This shift is especially significant in the current era, marked by global efforts towards achieving carbon neutrality and the widespread adoption of net-zero targets by governments and industries worldwide [162]. Saudi Arabia possesses considerable expertise and established capacities in industrial-scale hydrogen and chemical production, as well as in the capture and storage of CO_2. This proficiency forms the foundation for Saudi Arabia's vision of commercially viable hydrogen production and export, emphasizing the utilization of hydrocarbon resources with CCUS of emissions, commonly known as blue hydrogen. Furthermore, the prospect of hydrogen production from renewable sources opens up a new industrial frontier for Saudi Arabia, given its extensive sunbelt territory with vast expanses of flat land suitable for solar panel installations [163]. Despite current limitations in renewable capacity, with utility-scale photovoltaic capacity at 330 megawatts as of the first half of 2020, Saudi Arabia installed approximately 27 gigawatts of solar capacity in 2023 and has ambitious plans to install nearly 58 gigawatts by 2030 [164].

The push for a clean hydrogen industry, particularly through the production of blue hydrogen, aligns with Saudi Arabia's commitment to advancing a circular carbon economy. This framework involves managing and reducing emissions through carbon reduction, reuse, recycle, and removal [165]. The domestic use of clean hydrogen, especially in transportation, would contribute to lowering the carbon footprint of Saudi economic activities. Additionally, Saudi Arabia could position itself as a key player in the global energy transition by actively promoting its clean hydrogen supply to countries reliant on hydrocarbons and seeking lower-carbon energy sources. Nevertheless, uncertainties remain about the impact of developing a clean hydrogen industry on Saudi Arabia's natural gas sector. Questions arise about whether the kingdom's interest in hydrogen exports will supersede its previous focus on exporting natural gas, whether blue hydrogen will create a new market for its natural gas, and how these developments will shape the kingdom's overall energy strategy.

In October 2021, Saudi Arabia expressed the ambition to become the world's largest hydrogen producer while concurrently leveraging existing infrastructure and expertise to advance export decarbonization. A formal strategy or roadmap for hydrogen deployment is reportedly in the developmental stages in Saudi Arabia. While specific priorities, policies, and comprehensive targets with timelines are yet to be outlined, statements from political leaders and corporate executives offer insights into the potential direction of Saudi Arabia's hydrogen initiatives. The kingdom's hydrogen strategy is expected to be intricately linked with its CCUS strategy, as the progress in expanding the applications of CO_2 use beyond enhanced oil recovery can significantly impact the commercial viability of blue hydrogen production. Saudi Arabia is actively pursuing the goal of becoming the world's leading hydrogen supplier and has set ambitious clean hydrogen production targets, aiming for 2.9 million tons per year (t/yr) by 2030 and 4 million t/yr by 2035 [166]. The primary emphasis in the near term is on securing a substantial market share in blue hydrogen, particularly in the form of blue ammonia, which involves ammonia synthesis utilizing hydrocarbon CCUS, within the next decade [167]. A significant milestone was achieved in September 2020 when Saudi Aramco, the state oil firm, successfully shipped 40 tons of blue ammonia from Saudi Arabia to Japan [168]. This marked the world's first demonstration of blue ammonia supply chains, showcasing the production and international maritime transportation of blue ammonia. The success of this project reinforced Aramco's belief in existing technology solutions for the extraction, processing, and conversion of natural gas into hydrogen and ammonia, providing cost-effective and scalable low-emission solutions. Collaboration with South Korea is evident through a Memorandum of Understanding between Aramco and Hyundai OilBank Co. Hyundai OilBank plans to receive liquefied petroleum gas (LPG) cargoes from Saudi Aramco, convert the LPG into hydrogen, and return the emitted CO_2 to Saudi Arabia [169].

In the futuristic city of Neom, Saudi Arabia is actively exploring renewables-based hydrogen as a key focus of technological and economic experiments. A notable project is the $5 billion green hydrogen initiative, a joint venture between Neom, ACWA Power based in Riyadh, and Air Products headquartered in Pennsylvania. The project, valued at $8.4 billion, is expected to be operational in 2026. With a 4 GW renewable capacity, this project aims to be the world's largest renewable hydrogen-to-ammonia facility, producing 1.2 million tons per year of green hydrogen- equivalent to about 5 million barrels of oil per year in energy terms [170]. The kingdom is also keen on capitalizing on its abundant natural gas resources. A notable example is the plan to utilize a substantial portion of the natural gas reserves from the Jafurah field for the production of blue hydrogen. Situated approximately equidistant between Riyadh and the Saudi-Qatari border in the eastern part of the kingdom, the Jafurah field stands as Saudi Arabia's largest discovered non-associated gas reservoir, boasting 200 trillion cubic feet of gas. Aramco aims to commence gas production from this $110 billion project in 2024 [171]. Simultaneously, the development of the city of Neom, located in the northwest corner of the country along the Red Sea, is anticipated to play a pivotal role in Saudi Arabia's endeavors to establish a significant supply of green hydrogen.

Saudi Arabia also intends to explore hydrogen production through nuclear-powered electrolysis. Although the country currently lacks nuclear power generation infrastructure, it has outlined plans to incorporate approximately 17 GW of nuclear capacity by 2040. Additionally, there are ambitions to commence operations for two reactors with a combined capacity of 3.2 GW within the coming decade.

On the consumption side, Saudi Arabia is actively exploring hydrogen use in the transportation sector. Initiatives include testing Toyota Motor's Mirai FCEVs and Hyundai Motor's export of FCEVs to Saudi Arabia. There is also consideration for establishing manufacturing capacities for FCEVs in Saudi Arabia, involving companies from North America and Europe, such as Hyzon Motors and Gaussin. The initiatives align with the country's broader push to decarbonize and its commitment to sustainable development, as outlined in the Saudi Vision 2030. The total value of announced investments in hydrogen in the Middle East is set to hit $44 billion, with around $35 billion worth of the announced projects set to be operational by 2030 [172].

Abbreviations

APAC: Asia and the Pacific
AWE: Alkaline water electrolysis
CCS: Carbon capture and storage
CCUS: Carbon capture, utilization, and storage
DEWA: Dubai Electricity and Water Authority
DRI: Direct Reduced Iron
ECMC: Energy and Climate Change Ministerial Council
FCV: Fuel Cell Vehicle
FCEV: Fuel Cell Electric Vehicles
FH2R: Fukushima Hydrogen Energy Research Field
FID: Final Investment Decision
GCC: Gulf Cooperation Council
GHG: Greenhouse gas
HEM: Hydrogen Energy Ministerial Meeting
HFA: Hydrogen Flight Alliance
HRS: Hydrogen refueling stations
HTGR: High-Temperature Gas-Cooled Reactor
IOCL: Indian Oil Corporation Limited
KOGAS: Korea Gas Corporation
LCOE: Levelized costs of electricity
LNG: Liquefied natural gas
MCH: Methylcyclohexane
NDC: Nationally Determined Contributions
NDRC: National Development and Reform Commission
PEM: Proton exchange membrane
PFSR: Perfluorosulfonic Acid Resins
PRC: People’s Republic of China
PV: Photovoltaic
R&D: Research and development
REC: Renewable energy certificate

SOE: State-owned enterprise
SOEC: Solid oxide electrolyzer
STA: State Taxation Administration
UAE: United Arab Emirates

References

[1] CO_2 and Greenhouse Gas Emissions - Our World in Data n.d. https://ourworldindata.org/co2-and-greenhouse-gas-emissions (accessed January 20, 2024).

[2] The Future of Energy Transition: A Look at the Asia-Pacific Region's Hydrogen Strategies - PTR Inc. n.d. https://ptr.inc/asia-pacific-emerging-as-the-hydrogen-powerhouse/ (accessed January 21, 2024).

[3] Fortune India: Business News, Strategy, Finance and Corporate Insight n.d. https://www.fortuneindia.com/macro/green-hydrogen-costs-to-halve-to-160-170-per-kg-by-2030/107549 (accessed January 21, 2024).

[4] Seetha Ram K, Zhang H. Hydrogen in Decarbonization Strategies in Asia and the Pacific. Hydrogen in Decarbonization Strategies in Asia and the Pacific 2023. https://doi.org/10.56506/JPUT8568.

[5] Nnabuife SG, Oko E, Kuang B, Bello A, Onwualu AP, Oyagha S, et al. The prospects of hydrogen in achieving net zero emissions by 2050: A critical review. Sustainable Chemistry for Climate Action 2023;2:100024. https://doi.org/10.1016/J.SCCA.2023.100024.

[6] Griffiths S, Sovacool BK, Kim J, Bazilian M, Uratani JM. Industrial decarbonization via hydrogen: A critical and systematic review of developments, socio-technical systems and policy options. Energy Res Soc Sci 2021;80:102208. https://doi.org/10.1016/J.ERSS.2021.102208.

[7] Ministry of New and Renewable Energy | Ministry of New and Renewable Energy | India n.d. https://mnre.gov.in/ (accessed January 21, 2024).

[8] Kar SK, Sinha ASK, Harichandan S, Bansal R, Balathanigaimani MS. Hydrogen economy in India: A status review. Wiley Interdiscip Rev Energy Environ 2023;12:e459. https://doi.org/10.1002/WENE.459.

[9] Green hydrogen economy - predicted development of tomorrow: PwC n.d. https://www.pwc.com/gx/en/industries/energy-utilities-resources/future-energy/green-hydrogen-cost.html#qualifying-the-opportunity-in-green-hydrogen (accessed January 21, 2024).

[10] Current M, Qin T, Xiao G, Yu W, Yang L, Wang S, et al. Current Development Status, Policy Support and Promotion Path of China's Green Hydrogen Industries under the Target of Carbon Emission Peaking and Carbon Neutrality. Sustainability 2023, Vol 15, Page 10118 2023;15:10118. https://doi.org/10.3390/SU151310118.

[11] China unveils national 2025 target for green hydrogen and new strategies for further H2 growth | Recharge n.d. https://www.rechargenews.com/energy-transition/china-unveils-national-2025-target-for-green-hydrogen-and-new-strategies-for-further-h2-growth/2-1-1189877 (accessed January 25, 2024).

[12] China maps 2021-2035 plan on hydrogen energy development n.d. https://en.ndrc.gov.cn/news/pressreleases/202203/t20220329_1321487.html (accessed January 21, 2024).

[13] China's National Hydrogen Development Plan - Energy Iceberg n.d. https://energyiceberg.com/national-hydrogen-development-plan/ (accessed January 25, 2024).

[14] Global Hydrogen Review 2023 – Analysis - IEA n.d. https://www.iea.org/reports/global-hydrogen-review-2023 (accessed November 27, 2023).

[15] China's hydrogen roadmap: 4 things to know - Nikkei Asia n.d. https://asia.nikkei.com/Spotlight/Caixin/China-s-hydrogen-roadmap-4-things-to-know (accessed January 21, 2024).

[16] China sets green hydrogen target for 2025, eyes widespread use | Reuters n.d. https://www.reuters.com/world/china/china-produce-100000-200000-t-green-hydrogen-annually-by-2025-2022-03-23/ (accessed January 21, 2024).

[17] China Fuel Cell "Subsidy" Policy: Game Plan Breakdown - Energy Iceberg n.d. https://energyiceberg.com/china-fuel-cell-subsidy-design/ (accessed January 25, 2024).

[18] Mao S, Zhang Y, Bieker G, Rodriguez F. Zero-emission bus and truck market in China: A 2021 update 2023. https://theicct.org/wp-content/uploads/2023/01/china-hvs-ze-bus-truck-market-2021-jan23.pdf (accessed January 25, 2024).

[19] Hyundai opens 1st overseas hydrogen fuel cell plant in China - The Korea Times n.d. https://www.koreatimes.co.kr/www/tech/2023/07/129_352169.html (accessed January 25, 2024).

[20] Sinopec's first green hydrogen plant starts production in Xinjiang | Reuters n.d. https://www.reuters.com/business/energy/sinopecs-first-green-hydrogen-plant-xinjiang-starts-production-xinhua-2023-06-30/ (accessed January 25, 2024).

[21] China scaling up electrolyzer manufacturing base for domestic, export markets | S&P Global Commodity Insights n.d. https://www.spglobal.com/commodityinsights/en/market-insights/latest-news/energy-transition/120621-china-scaling-up-electrolyzer-manufacturing-base-for-domestic-export-markets (accessed January 22, 2024).

[22] International Energy Agency I. Global Hydrogen Review 2021 n.d. https://iea.blob.core.windows.net/assets/5bd46d7b-906a-4429-abda-e9c507a62341/GlobalHydrogenReview2021.pdf (accessed November 27, 2023).

[23] Ji M, Wang J. Review and comparison of various hydrogen production methods based on costs and life cycle impact assessment indicators. Int J Hydrogen Energy 2021;46:38612–35. https://doi.org/10.1016/J.IJHYDENE.2021.09.142.

[24] China tests hydrogen production through direct seawater electrolysis at Xinghua Bay OWF - Offshore Energy n.d. https://www.offshore-energy.biz/china-tests-hydrogen-production-through-direct-seawater-electrolysis-at-xinghua-bay-owf/ (accessed January 25, 2024).

[25] 4.0 kWh/Nm^3! LONGi Hydrogen launches new generation of electrolyzed water hydrogen production equipment ALK Hi1 - LONGi n.d. https://www.longi.com/in/news/hydrogen-alk-hi1-product-launch/ (accessed January 25, 2024).

[26] China's hydrogen roadmap: 4 things to know - Nikkei Asia n.d. https://asia.nikkei.com/Spotlight/Caixin/China-s-hydrogen-roadmap-4-things-to-know (accessed January 22, 2024).

[27] China's Solar Giants Make a Bid to Dominate Hydrogen Power - Bloomberg n.d. https://www.bloomberg.com/news/articles/2021-12-12/china-s-solar-giants-make-a-bid-to-dominate-hydrogen-power?sref=B2BBHw9t (accessed January 22, 2024).

[28] Nakano J. China Unveils its First Long-Term Hydrogen Plan n.d. https://www.csis.org/analysis/china-unveils-its-first-long-term-hydrogen-plan (accessed January 21, 2024).

[29] Sadik-Zada ER, Gatto A. Energy Security Pathways in South East Europe: Diversification of the Natural Gas Supplies, Energy Transition, and Energy Futures. From Economic to Energy Transition 2021:491–514. https://doi.org/10.1007/978-3-030-55085-1_17.

[30] Pandey RP, Shahi VK. Sulphonated imidized graphene oxide (SIGO) based polymer electrolyte membrane for improved water retention, stability and proton conductivity. J Power Sources 2015;299:104–13. https://doi.org/10.1016/J.JPOWSOUR.2015.08.093.

[31] Pedicini R, Carbone A, Saccà A, Gatto I, Di Marco G, Passalacqua E. Sulphonated polysulphone membranes for medium temperature in polymer electrolyte fuel cells (PEFC). Polym Test 2008;27:248–59. https://doi.org/10.1016/J.POLYMERTESTING.2007.11.002.

[32] Zhu Y, Li H, Tang J, Wang L, Yang L, Ai F, et al. Systematic stability investigation of perfluorosulfonic acid membranes with varying ion exchange capacities for fuel cell applications. RSC Adv 2014;4:6369–74. https://doi.org/10.1039/C3RA46182C.

[33] Jiang B, Pu H, Pan H, Chang Z, Jin M, Wan D. Proton conducting membranes based on semi-interpenetrating polymer network of fluorine-containing polyimide and perfluorosulfonic acid polymer via click chemistry. Electrochim Acta 2014;132:457–64. https://doi.org/10.1016/J.ELECTACTA.2014.03.135.

[34] Othman R, Dicks AL, Zhu Z. Non precious metal catalysts for the PEM fuel cell cathode. Int J Hydrogen Energy 2012;37:357–72. https://doi.org/10.1016/J.IJHYDENE.2011.08.095.

[35] Li A, Kong S, Guo C, Ooka H, Adachi K, Hashizume D, et al. Enhancing the stability of cobalt spinel oxide towards sustainable oxygen evolution in acid. Nature Catalysis 2022 5:2 2022;5:109–18. https://doi.org/10.1038/s41929-021-00732-9.

[36] Sun T, Xu L, Wang D, Li Y. Metal organic frameworks derived single atom catalysts for electrocatalytic energy conversion. Nano Research 2019 12:9 2019;12:2067–80. https://doi.org/10.1007/S12274-019-2345-4.

[37] Uncover the secrets of the proton exchange membrane! - iMedia n.d. https://min.news/en/news/92024a3aa3d870997e387c58143b42a5.html (accessed January 26, 2024).

[38] Atilhan S, Park S, El-Halwagi MM, Atilhan M, Moore M, Nielsen RB. Green hydrogen as an alternative fuel for the shipping industry. Curr Opin Chem Eng 2021;31:100668. https://doi.org/10.1016/J.COCHE.2020.100668.

[39] Yang X, Nielsen CP, Song S, McElroy MB. Breaking the hard-to-abate bottleneck in China's path to carbon neutrality with clean hydrogen. Nature Energy 2022 7:10 2022;7:955–65. https://doi.org/10.1038/s41560-022-01114-6.

[40] China to remain renewable energy leader with strong capacity growth in 2022, despite subsidies phase-out | South China Morning Post n.d. https://www.scmp.com/business/china-business/article/3161732/china-remain-renewable-energy-leader-strong-capacity-growth (accessed January 21, 2024).

[41] Cerniauskas S, Jose Chavez Junco A, Grube T, Robinius M, Stolten D. Options of natural gas pipeline reassignment for hydrogen: Cost assessment for a Germany case study. Int J Hydrogen Energy 2020;45:12095–107. https://doi.org/10.1016/J.IJHYDENE.2020.02.121.

[42] Zhu J, Pan J, Zhang Y, Li Y, Li H, Feng H, et al. Leakage and diffusion behavior of a buried pipeline of hydrogen-blended natural gas. Int J Hydrogen Energy 2023;48:11592–610. https://doi.org/10.1016/J.IJHYDENE.2022.10.185.

[43] Sinopec's long-distance pure hydrogen transmission pipeline project launched. Seetao 2023. https://www.seetao.com/details/210294.html (accessed January 25, 2024).

[44] China Plans 6,000-kilometer Integrated Pipeline Network to Meet Soaring Hydrogen Demand | Pipeline Technology Journal n.d. https://www.pipeline-journal.net/news/china-plans-6000-kilometer-integrated-pipeline-network-meet-soaring-hydrogen-demand (accessed January 25, 2024).

[45] Gao X, An R. Research on the coordinated development capacity of China's hydrogen energy industry chain. J Clean Prod 2022;377:134177. https://doi.org/10.1016/J.JCLEPRO.2022.134177.

[46] Okolie JA, Patra BR, Mukherjee A, Nanda S, Dalai AK, Kozinski JA. Futuristic applications of hydrogen in energy, biorefining, aerospace, pharmaceuticals and metallurgy. Int J Hydrogen Energy 2021;46:8885–905. https://doi.org/10.1016/J.IJHYDENE.2021.01.014.

[47] Zhang H, Yuan T, Tan J. Business model and planning approach for hydrogen energy systems at three application scenarios. Journal of Renewable and Sustainable Energy 2021;13. https://doi.org/10.1063/5.0031594/364102.

[48] Haslam GE, Jupesta J, Parayil G. Assessing fuel cell vehicle innovation and the role of policy in Japan, Korea, and China. Int J Hydrogen Energy 2012;37:14612–23. https://doi.org/10.1016/J.IJHYDENE.2012.06.112.

[49] Kovač A, Paranos M, Marciuš D. Hydrogen in energy transition: A review. Int J Hydrogen Energy 2021;46:10016–35. https://doi.org/10.1016/J.IJHYDENE.2020.11.256.

[50] Global Green Hydrogen Policy: China, A Giant Biding Its Time | Institut Montaigne n.d. https://www.institutmontaigne.org/en/expressions/global-green-hydrogen-policy-china-giant-biding-its-time (accessed January 26, 2024).

[51] Hosseini SE. Fundamentals of Hydrogen Production and Utilization in Fuel Cell Systems. Elsevier; 2023. https://doi.org/10.1016/C2020-0-03183-X.

[52] Ho T, Karri V, Madsen O. Prediction of hydrogen safety parameters using intelligent techniques. Int J Energy Res 2009;33:431–42. https://doi.org/10.1002/ER.1485.

[53] Burandt T. Analyzing the necessity of hydrogen imports for net zero emission scenarios in Japan. Appl Energy 2021;298:117265. https://doi.org/10.1016/J.APENERGY.2021.117265.

[54] Zhu D, Mortazavi SM, Maleki A, Aslani A, Yousefi H. Analysis of the robustness of energy supply in Japan: Role of renewable energy. Energy Reports 2020;6:378–91. https://doi.org/10.1016/J.EGYR.2020.01.011.

[55] Japan's Hydrogen Industrial Strategy n.d. https://www.csis.org/analysis/japans-hydrogen-industrial-strategy (accessed January 27, 2024).

[56] Australia's National Hydrogen Strategy | Hydrogen Knowledge Centre n.d. https://www.h2knowledgecentre.com/content/government622 (accessed January 27, 2024).

[57] Japan Solar Photovoltaic (PV) Market Analysis by Size, Installed Capacity, Power Generation, Regulations, Key Players and Forecast to 2035 n.d. https://www.globaldata.com/store/report/japan-solar-pv-market-analysis/ (accessed January 27, 2024).

[58] Basic Hydrogen Strategy The Ministerial Council on Renewable Energy, Hydrogen and Related Issues 2023. www.meti.go.jp/shingikai/enecho/shoene_shinene/suiso_seisaku/pdf/20230606_5.pdf (accessed January 27, 2024).

[59] World's first blue ammonia shipment opens new route to a sustainable future | Aramco n.d. https://www.aramco.com/en/news-media/news/2020/first-blue-ammonia-shipment (accessed January 27, 2024).

[60] Hydrogen Transportation in the form of MCH by Chemical Tanker 2022. www.chiyodacorp.com/media/20220208_E_R1.pdf (accessed January 28, 2024).

[61] Japan's Chiyoda says hydrogen pilot project begins to fuel power plant | Reuters n.d. https://www.reuters.com/article/idUSKBN23W1LQ/ (accessed January 27, 2024).

[62] Japan's Investments in Hydrogen and Its Derivatives in Southeast Asia - Center on Global Energy Policy at Columbia University SIPA | CGEP % n.d. https://www.energypolicy.columbia.edu/japans-investments-in-hydrogen-and-its-derivatives-in-southeast-asia/ (accessed January 27, 2024).

[63] Australia Exports World's First Shipment of Liquified Hydrogen to Japan | Global Australia n.d. https://www.globalaustralia.gov.au/news-and-resources/news-items/australia-exports-worlds-first-shipment-liquified-hydrogen-japan (accessed January 27, 2024).

[64] Japanese-funded $500m project to extract hydrogen from Victorian coal is at risk, sources say | Energy | The Guardian n.d. https://www.theguardian.com/australia-news/2023/apr/11/japanese-funded-500m-project-to-extract-hydrogen-from-victorian-coal-is-at-risk-sources-say (accessed January 27, 2024).

[65] First International Shipment of Liquid Hydrogen Arrives in Japan n.d. https://maritime-executive.com/article/first-international-shipment-of-liquid-hydrogen-arrives-in-japan (accessed January 27, 2024).

[66] Liquid Hydrogen: The Green Fuel That Isn't | TIME n.d. https://time.com/6139820/liquid-hdyrogen-shipment-hesc/ (accessed January 27, 2024).

[67] Japan to invest 15 tril. yen in hydrogen supply for decarbonization n.d. https://english.kyodonews.net/news/2023/06/c8162f931ea8-japan-to-invest-15-tril-yen-in-hydrogen-supply-for-decarbonization.html (accessed January 27, 2024).

[68] Japan promotes utilization of hydrogen with the framework of the Basic Hydrogen Strategy to contribute significantly to decarbonization n.d. www.env.go.jp/seisaku/list/ondanka_saisei/lowcarbon-h2-sc/PDF/Summary_of_Japan's_Hydrogen_Strategy.pdf (accessed January 27, 2024).

[69] Japan: Energy Country Profile - Our World in Data n.d. https://ourworldindata.org/energy/country/japan (accessed January 27, 2024).

[70] Japan looks to the Gulf as it bets big on hydrogen | Middle East Institute n.d. https://www.mei.edu/publications/japan-looks-gulf-it-bets-big-hydrogen (accessed January 27, 2024).

[71] Hosseini SE. Transition away from fossil fuels toward renewables: lessons from Russia-Ukraine crisis. Future Energy 2022;1:2–5. https://doi.org/10.55670/fpll.fuen.1.1.8.

[72] World Energy Investment 2020 – Analysis - IEA n.d. https://www.iea.org/reports/world-energy-investment-2020 (accessed January 7, 2024).

[73] Overview of Basic Hydrogen Strategy. Agency for Natural Resources and Energy 2023. www.meti.go.jp/shingikai/enecho/shoene_shinene/suiso_seisaku/pdf/20230606_4.pdf (accessed January 28, 2024).
[74] INTERVIEW: Japan's ENEOS shifts into high gear for hydrogen as new landscape emerges | S&P Global Commodity Insights n.d. https://www.spglobal.com/commodityinsights/en/market-insights/latest-news/energy-transition/072123-interview-japans-eneos-shifts-into-high-gear-for-hydrogen-as-new-landscape-emerges (accessed January 27, 2024).
[75] IEA – International Energy Agency n.d. https://www.iea.org/ (accessed March 11, 2022).
[76] Chapman A, Nguyen DH, Farabi-Asl H, Itaoka K, Hirose K, Fujii Y. Hydrogen penetration and fuel cell vehicle deployment in the carbon constrained future energy system. IET Electrical Systems in Transportation 2020;10:409–16. https://doi.org/10.1049/IET-EST.2020.0014.
[77] Stangarone T. South Korean efforts to transition to a hydrogen economy. Clean Technol Environ Policy 2021;23:509–16. https://doi.org/10.1007/S10098-020-01936-6/METRICS.
[78] Lee D, Kim K. Research and Development Investment and Collaboration Framework for the Hydrogen Economy in South Korea. Sustainability 2021, Vol 13, Page 10686 2021;13:10686. https://doi.org/10.3390/SU131910686.
[79] Lee JH, Woo J. Green New Deal Policy of South Korea: Policy Innovation for a Sustainability Transition. Sustainability 2020, Vol 12, Page 10191 2020;12:10191. https://doi.org/10.3390/SU122310191.
[80] The Korean New Deal National Strategy foraGreat Transformation 2020. https://faolex.fao.org/docs/pdf/kor203848.pdf (accessed January 28, 2024).
[81] GDP growth (annual %) - Korea, Rep. | Data n.d. https://data.worldbank.org/indicator/NY.GDP.MKTP.KD.ZG?locations=KR (accessed January 28, 2024).
[82] South Korea commits to lower renewables target in favour of nuclear n.d. https://renewablesnow.com/news/south-korea-commits-to-lower-renewables-target-in-favour-of-nuclear-811471/ (accessed January 28, 2024).
[83] South Korea outlines hydrogen roadmap to boost industry | Argus Media n.d. https://www.argusmedia.com/en/news/2389672-south-korea-outlines-hydrogen-roadmap-to-boost-industry (accessed January 28, 2024).
[84] POSCO Int'l to build 1.26 mn tons of green hydrogen infrastructure by 2035 - KED Global n.d. https://www.kedglobal.com/hydrogen-economy/newsView/ked202307070006 (accessed January 28, 2024).
[85] South Korea launches green H2 pilot project in Jeju | Argus Media n.d. https://www.argusmedia.com/en/news/2376600-south-korea-launches-green-h2-pilot-project-in-jeju (accessed January 28, 2024).
[86] Shin WJ, Lee Y, Yu Y, Ko M, Lee C, Song HH. Comparative life cycle greenhouse gas analysis of clean hydrogen pathways: Assessing domestic production and overseas import in South Korea. J Clean Prod 2023;425:138907. https://doi.org/10.1016/J.JCLEPRO.2023.138907.
[87] Hydrogen Developments | South Korea | Global Hydrogen Policy Tracker | Baker McKenzie Resource Hub n.d. https://resourcehub.bakermckenzie.com/en/resources/hydrogen-hcat-map/asia-pacific/south-korea/topics/hydrogen-developments (accessed January 28, 2024).
[88] Kogas will invest $4B in hydrogen infrastructure n.d. https://koreajoongangdaily.joins.com/news/article/article.aspx?aid=3062435 (accessed January 30, 2024).
[89] Hydrogen- Powering South Koreas future - Darren Choi n.d. https://www.eastspring.com/hk/insights/hydrogen-powering-south-koreas-future (accessed January 30, 2024).
[90] Hyundai Motor Group reveals "FCEV Vision 2030" n.d. https://www.hyundai.news/eu/articles/press-releases/hyundai-motor-group-reveals-fcev-vision-2030.html (accessed January 30, 2024).
[91] Hyundai Ioniq Electric News and Reviews | InsideEVs n.d. https://insideevs.com/hyundai/ioniq-electric/ (accessed January 30, 2024).

[92] Hyundai urges rivals to buy its fuel cell tech to boost sector n.d. https://www.ft.com/content/a9362bca-832a-11e9-9935-ad75bb96c849 (accessed January 30, 2024).
[93] Newsroom | Company - Hyundai Worldwide n.d. https://www.hyundai.com/worldwide/en/newsroom (accessed January 30, 2024).
[94] U.S. Department of Energy Joins Industry to Collaborate on Transportation Technology Validation and Assessment | Department of Energy n.d. https://www.energy.gov/articles/us-department-energy-joins-industry-collaborate-transportation-technology-validation-and (accessed January 30, 2024).
[95] Doosan Spins Out Fuel Cell Business And Announces Plans To Construct Korea's 1st Liquid Hydrogen Plant - FuelCellsWorks n.d. https://fuelcellsworks.com/news/doosan-spins-out-fuel-cell-business-and-announces-plans-to-construct-koreas-1st-liquid-hydrogen-plant/ (accessed January 30, 2024).
[96] World's first byproduct hydrogen power plant opened n.d. https://koreajoongangdaily.joins.com/2020/07/28/business/industry/world-first-byproduct-hydrogen-plant-Daesan-Green-Energy/20200728192100399.html#:%7e:text%3dKorea%20formally%20opened%20the%20world%e2%80%99s%2celectricity%20to%20supply%20160%252C000%20households (accessed January 30, 2024).
[97] PM launches National Hydrogen Mission n.d. https://www.projectstoday.com/News/PM-launches-National-Hydrogen-Mission (accessed January 30, 2024).
[98] National Green Hydrogen Mission | Ministry of New and Renewable Energy | India n.d. https://mnre.gov.in/national-green-hydrogen-mission/ (accessed January 30, 2024).
[99] India | Green Hydrogen Organisation n.d. https://gh2.org/countries/india (accessed January 30, 2024).
[100] India Introduces New Plans for Green Bunkering and Emission Reduction n.d. https://www.fleetmon.com/maritime-news/2023/41877/india-introduces-new-plans-green-bunkering-and-emi/ (accessed January 30, 2024).
[101] Cabinet approves Rs 19,744-cr National Green Hydrogen Mission, ET EnergyWorld n.d. https://energy.economictimes.indiatimes.com/news/renewable/cabinet-approves-rs-19744-cr-national-green-hydrogen-mission/96736053 (accessed January 30, 2024).
[102] MNRE Rolls Out Incentive Scheme Guidelines for Green Hydrogen Production and Electrolyzer Manufacturing - Asia Pacific | Energetica India Magazine n.d. https://www.energetica-india.net/news/mnre-rolls-out-incentive-scheme-guidelines-for-green-hydrogen-production-and-electrolyzer-manufacturing (accessed January 30, 2024).
[103] India – HyResource n.d. https://research.csiro.au/hyresource/policy/international/india/ (accessed January 30, 2024).
[104] Harnessing Green Hydrogen - RMI n.d. https://rmi.org/insight/harnessing-green-hydrogen/ (accessed August 20, 2023).
[105] Adani and TotalEnergies to create the world's largest green hydrogen ecosystem n.d. https://www.adani.com/newsroom/media-release/adani-and-totalenergies-to-create-the-worlds-largest-green-hydrogen-ecosystem (accessed February 1, 2024).
[106] New Energy – Reliance | Aim to Build World's Leading New Energy And New Materials Businesses n.d. https://www.ril.com/businesses/new-energy-materials (accessed February 1, 2024).
[107] Greenko to invest $6 b into pumped storage, green ammonia business in next 3 years - The Hindu BusinessLine n.d. https://www.thehindubusinessline.com/companies/greenko-to-invest-6-b-into-pumped-storage-green-hydrogen-business-in-next-3-years/article66506985.ece (accessed February 1, 2024).
[108] Avaada taps Brookfield for $1 bn | Mint n.d. https://www.livemint.com/companies/news/avaada-partners-with-brookfield-to-raise-1-07bn-for-green-hydrogen-ammonia-ventures-plans-to-reach-30-gw-by-2030-11682529032799.html (accessed February 1, 2024).
[109] India in talks to supply green hydrogen to EU, Singapore | Reuters n.d. https://www.reuters.com/sustainability/climate-energy/india-talks-supply-10-mln-tonnes-green-hydrogen-eu-2023-07-05/ (accessed February 1, 2024).

[110] Consultation hub | Review of the National Hydrogen Strategy - Climate n.d. https://consult.dcceew.gov.au/review-of-the-national-hydrogen-strategy (accessed February 1, 2024).

[111] Australia's Hydrogen Tipping Point | Deloitte Australia n.d. https://www.deloitte.com/au/en/Industries/power-utilities-renewables/perspectives/australia-hydrogen-tipping-point.html (accessed February 1, 2024).

[112] Australian hydrogen market study 2021. https://www.cefc.com.au/media/nhnhwlxu/australian-hydrogen-market-study.pdf (accessed February 2, 2024).

[113] The Fish That Got Away or the Fish in the Bucket? . Australian Hydrogen Research Conference 2023 2023. https://arena.gov.au/assets/2019/11/renewable-energy-options-for-industrial-process-heat.pdf (accessed February 1, 2024).

[114] Renewable energy options for industrial process heat 2019.

[115] Grange Resources Renewable Hydrogen Study (Archived) – HyResource n.d. https://research.csiro.au/hyresource/grange-resources-renewable-hydrogen-study/ (accessed February 1, 2024).

[116] Rio Tinto and Sumitomo to assess hydrogen pilot plant at Gladstone's Yarwun alumina refinery n.d. https://www.riotinto.com/en/news/releases/2021/rio-tinto-and-sumitomo-to-assess-hydrogen-pilot-plant-at-gladstones-yarwun-alumina-refinery (accessed February 1, 2024).

[117] Industry-first pilot using hydrogen in food manufacturing launches in NSW – pv magazine Australia n.d. https://www.pv-magazine-australia.com/2021/12/07/industry-first-pilot-using-hydrogen-in-food-manufacturing-launches-in-nsw/ (accessed February 1, 2024).

[118] South Australian Government Hydrogen Facility – HyResource n.d. https://research.csiro.au/hyresource/south-australian-government-hydrogen-facility/ (accessed February 1, 2024).

[119] Future-ready Fast Ferries: Developing pathways to Zero Emissions | Austal: Corporate n.d. https://www.austal.com/future-ready (accessed February 2, 2024).

[120] Yara Clean Ammonia and Pilbara Ports Authority team up to assess ammonia as a shipping fuel | Yara International n.d. https://www.yara.com/news-and-media/news/archive/news-2022/yara-clean-ammonia-and-pilbara-ports-authority-team-up-to-assess-ammonia-as-a-shipping-fuel/ (accessed February 2, 2024).

[121] Hydrogen Flight Alliance (HFA) – Accelerating Australian aviation towards an emission free future n.d. https://hfa.aero/ (accessed February 2, 2024).

[122] New partnerships target Australia's first test and commercial hydrogen-electric flights – GreenAir News n.d. https://www.greenairnews.com/?p=4636 (accessed February 2, 2024).

[123] Stralis Aircraft – Emission free hydrogen electric aircraft n.d. https://stralis.aero/ (accessed February 2, 2024).

[124] National Hydrogen Strategy Review. Australia Government | Department of Climate Change, Energy, the Environment and Wat 2023. https://storage.googleapis.com/files-au-climate/climate-au/p/prj277759bb48cda6df2c230/public_assets/National%20Hydrogen%20Strategy%20Review%20-%20Consultation%20Paper%20-%20July%202023.pdf (accessed February 1, 2024).

[125] Australia's international clean energy partnerships - DCCEEW n.d. https://www.dcceew.gov.au/climate-change/international-climate-action/international-partnerships (accessed February 2, 2024).

[126] Hydrogen: the energy resource of the future n.d. https://germany.embassy.gov.au/beln/hydrogen.html (accessed February 2, 2024).

[127] How Australia could help Europe solve its green energy gap | Reuters n.d. https://www.reuters.com/business/sustainable-business/how-australia-could-help-europe-solve-its-green-energy-gap-2022-04-21/ (accessed February 2, 2024).

[128] Australia and U.S. Join Forces on the Path to Net-Zero | Department of Energy n.d. https://www.energy.gov/articles/australia-and-us-join-forces-path-net-zero (accessed February 2, 2024).

[129] Australian Hydrogen Company Outlines US Expansion in New Mexico, Touts Research n.d. https://www.usnews.com/news/us/articles/2023-10-25/australian-hydrogen-company-outlines-us-expansion-in-new-mexico-touts-research (accessed February 2, 2024).
[130] Australia and Singapore to work together to accelerate low emissions technologies - DCCEEW n.d. https://www.dcceew.gov.au/about/news/australia-and-singapore-to-work-together-to-accelerate-low-emissions-technologies (accessed February 2, 2024).
[131] Home - HESC n.d. https://www.hydrogenenergysupplychain.com/ (accessed February 2, 2024).
[132] Japanese gov't awards $1.62 bln to Japan-Australia hydrogen energy supply JV | Reuters n.d. https://www.reuters.com/business/energy/japanese-govt-awards-162-bln-japan-australia-hydrogen-energy-supply-jv-2023-03-07/ (accessed February 2, 2024).
[133] Geopolitics of the Energy Transformation: The Hydrogen Factor n.d. https://www.irena.org/Digital-Report/Geopolitics-of-the-Energy-Transformation (accessed February 2, 2024).
[134] Australia in the Middle East: Enduring risks, interests, and opportunities | Lowy Institute n.d. https://www.lowyinstitute.org/publications/australia-middle-east-enduring-risks-interests-opportunities (accessed February 2, 2024).
[135] Green hydrogen in Africa: A continent of possibilities? | White & Case LLP n.d. https://www.whitecase.com/insight-our-thinking/africa-focus-winter-2023-green-hydrogen (accessed February 2, 2024).
[136] Green hydrogen offers a window to redefine the Africa-Australia and Africa-Australia-Europe partnerships | Institute for International Trade | University of Adelaide n.d. https://iit.adelaide.edu.au/news/list/2022/08/01/green-hydrogen-offers-a-window-to-redefine-the-africa-australia-and-africa (accessed February 2, 2024).
[137] International Collaboration is the Key to Accelerating Clean Hydrogen n.d. https://www.youngausint.org.au/post/international-collaboration-is-the-key-to-accelerating-clean-hydrogen (accessed February 2, 2024).
[138] Hydrogen: the energy resource of the future n.d. https://germany.embassy.gov.au/beln/hydrogen.html (accessed February 2, 2024).
[139] Memorandum of understanding between the governments of Australia and the Netherlands on cooperation in the field of hydrogen | Diplomatic statement | Government.nl n.d. https://www.government.nl/documents/diplomatic-statements/2023/01/30/memorandum-of-understanding-between-the-governments-of-australia-and-the-netherlands-on-cooperation-in-the-field-of-hydrogen (accessed February 2, 2024).
[140] FACT SHEET: Delivering on the Next Generation of Innovation and Partnership with Australia | The White House n.d. https://www.whitehouse.gov/briefing-room/statements-releases/2023/10/25/fact-sheet-delivering-on-the-next-generation-of-innovation-and-partnership-with-australia/ (accessed February 2, 2024).
[141] Vakulchuk R, Overland I, Scholten D. Renewable energy and geopolitics: A review. Renewable and Sustainable Energy Reviews 2020;122:109547. https://doi.org/10.1016/J.RSER.2019.109547.
[142] Ibrahim MD, Binofai FAS, Mohamad MOA. Transition to Low-Carbon Hydrogen Energy System in the UAE: Sector Efficiency and Hydrogen Energy Production Efficiency Analysis. Energies 2022, Vol 15, Page 6663 2022;15:6663. https://doi.org/10.3390/EN15186663.
[143] The Hurdles of Energy Transitions in Arab States - Disruptions and Dynamism in the Arab World - Carnegie Endowment for International Peace n.d. https://carnegieendowment.org/2023/05/03/hurdles-of-energy-transitions-in-arab-states-pub-89518 (accessed February 3, 2024).
[144] GCC well-positioned to capitalise on green hydrogen boom n.d. https://www.consultancy-me.com/news/3858/gcc-well-positioned-to-capitalise-on-green-hydrogen-boom (accessed February 4, 2024).
[145] International Energy Agency. Renewable Hydrogen from Oman A producer economy in transition n.d. https://iea.blob.core.windows.net/assets/338820b9-702a-48bd-b732-b0a43cda641b/RenewableHydrogenfromOman.pdf (accessed February 3, 2024).
[146] Hydrom - Home n.d. https://hydrom.om/ (accessed February 3, 2024).
[147] International Energy Agency n.d. http://www.iea.org/ (accessed January 19, 2017).

[148] Oman could be among the top green hydrogen exporters by 2030 n.d. https://renewablesnow.com/news/oman-could-be-among-the-top-green-hydrogen-exporters-by-2030-825903/ (accessed February 4, 2024).
[149] Oman to lead Middle East green hydrogen exports this decade: IEA | Hydrogen news and intelligence n.d. https://www.hydrogeninsight.com/production/oman-to-lead-middle-east-green-hydrogen-exports-this-decade-iea/2-1-1466092 (accessed February 4, 2024).
[150] Progress in Oman mega-projects – Ammonia Energy Association n.d. https://www.ammoniaenergy.org/articles/progress-in-oman-mega-projects/ (accessed February 4, 2024).
[151] U.S. ammonia prices rise in response to higher international natural gas prices - U.S. Energy Information Administration (EIA) n.d. https://www.eia.gov/todayinenergy/detail.php?id=52358 (accessed February 4, 2024).
[152] WAM - Emirates News Agency n.d. https://wam.ae/en/details/1395303086428 (accessed February 4, 2024).
[153] UAE's revised energy strategy includes big hydrogen plans | Reuters n.d. https://www.reuters.com/world/middle-east/uaes-revised-energy-strategy-includes-big-hydrogen-plans-2023-07-11/ (accessed February 4, 2024).
[154] Green Hydrogen Project | Green Hydrogen Project | Siemens Energy Middle East n.d. https://www.siemens-energy.com/mea/en/company/megaprojects/dewa-green-hydrogen-project/dewa-green-hydrogen-project.html (accessed February 4, 2024).
[155] United Arab Emirates and Hydrogen n.d. https://www.trade.gov/market-intelligence/united-arab-emirates-and-hydrogen (accessed February 4, 2024).
[156] Abu Dhabi government and HYCAP to build green hydrogen complex - Investment Monitor n.d. https://www.investmentmonitor.ai/news/abu-dhabi-government-and-hycap-to-build-green-hydrogen-complex/ (accessed February 4, 2024).
[157] United Arab Emirates – HyResource n.d. https://research.csiro.au/hyresource/policy/international/united-arab-emirates/ (accessed February 4, 2024).
[158] ADNOC joins Hydrogen Council, a global coalition of leading companies deploying hydrogen solutions - ADNOC n.d. https://www.adnoc.ae/en/news-and-media/press-releases/2021/adnoc-joins-hydrogen-council (accessed February 4, 2024).
[159] Gandhi K, Apostoleris H, Sgouridis S. Catching the hydrogen train: economics-driven green hydrogen adoption potential in the United Arab Emirates. Int J Hydrogen Energy 2022;47:22285–301. https://doi.org/10.1016/J.IJHYDENE.2022.05.055.
[160] Ruwais Refinery, Ruwais Refinery Expansion Project, UAE - Hydrocarbons Technology n.d. https://www.hydrocarbons-technology.com/projects/ruwais-refinery/ (accessed February 4, 2024).
[161] Saudi vision 2030 n.d. https://www.vision2030.gov.sa/ (accessed February 4, 2024).
[162] Saudi Arabia wants to be top supplier of hydrogen - energy minister | Reuters n.d. https://www.reuters.com/business/energy/saudi-arabia-wants-be-top-supplier-hydrogen-energy-minister-2021-10-24/ (accessed February 4, 2024).
[163] Saudi Arabian solar: how the state has invested in overseas solar projects - Power Technology n.d. https://www.power-technology.com/features/saudi-arabian-solar-how-the-state-has-invested-in-overseas-solar-projects/ (accessed February 4, 2024).
[164] Saudi Arabia's Hydrogen Industrial Strategy n.d. https://www.csis.org/analysis/saudi-arabias-hydrogen-industrial-strategy (accessed February 4, 2024).
[165] Circular Carbon Economy and sustainability | Aramco n.d. https://www.aramco.com/en/sustainability/climate-change/circular-carbon-economy (accessed February 4, 2024).
[166] Saudi Aramco plans new green hydrogen, ammonia project | Argus Media n.d. https://www.argusmedia.com/en/news/2267651-saudi-aramco-plans-new-green-hydrogen-ammonia-project (accessed February 4, 2024).
[167] Saudi Aramco Bets on Blue Hydrogen Exports Ramping Up From 2030 - Bloomberg n.d. https://www.bloomberg.com/news/articles/2024-01-30/aramco-ordered-not-to-raise-oil-production-capacity-further (accessed February 4, 2024).

[168] World's first blue ammonia shipment opens new route to a sustainable future | Aramco n.d. https://www.aramco.com/en/news-media/news/2020/first-blue-ammonia-shipment (accessed February 4, 2024).

[169] S Korea's Hyundai Heavy, Aramco sign blue hydrogen cooperation project | S&P Global Commodity Insights n.d. https://www.spglobal.com/commodityinsights/en/market-insights/latest-news/natural-gas/030421-s-koreas-hyundai-heavy-aramco-sign-blue-hydrogen-cooperation-project (accessed February 4, 2024).

[170] NEOM Green Hydrogen Company completes financial close at a total investment value of USD 8.4 billion in the world's largest carbon-free green hydrogen plant n.d. https://www.neom.com/en-us/newsroom/neom-green-hydrogen-investment (accessed February 4, 2024).

[171] Saudi's Blue Hydrogen Production to Use $110 Billion Gas Field in Jafurah - Bloomberg n.d. https://www.bloomberg.com/news/articles/2021-10-24/saudi-arabia-to-use-110-billion-gas-project-for-blue-hydrogen?sref=B2BBHw9t (accessed February 4, 2024).

[172] Saudi Arabia to develop hydrogen fuel cell-based transport, sustainable jet fuel | S&P Global Commodity Insights n.d. https://www.spglobal.com/commodityinsights/en/market-insights/latest-news/energy-transition/012022-saudi-arabia-to-develop-hydrogen-fuel-cell-based-transport-sustainable-jet-fuel (accessed February 4, 2024).

CHAPTER 10

Other main players in hydrogen technology

10.1 Russia's strategic influence on global hydrogen technology

Russia possesses 1,688 trillion cubic feet (Tcf) of confirmed natural gas reserves, representing approximately 24% of the global total (Figure 10.1). This places Russia as the foremost holder of proven gas reserves worldwide. These reserves amount to 102.3 times its yearly consumption, indicating that Russia has an estimated 102 years' worth of gas at present consumption rates. Furthermore, Russia holds the title of the world's largest gas exporter, having produced 762 billion cubic meters (bcm) of natural gas in 2021 and exporting around 210 bcm through pipelines [1]. Prior to the Ukraine conflict, Russia played a substantial role as a provider of fossil fuels to the European Union (EU). For instance, Russia supplied 45% of the EU's gas imports before the invasion of Ukraine in February 2022. However, in response to the conflict, the EU has initiated efforts to diversify its gas suppliers and decrease its reliance on Russian fossil fuels. By 2023, the EU is poised to significantly diminish its imports of Russian gas, with Norway and the United States emerging as major alternative suppliers. In 2020, over half of the EU's solid fossil fuel imports, primarily coal, originated from Russia, along with 43% of imported natural gas [2]. The EU's dependency on Russian fossil fuels has raised concerns due to the potential for supply disruptions and economic ramifications. Consequently, the EU has been implementing measures to mitigate this dependency and transition towards renewable energy sources. The EU has historically used natural gas, including that from Russia, to produce hydrogen, with the majority of hydrogen production being from natural gas [3]. Despite having the largest reserves of natural gas globally, a well-established nuclear sector, favorable geological conditions for carbon capture and storage (CCS), and considerable potential for renewable energy production, Russia is falling behind the leading nations in the pursuit of building a hydrogen economy [4].

DOI: https://doi.org/10.55670/fpll.book/1

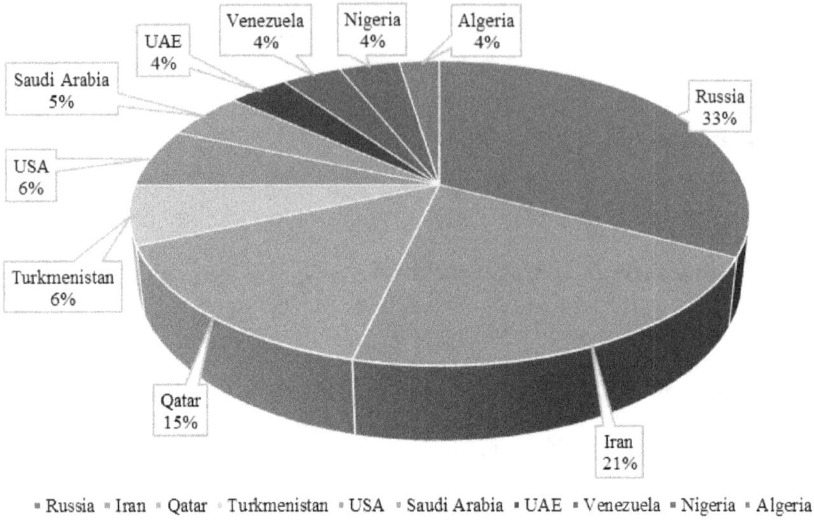

Figure 10.1 Countries by natural gas proven reserves

In June 2020, the Russian Federation unveiled its Energy Strategy for the period up to 2035. This blueprint delineates Russia's comprehensive, mid-term objectives for its energy sector, a critical component of its economy. Among its provisions, the strategy presents an outline for hydrogen, aiming to export 0.2 million metric tons of hydrogen by 2024, two million by 2035, and 15 million by 2050. Russian hydrogen production reached 3.4 million metric tons in ammonia and 2.7 million metric tons in refining in 2021. These estimates are anticipated to rise to 3.8 million metric tons and 3.1 million metric tons, respectively, by 2025 [5]. To provide context, global hydrogen production totaled approximately 70 million metric tons in 2019 [6]. Before the EU released its hydrogen strategy, other potential importers of Russian hydrogen had already published their national strategies or framework documents. Japan did so in 2017, followed by South Korea in 2019, and Germany and the Netherlands in the first half of 2020. Russia, on the other hand, took another year to release its "Concept on the development of hydrogen energy sector", a document that still lacks the focus of a proper comprehensive strategy [7]. Despite this delayed approach to developing its hydrogen sector, Russia expressed optimism about its potential position in the global market, aiming to capture up to 20 percent of it by 2030. While the exact size of this market remains uncertain, if

Russia's ambition were to be realized, it could potentially earn between $23 and $100 billion annually from hydrogen.

Further elaboration on Russia's hydrogen strategy came with the release of the Roadmap for Hydrogen Development by 2024 in October 2020. This roadmap offers a multi-year action plan, albeit in broad terms with some specifics, for establishing a hydrogen energy sector in Russia. In August 2021, the government released the Concept for the Development of Hydrogen Energy in Russia, marking the initial phase of the extensive action plan outlined in the roadmap [8]. Russia's energy strategy delineates two overarching objectives: (1) bolstering socioeconomic advancement and (2) maintaining a dominant position in global energy affairs. This strategy encompasses initiatives such as structural diversification, strengthening domestic fossil fuel energy infrastructure in critical regions, expanding Russia's presence in Asia-Pacific markets, and making non-committal references to climate action. Russia's hydrogen strategy aligns with these objectives and strategic facets, yet the precise implementation of this strategy remains ambiguous. Numerous critical inquiries- such as the prioritized areas, the allocated funding for sector development, the formation of clusters, the involvement of the private sector, and others- still lack definitive answers. Russia's major energy firms-initiated discussions regarding potential hydrogen supplies with their longstanding business partners from several leading hydrogen-producing nations even before the Hydrogen Concept received official approval from the government.

Russia perceives itself as possessing a competitive edge in hydrogen due to its abundant fossil fuel resources, complemented by a well-established global oil and gas industry. Given that natural gas presently serves as the primary feedstock for hydrogen production worldwide, Russia's position as one of the largest gas producers globally offers potential cost competitiveness. Moreover, conventional wisdom underscores the oil and gas sector (along with its skilled workforce) as highly transferable to hydrogen energy production. Additionally, Russia's existing expertise in hydrogen fuel, spanning production, transportation, and storage, is regarded as advantageous. However, Russia's hydrogen strategy is not solely reliant on its inherent strengths. All strategy documents acknowledge shifting global demand amidst the energy transition, indicating Russia's adaptation efforts. As a significant hydrocarbon producer, Russia faces compounded risks in an increasingly decarbonized global economy. Both the concept and the roadmap underscore the defensive elements of Russia's initiatives, acknowledging climate change, the global energy shift, and the burgeoning demand for low-carbon goods. Various action items outlined in the documents reflect an awareness that emissions from future hydrogen fuel production may face heightened scrutiny from international buyers. For instance, Russia's strategy explicitly outlines the establishment of low-carbon, export-focused hydrogen production facilities within the Northwest cluster, with a focus on catering to Europe's low-carbon demand. While the extent of emphasis on green hydrogen remains uncertain, it is frequently mentioned in both the roadmap and the concept as a probable priority. Being the foremost exporter of natural gas globally, Russia initially leaned towards prioritizing grey hydrogen to advance its hydrogen sector. However, it had to adapt its export strategies to align with the environmental demands associated with the decarbonization initiatives in Asia and Europe. Consequently, the several projects outlined for Russia's various regions were mandated to produce solely low-carbon or zero-carbon hydrogen and ammonia (Figure 10.2).

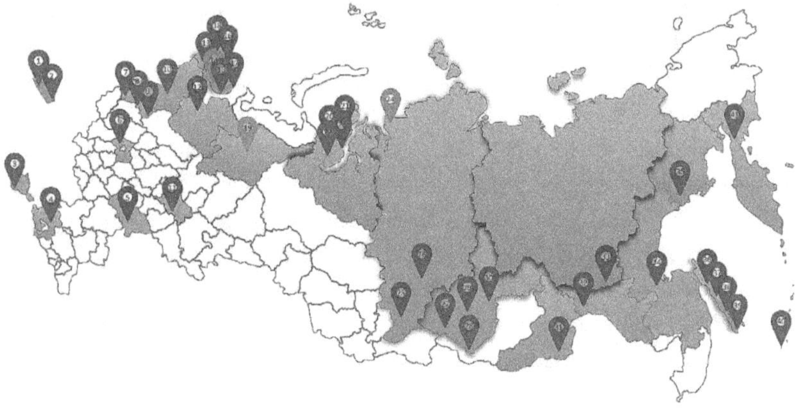

Figure 10.2 Russia's projects for hydrogen and ammonia production (the details of the projects are given in [9])

The Russia's hydrogen strategy outlines three distinct phases of development:

- Phase 1 (2021–2024): Aims to achieve 0.2 million metric tons of exports by establishing hydrogen clusters and fostering a research ecosystem. This phase involves advancing technologies, manufacturing industrial products for hydrogen, executing pilot projects, and stimulating domestic demand for hydrogen.

- Phase 2 (2025–2035): Seeks to attain 2 million metric tons of exports by launching commercial projects for hydrogen production, particularly focusing on large, export-oriented production facilities. It also aims to further expand and integrate hydrogen energy into the domestic market while scaling up the production and export of domestically manufactured hydrogen equipment.
- Phase 3 (2036–2050): Aims to establish and consolidate a significant presence in the global hydrogen energy market. It targets exporting 15–50 million metric tons to the global market by 2050.

The essence of the strategy appears to revolve around the development of hydrogen clusters. However, strategic documents have refrained from providing clear definitions of these clusters. Apart from implied export and production facilities and an explicit mention of utilizing university systems, it remains uncertain which other institutions and facilities referenced will be included in these clusters. Apart from these clusters, the concept document appears more like an extensive list of potential strategies rather than a cohesive framework with clear guiding principles, objectives, or directives. There is a lack of evident prioritization or overarching strategy evident in the extensive array of potential actions presented.

On the other hand, the roadmap provides a more structured approach, delineating specific steps and timelines for various aspects of the strategy. It encompasses eight distinct areas, including strategic planning, government support, production capacity, pilot projects, research and development, regulatory framework development, workforce training, and international engagement. However, the roadmap's timelines only extend to 2024, and much of its content essentially outlines a plan for further planning. Many steps involve government entities reporting on broad aspects of the strategy, such as workforce status, to clarify priorities, feasibility, and necessary actions. Responsibility for implementing these and future aspects remains somewhat unclear. While the roadmap identifies the entities involved in each step, it lacks a clear hierarchy or assignment for advancing the development of the hydrogen sector. The Ministry of Energy appears to play a pivotal role in this strategy, tasked with providing annual reports on sector development. The concept document frequently mentions mobilizing the private sector, although this aspect receives little attention in the roadmap. However, in July 2021, the government established a working group involving major companies such as Novatek, Sibur, and Sistema, along with state-owned enterprises like Rosatom, Rosneft, Gazprom, Gazpromneft, Kamaz, Rostec, and Rosnano [10]. Russia aims to assert itself as a global leader in hydrogen energy geopolitics. The concept underscores the importance of international cooperation, including collaboration on pilot projects, industry standards, and trade regulations with other countries. Notably, Russia's strategy advocates for technological neutrality, aiming to safeguard hydrogen energy derived from fossil fuels and nuclear power. The roadmap outlines potential international partners, including Germany, Japan, Denmark, Italy, Australia, the Netherlands, South Korea and Saudi Arabia. Additionally, Russia has signed bilateral agreements on hydrogen energy cooperation with the United Arab Emirates (UAE) and Japan, although the extent of collaboration with these countries remains somewhat unclear, with the UAE agreement involving a working group focused on hydrogen energy development issues. The partnership between Russia and the UAE includes a joint assessment of solutions for the manufacture, storage, and transportation of hydrogen fuel [11,12].

According to the concept, Russia aims to establish three hydrogen clusters, or hubs: a Northwest cluster, an Eastern cluster, and an Arctic cluster, with the possibility of a fourth Southern cluster remaining open. The Northwest and Eastern clusters are designed to cater to anticipated demand centers for hydrogen. Russia plans to utilize the Northwest cluster for exporting to European markets, emphasizing low-carbon (and potentially zero-carbon) products, while the Eastern cluster will focus on exporting to Asian markets. The Arctic cluster is aligned with supply considerations, likely corresponding with new gas resource development, and also aligns with Russia's broader strategic interest in maintaining a strong presence in the Arctic, both economically and militarily [13]. The potential for a Southern cluster is contingent upon demand growth for hydrogen fuel, but the concept provides limited information on its geographical or other aspects. The specific locations of these clusters are yet to be determined, as the concept lacks clarity in defining them and providing geographic details. However, there is a strong implication that the geography of these clusters will mirror Russia's existing natural gas sector. Based on current economic and infrastructure conditions, potential candidates for the Northwest, Eastern, and Arctic clusters could include the Leningrad Oblast, Primorsky Krai, and Yamalo-Nenets Autonomous Okrug, respectively. Furthermore, the concept discusses the establishment of a domestic research ecosystem and the development of local and regional markets, emphasizing the importance of creating domestic demand and research hubs throughout the strategy. However, beyond the clusters and hydrogen-powered urban transport, the concept does not provide a clear geographical framework for Russia's research ecosystem and domestic demand.

10.1.1 Effects of Russia-Ukraine war on Russia's hydrogen roadmap

While Russia entered the arena of hydrogen ambitions and initiatives relatively late, by early 2022, it possessed notable advantages over other nations, eyeing a substantial stake in the prospective hydrogen market, particularly within Europe. These advantages encompassed a well-established hydrocarbon sector and accompanying infrastructure, along with historical proficiency in nuclear energy and hydropower generation. Additionally, Russia boasted some of the most favorable geological conditions globally for CCS, alongside political backing at the highest echelons in numerous European and non-European capitals. In theory, these advantages positioned Russia to emerge as a predominant player in the future global hydrogen economy. However, Russia's invasion of Ukraine severely undermined its ambitions. This was not only due to facing the most stringent economic sanctions in its history and enduring complete political isolation from the West but also because mainstream European policymakers are now disinclined to entertain the notion of future Russian hydrogen supplies, given the analogous challenges associated with Russian gas supplies (Figure 10.3). Furthermore, with Japan and South Korea- two potentially significant hydrogen markets-also imposing sanctions on Russia, its potential Asian export trajectory is now uncertain [14].

Since the onset of the conflict, certain media reports within Russia have portrayed more modest forecasts regarding hydrogen from the Ministry of Energy. The country's aspirations for exports have diminished from an initial estimate of 2.2 million tons per year by 2030 to approximately 1.4 million tons. As a consequence of the war and subsequent market losses, Russia is anticipated to forfeit at least half of its long-term hydrogen export capacity, which is projected to be around 4.5 million tons per year by 2050 instead of surpassing 9.5 million tons. This decline is primarily attributed to key future consumers- such as Japan, Germany, and South Korea- being precluded from importing Russian hydrogen due to sanctions. These estimates were presented by the Energy Ministry as part of a "Preliminary 'stress scenario' of the program on the development of Russia's low-carbon economy" in June 2022. Given the absence of indications of an end to the conflict in the foreseeable future, these projections are likely to become a permanent reality.

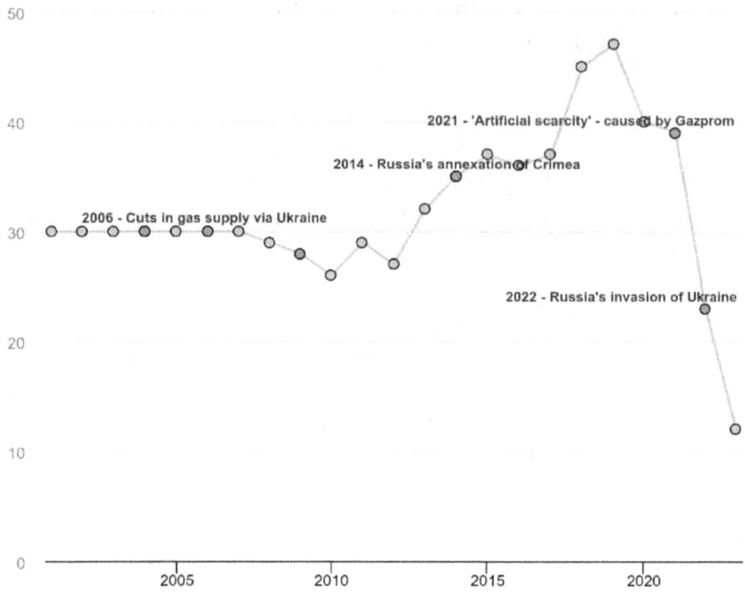

Figure 10.3 Share (%) of EU gas demand met by Russian supply, 2001-2023 [15]

With countries previously identified as primary consumers of its hydrogen now deemed unfriendly by Russia and excluded from cooperation, Russia will need to revise its hydrogen plans to target other nations, primarily in Asia, which exhibit less certainty about the potential of this fuel for their future growth or demonstrate a more pragmatic approach to engaging in business with the Russian government. This reorientation is expected to impact the technical aspects of Russia's eventual hydrogen sector, as its ambitions regarding blue and green hydrogen are likely to suffer significantly as a result. With its reliance on existing and planned natural gas infrastructure to supply hydrogen to Europe, Russia is facing a significant challenge as this option may no longer be viable in the aftermath of the conflict in Ukraine. The Nord Stream 2 pipeline, which could have been repurposed for hydrogen transport, remains unused despite being technically completed before the war due to the ongoing geopolitical situation [16]. Consequently, Russia will need to urgently develop entirely new facilities and networks for hydrogen storage and distribution to alternative destinations. While the anticipated profits remain substantial, estimated at approximately $12.8 billion, the investment required for production alone is expected to nearly double that amount, reaching at least $21.1 billion [17]. Moreover, the costs associated with the subsequent construction of hydrogen infrastructure could be even higher. With most investors from wealthy countries previously considered key partners unlikely to support these endeavors due to sanctions, achieving these goals will be exceedingly challenging. International sanctions, technology transfer bans, and loss of foreign direct investment due to the conflict with Ukraine have significantly impacted Russia's hydrogen strategy, necessitating the development of new infrastructure from scratch.

Under wartime conditions, Russia's prospective hydrogen industry will encounter challenges beyond end-user delivery. Specifically, meeting the projected quantities of blue and green hydrogen production may prove to be unrealistic. This challenge primarily stems from technological limitations, particularly in CCS and electrolyzer manufacturing, coupled with dependence on foreign suppliers. While CCS development could potentially benefit from collaboration with China, which could replace traditional leaders like Norway, expanding renewable-based hydrogen production is less feasible. This is because Europe leads in proton exchange membrane (PEM) electrolyzer production, which is well-suited for integration with intermittent renewable energy sources like wind and solar power plants [18]. Although China has emerged as a major manufacturer of alkaline electrolyzers, this technology is more compatible with stable-load power sources such as hydropower or nuclear plants [19]. Consequently, even with technological assistance from China, Russia's options for zero-carbon hydrogen production will remain significantly restricted in the short and medium terms. The question also arises regarding China's interest in establishing hydrogen-related collaboration with Russia. While Russia has identified China as a crucial potential hydrogen market, China has only announced one long-term hydrogen plan for the period of 2021–2035 [20]. China is adopting a phased approach to developing its domestic hydrogen industry and acquiring technologies and manufacturing capabilities, displaying conservatism in setting targets. Its goal of producing 100,000 to 200,000 tons of renewable hydrogen annually by 2025 pales in comparison to the EU's annual target of 1 Mt [21]. Germany's plan for a 14 terawatt-hours (TWh) production capacity by 2025 exceeds China's maximum goal of 6.7 TWh. Although the China Hydrogen Alliance has projected that hydrogen demand will reach 35 Mt in 2030 and 60 Mt in 2050, the extent of hydrogen imports required to meet this demand remains uncertain [22]. Despite being the world's largest hydrogen producer with an output of approximately 33 Mt, China primarily relies on locally sourced, cost-effective fossil fuels as feedstocks for production. Hence, Russia's proposal of a similar product (hydrogen derived from hydrocarbons) may not offer a significant price advantage considering the logistical challenges of long-distance transportation. The critical factor will be determining the quantity of hydrogen China requires and whether any portion needs to be imported. To enhance the appeal of its hydrogen to China, Russia must accelerate the development of energy infrastructure near the shared border. This entails expanding the current network of natural gas pipelines and potentially constructing new ones capable of transporting hydrogen or its derivatives. However, Russia's energy infrastructure for delivering hydrogen to China is notably less advanced than that serving Europe. Unlike the four major natural gas pipelines potentially adaptable for hydrogen transmission to Europe, the Power of Siberia pipeline is the sole conduit to Asia and has yet to achieve profitability since its inauguration in 2019 [23]. Given the current economic constraints resulting from significantly reduced state revenues, investing in a more expensive endeavor with uncertain prospects may not be justifiable.

10.2 United Kingdom's hydrogen horizon: pioneering diplomacy

Hydrogen represents one of several emerging low-carbon solutions crucial for the UK to achieve its ambitious emissions reduction targets, including Carbon Budget Six (CB6) [24] and net zero by 2050 [25]. Aligned with the Prime Minister's Ten Point Plan for a Green Industrial Revolution [26], the government has set a target to establish 10GW of low-carbon hydrogen production capacity by 2030 (Figure 10.4), subject to affordability and value for money, with at least half of this capacity coming from electrolytic hydrogen [27,28]. Given the current minimal production and utilization of low-carbon hydrogen, especially in energy applications, this goal necessitates a rapid and substantial scale-up from the present state. The Ten Point Plan introduces new funding initiatives and policies geared towards realizing this objective, such as the allocation of £240 million for government co-investment in production capacity through the Net Zero Hydrogen Fund (NZHF) [29]. Additionally, it outlines plans for a hydrogen business model to attract private sector investment, complemented by a revenue mechanism to bolster funding for this model.

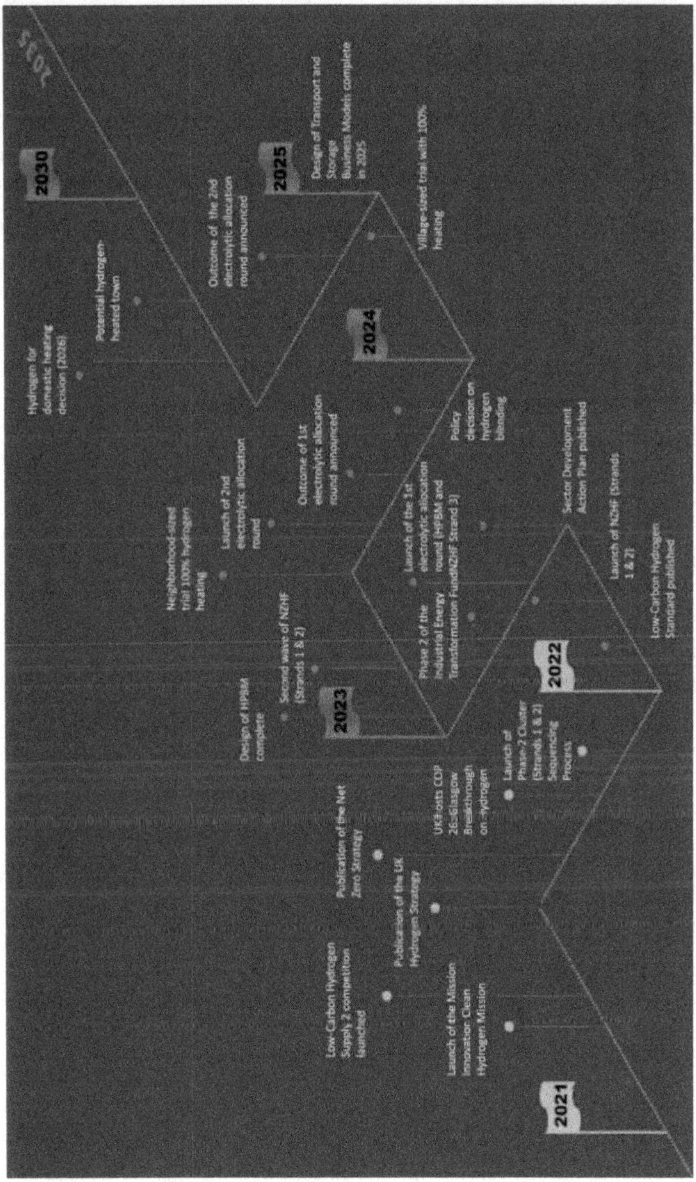

Figure 10.4 The UK's hydrogen roadmap by 2030 [30]

Ongoing advancements in hydrogen technologies, facilitated by innovative UK research, collaboration, and international partnerships, are deemed essential to this endeavor. Recognizing hydrogen's significance, the Ten Point Plan identifies it as a primary focus area within the Net Zero Innovation Portfolio, a £1 billion fund aimed at expediting the commercialization of low-carbon technologies for achieving net zero emissions. The forthcoming decade is pivotal for preparing energy users best suited to utilize hydrogen as a low-carbon solution. To this end, efforts are underway, including support for fuel switching to hydrogen in industry through initiatives like the £315 million Industrial Energy Transformation Fund and the £20 million Industrial Fuel Switching Competition. Additionally, plans encompass establishing the groundwork for hydrogen application and storage in the power sector, conducting demonstration competitions and trials for hydrogen utilization in transportation modes like road freight, shipping, and aviation, as well as pioneering trials of hydrogen heating.

These trials were scheduled to commence with a hydrogen neighborhood trial by 2023, followed by a large-scale hydrogen village trial by 2025, potentially culminating in a hydrogen town pilot before the decade's end. Collaborating with the Health and Safety Executive (HSE) and industry stakeholders, the government is assessing the feasibility of blending 20% hydrogen into the gas network and supporting the development of prototype "hydrogen-ready" appliances [31]. Furthermore, initiatives outlined in documents like the Energy White Paper [32], Industrial Decarbonization Strategy [33], and the recently published Transport Decarbonization Plan [34] complement these efforts by further stimulating hydrogen demand across various sectors. This comprehensive strategy delineates additional commitments and actions, showcasing the collaborative efforts between the government, industry, research and innovation community, and broader civil society toward realizing the vision of a UK hydrogen economy.

Statistics indicate that fostering a hydrogen economy in the UK could also create employment opportunities, with projections indicating the potential to support over 9,000 jobs by 2030 and up to 100,000 jobs by 2050, spanning the industrial hubs and regions throughout the UK [35]. Low-carbon hydrogen will play a crucial role in achieving net zero emissions and, prior to that, in meeting the UK's ambitious CB6 target to reduce emissions by 78 percent from 1990 levels by 2035. Analysis for CB6 indicates that between 250 and 460 TWh of hydrogen may be required by 2050, constituting 20 to 35 percent of the UK's final energy consumption [36]. The scale of the hydrogen economy in 2050 will be influenced by various factors, including the cost and accessibility of hydrogen and hydrogen-based technologies compared to alternatives such as electrification, biomass, and CCS utilization. Nevertheless, there is widespread agreement, including from the Climate Change Committee (CCC) and others, that substantial quantities of low-carbon hydrogen will be necessary in the energy system by 2050 (Figure 10.5) [37].

Figure 10.5 The existing hydrogen projects leading to the UK's net-zero target of 2050 [30]

Nevertheless, several strategic challenges exist across the hydrogen value chain that must be addressed to facilitate widespread production and utilization of hydrogen in the UK:

- Cost disparities between hydrogen and existing high-carbon fuels: Presently, hydrogen production and utilization are considerably more expensive compared to conventional fossil fuels. Although costs are anticipated to decline significantly with increased innovation and deployment, addressing this disparity remains crucial.
- Technological uncertainties: While certain hydrogen technologies are already operational, many applications require validation at a scale before widespread implementation becomes feasible.
- Policy and regulatory ambiguities: Given the nascent nature of hydrogen energy policy, industry stakeholders are reliant on governmental support for financial backing, regulatory frameworks, incentives, quality and safety standards, guidance on supply chain development and workforce skills, and overarching strategic direction.
- Infrastructure requirements: The adoption of hydrogen necessitates the establishment of new infrastructure for distribution and storage, as well as integration with carbon capture, utilization, and storage (CCUS) systems, gas networks, and electricity grids.
- Coordination of supply and demand: Developing a hydrogen economy entails overcoming the challenge of synchronizing the development of production facilities and end-use applications, as well as balancing supply and demand, which may involve implementing storage solutions over time.
- Investment and deployment challenges: Scaling up a low-carbon hydrogen economy requires addressing the barriers associated with being an early mover in the market and fostering an environment conducive to sustained investment and deployment in both initial and subsequent projects.

10.2.1 Hydrogen in Scotland, Wales, and Northern Ireland

Scotland is poised to play a pivotal role in advancing the UK's hydrogen economy, leveraging its abundant offshore and onshore wind resources, as well as wave and tidal power, along with CCUS technologies. With a robust industrial base and expertise in energy systems, including oil and gas and offshore wind, Scotland possesses the capability to produce substantial volumes of hydrogen on an industrial scale. Economic evaluations conducted for the Scottish Government indicate that by 2045, Scotland could potentially generate between 21 and 126 TWh of hydrogen annually, with the most ambitious scenario envisioning up to 96 TWh of hydrogen available for export to Europe and other parts of the UK. This endeavor is anticipated to yield significant employment opportunities and foster local economic growth [38]. In December 2020, the Scottish Government released a Hydrogen Policy Statement outlining its vision for Scotland's hydrogen economy and aspirations for renewable and low-carbon hydrogen production. Scotland's Hydrogen Action Plan is bolstered by a £100 million investment program spanning from 2021 to 2026 [39].

Scotland is at the forefront of pioneering hydrogen initiatives, hosting a range of groundbreaking demonstration projects aimed at elucidating the potential role of hydrogen in shaping Scotland and the UK's future energy landscape. Notably, the European Marine Energy Centre situated in the Orkney Islands spearheads a £65 million portfolio of renewable hydrogen projects, continually expanding its scope, thereby offering a scaled-down illustration of various facets of a hydrogen-based economy. Aberdeen serves as the testing ground for 25 hydrogen double-decker buses, which have paved the way for the establishment of infrastructure capable of supporting a diverse array of over 60 hydrogen-powered vehicles. This initiative has catalyzed the Aberdeen Hydrogen Hub endeavor, which is aimed at positioning the region as a leading exemplar of hydrogen adoption in Europe. Additionally, the H100 neighborhood trial project in Fife is underway, focusing on developing a 100% electrolytic hydrogen production and distribution network alongside the installation of hydrogen boilers in 300 residences to showcase the viability of hydrogen for domestic heating applications across the UK. Furthermore, in March 2021, both the UK and Scottish Governments unveiled plans to allocate £50 million each, as part of the Heads of Terms for the Islands Growth Deal, to foster the future economic prosperity of Orkney, Shetland, and the Outer Hebrides. These investments encompass several initiatives aimed at bolstering support for hydrogen-related endeavors in these regions [40].

Building Innovative Green Hydrogen Systems in Isolated Territories (BIG HIT) is a comprehensive six-year initiative based in Orkney aimed at establishing an integrated, environmentally friendly energy system tailored to local needs [41]. The project's primary objective is to develop a scalable model for hydrogen production, storage, distribution, and utilization across various sectors such as heating, electricity generation, and transportation. Supported by funding from the Fuel Cells and Hydrogen Joint Undertaking, BIG HIT builds upon the success of Orkney's Surf'n'Turf initiative, which leverages wind and tidal energy to produce hydrogen for community use. Cutting-edge PEM electrolyzers deployed in Eday and Shapinsay Islands facilitate hydrogen production through electrolysis, utilizing locally sourced renewable energy. The generated hydrogen is subsequently stored and utilized to meet heating, power, and transportation needs within the vicinity. By pioneering such initiatives, BIG HIT positions Orkney as a pioneering and scalable model for small-scale hydrogen deployment in isolated regions. The insights gained from BIG HIT will be instrumental in facilitating broader adoption and implementation of renewable energy solutions incorporating fuel cell and hydrogen technologies in similar isolated or constrained territories globally [42].

Wales boasts significant potential for the production and utilization of low carbon hydrogen, leveraging its offshore wind resources, tidal and wave power capabilities, robust infrastructure networks, and accessible ports, alongside its robust research and development capabilities and skilled workforce. This favorable environment lays the groundwork for the widespread adoption of hydrogen and fuel cell technologies, particularly in light of supportive policy frameworks. The Welsh Government released a hydrogen pathway report in December 2020 and published it in the Fall of 2021 [43]. Concurrently, the Welsh Hydrogen Business Research and Innovation for Decarbonization (H2BRID) initiative is being developed to complement the Welsh hydrogen pathway, with plans for launch around the same period. This initiative aims to address the challenges outlined in the pathway and invest in innovative hydrogen projects across Wales [44]. Welsh innovation in the hydrogen sector is evident through various pioneering companies, projects, and research clusters. For instance, Riversimple, a Welsh SME, is actively involved in designing, constructing, and testing innovative hydrogen fuel cell electric vehicles. Additionally, the Dolphyn FLOW has studied the feasibility of establishing a 100-300MW commercial hydrogen wind farm off South Wales, with future plans for expansion. This initiative also involves the development of hydrogen pipelines connecting strategic locations along the Milford Haven waterway for transportation and heating applications, potentially extending to Pembroke Dock for marine operations. The Hydrogen Centre, located at the Baglan Energy Park in Neath Port Talbot, serves as a collaborative hub for various projects between the University of South Wales and other academic and industrial partners. The center focuses on the experimental development of renewable hydrogen production, innovative hydrogen energy storage solutions, and further research and development of hydrogen-powered vehicles, fuel cell applications, and hydrogen energy systems [45]. Moreover, the UK Government announced capital funding for the Holyhead

Hydrogen Hub. This initiative aimed to establish a demonstration hydrogen production plant and fueling hub for heavy goods vehicles (HGVs) to serve freight traffic at Holyhead and port-side vehicles. The hub, operated in 2023, represents a significant step towards advancing hydrogen infrastructure and applications in Wales.

Northern Ireland is strategically positioned to drive forward hydrogen innovation and implementation, leveraging its abundant wind resources, modern gas network, connections to Ireland and Great Britain, access to salt cavern storage, and renowned engineering and manufacturing capabilities. Notably, Northern Ireland Water is set to acquire a new electrolyzer for one of its wastewater treatment facilities, marking a pioneering endeavor in the UK. Concurrently, the public transportation provider Translink is rolling out new hydrogen buses manufactured by local company Wrightbus in Ballymena, alongside plans to procure a new hydrogen refueling station.

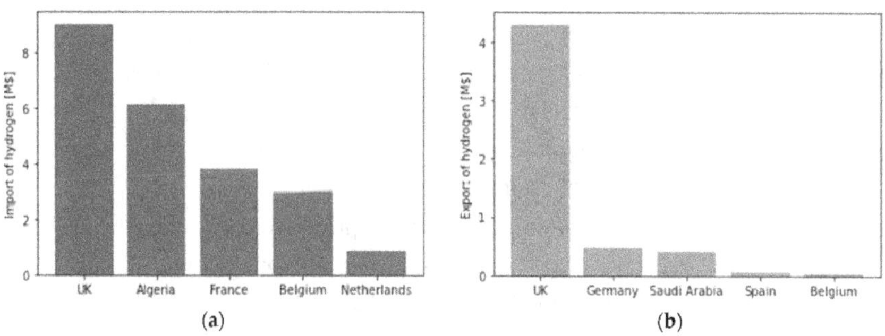

Figure 10.6 (a) Hydrogen imports from other countries to Ireland, (b) Ireland exports of hydrogen to other countries [46]

The GenComm initiative, spearheaded by Belfast Metropolitan College, has secured funding from both the EU and UK Government to conduct trials on hydrogen production via electrolysis for use in hydrogen-powered buses. Moreover, the Department for the Economy is currently seeking input on policy options for a new Energy Strategy encompassing hydrogen, which will outline Northern Ireland's energy priorities and trajectory up to 2050, with an anticipated publication by year-end. The UK Government is firmly committed to collaborating closely with devolved administrations, including through the joint government-industry

Hydrogen Advisory Council, to fully harness the UK's potential in establishing a leading hydrogen economy. This collaboration aims to ensure that low-carbon hydrogen plays a pivotal role in reducing emissions and fostering sustainable growth throughout the United Kingdom. Figure 10.6 demonstrates hydrogen imports from other countries to Ireland and Ireland exports of hydrogen to other countries.

10.2.2 The UK's hydrogen plans for 2030

The initial vision outlined in the Strategy, upon its release, aimed for the UK to emerge as a global frontrunner in hydrogen technology by 2030. This vision entails achieving 5 GW of low-carbon hydrogen production capacity, a target later revised upwards to 10 GW, as a catalyst for economy-wide decarbonization, with clear plans for further scaling up towards Carbon Budget 6 (CB6, spanning 2033-37) and eventual net zero emissions. The Strategy adopts a "twin track" approach, endorsing various technologies such as green electrolytic and blue carbon capture-enabled hydrogen production. By 2030, the envisioned outcomes encompass several strategic goals, including [47]:

- Progress towards the 2030 target of 10 GW low-carbon hydrogen production capacity, with an interim goal of 1 GW by 2025.
- Decarbonization of existing UK hydrogen supply through CCUS or through the integration of electrolytic hydrogen.
- Reduction in the cost of low-carbon hydrogen production through advancements driven by early projects, market maturity, and technological innovation.
- Establishment of an end-to-end hydrogen system catering to diverse users across various sectors and regions within the UK.
- Heightened public awareness and acceptance of hydrogen usage throughout the energy system.
- Promotion of UK economic growth and job opportunities, bolstered by a robust domestic supply chain and export potential, positioning the UK as a global leader and attractive destination for investment.
- Substantial contribution to emissions reduction targets under Carbon Budgets 4 and 5, propelling the UK towards achieving CB6.
- Preparation for future scaling beyond 2030, laying the groundwork for net zero emissions with requisite infrastructure and regulatory frameworks.
- Evidence-based policy development, informed by improved modeling and real-world data from innovation and deployment projects.

Additionally, the Strategy delineates a roadmap for the 2020s, outlining key policies and activities necessary to achieve different archetypes of a hydrogen economy by various timeframes, including the mid-2030s for CB6. It also includes a comprehensive array of key commitments spanning the hydrogen supply chain.

Project Union is an initiative by National Grid aimed at establishing a comprehensive hydrogen network across the UK, connecting industrial clusters throughout the country, potentially covering a distance of approximately 2000 kilometers by 2030 [48]. This project, funded through net zero development funding, plans to repurpose around 25 percent of the current gas transmission pipelines, capable of accommodating at least a quarter of the UK's current gas demand (Figure 10.7) [37]. The feasibility stage involves identifying pipeline routes, assessing the readiness of existing gas assets, and devising a transition plan for these assets. Additionally, National Grid intends to explore the phased conversion of pipelines starting from 2026.

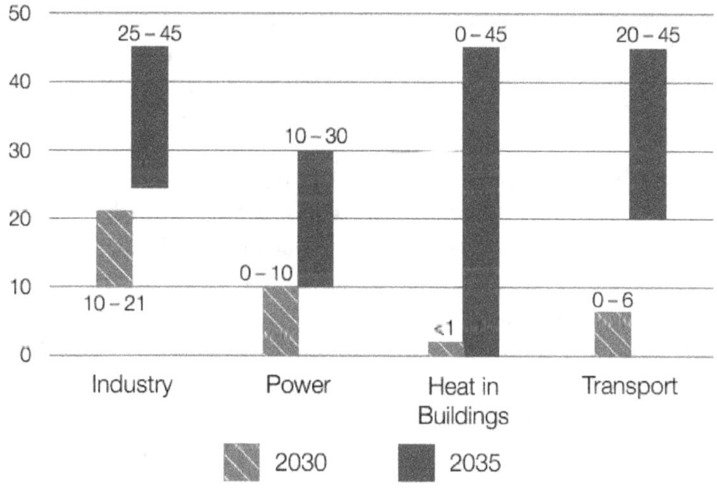

Figure 10.7 UK's hydrogen demand in 2030 and 2035 [49]

H21 comprises a series of industry-led projects funded by Ofgem, which focus on testing pure hydrogen in pipelines and associated infrastructure to gather evidence supporting hydrogen transport in dedicated pipelines. The outcomes of these initiatives will contribute to establishing safety frameworks for pipelines, which will undergo evaluation by the HSE's Science Division, informing government decisions on hydrogen's long-term role in heat supply by the mid-2020s [50]. FutureGrid aims to establish a representative transmission network for hydrogen trials, utilizing decommissioned transmission assets. This network will facilitate real-time testing and analysis of hydrogen transportation, with blends of hydrogen up to 100 percent being tested at transmission pressures [51]. The project, with funding primarily from Ofgem's Network Innovation Competition and project partners, will connect to Northern Gas Network's existing H21 distribution network facility and HyStreet homes to demonstrate the decarbonization potential of a complete beach-to-meter network. Future Billing Methodology, led by Cadent Gas, explores various options for future gas billing to adapt to potential changes in gas blends. This initiative aims to develop consumer gas billing methodologies that account for differences in calorific value between methane, biomethane, and hydrogen, facilitating their blending into the existing grid [52]. The Iron Mains Risk Reduction Programme, regulated by the HSE and initiated in 2002, involves decommissioning gas distribution iron mains and replacing them with new plastic ones. These plastic mains are considered suitable for transporting hydrogen within the existing gas grid over the long term [53].

10.2.3 The UK's hydrogen-powered transportation

Clean hydrogen presents a viable alternative to conventional fuels such as petrol, diesel, and kerosene, as it can be directly utilized in turbines, fuel cells, combustion engines, or as a precursor for the production of transportation fuels like ammonia and sustainable aviation fuels. Hydrogen adoption is expected to play a pivotal role in reducing emissions within the transportation sector, which stands as the largest single contributor to domestic greenhouse gas (GHG) emissions in the UK, approximately 24% of the total GHGs [37]. Transport represents a critical early market for hydrogen, spurring some of the initial low-carbon production endeavors in the UK. Presently, there are over 300 hydrogen vehicles operating on UK roads, predominantly comprising passenger cars and buses, and the government is actively fostering hydrogen utilization in transport through initiatives like the £23 million Hydrogen for Transport Programme [54]. The recent analysis underscores transport as a significant component of the prospective hydrogen economy, with projected demand for 2050 potentially reaching up to 140TWh.

It is anticipated that hydrogen's role in UK transportation will undergo significant evolution throughout the 2020s and beyond. Historically, road transport has spearheaded the early adoption of hydrogen in the UK. Fuel cell hydrogen buses boast comparable ranges to their diesel counterparts, and their back-to-depot operational model allows for more centralized hydrogen refueling infrastructure, which aligns well with the distributed hydrogen production anticipated during this period. By 2030, hydrogen will be integrated into various transport modes, including HGVs, buses, and rail, with initial applications in commercial shipping and aviation. The analysis indicates a potential demand of up to 6TWh for low-carbon hydrogen from the transport sector by 2030. Beyond this timeframe, a growing role for hydrogen in decarbonizing aviation and shipping is anticipated, which could constitute a significant portion of overall hydrogen demand in the long run. To meet the objectives of CB6 by 2035, it is estimated that demand from the transport sector could range from 20 to 45TWh. The longer-term role of hydrogen in transport decarbonization remains uncertain, but it is likely to be most effective in areas where energy density requirements, duty cycles, and refueling times make it the most suitable low-carbon energy source. Key challenges in this realm include technology uncertainties, limited existing hydrogen infrastructure, cost differentials, and low volumes of hydrogen-powered vehicles. Continued investment in research and innovation by both government and industry will be crucial in addressing these challenges [37]. The capability of hydrogen refueling stations influences the selection of hydrogen distribution methods, and conversely, distribution strategies impact refueling station choices. For instance, pipelines and liquid tankers are well-suited for refueling stations capable of dispensing over 500 kg of H_2 per day, whereas gas tube trailers are appropriate for smaller capacities ranging between 50 and 100 kg of H_2 per day. Meanwhile, on-site hydrogen production using methods like steam reforming and electrolysis can be employed for both large and small-scale refueling operations [55].

Approximately 2% of England's local operator bus fleet is now zero emission, either battery electric or hydrogen fuel cell. The National Bus Strategy, including the Zero Emission Bus Regional Areas (ZEBRA) scheme, aims to transition towards a fully zero-emission bus fleet. The government announced a budget of £525 million to assist operators with the upfront costs of buying zero-emission buses and refueling. The ZEBRA scheme provided £120 million in 2021/22 to deliver 4,000 new zero-emission buses and the infrastructure needed to support them [56]. The up-front cost of zero-emission buses remains higher, especially when the cost of upgrading and replacing infrastructure is included. The UK government has also pledged to introduce 4,000 zero-emission buses in England by

2025, supported by capital grants for new buses and infrastructure through the ZEBRA scheme [57].

The UK has made significant strides in the development and implementation of hydrogen trains. The first hydrogen-powered train, known as HydroFLEX successfully tested on the UK mainline, marking a major step towards decarbonizing Britain's railways [58]. The HydroFLEX project, a partnership between the University of Birmingham and Porterbrook, involves the conversion of an existing Class 319 train to run on hydrogen and battery power, with the aim of offering a cleaner alternative to current diesel trains [59]. Additionally, Alstom and Eversholt Rail signed an agreement to provide the UK's first-ever brand-new fleet of 10 hydrogen trains, further demonstrating the commitment to decarbonizing the rail sector [55]. The development of hydrogen train technology and infrastructure is seen as a crucial step in achieving a cleaner and more sustainable future for the UK's rail network.

In 2024, the UK government will invest up to £20 million in designing trials for electric road systems and hydrogen fuel cell heavy goods vehicles (HGVs), as well as running a battery electric trial to establish the feasibility, deliverability, costs, and benefits of these technologies in the UK. To support the shift away from fossil fuels, the government is also consulting on the phase-out date for the sale of new non-zero emission HGVs. Hydrogen vehicle systems (HVS) is inviting fleet operators to register expressions of interest in trialing its hydrogen-electric HGV prior to purchasing. The trials are set to start in Q4 2025 once the pre-production prototyping phase is completed and the HGVs are ready to enter volume production in the UK. The HGVs are designed to offer long ranges and quick refueling, making hydrogen a promising fuel for the haulage industry [60,61].

Research commissioned by the Department for Transport (DfT) projected that by 2050, there could be a demand of 75-95TWh for hydrogen-based fuels, primarily in the form of ammonia, for both domestic and international shipping in the UK. This adoption of hydrogen in shipping, alongside the decarbonization of road and rail freight, could potentially establish a comprehensive low-carbon freight system from ports to final destinations. While battery electric technology holds promise for domestic maritime navigation, its applicability is limited due to the size and weight constraints of batteries, making it unsuitable for larger ship types and longer distances. Hydrogen offers multiple pathways for decarbonizing ships, either through direct combustion or fuel cells, or by serving as a feedstock for methanol or ammonia. Ammonia, being more energy dense than hydrogen, requires less storage volume on vessels, making it a viable option for larger ships operating on long-distance routes. Furthermore, since ammonia is already transported internationally on ships, existing infrastructure and regulations partially support its use, although current ammonia production methods are not low carbon. Additionally, as outlined in DfT's Clean Maritime Plan, research suggests that the global market for alternative fuel production technologies, where the UK possesses competitive advantages such as upfront design, could potentially grow to approximately $11-15 billion per year (£8-£11 billion per year) by the mid-century. If the UK maintains its current export market share, estimated at around 5% of relevant global markets, this could result in economic benefits of around $490-690 (£360-£510) million per year by the middle of the century [37]. The study also revealed that while the UK possesses significant opportunities across all considered emission reduction options, it holds the most robust competitive edge in hydrogen and ammonia production technologies, as well as onboard batteries and electric engines. In March of 2021, the UK government launched the £20 million Clean Maritime Demonstration Competition with the aim of expediting the design and advancement of zero-emission vessels within the UK. This initiative established a framework for a series of technology demonstrations, hastening the process of maritime decarbonization. Additionally, the government is exploring the establishment of a dedicated unit within the Department for Transport, the UK Shipping Office for Reducing Emissions (UK-SHORE). UK-SHORE will concentrate on decarbonizing the maritime sector, leveraging insights gained from similar efforts in other transportation modes, such as the Office for Zero Emission Vehicles and the Future Fuels for Flight and Freight Competition. The objective of UK-SHORE is to position the UK as a global leader in the development and manufacturing of clean maritime technologies and fuels, including hydrogen and ammonia. The government intends to collaborate with industry stakeholders to assess how the implementation of this program, in partnership with UKRI and Innovate UK, could spur industry investment in clean maritime technologies. This collaborative effort aims to address both supply and demand-side obstacles while also fostering infrastructure development and enhancing consumer confidence [62].

The share of GHG emissions from aviation in the UK is anticipated to rise. The same as other sectors transition to low-carbon alternatives, it is imperative to address these emissions, leveraging the strengths of the UK's aerospace and aviation industries. To this end, the government has established the Jet Zero Council, a collaborative platform between industry and government aimed at accelerating the decarbonization efforts, with the goal of achieving zero-emission transatlantic flights within a generation. In July 2021, the "Jet Zero Consultation" was issued to gather feedback on the proposed strategy for achieving net-zero aviation [63]. While there are technological hurdles to overcome before hydrogen can be adopted in aviation, there is considerable interest from the aviation sector. Airbus has announced its ambition to develop and introduce a large commercial aircraft powered by hydrogen propulsion that is zero-emission by 2035. Concurrently, the UK government, through the Aerospace Technology Institute (ATI) program, is backing several projects in this domain. Furthermore, hydrogen can be utilized in the production of Sustainable Aviation Fuels (SAF). SAFs could play a crucial role in reducing emissions in the near to medium term, and the establishment of an SAF industry in the UK could generate numerous job opportunities nationwide. In March of 2021, the £15 million "Green Fuels, Green Skies" competition was launched to support the establishment of pioneering SAF plants in the UK. The government has outlined its proposed ambition for SAF uptake in the SAF blending mandate consultation, which was released for public review [64].

10.2.4 Hydrogen role in power generation in the UK

In the UK Energy White Paper, the government aims to achieve a fully decarbonized and cost-effective power system by 2050, necessitating the rapid expansion of renewables, which has been pivotal in reducing emissions thus far [65]. To meet CB6 on route to this goal, a predominantly decarbonized power sector is targeted by the mid-2030s. The deployment of renewables and other low-carbon generation forms is anticipated to escalate, driven by increasing demand for electricity as more sectors transition to electrification. Consequently, power generation will become more decentralized, variable, and intermittent, particularly with growing reliance on wind and solar energy. To facilitate this transition, there is a pressing need for more flexible, low-carbon generation and adaptable technologies like energy storage and demand-side response mechanisms to manage peak demands and balance electricity supply and demand. Low-carbon hydrogen can strategically contribute to meeting these future power system requirements, and its development and scaling in the power sector during the 2020s could alleviate pressure on other technologies such as renewables, CCUS, and nuclear power. Although not a singular solution, hydrogen could fulfill two key roles in the power system:

- Flexible power generation ("Gas to Power"): Clean hydrogen can serve as a flexible power generation source, including rapid-operating peaker plants and larger-scale but less flexible Combined Cycle Gas Turbines (CCGTs), aiding in meeting short- and long-term demand peaks. This hydrogen could be utilized either as a blend or at full capacity and would be supplied via pipelines or storage access. Initial projections suggest a modest yet significant role for clean hydrogen in power generation by 2030, with demand ranging from 0-10TWh. Expansion is expected beyond 2030, with projected hydrogen demand reaching 10-30TWh in 2035 and 25-40TWh by 2050. This utilization of hydrogen could also help secure demand for hydrogen production projects in the near future.

- System flexibility through electrolysis and storage ("Power to Gas", "Power to Gas to Power"): Electrolytic hydrogen production offers grid flexibility by utilizing surplus renewable or low-carbon electricity that would otherwise be constrained or curtailed. This excess electricity can be harnessed through electrolytic hydrogen production, enabling its flow across various parts of the system, including power, gas, transport, or industry, a concept often termed "sector coupling". This approach unlocks numerous system benefits and provides an additional revenue stream for both new and existing renewable capacity. When coupled with storage, including long-duration options where hydrogen is viable, electrolytic production can further integrate hydrogen into the power system by aiding grid balancing during periods of renewable generation surplus or deficit.

The use of hydrogen in the power sector must undergo rapid expansion through practical experience in the 2020s to facilitate further decarbonization by the 2030s and to fulfill its strategic role in achieving a fully decarbonized power sector in the long run. Starting from the mid-2020s, as the demand for flexible power generation rises, it is anticipated that hydrogen blends will dominate the use of hydrogen in the power industry, transitioning to fully hydrogen-powered turbines later in the decade. Additionally, at smaller scales, hydrogen fuel cells could emerge as an alternative, replacing high-carbon options like diesel generators to offer flexibility and backup power for off-grid areas and urban environments, building upon limited deployment thus far. By 2030, it is projected that clean hydrogen, potentially supplemented by ammonia, will increasingly contribute to peaking capacity and ensure supply security. As the demand for flexibility and the deployment of renewables increase leading up to 2030, there is an anticipation of a significant expansion in the deployment of electrolyzer capacity. This expansion will serve two primary purposes: firstly, it will support the achievement of the UK 10GW ambition; secondly, it will contribute to the decarbonization efforts in the power sector and other industries where it is economically feasible. The 2020s will concentrate on deploying a new generation of electrolyzers, which will be larger and more adaptable to variable operation in sync with renewable energy sources. Throughout the 2020s and leading up to 2030, the emergence and expansion of long-duration hydrogen storage, integrating hydrogen into the power grid and coupling it with flexible generation where necessary, is expected. To achieve this integration of hydrogen into the power sector by 2030, the primary barriers must be addressed.

10.2.5 Hydrogen fuel for heating buildings in the UK

Heating constitutes 74% of emissions from buildings in the UK and approximately 23% of the nation's total emissions [66]. While the transition to decarbonize electricity for lighting and appliances is progressing rapidly, the majority of buildings still rely on fossil fuels, predominantly natural gas, for heating, hot water, and cooking. Achieving the UK's net-zero goal by 2050 necessitates a shift to low-carbon alternatives to heating the UK's 30 million residential, commercial, industrial, and public sector buildings [67]. Initiating this transition now is crucial to meeting emission reduction targets cost-effectively, minimizing disruption, and ensuring continued reliability and comfort in heating systems for households. The UK's goal for the 2020s and early 2030s is to install low-carbon heating systems aligned with the net-zero objective exclusively while aligning with the natural replacement cycles of heating systems throughout the remainder of the 2030s and into the 2040s. The Heat and Buildings Strategy will outline the UK's approach to decarbonizing building heating in the UK. While the imperative to decarbonize heating for net-zero targets is clear, there remains uncertainty regarding the optimal approach to scale up heat decarbonization in the UK. Clean hydrogen fuel emerges as one of several key options for decarbonizing building heating alongside established technologies like electricity and heat networks. Though further assessment is needed to ascertain the feasibility of hydrogen utilization, it could serve as a direct substitute for natural gas in buildings presently reliant on grid-based heating. Given the diverse heat demands across different building types and regions in the UK, flexibility is essential in the decarbonization efforts.

Prior to considering hydrogen as a heating decarbonization option, comprehensive evidence on costs, benefits, safety, feasibility, air quality impacts, and consumer experience relative to established technologies is imperative. The 2020s will be pivotal for elucidating hydrogen's potential role, with the government collaborating with industry, network operators, and local stakeholders on extensive studies and testing endeavors to establish requisite evidence. Although the projected demand for low-carbon hydrogen for heating by 2030 is anticipated to be relatively modest (<1TWh), a compelling case for hydrogen heating could substantially elevate its significance as a future hydrogen demand source for building heating, impacting the design and timing of hydrogen production, storage, and network infrastructure. The analysis indicates that hydrogen demand for building heating could reach up to 45TWh by 2035 [37].

A broad spectrum of pertinent initiatives is already in progress. For instance, industry-led endeavors are actively investigating the distribution and transmission of hydrogen within gas networks, exemplified by projects like HyNet in the Northwest of England, the H21 project focusing on distribution across the North of England, and the HyNTS and LTS Futures projects centered on transmission, spearheaded by National Grid [68,69]. Moreover, the BEIS-funded £25 million Hy4Heat program has facilitated the development and demonstration of 100% hydrogen-ready appliances and components. This program has also established a framework for skills accreditation aimed at heating engineers engaged in hydrogen-related work [70]. As outlined in the Ten Point Plan, industry-led initiatives to conduct pioneering hydrogen heating trials are being backed, including a neighborhood trial scheduled for completion by 2023 and a village-scale trial by 2025. The village trial aims to leverage insights gained from the neighborhood trial, involving a larger and more diverse array of consumers and the conversion of existing local gas infrastructure to 100% hydrogen. These trials will furnish evidence on the practical, logistical, and technical challenges associated with building and appliance conversions. Particularly, they will assess and showcase consumer experiences with hydrogen heating in residential and commercial settings, the transition, operation, and performance of gas networks utilizing hydrogen, and the skills and training requisite for executing conversions. By 2025, plans will also be formulated for a potential hydrogen-heated town before the decade's end. This preparatory work will furnish crucial evidence regarding the feasibility and costs of transitioning from natural gas to hydrogen heating. It is anticipated that if the rationale for a widespread conversion of the gas grid to pure hydrogen is established, it will commence with the conversion of a pilot town in the late 2020s, accelerating from the early 2030s, taking into consideration practical implementation insights gleaned from the pilot program.

10.3 Canada: leading the way in hydrogen technology

Canada has played a significant role in shaping the burgeoning global hydrogen economy, tracing back over a century with advancements in hydrogen production technology and pioneering fuel cell technology four decades ago. Maintaining its position as a leader in research and development (R&D) and technology within the sector, Canada continues to drive innovation. Committed to the Paris Agreement, Canada has pledged to reduce GHG emissions by 30% below 2005 levels by 2030 and has set a target to achieve net-zero emissions by 2050, aligning with 72 other nations in this ambitious endeavor [71]. Envisioning a net-zero future, Canada aims to power its economy through electricity and low-carbon fuels, with hydrogen anticipated to fulfill up to 60% or more of the nation's energy requirements. Recognized as the lowest carbon fuel, hydrogen is deemed essential in decarbonizing Canada's most energy-intensive and challenging-to-abate end-use applications, necessitating substantial efforts to implement hydrogen on a large scale domestically. Over the course of three years, the Government of Canada, under the stewardship of NRCan, has collaborated with private sector stakeholders and governmental bodies at various levels to shape the Hydrogen Strategy for Canada. The unveiling of this strategy coincides with unprecedented circumstances as the world grapples with the impacts of COVID-19, and escalating evidence underscores the mounting risks of climate change to global economies, ecosystems, biodiversity, human health, and the future of civilization. Canada possesses all the requisite elements to foster a competitive and sustainable hydrogen economy. The modernization of the nation's energy systems toward a low-carbon economy presents a unique opportunity to harness Canadian expertise in constructing new infrastructure assets that will serve as the foundation for a low-carbon energy ecosystem spanning the country, with hydrogen assuming a pivotal role. Envisaging hydrogen contributing up to 30% of Canada's end-use energy by 2050, the nation stands poised to capitalize on this transformative shift.

Canada possesses distinctive competitive and relative advantages that position it to emerge as a global frontrunner in the production, utilization, and exportation of clean hydrogen alongside hydrogen-related technologies and services. A robust hydrogen economy promises to yield significant financial, environmental, and health advantages for Canadians. Abundant in feedstocks suitable for hydrogen production, Canada boasts some of the world's lowest carbon intensity (CI) electricity supplies, owing to its substantial hydroelectric generation capacity and recognition as a Tier-1 nuclear region. Moreover, the country holds abundant reserves of fossil fuels, boasts world-class CO_2 storage geology, exhibits potential for expanding variable renewables, possesses a large-scale biomass supply, and enjoys ample freshwater resources, all of which can be harnessed for hydrogen production.

Renowned for its leading-edge hydrogen and fuel cell technology companies and expertise, Canada hosts over 100 established companies as of 2017, employing over 2,100 individuals and generating revenues exceeding $200 million. The country also boasts significant proficiency in carbon capture technology, a critical element in producing low CI hydrogen from fossil fuels [72]. Canada's robust energy sector, which accounted for 832,500 direct and indirect jobs as of 2019, coupled with its substantial assets valued at $685 billion, including skilled labor and strategic infrastructure assets, positions the nation to swiftly transition to incorporating hydrogen on a large scale as an energy medium [73]. Engagement in international collaborations related to hydrogen by the Canadian government, industry, and academia underscores Canada's leadership on both innovation and commercial fronts. Benefitting from proximity to hydrogen import markets such as Japan, South Korea, California, the UK, Germany, and all of Europe, alongside export assets like deep-water ports, established pipeline networks, and a burgeoning LNG industry, Canada stands poised to serve as a hydrogen exporter as the global economy evolves. Canada's distinguished starting point in the hydrogen and fuel cell sector, recognized as a global hub for technical expertise, intellectual property, and premier products and services, solidifies its position as one of the top 10 hydrogen producers globally. Notably, Canada hosts the world's largest clean hydrogen production facility, specializing in producing hydrogen from natural gas with carbon capture and permanent storage for the ensuing CO_2 emissions. By capitalizing on these advantages to foster a dynamic and resilient clean hydrogen economy, Canadians stand to gain in several key areas. Seizing the hydrogen opportunity could potentially result in the creation of over 350,000 sector jobs and direct revenues exceeding $50 billion annually by 2050 [74]. Hydrogen plays a pivotal role in transitioning the oil and natural gas industries towards net-zero emissions, offering a chance to harness Canada's diverse talent pool, valuable energy reserves, and infrastructure assets in a carbon-free manner at the point of use, thus paving the way for the utilization of these assets in the future. With its unique ability to address hard-to-abate, energy-intensive applications such as long-range transportation, high-grade heat production, and industrial processes as a feedstock, hydrogen can reduce emissions of up to 190 million metric tons of CO_2 equivalent ($MtCO_2e$) [75].

Canada ranks as the fourth-largest producer of natural gas globally. Using natural gas and coal gasification, Canada produces approximately 3 million tons of grey hydrogen annually, primarily intended for industrial purposes. Moreover,

Alberta's Quest project produces steam methane reforming-based hydrogen using CCUS (blue hydrogen). The provinces with substantial natural gas and petroleum reserves, such as Alberta, British Columbia, Saskatchewan, and the Atlantic Provinces, are particularly well-positioned for hydrogen production derived from fossil fuels. Moreover, Canada possesses significant potential for hydrogen production from biomass sources. Air Liquide's electrolyzer plant, which has a capacity of 20 megawatts, is located in Becancour and produces green hydrogen. Additionally, there are projects underway in British Columbia aimed at establishing infrastructure to support a green hydrogen fueling network. Across most provinces, there is access to biomass residues sourced from the forestry and agricultural sectors. Hydrogen produced in Canada as a by-product of various industrial processes, such as chlor-alkali and sodium chlorate production, can be captured, purified, and utilized directly. The hydrogen released as a by-product from extensive industrial facilities may be adequate to meet certain short-term requirements but is restricted in its availability. Canada's hydrogen distribution infrastructure might encompass both sizable, centralized facilities situated in provinces abundant in natural gas or areas with extensive low-cost renewable energy sources, as well as smaller-scale decentralized electrolytic production facilities located close to areas of demand. As production scales up and investments are directed toward distribution infrastructure, it is anticipated that delivered hydrogen costs ranging from $1.50 to $3.50 per kilogram could be attained. Figure 10.8 illustrates the share of various hydrogen production methods in Canada by 2050 [76].

Canada ranks as the fourth-largest hydroelectricity producer globally after the United States, Brazil, and China [77]. The provinces leading in hydroelectric power generation include Manitoba with 96.8%, Quebec with 95%, Newfoundland and Labrador with 93.7%, Yukon with 92%, and British Columbia with 90% [78]. These provinces hold significant potential for electric utility providers to contribute to the hydrogen supply chain. By establishing electrolyzer farms at generation facilities, utilities can offer grid regulation services while generating hydrogen, presenting an alternative revenue stream. Moreover, decentralized electrolysis near demand hubs is feasible. Electrolyzers are flexible and scalable, with various manufacturers offering containerized solutions for easy deployment. As Canada incorporates more wind and solar energy into its energy portfolio, there is potential for increased production of low-carbon hydrogen and reduced costs associated with variable supply. Hydrogen, in turn, can enhance the economic viability of variable renewables by serving as a large-scale energy storage solution that optimizes the utilization of these renewable energy sources. For instance, Ontario experienced curtailment of approximately 7-8 TWh of renewable electricity in 2021, resulting in substantial lost revenue, which could have been mitigated by producing hydrogen instead [79].

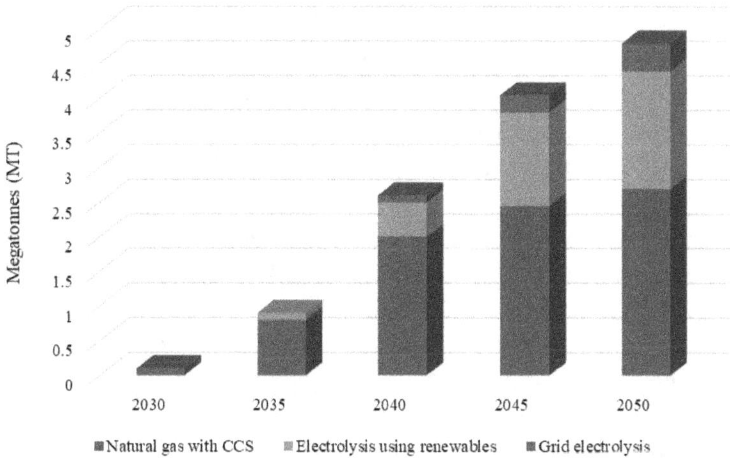

Figure 10.8 Hydrogen production in Canada by 2050

Canada ranks 9th globally in both wind and solar installations. Since 2005, generation from wind farms and solar photovoltaic panels has grown from negligible levels to approximately 5% of total electricity generation by 2018, with wind power capacity at 13.0 GW and solar power capacity at 2.9 GW. The majority of wind facilities are situated in Ontario, Quebec, and Alberta, while Ontario hosts over 98% of Canada's solar installations [80]. Nuclear reactors not only generate electricity but also produce process heat that can be utilized in the creation of low CI hydrogen. While large reactors are suitable for centralized hydrogen production on a large scale, small modular reactors are better suited for distributed hydrogen production. Electrolysis, a method of producing hydrogen using low-cost, off-peak electricity, can be employed at existing nuclear power plants. Efforts are currently underway to assess the economic viability of nuclear hydrogen production at the Bruce Nuclear Generating Station in Ontario. Presently, Ontario, where three of the four nuclear power plants are situated, along with New Brunswick, presents opportunities for nuclear hydrogen production. The development of small modular reactors is ongoing in Canada and globally. Certain designs of these reactors have the capability to produce high-temperature process heat, thereby enhancing the overall efficiency of hydrogen production. Although the commercial deployment of advanced reactors and small modular reactors is not anticipated in the near term, it presents a promising long-term opportunity for hydrogen production.

10.3.1 Hydrogen fuel utilization in Canada

Commercially available globally today are fuel cell light-duty passenger vehicles, albeit in limited numbers in Canada. The Canadian government has established federal targets for zero-emission vehicles (ZEV), aiming for them to constitute 10% of annual sales of light-duty vehicles by 2025, 30% by 2030, and 100% by 2035 [81]. Within the ZEV category, Canada recognizes battery electric vehicles (BEVs), fuel cell electric vehicles (FCEVs), and plug-in hybrid electric vehicles (PHEVs). Provinces like BC and Quebec have taken the lead by implementing consumer incentives and sales regulations to promote ZEV adoption, alongside initiating the deployment of hydrogen fueling infrastructure and light-duty FCEVs. While battery electric vehicles are anticipated to capture a significant share of the light-duty vehicle market in Canada, FCEVs provide an alternative for consumers seeking longer range and quicker refueling times, particularly suitable for larger passenger vehicle platforms.

Public transit agencies worldwide are transitioning to low- and zero-emission vehicles, with commercially available fuel cell electric buses (FCEBs) currently in service, including over 2000 globally, with about half powered by Canadian technology. Companies like New Flyer Industries and Ballard Power Systems are leading the North American market with their commercial FCEB deployments [82]. Canada's zero-emission bus (ZEB) initiative aims to support school boards and municipalities in procuring 5000 ZEBs over the next five years [83]. Leveraging the local supply chain could provide economic benefits if FCEBs are included. These buses are well-suited for longer routes and Canada's cold climate. Fuel cells are expected to have a significant role in medium- and heavy-duty trucks, rail, and ships with high power demands and lengthy duty cycles. For instance, heavy-duty trucks covering long distances may find heavy batteries impractical, reducing load capacity, while long charging times could disrupt operations. Fuel cell electric trucks, with their improved energy density and rapid refueling, are likely to be optimal for certain applications.

In Canada's oil and gas industry, utilizing low CI hydrogen can yield emission reduction advantages both in upstream extraction, where it can be utilized as a heat source, and in downstream refining, where it serves as a chemical feedstock. For instance, in upstream operations, low CI hydrogen has the potential to substitute natural gas used for steam production in steam-assisted gravity drainage (SAGD) processes for in-situ bitumen extraction. This substitution can effectively lower the carbon intensity of conventional refined petroleum products, offering a potential pathway for compliance with the federal Clean Fuel Standard. Various heavy industries in Canada, such as cement and steel manufacturing, the pulp and paper sector, and industrial processes reliant on high-grade heat production, can also achieve emission reductions by transitioning to blends of hydrogen and natural gas or pure hydrogen for heat generation.

Hydrogen holds promise for reducing emissions in heating applications within residential and commercial buildings. Natural gas utilities are exploring strategies to decarbonize the natural gas grid, including the incorporation of renewable natural gas and hydrogen as alternative low-carbon fuels. In Canada, where heating demands are substantial due to the cold climate, with heating representing nearly 80% of household energy usage, there is growing interest among utilities in utilizing hydrogen as a viable low-carbon option for both space and water heating. This interest extends to exploring hydrogen's potential as a blend with natural gas or as a standalone replacement fuel [84]. Various regions across Canada and globally are initiating pilot initiatives to assess the technical viability of integrating hydrogen into existing natural gas networks. However, to fully realize the potential of hydrogen blending, there will be a need for the development of appropriate codes and standards. Beyond certain blending thresholds, typically around 20% by volume, dedicated hydrogen pipelines may emerge as a preferable solution due to technical limitations. In a net-zero emissions future where the goal is to significantly reduce distributed combustion emissions, hydrogen could emerge as the preferred chemical heating fuel in Canada. Utilities are expected to play a pivotal role in leading this transition. Figure 10.9 demonstrates Canada's hydrogen demand by sector by 2050 [75].

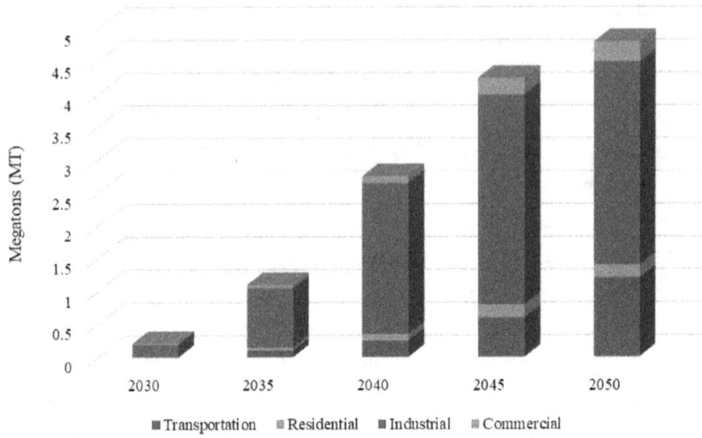

Figure 10.9 Canada's hydrogen demand by sector by 2050

10.3.2 Canada's international collaboration to deploy hydrogen technology

As a nation endowed with abundant energy resources and substantial capacity for low CI hydrogen production, coupled with well-established international trade alliances and vital infrastructure assets like deep-water ports and extensive pipeline networks, Canada is poised to emerge as a leading global provider of clean hydrogen. According to a 2019 study conducted in British Columbia, the province alone could potentially generate $15 billion in hydrogen exports by 2050 [85]. Furthermore, recent research suggests that Canada's hydrogen exports could soar to approximately $50 billion by 2050, effectively doubling the economic prospects envisioned for Canada's domestic market within the same timeframe. With importing nations increasingly focused on decarbonizing their energy systems, hydrogen stands to claim a substantial share of the global energy export market in the forthcoming decades. Just as Canada has made strides in capturing the global liquefied natural gas (LNG) export market, there exists an opportunity to leverage this experience in advancing a comprehensive hydrogen strategy. By implementing robust early initiatives and formulating a cohesive national plan that capitalizes on Canada's regional strengths, Canada can position itself as a frontrunner in the burgeoning hydrogen export arena.

Canada has established numerous bilateral and multilateral agreements, formalizing and enhancing cooperation with various countries and regions worldwide, including Germany [86], the EU [87], Portugal [88], and Japan [89]. Over the past three decades, Canada has been at the forefront of several international initiatives spanning the entire hydrogen value chain, leveraging these strategic partnerships to foster global collaboration on hydrogen-related endeavors. Canada played a pivotal role as a founding member of the IEA Hydrogen and Advanced Fuel Cell initiatives, which have since evolved into technology collaboration programs aimed at coordinating both public and private sector researchers to expedite research, development, and demonstrations on a global scale. Additionally, Canada has been a key participant and founding member of the International Partnership for Hydrogen and Fuel Cells in the Economy (IPHE), where member countries have committed to advancing the commercialization of fuel cell and hydrogen technologies while addressing critical issues such as awareness and the establishment of essential codes and standards. More recently, Canada spearheaded the development and launch of a Hydrogen Initiative within the Clean Energy Ministerial (CEM) framework. Serving as a co-lead alongside other member countries, Canada's initiative comprises over 20 nations, aiming to serve as a cornerstone for global collaboration on hydrogen. The objective is to underscore the pivotal role of hydrogen in the global energy transition during discussions among energy ministers worldwide [90].

Canadian industry has embarked on international collaborations to expedite and capitalize on research and development endeavors, as well as to exchange insights regarding the business feasibility and practical implementation considerations. An example of such collaboration is the memorandum of understanding signed between the Canadian Hydrogen and Fuel Cell Association (CHFCA) and the Australian Hydrogen Council (AHC), aimed at enhancing cooperation between Canada and Australia in the commercial deployment of hydrogen and fuel cell technologies, particularly in sectors such as mining and transportation [91]. Several of Canada's key energy partners, including Japan, South Korea, China, and the United States, have either released national strategies or announced substantial investments in their hydrogen economies. This surge of interest can be attributed to various factors, including the global drive towards decarbonization across all sectors, the escalating adoption of variable renewable energy sources, uncertainties surrounding future investments in the oil and gas sector, and the notable decrease in the costs associated with hydrogen production technologies. Unlike previous waves of enthusiasm surrounding hydrogen, the current interest is grounded in the understanding that hydrogen will be indispensable in addressing climate change. Although numerous challenges persist, the prevailing sentiment is unequivocal: hydrogen will play a pivotal role in achieving a carbon-neutral future, with many of the world's major economies already formulating strategies and making investments to actualize this vision.

10.4 Hydrogen in Africa: a vision for a cleaner future

As the global energy landscape shifts towards sustainability, Africa stands poised to emerge as a significant participant in the burgeoning realm of green hydrogen, thanks to its abundant and largely unexploited renewable energy resources [92]. Several African countries, including Egypt, Kenya, Mauritania, Morocco, Namibia, and South Africa, have shown a keen interest in developing green hydrogen projects. These countries have established the Africa Green Hydrogen Alliance to leverage the continent's vast potential for renewable energy and to pursue the development of green hydrogen as an alternative energy source. For instance, Kenya has set an ambitious target of achieving 100% clean energy by 2030 and has developed a Green Hydrogen Strategy and Roadmap to support this goal. Similarly, South Africa has published a Hydrogen Society Roadmap, which outlines national ambitions and prioritizes sectors for the deployment of the hydrogen economy, including the decarbonization of transport and energy-intensive industries, as well as the creation of an export market for green hydrogen. These initiatives demonstrate a growing momentum towards green hydrogen development in various African countries. In Africa, the projected electrolyzer projects could potentially yield 2 million tons of hydrogen by 2030. Over 20 projects spanning Kenya, Mauritania, Morocco, Namibia, and South Africa have been announced, boasting electrolyzer capacities of 100 MW or greater. Among these, nine projects are anticipated to have electrolyzer capacities exceeding 1,000 MW [92].

Africa has a unique opportunity to emerge as a major player in green hydrogen production. The continent boasts some of the world's most abundant solar and wind resources, much of which remains untapped. Despite having 60% of the world's best solar potential, Africa currently contributes only 1% to global solar generation capacity, as estimated by the International Energy Agency [2]. Moreover, Africa faces a significant energy deficit, with approximately 600 million people lacking access to electricity [93]. The anticipated rapid population and economic growth across the continent necessitates an increase in energy supply. By harnessing renewable resources for green hydrogen production, African countries can support this growth while promoting environmental sustainability. Additionally, African nations possess crucial resources like platinum and other minerals essential for green hydrogen production, including fuel cells and next-generation technologies.

There is significant potential for exporting African green hydrogen, with the European Investment Bank projecting a production capacity exceeding 50 million tons per year by 2035. This production is expected to be economically viable at €2 per kilogram, making it highly competitive with current global oil prices [94]. Countries like Morocco, Mauritania, and Egypt, due to their proximity to Europe, are well positioned to integrate their production into the European Hydrogen Backbone initiative. Additionally, other African nations can utilize existing sea transport infrastructure to export derivative products like green ammonia. Realizing this export potential could attract foreign investment and drive the development of essential power and water infrastructure across the continent, benefiting the entire population. It could also generate employment opportunities across various industries, including renewable energy, green hydrogen production, derivative products like green ammonia and fertilizers, as well as semi-finished and industrial goods such as green steel and cement. This diversification would support economic growth and resilience in African economies. Achieving the vision of Africa as a leading green hydrogen producer comes with considerable challenges. One of the major obstacles is the extensive infrastructure development needed to support large-scale production and export of green hydrogen. Many regions in Africa that are ideal for renewable energy generation lack adequate development. Thus, significant investments will be necessary in transmission infrastructure to link renewable energy sources to production facilities and export centers. Governments must also navigate their broader energy strategies alongside the rapid growth of intermittent renewables, which could strain underdeveloped national transmission grids. Furthermore, the need for clean water for hydrogen production presents additional hurdles, particularly in areas with limited water resources. Some projects are incorporating desalination into their overall design to meet clean water requirements.

Additionally, while certain countries can leverage existing gas transportation and export infrastructure, others must build such infrastructure from scratch to access key export markets. These diverse development needs increase the capital costs for large-scale green hydrogen projects and introduce significant project risks that require careful management to ensure viability and successful implementation.

The political and investment landscape across Africa varies widely and can be challenging. This diversity necessitates thorough consideration and management, given the substantial investments involved. While export-oriented projects benefit from selling to markets with hard currencies, some projects may still require domestic subsidies and price support to remain competitive. It's imperative that green hydrogen projects directly benefit local communities. Currently contributing only 4% of global carbon emissions, Africa should not solely serve as a provider of primary resources for wealthy foreign markets, replicating historical resource extraction patterns [95]. If green hydrogen development focuses primarily on exports, careful attention (and potential intervention from governments) is required to ensure that energy, infrastructure, and water resources also contribute to the country's and local communities' well-being. There is currently considerable interest in the potential of producing green hydrogen in Africa for export. Numerous large-scale projects have been announced, including the Aman green hydrogen project in Mauritania with a capacity of 30 GW (18 GW of solar and 12 GW of wind) [96], the Tsau Khaeb project in Namibia with 350,000 metric tons of green hydrogen production a year from approximately 7GW of renewable generation capacity and about 3GW electrolyzer [97], and the SCZONE project in Egypt with 4 GW [98]. Additionally, various cooperation agreements between African nations and European partners have been established, such as the financing initiatives for hydrogen investment in Namibia (SDG Namibia One) [99] and South Africa (SA-H_2) [100]. Export-oriented production offers several advantages, including attracting foreign investment, accessing external price support and incentives, and opportunities for concessional and export credit agency (ECA) financing. Moreover, the availability of cost-effective solar generation at scale provides a solid foundation for African projects to achieve cost competitiveness. However, there are significant challenges to overcome. The complex and multifaceted development required for these projects will increase capital costs and tariffs. Additionally, African nations face competition from other countries, such as Chile, vying to become leaders in hydrogen export, potentially leading to price competition for the same export markets with similar products. Moreover, the realization of large-scale planned projects remains uncertain. These projects are significantly larger than the current pilot projects and are being developed amidst a constrained supply environment for key components for green hydrogen production and high inflation rates. An alternative direction from export-oriented projects involves the establishment of captive green hydrogen initiatives, which may start off on a smaller scale and prioritize meeting domestic industrial demands. This approach could encompass various endeavors, including local airfields producing hydrogen for aviation fuel, especially as hydrogen-powered aircraft gain popularity. These initiatives are in line with domestic industrial progress and could offer a more sustainable model for growth. Figure 10.10 depicts green hydrogen production in various African countries [101].

Kenya: An illustration of this strategy is evident in Kenya's Green Hydrogen Strategy and Roadmap, which was introduced in September 2023 and places significant emphasis on producing domestic substitutes for imported hydrogen commodities like fertilizer or methanol. The objective is to mitigate supply risks and market price fluctuations, ultimately benefiting the domestic economy [102]. Kenya's hydrogen Roadmap, crafted in collaboration with EU specialists and integral to the EU's Global Gateway investment initiative, outlines Kenya's aspirations for nurturing its green hydrogen sector from the present through 2032 and beyond. Its primary emphasis lies on fostering domestic market expansion and progress, featuring distinct targets related to cutting emissions, fostering job opportunities, and attracting direct investments. Aligned with its revised Nationally Determined Contribution (NDC), Kenya aims to slash greenhouse gas emissions by one-third by 2030. As a frontrunner in transitioning to green energy, Kenya has achieved notable strides, with over 90% of its electricity sourced from renewable outlets. Kenya's collaboration with the EU centers on expediting investments in clean energy infrastructure, thereby bolstering Kenya's commitment to realizing a fully renewable electricity system by 2030 [103].

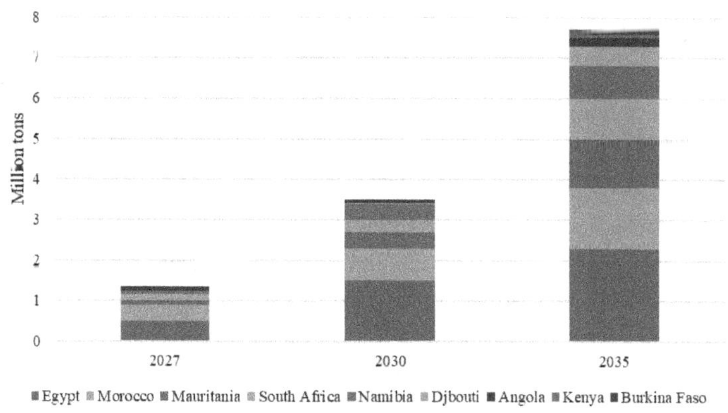

Figure 10.10 Green hydrogen production in various African countries

Egypt: Egypt is formulating a comprehensive strategy to position itself as a frontrunner in the global green hydrogen sector. Leveraging its vast renewable energy potential, particularly in wind and solar energy, Egypt aims to produce green hydrogen as a sustainable substitute for fossil fuels. With ambitious targets set by the government (up to $40bn hydrogen strategy) [104], including generating 42% of its electricity from renewable sources by 2035 and establishing a production capacity of 1.4GW by 2030, Egypt is channeling efforts into various areas such as policy interventions, renewable energy supply, hydrogen production, transportation and storage, water scarcity management, R&D capabilities enhancement, financial resources allocation, and regulatory framework development. Additionally, the country is actively seeking foreign investments and fostering local expertise to nurture a robust low-carbon hydrogen sector. Egypt's commitment to green hydrogen is underscored by its engagement in framework agreements and partnerships with international entities and companies to advance green hydrogen projects. These endeavors are anticipated to yield significant environmental, economic, and social benefits, including job creation, reduction of greenhouse gas emissions, and the establishment of domestic and export markets for green hydrogen and its by-products [105].

South Africa: The South African government has released the Hydrogen Society Roadmap (HSRM), outlining the nation's aspirations and sector prioritization for advancing the hydrogen economy [106]. The roadmap targets the production of 500,000 tons of green hydrogen annually by 2030 and the establishment of 10 GW of electrolysis capacity by the same year, with plans to expand to 15 GW by 2040 [107]. Its objectives include developing an export market for South African green hydrogen, transitioning to cleaner power generation methods, decarbonizing the transportation and heavy industry sectors, and fostering local manufacturing of components within the green hydrogen supply chain. Aligned with the country's Integrated Resource Plan, Integrated Energy Plan, and Renewable Energy Policy, the HSRM recognizes hydrogen's pivotal role in South Africa's equitable energy transition, which aims to achieve net-zero emissions by 2050. The roadmap signifies a critical milestone in South Africa's journey toward energy transition, emphasizing both economic growth and climate change mitigation efforts.

Namibia: The Namibia hydrogen roadmap outlines a strategic blueprint to harness the country's abundant renewable energy reservoirs, particularly solar and wind, to generate green hydrogen for both domestic utilization and export purposes. The roadmap entails establishing large-scale green hydrogen production facilities, with the potential to emerge as a major green hydrogen exporter. A pivotal initiative within this roadmap involves the deployment of up to 10 GW of offshore and onshore wind capacity for green hydrogen production, potentially inaugurating Africa's inaugural offshore wind farm. The initial phase of this endeavor aims to introduce 2 GW of renewable electricity generation capacity alongside electrolyzer infrastructure for green hydrogen production, subsequently destined for ammonia conversion. Subsequent expansion phases, projected for the late 2020s, anticipate a total investment value nearing USD 9.4 billion, nearly on par with Namibia's current GDP. This roadmap integrates with Namibia's overarching strategy to transition to renewable energy and diminish its carbon footprint. Leveraging the country's extensive renewable potential and cost-effective electricity production rates, Namibia is well-positioned to emerge as a noteworthy participant in the global green hydrogen market. The establishment of green hydrogen production facilities in Namibia is anticipated to unlock economic prospects, entice foreign investments, and contribute to the country's sustainable economic expansion. Nevertheless, the successful execution of the roadmap necessitates substantial investment, technological advancements, and harmonized efforts among diverse stakeholders. Addressing regulatory frameworks and infrastructure hurdles will be imperative to ensure the effective and sustainable advancement of the green hydrogen sector in Namibia [108].

Morocco: The Moroccan government has outlined a strategic roadmap for hydrogen production and export as part of its Power-to-X (PtX) strategy. This roadmap aims to harness the country's abundant wind and solar resources to generate hydrogen via water electrolysis, with the objective of establishing a significant presence in the global PtX market. Alongside, the roadmap includes provisions for integrating large-scale renewable energy sources like wind and solar to address Morocco's growing demand for primary energy. Moreover, Morocco seeks to implement sustainability criteria within its hydrogen/PtX roadmap to align with evolving European market regulations, particularly concerning carbon levies on goods. The nation aspires to emerge as a green hydrogen exporter, prioritizing domestic market needs such as green ammonia production. Additionally, the Moroccan government is actively forging energy partnerships with European nations and engaging in international hydrogen agreements to bolster the development of a green hydrogen sector within the country [109]. The PtX roadmap entails the formulation of a comprehensive 2050 energy and climate strategy, an infrastructure blueprint for hydrogen and PtX product distribution, and the establishment of sustainability standards. It also involves exploring governance frameworks for Morocco's hydrogen and ammonia sectors alongside R&D initiatives and demonstration plant development. Leveraging its abundant renewable energy potential, particularly in wind and solar, Morocco stands as a promising contender for green hydrogen and derivative production and export. This roadmap aligns with Morocco's broader objectives to enhance the share of renewable energy in its electricity generation mix and reduce energy consumption through efficiency enhancements.

Looking further ahead, sustainable energy and green hydrogen could serve as the foundation for domestically oriented industrial hubs, supporting industrial advancement in regions abundant in sunlight, wind, and water to meet the rising demands of population and economy in an environmentally conscious manner from the outset. As technology costs associated with green hydrogen production continue to decrease due to export development, this presents a promising opportunity for countries with substantial renewable energy capacities but limited access to foreign export markets. Emphasizing the production of semi-finished and downstream products, such as green steel or green cement, could further enhance the positioning of African green hydrogen economies along the value chain for long-term growth. However, a significant challenge for such industrial development lies in achieving cost competitiveness, particularly when operating within a smaller market and navigating a context where securing hard currency financing and external price support is more complex. Access to suitable financing will be crucial, especially for medium-sized local enterprises pivotal to the development of industrial hubs. These entities often face limitations in accessing existing financing mechanisms in the green hydrogen sector due to their smaller transaction volumes and primarily domestic customer base.

There exist numerous opportunities for the advancement of green hydrogen initiatives across Africa, fostering the emergence of significant green hydrogen markets and serving as pivotal contributors to the global energy transition, benefitting both Africa and the broader international community. However, it would be erroneous to presume that a singular, universally applicable approach exists that all African nations could or should adopt. Development efforts will inherently need to capitalize on and be tailored to the specific dynamics of local markets and conditions, charting distinctive trajectories in different geographical contexts. Prominent markets in North Africa are evidently focused on export-oriented development, with a compelling case for direct integration into expanding European pipeline networks and the utilization of maritime transport routes from Southern Africa. Nonetheless, this emphasis on exportation cannot be the sole focus, necessitating that other nations identify alternative pathways for development to reap rewards. This may involve concentrating on domestic-oriented production geared towards import substitution, fostering the broader evolution of green industries like green steel or green cement, and establishing industrial clusters centered around abundant renewable energy sources. Additionally, there are prospects for blue hydrogen adoption, integrating carbon capture technologies in countries endowed with substantial and well-developed oil and gas reserves. Unleashing this vast potential presents formidable challenges as initiatives traverse various sectors and demand substantial coordination and extensive financial backing. The regulatory framework for green hydrogen ventures will be intricate, given the nascent nature of the green hydrogen industry itself, and the multifaceted development requisites will inevitably intersect with diverse sectors such as power generation, water, industrial processes, transportation, and export, each governed by its distinct regulatory frameworks and practices. Ensuring compliance with evolving green standards in export markets will pose an ongoing and pivotal challenge. Nonetheless, as projects advance and African nations continue to explore and enhance their capacities, the prospects for catalyzing an economic and environmental metamorphosis across the continent and beyond grow increasingly promising [110].

10.5 South America's hydrogen ambitions

The surge of green hydrogen development in Latin America is becoming increasingly pronounced, with numerous countries in the region actively incorporating green hydrogen technology into their national energy policies and strategies for reducing greenhouse gas emissions. For instance, Paraguay's green hydrogen roadmap focuses on leveraging green hydrogen primarily in the transportation sector, while Chile and Peru have jointly unveiled a roadmap outlining green hydrogen's role in mining. Recognizing transportation as one of the most environmentally polluting sectors in Latin America, the Inter-American Development Bank has pinpointed clean hydrogen technology as a viable solution [111]. Latin America's capacity to generate substantial volumes of competitively priced low-carbon hydrogen and export it, alongside its steadfast commitment to renewable energy and global sustainability goals, is propelling the ascent of green hydrogen in the region. Several countries, such as Chile, Colombia, Uruguay, and

Brazil, are actively incorporating green hydrogen technology into their national energy strategies and plans for reducing greenhouse gas emissions [112]. This momentum for low-carbon hydrogen is gaining traction, with numerous countries in Latin America currently formulating long-term hydrogen strategies (so far 11 countries) and boasting a project pipeline comprising over 25 low-carbon hydrogen projects in their nascent stages [113]. The surge of green hydrogen in Latin America is anticipated to position the region as a significant player in the global green hydrogen industry by 2050 (Figure 10.11). Moreover, the region is gearing up for large-scale green hydrogen production, with several projects already underway in countries like Brazil, Mexico, Chile, and Costa Rica. The extensive implementation of low-carbon hydrogen production and applications presents a multifaceted challenge for Latin America, necessitating simultaneous and coordinated efforts across various fronts. Achieving this goal demands sustained engagement from all pertinent stakeholders, encompassing governmental bodies, industry players, research institutions, financial entities, labor unions, and civil society.

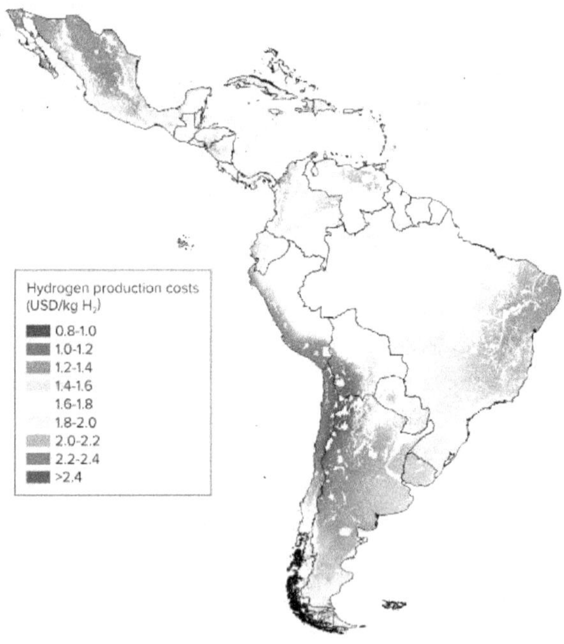

Figure 10.11 Hydrogen production cost in South America by 2050 [2]

In contrast to the recent integration of variable renewables in the region, the adoption of low-carbon hydrogen production and utilization entails the utilization of numerous technologies that are presently not fully developed. Consequently, policymakers must devise tailored measures to support the emergence of these sustainable technologies as they enter the market, alongside broader policies like carbon pricing to provide enduring economic incentives. Latin America can leverage its past experiences in developing and deploying clean energy technologies, such as biofuels for transportation in Brazil. While the recent incorporation of variable renewables in the region benefited from existing electricity markets and supportive infrastructure facilitating product distribution to consumers, the uptake of low-carbon hydrogen will hinge on the concurrent and coordinated adoption of emerging end-use technologies and enabling infrastructure, resembling the introduction of natural gas as a fuel in the region. The expansion of low-carbon hydrogen production and utilization will necessitate timely investments in enabling infrastructure, encompassing the establishment of new transmission lines (to deliver low-carbon electricity to electrolyzers for on-grid projects), hydrogen transport and storage infrastructure, and port terminals. The scaling up of these endeavors will require the development of new value chains, such as the establishment of electrolyzer manufacturing facilities in the region, which will generate employment opportunities and economic growth. To achieve this, there must be an early emphasis on education to cultivate the requisite skills and expertise needed in this sector.

Chile: Chile has set forth ambitious objectives for fostering the growth of a green hydrogen industry through its National Green Hydrogen Strategy. This strategy identifies six prioritized sectors that are pivotal in driving robust domestic demand while concurrently aiming to supply hydrogen and its derivatives to an increasingly expanding global market. The establishment of designated hydrogen hubs, referred to as H_2 hubs, is envisioned to unlock synergies in catering to both domestic and international markets. These H_2 hubs represent comprehensive ecosystems encompassing hydrogen production, distribution, utilization, innovation, and, potentially, export activities. Renewable energy production for hydrogen generation will be strategically distributed across expansive areas to form broader hydrogen valleys in conjunction with the hydrogen hub. To capitalize on Chile's world-leading levelized costs of energy, suitable locations should ideally be selected from two regions: the Antofagasta region and the broader Atacama desert in the north, boasting solar potentials reaching up to 1400 GW, and the Magallanes region in the south, with onshore wind potentials of approximately 130 GW. The identification of H_2 hub locations will be based on various factors, including existing infrastructure such as ports, proximity to domestic demand centers, synergies with existing industries, local capabilities such as the availability of skilled labor, hazard risks, and the ongoing development of hydrogen projects, among other considerations [114].

In the Antofagasta region, it may be advantageous to establish corridors comprising two or more H_2 hubs along the coastline, for instance, in Mejillones and Tocopilla, to leverage synergies and streamline regional participation and infrastructure requirements, provided that the critical mass necessary for self-sustaining hub development can be attained. The Mejillones Bay in the Antofagasta region and Cabo Negro Bay in the Magallanes region emerge as highly promising sites for an H_2 hub based on the evaluated criteria and are thus identified as prospective locations for further examination. Upon exploring various export scenarios for hydrogen and its derivatives, it is determined that both H_2 hubs can offer substantial quantities of hydrogen for export to global markets in Asia, North America, and Europe. Under a high-export scenario, the Magallanes region can fully capitalize on its green hydrogen potential, while the renewable potential in Antofagasta exceeds the requirements for hydrogen production scenarios [115]. Detailed development pathways until 2050 have been outlined for the two H_2 hubs in Antofagasta and Magallanes based on the low-export scenario outlined in the National Green Hydrogen Strategy. In this trajectory, Antofagasta will see the installation of 38 GW of electrolyzer capacity by 2050, necessitating approximately 77 GW of installed solar power generation. Concurrently, the Magallanes region is projected to achieve 31 GW of electrolyzer capacity and 61 GW of installed wind power generation. Antofagasta is estimated to produce around 2.8 million tons of hydrogen annually, while Magallanes is expected to reach production volumes of 3.1 million tons per year. Derivatives of hydrogen such as ammonia, methanol, or synthetic fuels may serve as attractive export products from the outset, while the technology for the safe and cost-effective transportation of liquefied hydrogen (LH_2) is anticipated to become pertinent on a large scale at the beginning of the next decade. Production estimates indicate 5.6 million tons of ammonia per year in Antofagasta by 2050 and 6.9 million tons of ammonia in Magallanes [114]. To accomplish this objective, a significant expansion of the existing infrastructure will be necessary. Particularly, the current electrical grid must undergo substantial growth to transmit the requisite renewable electricity to the H_2 hubs. These infrastructure enhancements offer considerable potential for synergies, notably in northern Chile, where the existing grid and expansions can cater to both existing and additional electricity demand from renewable sources. The establishment of a green hydrogen industry presents substantial opportunities but also entails risks. According to the development pathways outlined, cumulative investments exceeding \$70 billion by 2050 in Antofagasta and over \$100 billion in Magallanes will be essential to establish and cultivate the H_2 hubs to the projected capacities. Renewable capacities are the primary drivers of investments, comprising 40-65% of the total investments, contingent on the location and installation year. Hydrogen generation necessitates approximately 13-32% of total investments, with hydrogen conditioning and derivatives synthesis accounting for 9-17%. Additionally, investments in electrical infrastructure constitute 3-13%, while expenditure on export ships ranges from 2-11%. Both regions are anticipated to experience an investment peak around 2035, attributed to further declines in investment costs of renewable electricity plants and hydrogen technologies, alongside a surge in new installations during that period. By the year 2050, the sole hydrogen generation is projected to yield annual tax revenues of approximately \$700 million in both the Antofagasta and Magallanes regions. Furthermore, additional value will be generated by companies participating in the green hydrogen value chain, encompassing equipment logistics, electricity generation, water supply, electrolysis, conditioning, and both national and international transport [116]. In both regions, Antofagasta and Magallanes, at least a quarter of existing companies stand to potentially benefit from green hydrogen production. The nascent industry offers the highest potential for job creation in the construction of renewable energy generation plants, electrical infrastructure, and hydrogen generation, with an estimated 200,000 jobs expected by around 2035 and beyond, including indirect job creation through supportive activities. While construction-related jobs align with the trend of new capacity installations, employment in operation and maintenance steadily grows over time alongside the expanding fleet of operational capacities. Antofagasta is anticipated to exceed 130,000 full-time equivalents by 2050, while Magallanes is expected to surpass 80,000. Chile offers various instruments for territorial planning, crucial for both the layout of the H_2 hubs and renewable energy generation [117]. In areas of interest, such as Mejillones, planning details extend to dedicated zones for industrial development, accentuating potential synergies with the phasing-out of coal-fired power plants. Conversely, the Cabo Negro Bay area only offers communal planning, which does not encompass the H_2 hub site. Existing planning instruments need adjustment to meet H_2 hub requirements, particularly regarding the substantial expansion of renewable energy production to supply the hubs, which must be incorporated into respective planning instruments, such as the long-term energy plan, focusing on Chile's transition to renewable energy.

Colombia: Colombia possesses abundant and diverse natural resources, positioning it favorably for meeting its current energy demands and paving the way for future low-carbon hydrogen production. Firstly, Colombia boasts significant reserves of oil, natural gas, and coal, contributing to its near self-sufficiency in fuel and providing potential for the production of blue hydrogen through the capture, storage, and utilization of CO_2 emissions. The country's natural gas reserves are estimated at approximately 2,949 Gpc (equivalent to 7.7 years), concentrated in production fields located in the Llanos Orientales, the Lower Magdalena Valley, and La Guajira. Additionally, Colombia has the capability to import natural gas via the port of Cartagena, with another plant set to commence operations soon in the port of Buenaventura, ensuring a secure long-term gas supply [118]. As for coal, Colombia ranks as the largest producer in Latin America and a major global exporter, boasting proven reserves exceeding 4,500 Mt, sufficient to meet domestic demand for over 50 years. These coal reserves are primarily situated in regions such as La Guajira, Cesar, and select areas in the central part of the country [119]. Considering the annual production rates of both energy sources, Colombia stands well-positioned to meet national low-carbon hydrogen demands in both the medium and long term, particularly as technologies for CO_2 capture and storage become increasingly pivotal in leveraging these existing resources. Furthermore, Colombia is exploring the potential of developing natural hydrogen reservoirs, albeit at an early stage where determination of their locations and potential remains preliminary. Secondly, Colombia possesses renewable energy resources comparable to global benchmarks, notably in hydroelectric generation. With water supplies six times the world average and three times that of Latin America, Colombia's topography and precipitation patterns offer conducive conditions for hydroelectric power generation, which presently accounts for 70% of the country's electricity production [120]. Conversely, the extensive wind potential in the northern regions of Colombia, coupled with favorable solar conditions across much of the nation's territory and the substantial decline in costs associated with non-conventional renewable technologies, will facilitate the swift deployment of Non-Conventional Renewable Energy Sources (NCRE). The proportion of NCRE in Colombia's electricity grid, which stood at under 1% in 2018, surged to over 12% by 2022, with an installed capacity of 2,800 MW. It is anticipated that by 2027, the nation will achieve a cumulative installed capacity of nearly 32 GW, with over 11 GW derived from renewable energy sources [121]. Particularly, certain coastal areas in northern Colombia, such as La Guajira, Magdalena, or Atlántico, possess a blend of renewable resources conducive to cost-effective green hydrogen production. Notably, wind resources along the coast exhibit intensity and consistency comparable to offshore conditions, boasting plant capacity factors exceeding 60%. Similarly, solar resources in these regions are abundant, with solar photovoltaic installations achieving plant capacity factors surpassing 19-20%. These favorable conditions are poised to yield green hydrogen production costs akin to those derived from fossil sources by the decade's end, positioning them competitively on a global scale. Colombia is located in a strategically important geographic position as the gateway between Central and South America, with established transport, distribution, and trade networks connecting it to five neighboring countries. Its electrical grid is interconnected with those of neighboring nations and is set to expand in the coming decade, including new links with Ecuador and Panama. The country boasts a substantial natural gas pipeline network spanning over 7,500 kilometers and a distribution system meeting a national demand of 911 million cubic feet per day. If necessary, this network could be adapted for the transport and distribution of hydrogen between production and demand centers. Furthermore, Colombia's location between two oceans positions it as a pivotal maritime trade and export hub, with 10 port zones across the country. Ports like Cartagena in the Caribbean and Buenaventura in the Pacific are particularly significant, boasting developed infrastructure capable of handling crude oil and gas imports/exports, which could be repurposed for hydrogen and its derivatives. Consequently, Colombia has the potential to emerge as a key hydrogen logistics center, facilitating both maritime and land-based transportation to supply neighboring countries. Additionally, the anticipated competitive costs of low-carbon hydrogen production in Colombia (Figure 10.12) are expected to drive exports, thereby stimulating trade in various products across the entire hydrogen value chain [122]. To establish itself in this arena, Colombia must attract both domestic and international investment to kickstart commercial projects at both production and local demand levels. Recognizing the strategic importance of low-carbon hydrogen in its decarbonization strategy, Colombia has integrated it into its Energy Transition Law alongside green and blue hydrogen. This move is expected to further propel the development of renewable generation projects aimed at green hydrogen production, building on the momentum initiated in 2014 with the approval of Law 1715 and the organization of auctions for renewable energy. Additionally, the government's support extends to domestic mining and gas operations, which will play a crucial role in exploring and exploiting essential resources along the low-carbon hydrogen value chain.

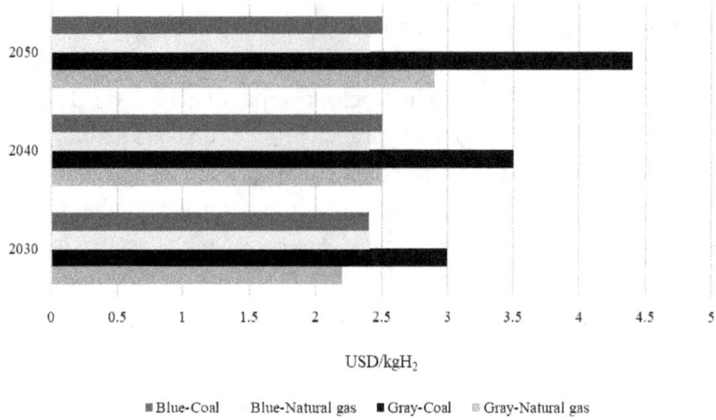

(a)

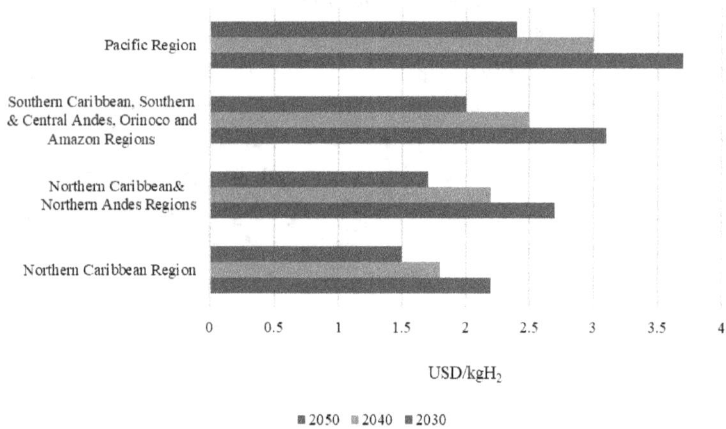

(b)

Figure 10.12 Low carbon hydrogen production in Colombia by 2050 (a) gray and blue hydrogen (b) green hydrogen

Encouraging a just transition, Colombia plans to repurpose mining and hydrocarbon projects for low-carbon hydrogen generation, incorporating CO2 capture, storage, and potential utilization. Implementing Colombia's low-carbon hydrogen roadmap will necessitate significant public-private investments in production, transportation, and end-use infrastructure. To incentivize these investments, a framework of attractive, stable, and predictable support mechanisms has been established, aiming to provide investors with the necessary assurances.

Brazil: Brazil is poised to emerge as a significant player in the global hydrogen market, leveraging its abundant clean energy resources to drive the growth of a low-carbon hydrogen sector. This initiative aims to facilitate a transition towards a low-carbon economy within the country, particularly in industries facing challenges in reducing emissions. Green hydrogen is expected to play a pivotal role in realizing Brazil's vision for energy transition and achieving a net-zero economy. By prioritizing low-carbon hydrogen, Brazil can explore projects integrating diverse technological approaches and business models, offering additional avenues to promote the adoption of green hydrogen. Furthermore, Brazil boasts a substantial domestic market potential for low-carbon hydrogen and possesses robust logistics for exporting to key international markets. Against this backdrop, Brazil aspires to cultivate a competitive low-carbon hydrogen economy. To become a prominent player in the low-carbon hydrogen economy, Brazil has created a conducive marketplace and an appealing business environment, surmounting various challenges. This endeavor presents significant opportunities for the country, unlocking latent potential across the entire low-carbon hydrogen value chain. Acknowledging both challenges and opportunities, the Brazilian Government introduced the Guidelines for the National H2 Program (PNH2) in August 2021, with related initiatives currently underway and coordinated by the Brazilian Ministry of Mines and Energy (MME) [123]. The PNH2 encompasses six key focus areas: reinforcing research, development, and innovation (R&DI) efforts; enhancing capacity building and human capital formation; refining energy planning strategies; establishing a sound legal and regulatory framework; fostering market development and competitiveness; and promoting international partnerships and cooperation. The proposed concepts and framework were subject to discussions with stakeholders to gather feedback and recommendations. This participatory process involved engagements with stakeholders across the hydrogen value chain, energy sector, potential consumers and investors, energy consultants, legal experts, academic institutions, and civil society organizations. The primary goal of this program is to significantly enhance the competitiveness and integration of hydrogen within the Brazilian energy matrix, recognizing the pivotal role that the production and utilization of low-carbon hydrogen can play in achieving a net-zero economy by 2050 [124]. The Brazilian government's dedication to fostering a low-carbon hydrogen economy is demonstrated through several key initiatives. Firstly, a commitment to support research and development in low-carbon hydrogen has been upheld for two decades. Additionally, the 2021 National Energy Policy Council (CNPE) Resolution designates hydrogen as a priority area for public and publicly oriented investment in energy R&DI, further strengthening support for research and development. Moreover, the establishment of the Brazilian Energy Compact on Hydrogen at the UN High-Level Dialogue on Energy in 2021 underscores Brazil's commitment to advancing hydrogen-related initiatives [125]. Furthermore, the launch of the PNH2 in 2021 aims to organize stakeholders' governance to drive the development of a low-carbon hydrogen economy. Furthermore, Brazil has implemented public policies supporting renewables, which will contribute to the advancement of the low-carbon hydrogen economy. These policies include tax incentives or differential taxes for renewables, especially within Export Processing Zones (EPZs), as well as investments in infrastructure, green finance, and dedicated funds for renewables and infrastructure projects. Moreover, ongoing discussions include the introduction of new programs such as Fuel for the Future (incorporating fuel cell technology, advanced biofuels, e-fuels, etc.) and carbon pricing, complementing existing initiatives like Route 2030 and RenovaBio. Additionally, Brazil actively participates in international cooperation efforts related to low-carbon hydrogen, collaborating with countries and organizations such as Germany, the United States, the United Kingdom, India, and other BRICS nations. These partnerships aim to promote knowledge exchange and joint initiatives in the field of low-carbon energy. Regarding impact targets, Brazil's approach to low-carbon hydrogen development does not set specific targets for production or end-use consumption. Instead, these impacts will naturally emerge from market dynamics influenced by energy and climate policies. Nonetheless, the Brazilian government anticipates a significant reduction in the country's carbon emissions, particularly as approximately two-thirds of greenhouse gas emissions associated with energy are from hard-to-abate sectors with considerable potential for low-carbon hydrogen utilization.

Abbreviations

AHC: Australian Hydrogen Council
ATI: Aerospace Technology Institute
Bcm: Billion cubic meters
BEV: Battery electric vehicle
BIG HIT: Building Innovative Green Hydrogen Systems in Isolated Territories
CB6: Carbon Budget Six
CCC: Climate Change Committee
CCGT: Combined Cycle Gas Turbine
CCS: Carbon capture and storage
CCUS: Carbon capture, utilization, and storage
CEM: Clean Energy Ministerial
CHFCA: Canadian Hydrogen and Fuel Cell Association
CI: Carbon intensity
CNPE: National Energy Policy Council
DfT: Department for Transport
EPZ: Export Processing Zone
EU: European Union
FCEV: Fuel cell electric vehicle
FCEB: Fuel cell electric bus
GHG: Greenhouse gas
H2BRID: Hydrogen Business Research and Innovation for Decarbonization
HGV: heavy goods vehicle
HSE: Health and Safety Executive
HVS: Hydrogen vehicle system
IPHE: International Partnership for Hydrogen and Fuel Cells in the Economy
LH2: Liquefied hydrogen
LNG: Liquefied natural gas
MME: Brazilian Ministry of Mines and Energy
NCRE: Non-Conventional Renewable Energy Sources
NDC: Nationally Determined Contribution
NZHF: Net Zero Hydrogen Fund
PEM: Proton exchange membrane
PHEV: Plug-in hybrid electric vehicle
PNH2: Guidelines for the National H2 Program

PtX: Power-to-X
R&D: Research and development
R&DI: Research, development, and innovation
SAF: Sustainable Aviation Fuel
SAGD: Steam-assisted gravity drainage
Tcf: Trillion cubic feet
TWh: terawatt-hours
UAE: United Arab Emirates
UK-SHORE: UK Shipping Office for Reducing Emissions
ZEB: zero-emission bus
ZEBRA: Zero Emission Bus Regional Area
ZEV: zero-emission vehicles

References

[1] Russia Natural Gas Reserves, Production and Consumption Statistics - Worldometer n.d. https://www.worldometers.info/gas/russia-natural-gas/ (accessed February 8, 2024).

[2] IEA – International Energy Agency n.d. https://www.iea.org/ (accessed March 11, 2022).

[3] Hydrogen n.d. https://energy.ec.europa.eu/topics/energy-systems-integration/hydrogen_en (accessed February 8, 2024).

[4] Top five countries with the biggest natural gas reserves n.d. https://www.nsenergybusiness.com/features/biggest-natural-gas-reserves-countries/ (accessed February 9, 2024).

[5] Insight from Moscow: Russia aiming to take major role in global hydrogen markets | S&P Global Commodity Insights n.d. https://www.spglobal.com/commodityinsights/en/market-insights/blogs/natural-gas/012022-russia-hydrogen-gas-reserves (accessed February 8, 2024).

[6] Barlow I, Tsafos N. Russia's Hydrogen Energy Strategy 2021. https://www.csis.org/analysis/russias-hydrogen-energy-strategy (accessed February 8, 2024).

[7] Patonia A. How Its War on Ukraine Killed Russia's Hydrogen Ambitions | German Marshall Fund of the United States 2022. https://www.gmfus.org/news/how-its-war-ukraine-killed-russias-hydrogen-ambitions (accessed February 9, 2024).

[8] Hydrogen Developments | Russia | Global Hydrogen Policy Tracker | Baker McKenzie Resource Hub n.d. https://resourcehub.bakermckenzie.com/en/resources/hydrogen-heat-map/emea/russia/topics/hydrogen-developments (accessed February 8, 2024).

[9] Vechkinzova E, Steblyakova LP, Roslyakova N, Omarova B. Prospects for the Development of Hydrogen Energy: Overview of Global Trends and the Russian Market State. Energies 2022, Vol 15, Page 8503 2022;15:8503. https://doi.org/10.3390/EN15228503.

[10] Russia gathers firms to develop hydrogen industry | Argus Media n.d. https://www.argusmedia.com/en/news/2236671-russia-gathers-firms-to-develop-hydrogen-industry (accessed February 9, 2024).

[11] Russia, UAE plan to work on hydrogen and manufacturing related equipment - Business & Economy - TASS n.d. https://tass.com/economy/1362893 (accessed February 9, 2024).

[12] UAE and Russia agreement to collaborate on hydrogen development – Policies - IEA n.d. https://www.iea.org/policies/14745-uae-and-russia-agreement-to-collaborate-on-hydrogen-development (accessed February 9, 2024).
[13] Russia in the Global Hydrogen Race - Stiftung Wissenschaft und Politik n.d. https://www.swp-berlin.org/10.18449/2021C34/ (accessed February 9, 2024).
[14] South Korea to Join Russia Sanctions, But Won't Lodge Its Own n.d. https://www.voanews.com/a/south-korea-to-join-russia-sanctions-but-won-t-lodge-its-own/6459138.html (accessed February 9, 2024).
[15] Share of European Union gas demand met by Russian supply, 2001-2023 – Charts – Data & Statistics - IEA n.d. https://www.iea.org/data-and-statistics/charts/share-of-european-union-gas-demand-met-by-russian-supply-2001-2023 (accessed April 15, 2024).
[16] The Nord Stream 2 pipeline lies abandoned after Russia invaded Ukraine n.d. https://www.cnbc.com/2022/03/31/the-nord-stream-2-pipeline-lies-abandoned-after-russia-invaded-ukraine.html (accessed February 9, 2024).
[17] No looking back: Energy transition in Russia - n.d. https://etn.news/buzz/no-looking-back-energy-transition-in-russia (accessed February 9, 2024).
[18] Patonia A, Poudineh R. Cost-competitive green hydrogen: how to lower the cost of electrolysers? - Oxford Institute for Energy Studies 2022. https://www.oxfordenergy.org/publications/cost-competitive-green-hydrogen-how-to-lower-the-cost-of-electrolysers/ (accessed February 9, 2024).
[19] China scaling up electrolyzer manufacturing base for domestic, export markets | S&P Global Commodity Insights n.d. https://www.spglobal.com/commodityinsights/en/market-insights/latest-news/energy-transition/120621-china-scaling-up-electrolyzer-manufacturing-base-for-domestic-export-markets (accessed February 9, 2024).
[20] China maps 2021-2035 plan on hydrogen energy development 2022. https://english.news.cn/20220323/428eaeae2c0a41b98ffb8d5ef4e91190/c.html (accessed February 9, 2024).
[21] China's nascent green hydrogen sector: How policy, research and business are forging a new industry | Merics n.d. https://merics.org/en/report/chinas-nascent-green-hydrogen-sector-how-policy-research-and-business-are-forging-new (accessed February 9, 2024).
[22] Nakano J. China Unveils its First Long-Term Hydrogen Plan n.d. https://www.csis.org/analysis/china-unveils-its-first-long-term-hydrogen-plan (accessed January 21, 2024).
[23] What Russia's First Gas Pipeline to China Reveals About a Planned Second One - Carnegie Endowment for International Peace n.d. https://carnegieendowment.org/politika/89552 (accessed February 9, 2024).
[24] UK Government Adopts Ambitious Sixth Carbon Budget | Jones Day n.d. https://www.jonesday.com/en/insights/2021/07/uk-government-adopts-ambitious-sixth-carbon-budget (accessed February 9, 2024).
[25] Watson N, Burnett N, Edwards T. The UK's plans and progress to reach net zero by 2050 n.d. https://assets.publishing.service.gov.uk/media/659c04aad7737c000df3356d/hydrogen-production-delivery-roadmap.pdf (accessed February 9, 2024).
[26] The ten point plan for a green industrial revolution - GOV.UK n.d. https://www.gov.uk/government/publications/the-ten-point-plan-for-a-green-industrial-revolution (accessed February 9, 2024).
[27] Department for Energy Security & Net Zero. Hydrogen Production Delivery Roadmap 2023.
[28] UK hydrogen policy takes shape (5): the Hydrogen Investor Roadmap - Global Energy Blog n.d. https://www.globalenergyblog.com/uk-hydrogen-policy-takes-shape-5-the-hydrogen-investor-roadmap/ (accessed February 9, 2024).

[29] Competition overview - Net Zero Hydrogen Fund: Strand 2 Capital Expenditure Round 2 - Innovation Funding Service n.d. https://apply-for-innovation-funding.service.gov.uk/competition/1541/overview/51b5d1d1-dc76-494c-9cb3-2b1dacf41199 (accessed February 9, 2024).

[30] Li H, Yazdi M, Moradi R, Pirbalouti RG, Nedjati A. Synergistic Integration of Hydrogen Energy Economy with UK's Sustainable Development Goals: A Holistic Approach to Enhancing Safety and Risk Mitigation. Fire 2023, Vol 6, Page 391 2023;6:391. https://doi.org/10.3390/FIRE6100391.

[31] Hydrogen blending in GB distribution networks: strategic decision - GOV.UK n.d. https://www.gov.uk/government/publications/hydrogen-blending-in-gb-distribution-networks-strategic-decision (accessed February 9, 2024).

[32] Energy white paper: Powering our net zero future - GOV.UK n.d. https://www.gov.uk/government/publications/energy-white-paper-powering-our-net-zero-future (accessed February 9, 2024).

[33] Industrial decarbonisation strategy - GOV.UK n.d. https://www.gov.uk/government/publications/industrial-decarbonisation-strategy (accessed February 9, 2024).

[34] Transport decarbonisation plan - GOV.UK n.d. https://www.gov.uk/government/publications/transport-decarbonisation-plan (accessed February 9, 2024).

[35] UK government launches plan for a world-leading hydrogen economy - GOV.UK n.d. https://www.gov.uk/government/news/uk-government-launches-plan-for-a-world-leading-hydrogen-economy (accessed February 9, 2024).

[36] The Carbon Budget Order 2021 - Impact Assessment n.d. https://www.legislation.gov.uk/ukdsi/2021/9780348222616/impacts (accessed February 10, 2024).

[37] HM Government. UK Hydrogen Strategy 2021. https://assets.publishing.service.gov.uk/media/64c7e8bad8b1a70011b05e38/UK-Hydrogen-Strategy_web.pdf (accessed February 10, 2024).

[38] Scottish hydrogen: assessment report - gov.scot n.d. https://www.gov.scot/publications/scottish-hydrogen-assessment-report/ (accessed February 10, 2024).

[39] Scottish Government Hydrogen Policy Statement - gov.scot n.d. https://www.gov.scot/publications/scottish-government-hydrogen-policy-statement/ (accessed February 10, 2024).

[40] Islands Growth Deal worth £335 million signed - GOV.UK n.d. https://www.gov.uk/government/news/islands-growth-deal-worth-335-million-signed (accessed February 10, 2024).

[41] BIG HIT n.d. https://www.bighit.eu/ (accessed February 10, 2024).

[42] BIG HIT n.d. https://www.orkney.gov.uk/Service-Directory/Renewable/big-hit.htm (accessed February 10, 2024).

[43] Hydrogen in Wales - Energy Systems Catapult n.d. https://es.catapult.org.uk/consultation/hydrogen-in-wales-a-pathway-and-next-steps-for-developing-the-hydrogen-energy-sector-in-wales/ (accessed February 10, 2024).

[44] Hydrogen Business Research & Innovation for Decarbonisation (HyBRID) - Innovate UK KTN n.d. https://iuk.ktn-uk.org/opportunities/hydrogen-business-research-innovation-for-decarbonisation-hybrid/ (accessed February 10, 2024).

[45] H2 Wales - Renewable Hydrogen Research and Development Centre | Hydrogen Centre n.d. http://www.h2wales.org.uk/pages/hydrogen-centre/hydrogen-centre.html (accessed February 10, 2024).
[46] De Tommasi L, Lyons P. Towards the Integration of Flexible Green Hydrogen Demand and Production in Ireland: Opportunities, Barriers, and Recommendations. Energies 2023, Vol 16, Page 352 2022;16:352. https://doi.org/10.3390/EN16010352.
[47] United Kingdom – HyResource n.d. https://research.csiro.au/hyresource/policy/international/united-kingdom/ (accessed February 10, 2024).
[48] National Grid explores plans for UK hydrogen 'backbone' | National Grid Group n.d. https://www.nationalgrid.com/national-grid-explores-plans-uk-hydrogen-backbone (accessed February 10, 2024).
[49] Jahanbakhsh A, Louis Potapov-Crighton A, Mosallanezhad A, Tohidi Kaloorazi N, Maroto-Valer MM. Underground hydrogen storage: A UK perspective. Renewable and Sustainable Energy Reviews 2024;189:114001. https://doi.org/10.1016/J.RSER.2023.114001.
[50] H21 | About H21 n.d. https://h21.green/about/ (accessed February 10, 2024).
[51] FutureGrid | National Gas n.d. https://www.nationalgas.com/insight-and-innovation/transmission-innovation/futuregrid (accessed February 10, 2024).
[52] Industry Information - Future Billing Methodology n.d. https://futurebillingmethodology.co.uk/industry-information/ (accessed February 10, 2024).
[53] Iron mains risk reduction n.d. https://www.hse.gov.uk/gas/supply/mainsreplacement/index.htm (accessed February 10, 2024).
[54] Hydrogen Powered Cars in the UK - ASM Auto Recycling n.d. https://www.asm-autos.co.uk/blog/hydrogen-powered-cars-in-the-uk/ (accessed February 11, 2024).
[55] Ling-Chin J, Giampieri A, Wilks M, Lau SW, Bacon E, Sheppard I, et al. Technology roadmap for hydrogen-fuelled transportation in the UK. Int J Hydrogen Energy 2024;52:705–33. https://doi.org/10.1016/J.IJHYDENE.2023.04.131.
[56] Greater support needed to realise ambitions of the National Bus Strategy, Transport Committee tells Government - Committees - UK Parliament n.d. https://committees.parliament.uk/committee/153/transport-committee/news/194531/greater-support-needed-to-realise-ambitions-of-the-national-bus-strategy-transport-committee-tells-government/ (accessed February 11, 2024).
[57] Apply for zero emission bus funding (ZEBRA 2) - GOV.UK n.d. https://www.gov.uk/government/publications/apply-for-zero-emission-bus-funding-zebra-2/apply-for-zero-emission-bus-funding-zebra-2 (accessed February 11, 2024).
[58] UK's first hydrogen train takes to the mainline marking major step towards decarbonising Britain's railways University of Birmingham n.d. https://www.birmingham.ac.uk/research/railway/research/rail-decarbonisation/power-electronics-energy/hydrogen-train-mainline.aspx (accessed February 11, 2024).
[59] HydroFLEX · Porterbrook n.d. https://www.porterbrook.co.uk/innovation/hydroflex-cop (accessed February 11, 2024).
[60] UK: testing the first Zero-Emission Hydrogen-Electric heavy vehicle - Electric Motor Engineering n.d. https://www.electricmotorengineering.com/uk-testing-the-first-zero-emission-hydrogen-electric-heavy-vehicle/ (accessed February 11, 2024).
[61] HVS appeals to fleets to get involved in hydrogen truck trials n.d. https://www.fleetnews.co.uk/news/truck-news/2023/05/10/hvs-appeals-to-fleets-to-get-involved-in-hydrogen-truck-trials (accessed February 11, 2024).

[62] UK Shipping Office for Reducing Emissions - GOV.UK n.d. https://www.gov.uk/government/speeches/uk-shipping-office-for-reducing-emissions (accessed February 11, 2024).
[63] Jet Zero: our strategy for net zero aviation - GOV.UK n.d. https://www.gov.uk/government/consultations/achieving-net-zero-aviation-by-2050 (accessed February 11, 2024).
[64] UK launches Green Fuel, Green Skies competition to support SAF | Biomass Magazine n.d. https://biomassmagazine.com/articles/uk-launches-green-fuel-green-skies-competition-to-support-saf-17840 (accessed February 11, 2024).
[65] HM Government. Energy White Paper | Powering our Net Zero Future 2020. https://assets.publishing.service.gov.uk/media/5fdc61e2d3bf7f3a3bdc8cbf/201216_BEIS_EWP_Command_Paper_Accessible.pdf (accessed February 12, 2024).
[66] Household projections for England - Office for National Statistics n.d. https://www.ons.gov.uk/peoplepopulationandcommunity/populationandmigration/populationprojections/bulletins/householdprojectionsforengland/2018based (accessed February 13, 2024).
[67] National Energy Efficiency Data-Framework (NEED) - GOV.UK n.d. https://www.gov.uk/government/collections/national-energy-efficiency-data-need-framework (accessed February 13, 2024).
[68] Hydrogen in the UK - Hydrogen UK n.d. https://hydrogen-uk.org/hydrogen-in-the-uk/ (accessed February 13, 2024).
[69] Major boost for hydrogen as UK unlocks new investment and jobs - GOV.UK n.d. https://www.gov.uk/government/news/major-boost-for-hydrogen-as-uk-unlocks-new-investment-and-jobs (accessed February 13, 2024).
[70] HydrogenAppliances — Hy4Heat n.d. https://www.hy4heat.info/appliances (accessed February 13, 2024).
[71] Paris Agreement - Canada.ca n.d. https://www.canada.ca/en/environment-climate-change/services/climate-change/paris-agreement.html (accessed February 13, 2024).
[72] The Canadian Hydrogen and Fuel Cell Association n.d. https://www.chfca.ca/ (accessed February 13, 2024).
[73] Hydrogen Strategy for Canada 2020. https://natural-resources.canada.ca/sites/nrcan/files/environment/hydrogen/NRCan_Hydrogen-Strategy-Canada-na-en-v3.pdf (accessed February 13, 2024).
[74] Canada hydrogen strategy – Policies - IEA n.d. https://www.iea.org/policies/17710-canada-hydrogen-strategy (accessed February 13, 2024).
[75] Canada's clean hydrogen gets set to power the world - Financial Times - Partner Content by Invest in Canada n.d. https://canada-next-best-place-to-home.ft.com/canadas-clean-hydrogen-gets-set-to-power-the-world (accessed February 13, 2024).
[76] CER – Canada's Energy Future 2021 - Results n.d. https://www.cer-rec.gc.ca/en/data-analysis/canada-energy-future/2021/results.html (accessed February 13, 2024).
[77] International hydropower association. Hydropower Status Report | Sector trends and insights 2021.
[78] CER – Provincial and Territorial Energy Profiles – Canada n.d. https://www.cer-rec.gc.ca/en/data-analysis/energy-markets/provincial-territorial-energy-profiles/provincial-territorial-energy-profiles-canada.html (accessed February 13, 2024).
[79] Ontario Continues Wasting Clean Electricity for 7th Consecutive Year: Engineers n.d. https://www.newswire.ca/news-releases/ontario-continues-wasting-clean-electricity-for-7th-consecutive-year-engineers-824940352.html (accessed February 13, 2024).

[80] By the Numbers – Canadian Renewable Energy Association n.d. https://renewablesassociation.ca/by-the-numbers/ (accessed February 13, 2024).
[81] Canada's Electric Vehicle Availability Standard (regulated targets for zero-emission vehicles) - Canada.ca n.d. https://www.canada.ca/en/environment-climate-change/news/2023/12/canadas-electric-vehicle-availability-standard-regulated-targets-for-zero-emission-vehicles.html (accessed February 13, 2024).
[82] Bus Transit - Fuel Cell Electric Buses | Ballard Power n.d. https://www.ballard.com/markets/transit-bus (accessed February 13, 2024).
[83] Zero Emission Transit Fund | Municipal Climate Change Action Centre n.d. https://mccac.ca/funding-guide/zero-emission-transit-fund/ (accessed February 13, 2024).
[84] Karaca AE, Dincer I. An updated overview of Canada's hydrogen related research and development activities. Int J Hydrogen Energy 2021;46:34515–25. https://doi.org/10.1016/J.IJHYDENE.2021.07.235.
[85] British Columbia Hydrogen Study | Executive Summary n.d. https://assets-global.website-files.com/5f749e4b9399c80b5e421384/60c2207c71746c499c0cd297_2021%20Hydropower%20Status%20Report%20-%20International%20Hydropower%20Association%20Reduced%20file%20size.pdf (accessed February 13, 2024).
[86] Canada and Germany Sign Agreement to Enhance German Energy Security with Clean Canadian Hydrogen - Canada.ca n.d. https://www.canada.ca/en/natural-resources-canada/news/2022/08/canada-and-germany-sign-agreement-to-enhance-german-energy-security-with-clean-canadian-hydrogen.html (accessed February 13, 2024).
[87] EU-Canada Summit n.d. https://ec.europa.eu/commission/presscorner/detail/en/ac_23_6059 (accessed February 13, 2024).
[88] Canadian-Portuguese joint venture plans $1.04 bln hydrogen plant in Sines | Reuters n.d. https://www.reuters.com/business/energy/canadian-portuguese-joint-venture-plans-104-bln-hydrogen-plant-sines-2022-11-14/ (accessed February 13, 2024).
[89] Minister Wilkinson positions Canada as a clean energy and technology supplier of choice to Japan - Canada.ca n.d. https://www.canada.ca/en/natural-resources-canada/news/2023/01/minister-wilkinson-positions-canada-as-a-clean-energy-and-technology-supplier-of-choice-to-japan.html (accessed February 13, 2024).
[90] CEM Hydrogen Initiative – Programmes - IEA n.d. https://www.iea.org/programmes/cem-hydrogen-initiative (accessed February 13, 2024).
[91] Canada and Australia sign MOU to increase collaboration in hydrogen and fuel cell technologies - CHFCA n.d. https://www.chfca.ca/2020/07/07/canada-and-australia-sign-mou-to-increase-collaboration-in-hydrogen-and-fuel-cell-technologies/ (accessed February 13, 2024).
[92] The African economies seeking a slice of the $130 billion green hydrogen market | The Habari Network n.d. https://www.thehabarinetwork.com/the-african-economies-seeking-a-slice-of-the-130-billion-green-hydrogen-market (accessed February 16, 2024).
[93] Global Hydrogen Review 2023 – Analysis - IEA n.d. https://www.iea.org/reports/global-hydrogen-review-2023 (accessed November 27, 2023).
[94] Mo Ibrahim Foundation. Addressing Africa's energy deficit | Climate change, renewables, and gas n.d. https://energia.gob.cl/sites/default/files/documentos/green_h2_strategy_chile.pdf (accessed February 15, 2024).
[95] New study confirms €1 trillion Africa's extraordinary green hydrogen potential n.d. https://www.eib.org/en/press/all/2022-574-new-study-confirms-eur-1-trillion-africa-s-extraordinary-green-hydrogen-potential (accessed February 15, 2024).

[96] How much does Africa contribute to global carbon emissions? | Climate News | Al Jazeera n.d. https://www.aljazeera.com/news/2023/9/4/how-much-does-africa-contribute-to-global-carbon-emissions (accessed February 16, 2024).
[97] The AMAN Project: Accelerating the green energy transition in Mauritania • RSK • Engineering and Environmental Consultancy n.d. https://rskgroup.com/projects/the-aman-project-accelerating-the-green-energy-transition-in-mauritania/ (accessed February 16, 2024).
[98] Projects - Hyphen Hydrogen Energy n.d. https://hyphenafrica.com/projects/ (accessed February 16, 2024).
[99] Egypt's SCZone signed an $6.75 billion agreement with ENERGY CHINA to build 210k ton green hydrogen capacity n.d. https://chinahydrogen.substack.com/p/egypts-sczone-signed-an-675-billion (accessed February 16, 2024).
[100] SDG Namibia One Fund Acquires 24% Stake in Nation's First Gigawatt Green Hydrogen Project - ESG News n.d. https://esgnews.com/sdg-namibia-one-fund-acquires-24-stake-in-nations-first-gigawatt-green-hydrogen-project/ (accessed February 16, 2024).
[101] Unveiling the 'SA-H2 Fund': South Africa's Dedicated Green Hydrogen Fund | Development Bank of Southern Africa n.d. https://www.dbsa.org/press-releases/unveiling-sa-h2-fund-south-africas-dedicated-green-hydrogen-fund (accessed February 16, 2024).
[102] Africa and Europe: green hydrogen economy n.d. https://www.rystadenergy.com/news/africa-and-europe-set-to-be-the-dynamos-for-the-global-green-hydrogen-economy (accessed February 16, 2024).
[103] Green Hydrogen Strategy and Roadmap for Kenya n.d. https://ec.europa.eu/commission/presscorner/detail/es/ip_23_4324 (accessed February 16, 2024).
[104] Kenya's Green Hydrogen Strategy launches at Africa Climate Summit n.d. https://african.business/2023/09/resources/kenyas-green-hydrogen-strategy-launched-at-africa-climate-summit (accessed February 16, 2024).
[105] Egypt to launch $40bn hydrogen strategy before June 2022 | Energy & Utilities n.d. https://energy-utilities.com/egypt-to-launch-40bn-hydrogen-strategy-before-news116149.html (accessed February 16, 2024).
[106] Positioning Egypt as a Global Green Hydrogen Leader - Meeting Egypt's Environmental Challenges - Carnegie Endowment for International Peace n.d. https://carnegieendowment.org/2023/12/05/positioning-egypt-as-global-green-hydrogen-leader-pub-90716 (accessed February 16, 2024).
[107] Department of Science and Innovation. Hydrogen Society Roadmap for South Africa 2021.
[108] Hydrogen Society Roadmap – Policies - IEA n.d. https://www.iea.org/policies/16974-hydrogen-society-roadmap (accessed February 16, 2024).
[109] Mukelabai MD, Wijayantha KGU, Blanchard RE. Hydrogen technology adoption analysis in Africa using a Doughnut-PESTLE hydrogen model (DPHM). Int J Hydrogen Energy 2022;47:31521–40. https://doi.org/10.1016/J.IJHYDENE.2022.07.076.
[110] Boulakhbar M, Lebrouhi B, Kousksou T, Smouh S, Jamil A, Maaroufi M, et al. Towards a large-scale integration of renewable energies in Morocco. J Energy Storage 2020;32:101806. https://doi.org/10.1016/J.EST.2020.101806.
[111] Green hydrogen in Africa: A continent of possibilities? | White & Case LLP n.d. https://www.whitecase.com/insight-our-thinking/africa-focus-winter-2023-green-hydrogen (accessed February 16, 2024).
[112] IDB | The IDB launches clean hydrogen certification system and sponsors decarbonization declaration for heavy industry n.d. https://www.iadb.org/en/news/idb-launches-clean-

hydrogen-certification-system-and-sponsors-decarbonization-declaration (accessed February 16, 2024).
[113] How the hydrogen market has evolved by 2023 in Latin America & the Caribbean – HINICIO n.d. https://hinicio.com/h2_latin_america_market_2023_article_eng/ (accessed February 17, 2024).
[114] Hydrogen in Latin America – Analysis - IEA n.d. https://www.iea.org/reports/hydrogen-in-latin-america (accessed February 16, 2024).
[115] Matthias Altmann LRWMZCVNMFDÁZFRFM. Development Pathways for "Hydrogen Hubs" in Chile 2022.
[116] Chavez-Angel E, Castro-Alvarez A, Sapunar N, Henríquez F, Saavedra J, Rodríguez S, et al. Exploring the Potential of Green Hydrogen Production and Application in the Antofagasta Region of Chile. Energies (Basel) 2023;16:4509. https://doi.org/10.3390/EN16114509/S1.
[117] Why the World Bank should be cautious on green hydrogen n.d. https://www.energymonitor.ai/opinion/beyond-the-h2ype-why-the-world-bank-should-be-cautious-on-green-hydrogen/ (accessed February 17, 2024).
[118] Chile's Green Hydrogen Strategy and investment opportunities 2021.
[119] Colombia Exploration & Production, 2021 n.d. https://www.globaldata.com/store/report/colombia-oil-and-gas-exploration-and-production-analysis/ (accessed February 17, 2024).
[120] Colombia Coal Reserves and Consumption Statistics - Worldometer n.d. https://www.worldometers.info/coal/colombia-coal/ (accessed February 17, 2024).
[121] Mesa O, Urrea V, Ochoa A. Trends of hydroclimatic intensity in Colombia. Climate 2021;9:120. https://doi.org/10.3390/CLI9070120/S1.
[122] Colombia eyeing 11GW of renewables capacity by 2027 - BNamericas n.d. https://www.bnamericas.com/en/news/colombia-eyeing-11gw-of-renewables-capacity-by-2027 (accessed February 17, 2024).
[123] Colombia's hydrogen roadmap n.d. https://www.minenergia.gov.co/documents/5862/Colombias_Hydrogen_Roadmap_2810.pdf (accessed February 17, 2024).
[124] Brazil – HyResource n.d. https://research.csiro.au/hyresource/policy/international/brazil/ (accessed February 17, 2024).
[125] Brazil | Green Hydrogen Organisation n.d. https://gh2.org/countries/brazil (accessed February 17, 2024)
[126] Energy Compacts Registry | United Nations n.d. https://www.un.org/en/energycompacts/page/registry#Brazil (accessed February 17, 2024).

www.ingramcontent.com/pod-product-compliance
Lightning Source LLC
LaVergne TN
LVHW052022200926
848481LV00005B/1557